INTRODUCTION TO PARTIAL DIFFERENTIAL EQUATIONS AND BOUNDARY VALUE PROBLEMS

INTERNATIONAL SERIES IN
PURE AND APPLIED MATHEMATICS

William Ted Martin and E. H. Spanier
CONSULTING EDITORS

AHLFORS · Complex Analysis
BELLMAN · Stability Theory of Differential Equations
BUCK · Advanced Calculus
BUSACKER AND SAATY · Finite Graphs and Networks
CHENEY · Introduction to Approximation Theory
CODDINGTON AND LEVINSON · Theory of Ordinary Differential Equations
COHN · Conformal Mapping on Riemann Surfaces
DENNEMEYER · Introduction to Partial Differential Equations and Boundary
    Value Problems
DETTMAN · Mathematical Methods in Physics and Engineering
EPSTEIN · Partial Differential Equations
GOLOMB AND SHANKS · Elements of Ordinary Differential Equations
GRAVES · The Theory of Functions of Real Variables
GREENSPAN · Introduction to Partial Differential Equations
GRIFFIN · Elementary Theory of Numbers
HAMMING · Numerical Methods for Scientists and Engineers
HILDEBRAND · Introduction to Numerical Analysis
HOUSEHOLDER · Principles of Numerical Analysis
KALMAN, FALB, AND ARBIB · Topics in Mathematical System Theory
LASS · Elements of Pure and Applied Mathematics
LASS · Vector and Tensor Analysis
LEPAGE · Complex Variables and the Laplace Transform for Engineers
McCARTY · Topology: An Introduction with Applications to
    Topological Groups
MOORE · Elements of Linear Algebra and Matrix Theory
MOURSUND AND DURIS · Elementary Theory and Application
    of Numerical Analysis
NEF · Linear Algebra
NEHARI · Conformal Mapping
NEWELL · Vector Analysis
RALSTON · A First Course in Numerical Analysis
RITGER AND ROSE · Differential Equations with Applications
ROSSER · Logic for Mathematicians
RUDIN · Principles of Mathematical Analysis
SAATY AND BRAM · Nonlinear Mathematics
SIMMONS · Introduction to Topology and Modern Analysis
SNEDDON · Elements of Partial Differential Equations
SNEDDON · Fourier Transforms
STOLL · Linear Algebra and Matrix Theory
STRUBLE · Nonlinear Differential Equations
WEINSTOCK · Calculus of Variations
WEISS · Algebraic Number Theory
ZEMANIAN · Distribution Theory and Transform Analysis

INTRODUCTION TO PARTIAL DIFFERENTIAL EQUATIONS AND BOUNDARY VALUE PROBLEMS

Rene Dennemeyer
California State College at San Bernardino

McGraw-Hill Book Company
New York, St. Louis, San Francisco, Toronto, London, Sydney

**Introduction to Partial Differential Equations and
Boundary Value Problems**

32212

PREFACE

This book originated from an introductory course in partial differential equations which the author has given for a number of years. The course is at the senior level and is intended for mathematics, physics, or engineering majors who have completed a semester of ordinary differential equations and vector analysis or, alternatively, a year course at the junior level of the type often called "applied mathematics," or "advanced mathematics for engineers and physicists." No previous knowledge of partial differential equations is required. The book gives an introduction to the theory of linear partial differential equations and some methods of solution of the classical boundary- and initial-value problems of mathematical physics.

Chapter 1 is devoted to first-order equations. Following the first several sections on notation conventions, classification of equations, preliminary definitions and linear first-order equations in two independent variables, the method of Lagrange for quasilinear first-order equations is described. Characteristic curves are discussed, and their role in the Cauchy problem for quasilinear first-order equations is emphasized in Section 1-6. This section also contains an existence and uniqueness proof for the Cauchy problem for such equations. The ideas are illustrated by examples and counterexamples.

Chapter 2 is concerned with linear second-order equations. Some formal methods of obtaining solutions (mainly for constant-coefficient equations) are described, and functionally invariant solutions are illustrated in Section 2-4. Classification as to type and reduction to normal form is discussed in Section 2-4. The next two sections are devoted to the Cauchy problem, first for the case of two independent variables and then for equations in n independent variables. Characteristics of linear second-order equations are discussed and illustrated in these sections. The idea of a tangential operator is used to show that if the initial curve (or surface) is characteristic, then the differential equation itself implies a relation between the data of the Cauchy problem. The final section contains the definitions of the adjoint operator corresponding to a given operator, a self-adjoint operator, and concludes with a derivation of Green's formula. This material is useful in subsequent chapters.

The remaining three chapters are mainly devoted to boundary-value problems for Laplace's and Poisson's equation and to initial-value and boundary- and initial-value problems for the wave equation and the heat equation. Chapter 3 contains a discussion of properties of harmonic functions (including the maximum-minimum principle), separation of variables, and spherical harmonics. The last section is devoted to self-adjoint elliptic-type boundary-value problems and the eigenfunction representation of the Green's function of such a problem. The initial-value problems and the boundary- and initial-value problems for the wave equation in one, two, and three dimensions are discussed in Chapter 4. Poisson's formula for the solution of the initial-value problem is derived in Section 4-6. Examples of the series representation of the Green's function for a boundary- and initial-value problem are given in Sections 4-5 and 4-7. Similarly in Chapter 5 initial-value and boundary- and initial-value problems for the heat equation are discussed. The maximum-minimum principle for solutions of the heat equation together with an existence and uniqueness theorem is proved in the final section.

In addition there are several appendixes. Appendix 1 contains a proof of a special case of the Cauchy-Kowalewski theorem used in Chapter 2, namely, the Cauchy problem for a linear second-order equation in two independent variables where the initial curve is a segment of the x axis. In Appendix 2 there is brief discussion of Sturm-Liouville problems and orthogonal systems of functions together with formulas and properties of Fourier series and Fourier-Legendre and Fourier-Bessel expansions. This material is pertinent to the separation-of-variables technique and the special functions used in Chapters 3 to 5. Appendix 3 contains the references cited in the text, together with an additional bibliography on partial differential equations and related subjects.

Examples which illustrate the theory as well as the formal methods of solution are given in each chapter. At the end of each chapter there is a list of problems subdivided according to section. There are problems involving extensions of the theory or additional results, and there are exercises in formal techniques. Answers to many of the problems are given in Appendix 4.

Of necessity some topics have been omitted from the text. Among these are numerical methods and integral-transform and conformal-mapping techniques. Numerical and approximate methods of solution are of course very important; however, these topics are frequently taken up in separate courses in numerical analysis. Integral-transform or conformal-mapping techniques require of the student a more extensive background in complex analysis than is assumed here.

Acknowledgment is made to the various works consulted in the development of the textbook, which are included in the references and bibliography. Also the author wishes to express appreciation to Dr. E. David Callender, for encouragement given during the preparation of the book, and to Miss Nancy Fisher and Miss Helen Kuntz, for their patience and help in typing the manuscript.

Rene Dennemeyer

CONTENTS

INTRODUCTION TO PARTIAL DIFFERENTIAL EQUATIONS; FIRST-ORDER EQUATIONS

1-1 INTRODUCTION

The study of partial differential equations is a classical branch of analysis. There are many important applications of the subject in the physical sciences and engineering. Moreover it is an active field of modern mathematical research. An extensive literature deals with the many aspects of this interesting subject, and a number of references for further study are listed in Appendix 3. The subsequent chapters constitute an introduction to linear partial differential equations (except that quasilinear first-order equations are considered in this chapter) together with some applications. This chapter is primarily devoted to linear first-order equations.

In the general solution of an ordinary differential equation one or more arbitrary constants appear. There is only one independent variable, and the arbitrary constants serve to fix the initial values of the solution and its derivatives at a given point. With partial differential equations the situation is basically different. Several independent variables are present, and the general solution involves arbitrary functions of the independent variables rather than arbitrary constants. Thus there is a much greater degree of generality of *form* of solution. Also, instead of specifying the value of a solution (and possibly its derivatives) at a point, the values are prescribed along a curve or over a surface. Particularly with regard to the applications the principal problem is to obtain from the wealth of solutions of a given linear partial differential equation one which will fit prescribed initial and boundary conditions.

In order to obtain geometric interpretations of the concepts and theorems in a familiar setting, namely, three-dimensional euclidean space, the discussion for the most part is restricted to equations involving two independent variables and one dependent variable. However, the analysis extends readily to the general case of n independent variables. This extension is indicated in some of the problems at the end of the chapter.

1-2 CLASSIFICATION OF EQUATIONS; NOTATION

A partial differential equation is an equation which involves two or more independent variables together with partial derivatives of one or more dependent variables with respect to the independent variables. The general form of a partial differential equation in two independent variables x, y and one dependent variable z is

$$F(x, y, z, z_x, z_y, z_{xx}, z_{xy}, z_{yy}, \ldots) = 0$$

where $z_x = \partial z/\partial x$, $z_y = \partial z/\partial y$, etc., denote partial derivatives. The general form of a partial differential equation in n independent variables and one dependent variable is

$$F(x_1, \ldots, x_n, u, u_{x_1}, \ldots, u_{x_n}, u_{x_1 x_1}, u_{x_1 x_2}, \ldots) = 0$$

where $x_1, \ldots, x_n$ are the independent variables and $u_{x_1} = \partial u/\partial x_1$, $u_{x_2} = \partial u/\partial x_2, \ldots, u_{x_n} = \partial u/\partial x_n$ denote the partial derivatives of the dependent variable u. Following are some examples of partial differential equations:

$$xz_x - yz_y = \sin xy \tag{1-1}$$

$$\frac{\partial^2 u}{\partial t^2} = \frac{\partial^2 u}{\partial x^2} + \frac{\partial^2 u}{\partial y^2} \tag{1-2}$$

$$uu_y u_{xxx} + u_{yy}{}^2 = \sin u \tag{1-3}$$

$$\left(\frac{\partial^2 u}{\partial x^2}\right)^2 + \frac{\partial^2 u}{\partial y^2} = \frac{\partial u}{\partial z} + z^3 \tag{1-4}$$

$$\begin{cases} w\dfrac{\partial v}{\partial x} + v\dfrac{\partial w}{\partial y} = x + y \\[2mm] v\dfrac{\partial v}{\partial x} + w\dfrac{\partial w}{\partial y} = x - y \end{cases} \tag{1-5}$$

$$x\frac{\partial^2 u}{\partial t^2} + t\frac{\partial^2 u}{\partial y^2} + u^3\left(\frac{\partial u}{\partial x}\right)^2 = t + 1 \tag{1-6}$$

The classification of partial differential equations as to order, linearity, and nonlinearity is similar to the classification of ordinary differential equations. The *order* of a partial differential equation is the order of the partial derivative (or derivatives) of highest order which appear in the equation. Equations (1-1) to (1-6) are of the following orders, respectively: 1, 2, 3, 2, 1, 2. In Eq. (1-5) there appears a system of two first-order equations in two independent variables. A partial differential equation is said to be *linear* if it is of the first degree in the dependent variable (or dependent variables when there is more than one dependent variable present) and the partial derivatives which occur in the equation. A partial differential equation is called *nonlinear* if it is not linear. The partial differential equations (1-1) and (1-2) are linear, while the remaining equations are nonlinear. The general form of a linear first-order equation in two independent variables is

$$P(x,y)z_x + Q(x,y)z_y + R(x,y)z = S(x,y) \tag{1-7}$$

and the general form of a linear second-order equation in two independent variables is

$$A(x,y)z_{xx} + 2B(x,y)z_{xy} + C(x,y)z_{yy} + D(x,y)z_x$$
$$+ E(x,y)z_y + F(x,y)z = G(x,y) \quad (1\text{-}8)$$

There is a further classification of partial differential equations. A partial differential equation is called *quasilinear* if it is linear in the highest-order derivatives which appear in the equation (regardless of the manner in which lower-order derivatives and dependent variable occur in the equation). All the partial differential equations (1-1) to (1-6) are quasilinear except Eq. (1-4). The general form of a quasilinear first-order equation in two independent variables is

$$P(x,y,z)z_x + Q(x,y,z)z_y = R(x,y,z) \quad (1\text{-}9)$$

Observe that a linear equation is certainly quasilinear, but in general the converse is not true. A class of partial differential equations intermediate between linear and quasilinear is defined as follows. A partial differential equation is said to be *almost linear* if it is quasilinear and the coefficients of the highest-order derivatives which appear in the equation are functions of the independent variables only. The general form of an almost-linear equation of the second order in two independent variables is

$$A(x,y)z_{xx} + 2B(x,y)z_{xy} + C(x,y)z_{yy} + D(x,y,z,z_x,z_y) = 0 \quad (1\text{-}10)$$

Equations (1-1), (1-2), and (1-6) are almost linear. At this point it should be mentioned that in the literature the term quasilinear is sometimes reserved for what here has been called almost linear.

In writing partial differential equations involving two independent variables the following notation for partial derivatives is used:

$$p = \frac{\partial z}{\partial x} \qquad q = \frac{\partial z}{\partial y} \qquad r = \frac{\partial^2 x}{\partial x^2} \qquad s = \frac{\partial^2 z}{\partial x \, \partial y} \qquad t = \frac{\partial^2 z}{\partial y^2} \quad (1\text{-}11)$$

Accordingly the general form of a quasilinear equation of the second order in two independent variables can be written

$$A(x,y,z,p,q)r + 2B(x,y,z,p,q)s + C(x,y,z,p,q)t + D(x,y,z,p,q) = 0 \quad (1\text{-}12)$$

1-3 FORMATION OF EQUATIONS; GEOMETRIC EXAMPLES

Partial differential equations occur in diverse fields of study, mathematical and physical. The classical equations of mathematical physics are considered in subsequent chapters. In this section the formation of partial differential equations by elimination of arbitrary functions is illustrated, and several examples of geometric problems which lead to partial differential equations are presented. The geometric examples furnish an opportunity to review some ideas of analytic geometry. These ideas are useful in the following sections.

Regions.   Surfaces

In the sequel frequent use is made of the concept of a region. Here the pertinent definitions are phrased in terms of sets of points in the xy plane. Recall that a *point* in real two-dimensional euclidean space is an ordered pair (x,y) of real numbers. Let (x_0,y_0) be a given point, and ϵ a real positive number. Then the ϵ *neighborhood* of (x_0,y_0) is the set of all points (x,y) such that

$$[(x - x_0)^2 + (y - y_0)^2]^{1/2} < \epsilon$$

Visualize a circle of radius $\epsilon > 0$ with center at the point (x_0,y_0). Then the ϵ neighborhood of (x_0,y_0) is just the set of all points lying inside the circle. Note that points on the circle are not in the ϵ neighborhood. Suppose now that S is a set of points in the plane. A point (x_0,y_0) is called an *interior point of* S if there exists an ϵ neighborhood of (x_0,y_0) which is contained within S; that is, every point in the ϵ neighborhood is also a point of S. Observe that if (x_0,y_0) is an interior point of S, then certainly (x_0,y_0) itself is a point of S. As an example, let S be the set of all points in the first quadrant which do not lie on one of the coordinate axes. Then every point of S is an interior point of S. A set having the property that every member of the set is an interior point of the set is called an *open set*. Let $\mathcal{D}$ denote the set of all points lying on or within the circle of unit radius about the origin. Then each point within the circle is an interior point of $\mathcal{D}$. However, the points of $\mathcal{D}$ which lie on the circle are not interior points of $\mathcal{D}$ since no matter how small $\epsilon > 0$ is chosen, the ϵ neighborhood of a point on the circle will contain points which do not belong to $\mathcal{D}$. Thus $\mathcal{D}$ is not open.

A set $\mathcal{R}$ is called a *region* if (1) $\mathcal{R}$ is an open set and (2) each pair (x_1,y_1), (x_2,y_2) of distinct points of $\mathcal{R}$ can be joined by a polygonal line (broken-line path) which lies within $\mathcal{R}$. The set S of all points (x,y) such that $x > 1$ and $y > 2$ is an example of a region. Another example of a region is the set of all (x,y) such that $y < x^2$. An example of an open set which is not a region is furnished by the set of all points such that $x < 0$ or $x > 1$. Observe that $(-1,2)$ and $(2,3)$ are points in the set, but they cannot be joined by a polygonal line which lies entirely in the set.

Analogous definitions are made in higher-dimensional spaces once the concepts of neighborhood and path are clearly understood.

Let $\mathcal{R}$ be a region in the plane, and let f be a function which is defined and differentiable on $\mathcal{R}$. The continuum of points (x,y,z) in three-dimensional space defined by $z = f(x,y)$ whenever (x,y) belongs to $\mathcal{R}$ is visualized as a smooth surface. Such a surface may also be defined by an equation

$F(x,y,z) = 0$ provided the equation implicitly defines a differentiable function $z = f(x,y)$ on $\mathscr{R}$. Another mode of analytic representation of surfaces in three dimensions is the parametric form

$$x = f(u,v) \qquad y = g(u,v) \qquad z = h(u,v)$$

where u, v are parameters. Here it is assumed that the functions f, g, h are continuously differentiable and the Jacobians

$$J_1 = \frac{\partial(g,h)}{\partial(u,v)} = \frac{\partial g}{\partial u}\frac{\partial h}{\partial v} - \frac{\partial h}{\partial u}\frac{\partial g}{\partial v}$$

$$J_2 = \frac{\partial(h,f)}{\partial(u,v)} \qquad J_3 = \frac{\partial(f,g)}{\partial(u,v)}$$

are not simultaneously zero in some region of the uv plane.

Elimination of Arbitrary Functions

Recall the method used to derive an ordinary differential equation from a relation $\Phi(x,y,c) = 0$ containing an arbitrary constant c. Differentiation with respect to x yields another equation involving c, and then c is eliminated between the two relations to obtain a first-order differential equation $F(x,y,y') = 0$. If the arbitrary constants c_1, c_2 are eliminated from a given relation $\Phi(x,y,c_1,c_2) = 0$, a second-order equation $F(x,y,y',y'') = 0$ is obtained. In this case the given relation represents geometrically a two-parameter family of plane curves. The equation of any particular curve in the family satisfies the derived differential equation, and the differential equation states a property common to each curve of the family. For example, elimination of c_1, c_2 from $y - c_1 x - c_2 = 0$ results in the differential equation $y'' = 0$. The property common to each curve of the family in this case is that of zero curvature. Partial differential equations may also result from the elimination of arbitrary constants from a given relation. However, the degree of generality of form of solution of a partial differential is better shown by illustrating how such equations result from the elimination of arbitrary functions.

EXAMPLE 1-1 Eliminate the arbitrary functions f_1, f_2 in the relation $z = f_1(x) + f_2(y)$ and obtain a partial differential equation.

Consider the class of all functions of the independent variables x, y which are expressible

$$z = f_1(x) + f_2(y)$$

on the domain of definition, where f_1, f_2 are differentiable. For any such function differentiation yields $z_x = f_1'(x)$, and

$$z_{xy} = \frac{\partial}{\partial y}[f_1'(x)] = 0$$

Thus the second-order equation $z_{xy} = 0$ results.   Corresponding to each choice of functions f_1, f_2 a surface is obtained, the equation of which satisfies $z_{xy} = 0$.   Examples are

$$z = 6e^x + \log (1 + \cos^2 y)$$
$$z = x^4 - x + 3 \cosh^3 y$$

Note that if c_1, c_2 are arbitrary constants, then the function defined by

$$z = 6c_1 e^x + x^4 - x + c_2 \log (1 + \cos^2 y) + 3 \cosh^3 y$$

satisfies $z_{xy} = 0$ on its domain.   But so does

$$z = 6c_1 e^x + c_2(x^4 - x) + c_3 \log (1 + \cos^2 y) + 3c_4 \cosh^3 y$$

for arbitrary choice of constants c_1, c_2, c_3, c_4.   Clearly any finite number of arbitrary constants can be made to appear.   The partial differential equation $z_{xy} = 0$ expresses a geometric property common to each surface in the family.   The property is that the principal sections (the sections furnishing maximum and minimum values of curvature relative to all those curvatures obtained by taking normal sections through the surface at the point) at each point on the surface are obtained by taking a section either parallel to the yz plane or parallel to the xz plane.

EXAMPLE 1-2   Find the differential equation of the family of all surfaces of revolution about the z axis.

A surface of this type has an equation of the form $z = f(x^2 + y^2)$.   Let $t = x^2 + y^2$; then $z_x = 2xf'(t)$, and $z_y = 2yf'(t)$.   If $f'(t)$ is eliminated, the first-order linear equation $yp - xq = 0$ is obtained.   Recall that a normal $\mathbf{N}$ to a surface $z = f(x,y)$ is given by $\mathbf{N} = p\mathbf{i} + q\mathbf{j} - \mathbf{k}$, where $\mathbf{i}$, $\mathbf{j}$, $\mathbf{k}$ are the usual unit vectors along the coordinate axes.   Let $\mathbf{r} = x\mathbf{i} + y\mathbf{j} + z\mathbf{k}$ be the variable position vector.   Then the partial differential equation of the family states that $\mathbf{N} \times \mathbf{r} \cdot \mathbf{k} = 0$ at each point of the surface; that is, $\mathbf{N} \times \mathbf{r}$ is perpendicular to the z axis.

EXAMPLE 1-3   Find the differential equation satisfied by all separable functions of the form $z = f(x)g(y)$.

Observe that $p = f'(x)g(y)$, $q = f(x)g'(y)$, and $s = f'(x)g'(y)$.   Thus

$$\frac{pq}{s} = f(x)g(y)$$

and hence such functions satisfy the almost-linear second-order equation $zs - pq = 0$.

EXAMPLE 1-4   Let $u = f(x,y,z)$, $v = g(x,y,z)$ be given differentiable functions of the independent variables x, y, z.   Find the differential equation of lowest order which is satisfied by the class of all functions defined implicitly by a relation of the form $F(u,v) = 0$, where the derivatives F_u, F_v are not both zero.

For each suitable choice of function F a function $z = \varphi(x,y)$ is implicitly defined.   It is desired to eliminate the arbitrary function F.   Differentiate the relation $F(u,v) = 0$, first with respect to x and with respect to y; then

$$F_u(u_x + u_z z_x) + F_v(v_x + v_z z_x) = 0$$

$$F_u(u_y + u_z z_y) + F_v(v_y + v_z z_y) = 0$$

Elimination of F_u and F_v between the two equations yields the quasilinear first-order equation

$$p \frac{\partial(u,v)}{\partial(z,y)} + q \frac{\partial(u,v)}{\partial(x,z)} = \frac{\partial(u,v)}{\partial(y,x)}$$

Partial differential equations may also result from the elimination of arbitrary constants appearing in a given relation. If two or more arbitrary constants are involved, then, in general, several distinct partial differential equations may be obtained. As an example consider

$$z = c_1 \cos x + c_2 \sin y$$

where c_1, c_2 are arbitrary constants. Then

$$z_{xx} = -c_1 \cos x \qquad z_{yy} = -c_2 \sin y$$

so that

$$z_{xx} + z_{yy} + z = 0$$

However observe also that $z_{xy} = 0$.

Geometric Examples

If f is a given differentiable function of x and y in some region in the xy plane, then the corresponding surface S in xyz space has a *tangent plane* at each point $P_0(x_0, y_0, z_0)$ on S. The equation of the tangent plane is

$$p_0(x - x_0) + q_0(y - y_0) - (z - z_0) = 0 \tag{1-13}$$

where $p_0 = f_x(x_0, y_0)$ and $q_0 = f_y(x_0, y_0)$. The unit vector

$$\mathbf{n} = \frac{p_0 \mathbf{i} + q_0 \mathbf{j} - \mathbf{k}}{(p_0^2 + q_0^2 + 1)^{1/2}} \tag{1-14}$$

where $\mathbf{i}$, $\mathbf{j}$, and $\mathbf{k}$ are the unit vectors along the x, y, and z axes, respectively, is called the *unit normal* to S at P_0. The vector $\mathbf{n}$ is perpendicular to the tangent plane at P_0. If the equation of S is in the implicit form $F(x,y,z) = 0$, and if $F_x^2 + F_y^2 + F_z^2 \neq 0$, the equation of the tangent plane to S at P_0 is

$$F_x|_{P_0}(x - x_0) + F_y|_{P_0}(y - y_0) + F_z|_{P_0}(z - z_0) = 0 \tag{1-15}$$

The unit normal at P_0 is given by

$$\mathbf{n} = \frac{F_x|_{P_0}\mathbf{i} + F_y|_{P_0}\mathbf{j} + F_z|_{P_0}\mathbf{k}}{(F_x^2 + F_y^2 + F_z^2)|_{P_0}} \tag{1-16}$$

Suppose that the surface S is given in parametric form $x = f(u,v)$, $y = g(u,v)$, $z = h(u,v)$, where the Jacobians J_1, J_2, J_3 defined previously are not simultaneously zero. Then the equation of the tangent plane to S at $P_0(x_0, y_0, z_0)$ is

$$J_1(x - x_0) + J_2(y - y_0) + J_3(z - z_0) = 0 \qquad (1\text{-}17)$$

and the normal

$$\mathbf{n} = \frac{J_1 \mathbf{i} + J_2 \mathbf{j} + J_3 \mathbf{k}}{J_1^2 + J_2^2 + J_3^2} \qquad (1\text{-}18)$$

In Eqs. (1-17) and (1-18) the Jacobians are evaluated at u_0, v_0 corresponding to the values x_0, y_0, z_0.

EXAMPLE 1-5 Find the differential equation of the family of all tangent planes to the ellipsoid $x^2 + 4y^2 + 4z^2 = 4$ which are not perpendicular to the xy plane.

If $P_0(x_0, y_0, z_0)$ is a point on the ellipsoid, $z_0 \neq 0$, the equation of the tangent plane to the surface at P_0 is

$$x_0(x - x_0) + 4y_0(y - y_0) + 4z_0(z - z_0) = 0$$

or

$$x_0 x + 4y_0 y + 4z_0 z = 4$$

As P_0 varies over the ellipsoid, this equation yields all tangent planes in the family. Observe that only two arbitrary constants are present, since P_0 must lie on the surface. Differentiation gives

$$x_0 + 4z_0 p = 0 \qquad 4y_0 + 4z_0 q = 0$$

Substitute $-4z_0 p$ for x_0 and $-z_0 q$ for y_0 in the equation of the family of tangent planes. The result is

$$px + qy - z = -\frac{1}{z_0}$$

Now

$$\left(\frac{x_0}{z_0}\right)^2 + 4\left(\frac{y_0}{z_0}\right)^2 + 4 = \frac{4}{z_0^2}$$

Hence

$$16p^2 + 4q^2 + 4 = \frac{4}{z_0^2}$$

The differential equation of the family of tangent planes is

$$xp + yq - z = -(1 + 4p^2 + q^2)^{\frac{1}{2}}$$

1-4 LINEAR FIRST-ORDER EQUATIONS

A linear first-order partial differential equation in two independent variables x, y and dependent variable z has the form

$$A(x,y)\frac{\partial z}{\partial x} + B(x,y)\frac{\partial z}{\partial y} + C(x,y)z = G(x,y) \qquad (1\text{-}19)$$

It is assumed in this section that the coefficients A, B, C and given function G have continuous first derivatives with respect to x and y in some region $\mathscr{R}$ of the xy plane and that at least one of the coefficients A, B does not vanish on $\mathscr{R}$. In Eq. (1-19) there appears the differential operator

$$L = A\frac{\partial}{\partial x} + B\frac{\partial}{\partial y} + C \qquad (1\text{-}20)$$

The operator is *linear*; that is, L has the characteristic property

$$L(c_1 z_1 + c_2 z_2) = c_1 L z_1 + c_2 L z_2 \qquad (1\text{-}21)$$

which holds for every pair of constants c_1, c_2 and for every pair of functions z_1, z_2 which are differentiable on $\mathscr{R}$. Equation (1-19) can be written briefly

$$Lz = G$$

The *homogeneous equation* corresponding to Eq. (1-19) is

$$Lz = 0 \qquad (1\text{-}22)$$

By a *solution* of Eq. (1-19) on $\mathscr{R}$ is meant a function $z = \varphi(x,y)$ with continuous first partial derivatives in $\mathscr{R}$ such that if φ and the derivatives φ_x, φ_y are substituted into Eq. (1-19), an identity in x and y on $\mathscr{R}$ results. The corresponding surface in xyz space is called an *integral surface*. The phrase *general solution of the homogeneous equation* means a relation involving an arbitrary function such that each choice for the arbitrary function yields a solution of Eq. (1-22). If z_h denotes the general solution of the homogeneous equation, and if z_p denotes a particular solution of the inhomogeneous equation (1-19), then

$$z = z_h + z_p \qquad (1\text{-}23)$$

is called the *general solution of the inhomogeneous equation*. It is a consequence of the linearity property of the operator L that if z_p is any particular solution of Eq. (1-19) and z is given by Eq. (1-23), then

$$Lz = L(z_h + z_p) = Lz_h + Lz_p = Lz_p = G$$

Hence z is a solution of the inhomogeneous equation.

If the general solution z_h includes all solutions of the homogeneous equation on $\mathscr{R}$, the relation (1-23) obtains all solutions of the inhomogeneous equation on $\mathscr{R}$. Again this is a consequence of the linearity of L. Suppose that v is a solution of Eq. (1-19) on $\mathscr{R}$. Let $w = v - z_p$. Then

$$Lw = L(v - z_p) = Lv - Lz_p = G - G = 0$$

so that w is a solution of the homogeneous equation on $\mathscr{R}$. By the assumption made on z_h, $v - z_p = z_h$ for some choice of the arbitrary function which appears in z_h. Thus $v = z_h + z_p$. The following examples illustrate the foregoing ideas.

EXAMPLE 1-6

$$\frac{\partial z}{\partial x} + z = x$$

Here $A = 1$, $B = 0$, $C = 1$, and $G(x,y) = x$ in Eq. (1-19). The corresponding homogeneous equation is $p + z = 0$. Hold y constant and integrate with respect to x. Since

$$e^x \frac{\partial z}{\partial x} + e^x z = \frac{\partial}{\partial x}(e^x z) = 0$$

it follows that

$$e^x z = f(y) \qquad \text{or} \qquad z = e^{-x} f(y)$$

where f is an arbitrary function. Thus the general solution of the homogeneous equation is $z_h = e^{-x} f(y)$, f an arbitrary continuously differentiable function. All solutions (in the sense stated in the text) of the homogeneous equation are included in this relation. This follows from the manner in which it was derived. Corresponding to each choice of continuously differentiable function f, an integral surface of the homogeneous equation is obtained (see Fig. 1-1). Note that the general solution can also be written

$$F(e^x z, y) = 0$$

where F is an arbitrary differentiable function $F(u,v)$ such that $F_u \neq 0$ in some region of the uv plane. For if F is such a function, it is possible to solve $F(u,v) = 0$ and obtain $u = f(v)$, that is, $z = e^{-x} f(y)$, for some function f.

By inspection a particular solution of the inhomogeneous equation is $z_p = x - 1$. Accordingly the general solution of the given partial differential equation is

$$z = e^{-x} f(y) + x - 1$$

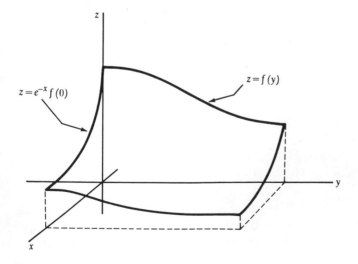

$z = e^{-x} f(0)$

$z = f(y)$

Figure 1-1

where f is an arbitrary continuously differentiable function. This relation represents a class of integral surfaces of the partial differential equation.

EXAMPLE 1-7   Obtain the general solution of

$$Ap + Bq + Cz = G$$

where A, B, and C are constants such that $A^2 + B^2 \neq 0$, and G is a given continuously differentiable function of x and y.

The corresponding homogeneous equation is

$$Ap + Bq + Cz = 0 \tag{1-24}$$

A technique often used to simplify the form of a given differential equation is to make a change of variable. Since in this case the differential equation has constant coefficients, consider a homogeneous affine transformation of the independent variables

$$\xi = c_{11}x + c_{12}y \qquad \eta = c_{21}x + c_{22}y$$

The coefficients c_{ij} are chosen subsequently. Now

$$z_x = c_{11}z_\xi + c_{21}z_\eta \qquad z_y = c_{12}z_\xi + c_{22}z_\eta$$

If these expressions for z_x, z_y are substituted into Eq. (1-24), one obtains the equation

$$(Ac_{11} + Bc_{12})z_\xi + (Ac_{21} + Bc_{22})z_\eta + Cz = 0$$

Assume $A \neq 0$, and choose $c_{11} = 1$, $c_{12} = 0$, $c_{21} = B$, and $c_{22} = -A$. Then

$$\xi = x \qquad \eta = Bx - Ay$$

and the transformed differential equation is

$$z_\xi + \frac{Cz}{A} = 0$$

Hold η fixed and integrate with respect to ξ. The result is

$$z = e^{-C\xi/A}f(\eta)$$

where f is an arbitrary function. Thus the general solution of the homogeneous equation is

$$z_h = e^{-Cx/A}f(Bx - Ay) \tag{1-25}$$

where f is an arbitrary continuously differentiable function. The general solution can also be written

$$F(ze^{Cx/A}, Bx - Ay) = 0$$

where $F(u,v)$ is an arbitrary continuously differentiable function such that $F_u \neq 0$. If $A = 0$, then $B \neq 0$, and the general solution of (1-24) is

$$z_h = e^{-Cy/B}f(Bx) \tag{1-26}$$

or

$$F(ze^{Cy/B}, Bx) = 0$$

If $A \neq 0$, and if z_p is a particular solution of the original differential equation, then the general solution of the inhomogeneous equation is

$$z = e^{-Cx/A}f(Bx - Ay) + z_p \tag{1-27}$$

To find particular solutions many of the techniques used to obtain particular solutions of inhomogeneous linear ordinary differential equations are applicable. If G is a sum of terms, and if a solution of the differential equation corresponding to each separate term is found, the superposition of these solutions yields a particular solution z_p. In the special case where G involves but one independent variable, say x, the term $B \, \partial z / \partial y$ is omitted, and the resulting differential equation is integrated with respect to x to obtain a particular solution. For example, if $G(x) = 3x^2$, integration of the ordinary differential equation

$$A \frac{dz}{dx} + Cz = 3x^2$$

gives the particular solution $z_p = (3C^2 x^2 - 6ACx + 6A^2)/C^3$ of the partial differential equation. In further illustration, suppose that $G(x,y) = 4e^{ax+by}$, where a and b are constants. Then a particular solution $z = ce^{ax+by}$ is assumed, where c is a constant to be determined. Substitution into the differential equation gives

$$c(Aa + Bb + C) = 4$$

Accordingly if $Aa + Bb + C \neq 0$, then a particular solution is

$$z_p = \frac{4e^{ax+by}}{Aa + Bb + C}$$

If $Aa + Bb + C = 0$, assume a particular solution of the form $z_p = cxe^{ax+by}$, where c is a constant to be determined.

To derive the form of the general solution of the linear first-order homogeneous equation

$$A(x,y) \frac{\partial z}{\partial x} + B(x,y) \frac{\partial z}{\partial y} + C(x,y)z = 0 \tag{1-28}$$

let

$$\xi = \xi(x,y) \qquad \eta = \eta(x,y) \tag{1-29}$$

be a transformation on $\mathscr{R}$ with Jacobian

$$\frac{\partial(\xi,\eta)}{\partial(x,y)} \neq 0 \tag{1-30}$$

Since $z_x = z_\xi \xi_x + z_\eta \eta_x$ and $z_y = z_\xi \xi_y + z_\eta \eta_y$, the transformed equation is

$$(A\xi_x + B\xi_y)z_\xi + (A\eta_x + B\eta_y)z_\eta + Cz = 0 \tag{1-31}$$

It is understood that the coefficients are now expressed in terms of the variables ξ, η. The transformation (1-29) is arbitrary to within the requirement (1-30). In order to simplify Eq. (1-31) choose η such that

$$A\eta_x + B\eta_y = 0 \tag{1-32}$$

This can be accomplished as follows. Assume $A(x,y) \neq 0$ and consider the ordinary differential equation

$$\frac{dy}{dx} = \frac{B(x,y)}{A(x,y)} \qquad (1\text{-}33)$$

Let the general solution of Eq. (1-33) be

$$\eta(x,y) = c \qquad (1\text{-}34)$$

where $\eta_y \neq 0$ and c is an arbitrary constant. Then the function $\eta(x,y)$ determined in this manner satisfies Eq. (1-32), since

$$\eta_x \, dx + \eta_y \, dy = 0$$

and so

$$\frac{B}{A} = -\frac{\eta_x}{\eta_y}$$

Now choose $\xi(x,y) = x$. Then $\xi_x \eta_y - \xi_y \eta_x \, \eta_y \neq 0$, and the transformation constructed in this manner is invertible. Equation (1-31) becomes simply

$$Az_\xi + Cz = 0 \qquad (1\text{-}35)$$

Hold η fixed and integrate with respect to ξ. The result is

$$z(\xi,\eta) = f(\eta) \exp\left[-\int \frac{C}{A} \, d\xi \right]$$

where f is an arbitrary function. The function

$$\psi(\xi,\eta) = \exp\left[-\int \frac{C(\xi,\eta)}{A(\xi,\eta)} \, d\xi \right]$$

satisfies Eq. (1-35). Hence

$$u(x,y) = \psi[x,\eta(x,y)] \qquad (1\text{-}36)$$

is a particular solution of Eq. (1-28). This can be established directly by differentiation. The *general solution* of Eq. (1-28) is

$$z_h = u(x,y)f\,[\eta(x,y)] \qquad (1\text{-}37)$$

To verify this, let f be a given continuously differentiable function. Then

$$z_x = u_x f(\eta) + uf'(\eta)\eta_x \qquad z_y = u_y f(\eta) + uf'(\eta)\eta_y$$

and so

$$Az_x + Bz_y + Cz = (Au_x + Bu_y + Cu)f(\eta) + (A\eta_x + B\eta_y)f'(\eta) = 0$$

Recall that the coefficients in Eq. (1-28) are assumed to be continuously differentiable in $\mathcal{R}$. All solutions of Eq. (1-28) are embodied in the general solution (1-37) in the following local sense.   If v is a solution of Eq. (1-28) on $\mathscr{R}$, and if P_0 is a point of $\mathscr{R}$, then there exists a neighborhood of P_0 and a continuously differentiable function f such that

$$v = u(x,y)f\,[\eta(x,y)]$$

on the neighborhood, where the functions u and η are those defined above. To establish this observe that the equations

$$Av_x + Bv_y + Cv = 0$$

$$Au_x + Bu_y + Cu = 0$$

hold simultaneously in a neighborhood of P_0.   Multiply the first equation by u, the second by v, and subtract.   The result is

$$A(uv_x - vu_x) + B(uv_y - vu_y) = 0$$

Let $w = v/u$.   Then the equations

$$Aw_x + Bw_y = 0$$

$$A\eta_x + B\eta_y = 0$$

hold simultaneously.   Since $A^2 + B^2 \neq 0$ in $\mathscr{R}$, it follows that the Jacobian

$$\frac{\partial(w,\eta)}{\partial(x,y)} = w_x\eta_y - \eta_x w_y = 0$$

Now (see Ref. 11) the identical vanishing of the Jacobian, together with the fact that $\eta_y \neq 0$, implies that there exists a neighborhood of P_0 and a continuously differentiable function f such that

$$w(x,y) = f\,[\eta(x,y)]$$

on the neighborhood; that is,

$$v = u(x,y)f\,[\eta(x,y)]$$

Let z_p be a particular solution of the inhomogeneous equation

$$Lz = Az_x + Bz_y + Cz = G$$

where G is a given continuously differentiable function on $\mathscr{R}$.   Then the general solution of the inhomogeneous equation is

$$z = z_p + uf(\eta)$$

where u and η are those defined previously.   First, the linearity of the operator L implies that for each choice of continuously differentiable function f, the function z so defined satisfies the inhomogeneous equation.   Next, let φ be any solution of the inhomogeneous equation on $\mathscr{R}$.   Let $v = \varphi - z_p$.

Then

$$Lv = L(\varphi - z_p) = L\varphi - Lz_p = G - G = 0$$

Hence v is a solution of the homogeneous equation. From the results of the preceding paragraph it follows that, locally at least, there exists a function f such that

$$v = u(x,y)f[\eta(x,y)]$$

and thus

$$\varphi = z_p + u(x,y)f[\eta(x,y)]$$

EXAMPLE 1-8 $x^2p - xyq + yz = 0$
The coefficients are $A(x,y) = x^2$, $B(x,y) = -xy$, and $C(x,y) = y$. Equation (1-33) takes the form

$$\frac{dy}{dx} = -\frac{y}{x}$$

Thus $x\,dy + y\,dx = 0$, and

$$\eta(x,y) = xy = c$$

The appropriate transformation is $\xi = x$, $\eta = xy$. The coefficient $A(\xi,\eta) = \xi^2$, and $C(\xi,\eta) = \eta/\xi$. Hence Eq. (1-35) is

$$\xi^2 z_\xi + \frac{\eta z}{\xi} = 0 \qquad \text{or} \qquad z_\xi + \frac{\eta z}{\xi^3} = 0$$

Then

$$\int \frac{C(\xi,\eta)}{A(\xi,\eta)}\,d\xi = \int \frac{\eta}{\xi^3}\,d\xi = -\frac{\eta}{2\xi^2}$$

and $\psi(\xi,\eta) = e^{\eta/2\xi^2}$. A particular solution of the partial differential equation is

$$u(x,y) = e^{y/2x}$$

The general solution of the partial differential equation is

$$z = e^{y/2x}f(xy)$$

where f is an arbitrary function.

1-5   QUASILINEAR FIRST-ORDER EQUATIONS; METHOD OF LAGRANGE

In this section quasilinear equations of the form

$$P(x,y,z)\frac{\partial z}{\partial x} + Q(x,y,z)\frac{\partial z}{\partial y} = R(x,y,z) \tag{1-38}$$

in the two independent variables x, y and the dependent variable z are considered. A technique called the *method of Lagrange* enables one to derive solutions of such an equation in many cases. The technique encompasses the method for linear first-order equations with variable coefficients set forth in Sec. 1-4.   The extension of Lagrange's method to quasilinear first-order equations in n independent variables is indicated in the problems at the end of the chapter.

Let $\mathscr{T}$ be a region in three-dimensional xyz space.   Assume the coefficients P, Q, R in Eq. (1-38) are continuously differentiable functions of the variables x, y, z in $\mathscr{T}$ and that P and Q do not vanish in $\mathscr{T}$.   By the *projection of $\mathscr{T}$ onto the xy* plane is meant the set of all points (x,y) in the plane such that for some value z the point (x,y,z) belongs to $\mathscr{T}$ (see Fig. 1-2).   Let $\mathscr{R}$ denote the projection of $\mathscr{T}$ onto the xy plane.   A *solution* of Eq. (1-38) on $\mathscr{R}$ is a function $z = \varphi(x,y)$ with continuous first partial derivatives such that (1) if (x_0,y_0) is a point of $\mathscr{R}$ and $z_0 = \varphi(x_0,y_0)$, then (x_0,y_0,z_0) lies in $\mathscr{T}$, and (2) the relation

$$P[x,y,\varphi(x,y)]\,\frac{\partial \varphi}{\partial x} + Q[x,y,\varphi(x,y)]\,\frac{\partial \varphi}{\partial y} = R[x,y,\varphi(x,y)]$$

holds identically on $\mathscr{R}$.   It is shown in Sec. 1-6 that under the preceding hypotheses Eq. (1-38) always has a solution, and indeed infinitely many

Figure 1-2

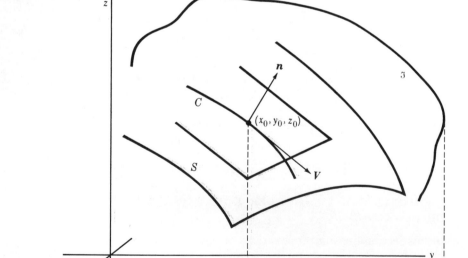

distinct solutions, in a neighborhood of a given point of $\mathcal{R}$. It should be noted, however, that Eq. (1-38) need not have any solutions, in the sense stated above, if only continuity is required of the coefficients P, Q, and R. In this connection see Ref. 6.

Method of Lagrange

Associated with Eq. (1-38) is the system of first-order ordinary differential equations

$$\frac{dx}{P} = \frac{dy}{Q} = \frac{dz}{R} \tag{1-39}$$

called the *subsidiary equations*. An equivalent system is

$$\frac{dy}{dx} = \frac{Q}{P} \qquad \frac{dz}{dx} = \frac{R}{P} \tag{1-40}$$

wherein x is the independent variable. The general solution of the system (1-40) has the form

$$y = y(x,c_1,c_2) \qquad z = z(x,c_1,c_2) \tag{1-41}$$

where c_1, c_2 are arbitrary constants. If these equations are solved for c_1 and c_2, then the general solution of the subsidiary equations can be written in the form

$$u(x,y,z) = c_1 \qquad v(x,y,z) = c_2 \tag{1-42}$$

Each relation $u = c_1$, $v = c_2$ is called an *integral* of the subsidiary equations. It is assumed that the functions u, v are functionally independent in $\mathcal{T}$; that is, the Jacobians

$$\frac{\partial(u,v)}{\partial(x,y)} \qquad \frac{\partial(u,v)}{\partial(x,z)} \qquad \frac{\partial(u,v)}{\partial(y,z)}$$

are not all zero at any point of $\mathcal{T}$. The relation

$$F(u,v) = 0 \tag{1-43}$$

where F is arbitrary, is called the *general solution of* Eq. (1-38), for reasons set forth below. The general solution can be written in the alternative forms $u = f(v)$ and $v = g(u)$, where f and g are arbitrary functions. Before discussing the theory of the method of Lagrange several examples are given to illustrate the technique.

EXAMPLE 1-9 $xzp + yzq = -(x^2 + y^2)$

Here $P(x,y,z) = xz$, $Q(x,y,z) = yz$, $R(x,y,z) = -(x^2 + y^2)$. These are continuously differentiable functions of the variables x, y, z everywhere. The subsidiary equations are

$$\frac{dx}{xz} = \frac{dy}{yz} = \frac{dz}{-(x^2 + y^2)}$$

Since $dx/x = dy/y$, integration gives $\log y = \log x + c_1$. The first integral of the subsidiary equations is then

$$u(x,y,z) = \frac{y}{x} = c_1$$

This relation is used to eliminate y from the equation

$$\frac{dx}{xz} = -\frac{dz}{x^2 + y^2}$$

Thus

$$\frac{dx}{x} = -\frac{z\,dz}{x^2(1 + c_1{}^2)} \qquad \text{or} \qquad x\,dx = -\frac{z\,dz}{1 + c_1{}^2}$$

Integration gives

$$(1 + c_1{}^2)x^2 + z^2 = c_2$$

Now replace c_1 by y/x. The second integral of the subsidiary equations is

$$v(x,y,z) = x^2 + y^2 + z^2 = c_2$$

The Jacobians are

$$\frac{\partial(u,v)}{\partial(x,y)} = -2\frac{x^2 + y^2}{x^2} \qquad \frac{\partial(u,v)}{\partial(x,z)} = -2\frac{zy}{x^2} \qquad \frac{\partial(u,v)}{\partial(y,z)} = 2\frac{z}{x}$$

They are different from zero in any region of space in which $xyz \neq 0$. If $\mathscr{T}$ is such a region, then the general solution of the quasilinear equation is

$$F\left(\frac{y}{x},\, x^2 + y^2 + z^2\right) = 0$$

To see why this relation is called the general solution in this particular example let $F(u,v)$ be a given continuously differentiable function of u, v such that $F_v \neq 0$. Then $F(u,v) = 0$ implicitly defines a differentiable function $z = \varphi(x,y)$. The function φ is a solution of the partial differential equation, since by the results of Example 1-4, elimination of F from the relation $F(u,v) = 0$ yields

$$p\,\frac{\partial(u,v)}{\partial(z,y)} + q\,\frac{\partial(u,v)}{\partial(x,z)} = -\frac{\partial(u,v)}{\partial(x,y)}$$

where now $p = \varphi_x$, $q = \varphi_y$. Substitution of the previously derived expressions for the Jacobians gives

$$-\frac{2z}{x}p - \frac{2zy}{x^2}q = \frac{2(x^2 + y^2)}{x^2}$$

or

$$xzp + yzq = -(x^2 + y^2)$$

The general solution can also be written in the alternative forms

$$\frac{y}{x} = f(x^2 + y^2 + z^2) \qquad x^2 + y^2 + z^2 = g\left(\frac{y}{x}\right)$$

where f, g are arbitrary functions.

A useful technique for integrating a system of first-order equations is that of multipliers. Recall from algebra that if $a/b = c/d$, then

$$\frac{\lambda a + \mu c}{\lambda b + \mu d} = \frac{a}{b} = \frac{c}{d}$$

for arbitrary values of multipliers λ, μ. Hence from Eq. (1-39) it follows that

$$\frac{\lambda\, dx + \mu\, dy + \nu\, dz}{\lambda P + \mu Q + \nu R} = \frac{dx}{P} = \frac{dy}{Q} = \frac{dz}{R}$$

for arbitrary multipliers λ, μ, ν. In this manner related differential equations can be formed, some of which may be readily integrated. In particular if λ, μ, ν are chosen such that $\lambda P + \mu Q + \nu R = 0$, then $\lambda\, dx + \mu\, dy + \nu\, dz = 0$. Now if there exists a function u such that

$$du = \lambda\, dx + \mu\, dy + \nu\, dz$$

then $u(x,y,z) = c_1$ is an integral of the subsidiary equations.

EXAMPLE 1-10 $\quad (y - x)p + (y + x)q = \dfrac{x^2 + y^2}{z}$

The subsidiary equations are

$$\frac{dx}{y - x} = \frac{dy}{y + x} = \frac{z\, dz}{x^2 + y^2}$$

Since $P + Q = 2y$, take $\lambda = 1$, $\mu = 1$, and $\nu = 0$. Then

$$\frac{dx + dy}{2y} = \frac{dy}{y + x} \qquad \text{or} \qquad (y + x)(dx + dy) = 2y\, dy$$

Integration gives $(y + x)^2/2 = y^2 + c_1$. Thus the first integral is

$$u(x,y,z) = x^2 + xy - y^2 = c_1$$

Now $xP - yQ = -(x^2 + y^2) = -zR$. Hence

$$x\, dx - y\, dy + z\, dz = 0$$

A second independent integral is

$$v(x,y,z) = x^2 - y^2 + z^2 = c_2$$

The general solution of the partial differential equation is

$$F(x^2 + xy - y^2, x^2 - y^2 + z^2) = 0$$

Note that the two independent integrals obtained above are not the only pair which can be used in writing down the general solution. Observe that $w(x,y,z) = z^2 + 2xy = c_1$ and $v(x,y,z) = c_2$ also constitute a pair of independent integrals of the subsidiary equations. Accordingly the general solution can also be written

$$F(z^2 + 2xy, x^2 - y^2 + z^2) = 0$$

where F is arbitrary.

EXAMPLE 1-11 $(z^2 - 2yz - y^2)p + (xy + xz)q = xy - xz$
The subsidiary equations are

$$\frac{dx}{z^2 - 2yz - y^2} = \frac{dy}{xy + xz} = \frac{dz}{xy - xz}$$

Since $Q/(y + z) = R/(y - z)$, it follows that $dy/(y + z) = dz/(y - z)$. Hence $y \, dy - z \, dz = z \, dy + y \, dz$. Integration yields the first integral

$$u(x,y,z) = z^2 - y^2 + 2zy = c_1$$

Also, $xP + yQ + zR = 0$, so $x \, dx + y \, dy + z \, dz = 0$. A second independent integral is

$$v(x,y,z) = x^2 + y^2 + z^2 = c_2$$

The general solution is

$$F(z^2 - y^2 + 2xy, \, x^2 + y^2 + z^2) = 0$$

Characteristic Curves

From a geometric as well as an analytic viewpoint there are many connections between the system of ordinary differential equations (1-39) and the partial differential equation (1-38).   Of importance in this regard is the vector field

$$\mathbf{V}(x,y,z) = P(x,y,z)\mathbf{i} + Q(x,y,z)\mathbf{j} + R(x,y,z)\mathbf{k} \tag{1-44}$$

called the *characteristic vector field* associated with Eq. (1-38).   From each point in the region $\mathscr{T}$ there emanates a characteristic vector.   Let $z = \varphi(x,y)$ be a solution of Eq. (1-38) in $\mathscr{R}$.   Geometrically the solution is visualized as a smooth surface S lying in $\mathscr{T}$, called an *integral surface*.   Let $\mathbf{n}$ be the normal vector to S [see Eq. (1-14)].   If $\mathbf{V}$ is the characteristic vector at a point on S, then

$$\mathbf{V} \cdot \mathbf{n} = \frac{Pp + Qq - R}{(p^2 + q^2 + 1)^{1/2}} = 0$$

Accordingly $\mathbf{V}$ lies in the tangent plane to S at each point on S.   Conversely, if S is a smooth surface with equation $z = \varphi(x,y)$, and if at each point on S the characteristic vector $\mathbf{V}$ at that point lies in the tangent plane to S, then S is an integral surface, and $z = \varphi(x,y)$ is a solution of Eq. (1-38) (see Fig. 1-2).

A smooth curve C lying in $\mathscr{T}$ and such that the characteristic vector $\mathbf{V}$ at each point on C is tangent to C is called a *characteristic curve* of Eq. (1-38). Accordingly the characteristic curves are just the field lines of the vector field $\mathbf{V}(x,y,z)$.   A necessary and sufficient condition for a smooth curve C to be a characteristic curve is that the subsidiary equations (1-39) hold along C. This follows from the fact that a set of direction numbers of the tangent to C is the set of differentials dx, dy, dz, and the subsidiary equations state that these direction numbers are proportional to the set of direction numbers P, Q, R of the vector $\mathbf{V}$; that is, the tangent has the direction of $\mathbf{V}$.

The characteristic curves constitute a two-parameter family of space curves. Exactly one characteristic passes through each point of $\mathscr{T}$. To see this, take x as the independent variable in Eq. (1-40). The system (1-40) satisfies the hypotheses of the fundamental existence theorem for a system of first-order equations (see Ref. 2). Thus, if x_0 is fixed and y_0, z_0 are chosen values, then there exists a unique solution

$$y = y(x,y_0,z_0) \qquad z = z(x,y_0,z_0) \tag{1-45}$$

in a neighborhood of x_0 such that

$$y(x_0,y_0,z_0) = y_0 \qquad z(x_0,y_0,z_0) = z_0 \tag{1-46}$$

In xyz space each function in Eq. (1-45) represents a cylinder, and relation (1-46) states that these cylinders intersect at (x_0,y_0,z_0). In a neighborhood of the point (x_0,y_0,z_0) the cylinders intersect in a smooth curve C which passes through (x_0,y_0,z_0). Since the system (1-40) is satisfied along C, it follows that C is a characteristic curve. By the uniqueness of solution of the system (1-40) subject to conditions (1-46), C is the unique characteristic through (x_0,y_0,z_0). Now regard y_0 and z_0 as parameters. Then the family of curves defined in Eq. (1-45) constitute a two-parameter family of characteristic curves such that exactly one member passes through each point of $\mathscr{T}$. Moreover every characteristic curve of the partial differential equation belongs to this family. Note that the preceding statements imply that distinct characteristic curves cannot intersect.

If S is an integral surface of Eq. (1-38) and C is a characteristic curve, either C does not intersect S, or else C is embedded in S. To show this, suppose $z = \varphi(x,y)$ is the equation of the integral surface S. Let C be a characteristic curve which intersects S at (x_0,y_0,z_0). In a neighborhood of $x = x_0$ there is a unique solution $y = y(x)$ of the differential equation

$$\frac{dy}{dx} = \frac{Q[x,y,\varphi(x,y)]}{P[x,y,\varphi(x,y)]} \tag{1-47}$$

such that $y(x_0) = y_0$. The space curve C' defined by the equations $y = y(x)$, $z = \varphi[x,y(x)]$ passes through the point (x_0,y_0,z_0) and lies in the surface S. Also

$$\frac{dz}{dx} = \varphi_x + \varphi_y y' = \varphi_x + \varphi_y \frac{Q}{P} = \frac{P\varphi_x + Q\varphi_y}{P} = \frac{R}{P}$$

Thus the functions which define C' satisfy the system (1-40); moreover $y(x_0) = y_0$, and $z(x_0) = z_0$. But the functions which define the characteristic C satisfy exactly the same conditions. There can be but one set of functions $y(x)$, $z(x)$ with these properties. Accordingly the characteristic C and the embedded curve C' are one and the same.

Geometric and Analytic Justification of Lagrange's Method

A heuristic geometric argument based on the preceding result may be given in support of the fact that under the present hypotheses on P, Q, R each solution of Eq. (1-38) is defined by a relation of the form in Eq. (1-43), at least in the small. Let S be an integral surface of Eq. (1-38), and let (x_0, y_0, z_0) be a point on S. Choose a smooth curve Γ such that (1) Γ lies in S and passes through (x_0, y_0, z_0), (2) Γ is *noncharacteristic*; i.e., the tangent to Γ is nowhere parallel to the characteristic vector $\mathbf{V}$ along Γ. Through each point of Γ there passes exactly one characteristic C, and C is embedded in S. Also, the tangent to such a characteristic C does not coincide with the tangent to Γ at the point where C intersects Γ. In this manner a subfamily of the family of characteristics is singled out, and in a neighborhood of (x_0, y_0, z_0) this sub-family of characteristic curves generates, or sweeps out, the integral surface S (see Fig. 1-3). The particular subfamily in question is obtained from the two-parameter family of all characteristics by imposing a functional relation $F(c_1, c_2) = 0$ on the parameters c_1, c_2 appearing in Eq. (1-42).

Assume that

$$u(x,y,z) = c_1 \qquad v(x,y,z) = c_2 \tag{1-48}$$

are functionally independent integrals of the subsidiary equations in $\mathcal{T}$. Let (x_0, y_0, z_0) be a fixed but otherwise arbitrarily chosen point of $\mathcal{T}$. Then in a neighborhood of (x_0, y_0, z_0) the surfaces

$$u(x,y,z) = u(x_0, y_0, z_0) \qquad v(x,y,z) = v(x_0, y_0, z_0) \tag{1-49}$$

intersect in a smooth curve C, which passes through the point. The curve C is the characteristic passing through (x_0, y_0, z_0). Choose x as the parameter

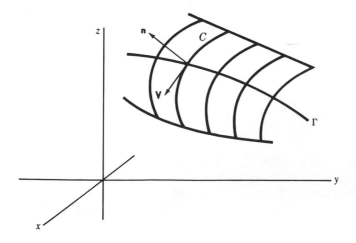

Figure 1-3

along C; then differentiation of Eqs. (1-49) yields

$$u_x + u_y \frac{dy}{dx} + u_z \frac{dz}{dx} = 0 \qquad v_x + v_y \frac{dy}{dx} + v_z \frac{dz}{dx} = 0$$

Hence, from the subsidiary equations (1-40), it follows that

$$Pu_x + Qu_y + Ru_z = 0 \qquad Pv_x + Qv_y + Rv_z = 0 \qquad (1\text{-}50)$$

These equations hold simultaneously along C.   In vector form they are

$$\mathbf{V} \cdot \nabla u = 0 \qquad \mathbf{V} \cdot \nabla v = 0$$

Thus (1-50) is just a restatement of the fact that at each point of C the characteristic vector $\mathbf{V}$ is perpendicular to each of the vectors ∇u, ∇v and so is parallel to $\nabla u \times \nabla v$.   Recall $\nabla u \times \nabla v$ has the same direction as the tangent to C at the point.

The term general solution applied to the relation (1-43) can be justified analytically.   Let u, v be as described in the preceding paragraph, and let F be any continuously differentiable function such that $F(u,v) = 0$ implicitly defines a function

$$z = \varphi(x,y)$$

having continuous partial derivatives in some neighborhood of a point (x_0,y_0,z_0) of $\mathscr{T}$.   Exactly as in Example 1-4, differentiation of $F = 0$ with respect to x and y and elimination of F lead to the equation

$$\frac{\partial(u,v)}{\partial(y,z)} \varphi_x + \frac{\partial(u,v)}{\partial(z,x)} \varphi_y = \frac{\partial(u,v)}{\partial(x,y)} \qquad (1\text{-}51)$$

Now

$$\frac{\partial(u,v)}{\partial(z,x)} = \frac{Q}{P} \frac{\partial(u,v)}{\partial(y,z)} \qquad \frac{\partial(u,v)}{\partial(x,y)} = \frac{R}{P} \frac{\partial(u,v)}{\partial(y,z)} \qquad (1\text{-}52)$$

The preceding relations are obtained from the simultaneous equations (1-50). If the expressions for the Jacobians are inserted in Eq. (1-51), the result is

$$P\varphi_x + Q\varphi_y = R \qquad (1\text{-}53)$$

Hence φ is a solution of the partial differential equation (1-38).   It is emphasized here that the preceding results are local, since the reasoning is based upon the existence theorem for the system of ordinary differential equations (1-40).   Thus the properties have been shown to hold only in some neighborhood of a point (x_0,y_0,z_0) of $\mathscr{T}$.

EXAMPLE 1-12 $xzp + yzq = -(x^2 + y^2)$
From Example 1-9 two independent integrals of the corresponding subsidiary equations are

$$u(x,y,z) = \frac{y}{x} = c_1 \qquad v(x,y,z) = x^2 + y^2 + z^2 = c_2$$

If c_1, c_2 are assigned positive values, the corresponding surfaces are a plane and a sphere. In the first octant these surfaces intersect in a segment of a circle (see Fig. 1-4).   To show directly that the curve of intersection is a characteristic, recall that the tangent to C has the direction of the vector $\nabla u \times \nabla v$.   Now

$$\nabla u = \left(\frac{-y}{x^2}\right)\mathbf{i} + \frac{1}{x}\mathbf{j} \qquad \nabla v = 2(x\mathbf{i} + y\mathbf{j} + z\mathbf{k})$$

so that

$$\nabla u \times \nabla v = \frac{2[xz\mathbf{i} + zy\mathbf{j} - (x^2 + y^2)\mathbf{k}]}{x^2} = \frac{2\mathbf{V}}{x^2}$$

Every characteristic in the first octant is a segment of such a circle.   Suppose F is a function such that $F(u,v) = 0$ can be solved for z to obtain

$$z = \varphi(x,y) = \left[f\left(\frac{y}{x}\right) - (x^2 + y^2)\right]^{1/2}$$

where φ is continuously differentiable.   Then φ satisfies

$$xz\varphi_x + yz\varphi_y = -(x^2 + y^2)$$

1-6 CAUCHY PROBLEM FOR QUASILINEAR FIRST-ORDER EQUATIONS

A fundamental problem in the study of ordinary differential equations is to determine a solution of a first-order equation $y' = f(x,y)$ which passes

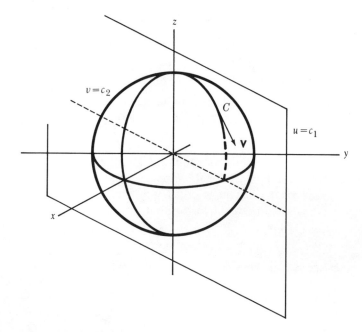

Figure 1-4

through a prescribed point in the xy plane.   Under quite general conditions a unique solution to the problem exists.   An analogous problem in the study of first-order partial differential equations in two independent variables x, y is to determine an integral surface such that the surface passes through a prescribed curve in xyz space.   Such a problem is termed a *Cauchy problem*. In this section a method of solving the Cauchy problem for Eq. (1-38) is described.

Before proceeding with the general case consider the following example.

EXAMPLE 1-13   Find a solution $z = \varphi(x,y)$ of $yp - xq = 0$ such that $\varphi(x,0) = x^4$.
The problem is to find an integral surface which passes through the curve Γ defined by the simultaneous equations $z = x^4$ and $y = 0$.   This curve lies in the xz plane.   By the method of Lagrange one obtains the general solution of the differential equation as

$$z = f(x^2 + y^2)$$

where f is arbitrary.   Every integral surface is a surface of revolution about the z axis. The condition that such a surface contains Γ is $f(x^2) = x^4$.   Thus $f(t) = t^2$.   The solution of the Cauchy problem is $z = (x^2 + y^2)^2$ (see Fig. 1-5).   There is but one surface of revolution about the z axis which contains Γ, and so the solution obtained is unique.

In general there may or may not exist a solution of the Cauchy problem for Eq. (1-38).   There is also the possibility that infinitely many distinct solutions exist.   All three cases are illustrated, with the aid of the partial

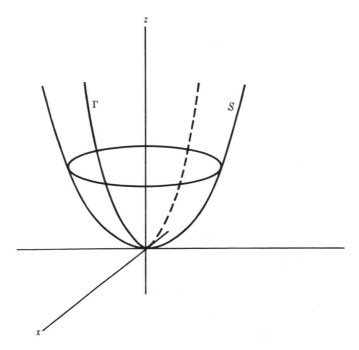

Figure 1-5

differential equation of the preceding example. If Γ is the curve stated in the example, there exists exactly one solution. Suppose instead Γ is the circle $x^2 + y^2 = 1$, $z = 1$. Choose any function $h(t)$ such that $h(1) = 1$ and consider the function $z = h(x^2 + y^2)$. The corresponding surface is an integral surface which contains Γ. Clearly there are infinitely many solutions in this case. Observe that the given curve Γ is now a characteristic curve of the partial differential equation. At the opposite extreme of circumstances let Γ be the ellipse $x^2 + y^2 = 1$, $z = y$. If $z = f(x^2 + y^2)$ is a solution of the Cauchy problem, then on the circle $x^2 + y^2 = 1$ one has $z = f(1)$, a constant. However this is incompatible with the requirement that $z = y$ whenever $x^2 + y^2 = 1$. Thus no solution exists. Note that in the last case the given curve is such that its projection on the xy plane coincides with the projection on the xy plane of a characteristic curve, but Γ itself is non-characteristic.

A Method of Solution

Let Γ be a given smooth curve defined parametrically by

$$x = f(t) \qquad y = g(t) \qquad z = h(t) \tag{1-54}$$

for $a < t < b$. Assume also that Γ is noncharacteristic. To construct an integral surface of Eq. (1-38) which contains Γ one can proceed as follows. Let $u = c_1$, $v = c_2$ be two independent integrals of the subsidiary equations (1-39). Write down the equations

$$u[f(t),g(t),h(t)] = c_1 \qquad v[f(t),g(t),h(t)] = c_2 \tag{1-55}$$

Eliminate t from the pair of equations so as to derive a functional relation

$$F(c_1,c_2) = 0 \tag{1-56}$$

between c_1 and c_2. Then the solution of the Cauchy problem is

$$F[u(x,y,z),v(x,y,z)] = 0 \tag{1-57}$$

EXAMPLE 1-14   Find an integral surface of

$$(y + xz)p + (x + yz)q = z^2 - 1$$

which passes through the parabola $x = t$, $y = 1$, $z = t^2$.
    The subsidiary equations are

$$\frac{dx}{y + xz} = \frac{dy}{x + yz} = \frac{dz}{z^2 - 1}$$

Since $P + Q = (x + y)(z + 1)$ and $P - Q = (x - y)(z - 1)$,

$$\frac{dx + dy}{x + y} = \frac{dz}{z - 1} \qquad \frac{dx - dy}{x - y} = \frac{dz}{z + 1}$$

Two independent integrals of this system are

$$u = \frac{z-1}{x+y} = c_1 \qquad v = \frac{z+1}{x-y} = c_2$$

Note that the characteristic curves here are the straight lines determined by the intersection of the planes

$$z - 1 - c_1(x + y) = 0 \qquad z + 1 - c_2(x - y) = 0$$

The given curve Γ is noncharacteristic.   Equation (1-55) takes the form

$$\frac{t^2-1}{t+1} = c_1 \qquad \frac{t^2+1}{t-1} = c_2$$

From the first equation, $t = c_1 + 1$.   Insert this into the second equation; then $(c_1 + 1)^2 = c_1 c_2$.   The integral surface which contains Γ has the equation

$$\left(\frac{z-1+x+y}{x+y} \right)^2 + 1 = \frac{z^2-1}{x^2-y^2}$$

An Existence and Uniqueness Theorem

From a geometric viewpoint a functional relation $F(c_1,c_2) = 0$ imposed on the arbitrary constants c_1, c_2 singles out a one-parameter subfamily of the two-parameter family of characteristic curves in Eq. (1-42).   Analytically, the condition that the resulting surface contains the given curve Γ is that the relation

$$F\{u[f(t),g(t),h(t)],v[f(t),g(t),h(t)]\} = 0 \tag{1-58}$$

holds identically in t, for $a < t < b$.   But if F is obtained by eliminating t between the pair of equations in (1-55), then clearly the identity holds.   The success of the method hinges on the ability to eliminate t in Eq. (1-55).   The following local existence and uniqueness theorem for the Cauchy problem shows that this is always possible provided Γ is noncharacteristic.

THEOREM 1-1   Let $\mathcal{T}$ be a region of xyz space and $\mathcal{R}$ the projection of $\mathcal{T}$ on the xy plane.   Let the following properties hold: (1) the coefficients P, Q, R in Eq. (1-38) are continuously differentiable functions of x, y, z, and P, Q do not vanish in $\mathcal{T}$; (2) Γ is a given space curve lying in $\mathcal{T}$ and is defined parametrically by Eq. (1-54), where the functions f, g, h are invertible and have continuous first derivatives; (3) $[f'(t)]^2 + [g'(t)]^2 \neq 0$, $a < t < b$; (4) (x_0,y_0,z_0) is a point on Γ corresponding to $t = t_0$; (5) $P(x_0,y_0,z_0)g'(t_0) - Q(x_0,y_0,z_0)f'(t_0) \neq 0$.   Then there exists a neighborhood N of (x_0,y_0) in $\mathcal{R}$, a neighborhood $|t - t_0| < \delta$ of t_0, and a unique function $z = \varphi(x,y)$ such that φ is a solution of Eq. (1-38) on N and

$$h(t) = \varphi[f(t),g(t)]$$

holds identically in t for $|t - t_0| < \delta$.

Proof The proof is based on the following property of the system (1-40). Under the present hypotheses on P, Q, and R there exists a two-parameter family of solutions

$$y = y(x,c_1,c_2) \qquad z = z(x,c_1,c_2) \tag{1-59}$$

of Eq. (1-40), where the functions y and z are continuous and have continuous first partial derivatives with respect to the parameters c_1, c_2 in a certain range of values of these parameters which includes y_0, z_0.   Moreover

$$y(x_0,c_1,c_2) = c_1 \qquad z(x_0,c_1,c_2) = c_2 \tag{1-60}$$

for each pair of values c_1, c_2.   The proof of this property is given in Ref. 2. From Eq. (1-60) one obtains

$$\frac{\partial y}{\partial c_1} = 1 \qquad \frac{\partial y}{\partial c_2} = 0 \qquad \frac{\partial z}{\partial c_1} = 0 \qquad \frac{\partial z}{\partial c_2} = 1$$

at $x = x_0$.   Hence the Jacobian $\partial(y,z)/\partial(c_1,c_2) \neq 0$ in some neighborhood of (x_0,y_0,z_0).   In turn this implies that Eq. (1-59) can be solved for c_1, c_2 to obtain the functions

$$c_1 = u(x,y,z) \qquad c_2 = v(x,y,z) \tag{1-61}$$

and the functions u and v have the property that

$$y_0 = u(x_0,y_0,z_0) \qquad z_0 = v(x_0,y_0,z_0)$$

The Jacobian $\partial(u,v)/\partial(y,z)$ is different from zero in some neighborhood of (x_0,y_0,z_0) since it is the reciprocal of $\partial(y,z)/\partial(c_1,c_2)$.   With u and v constructed in this manner, define the functions $c_1(t)$ and $c_2(t)$ by

$$c_1(t) = u[f(t),g(t),h(t)] \qquad c_2(t) = v[f(t),g(t),h(t)]$$

Then $c_1(t_0) = y_0$, and $c_2(t_0) = z_0$.   Moreover c_1 and c_2 have continuous first derivatives with respect to t in some neighborhood of $t = t_0$, and, by the chain rule,

$$\frac{dc_1}{dt} = \nabla u \cdot \frac{d\mathbf{r}}{dt} \qquad \frac{dc_2}{dt} = \nabla v \cdot \frac{d\mathbf{r}}{dt} \tag{1-62}$$

where

$$\frac{d\mathbf{r}}{dt} = f'(t)\mathbf{i} + g'(t)\mathbf{j} + h'(t)\mathbf{k}$$

Recall that the vector $d\mathbf{r}/dt$ is tangent to Γ.   Now at least one of the values $c_1'(t_0)$, $c_2'(t_0)$ is different from zero.   For suppose that $c_1'(t_0) = c_2'(t_0) = 0$. Then at $t = t_0$, $x = x_0$, $y = y_0$, $z = z_0$ the vector $d\mathbf{r}/dt$ is perpendicular to the vector ∇u and also perpendicular to the vector ∇v.   Since the Jacobian $\partial(u,v)/\partial(y,z) \neq 0$, at (x_0,y_0,z_0) the vector $\nabla u \times \nabla v$ has a nonzero x component and so is not the zero vector at that point.   Also $d\mathbf{r}/dt$ is not the zero vector at $t = t_0$, by virtue of hypothesis 3.   Hence $d\mathbf{r}/dt$ has the same direction as $\nabla u \times \nabla v$.   But $\nabla u \times \nabla v$ has the same direction as the characteristic vector $\mathbf{V}$ (see Sec. 1-5).   Thus Γ is characteristic at (x_0,y_0,z_0).   This, however, is impossible by virtue of hypothesis 5, which asserts that $\mathbf{V} \times d\mathbf{r}/dt \cdot \mathbf{k} \neq 0$ at that point.   Thus at least one of the derivatives $c_1'(t_0)$, $c_2'(t_0)$ must be different from zero.   For definiteness assume $c_1'(t_0) \neq 0$.   By a basic theorem of calculus the equation $c_1 = c_1(t)$ can be solved in a neighborhood of t_0 to obtain the inverse function $t = t(c_1)$, and moreover $t(y_0) = t_0$.   Substitute $t(c_1)$ for t in the equation $c_2 = c_2(t)$.   Then

$$c_2 = c_2[t(c_1)] = \psi(c_1)$$

With the function ψ constructed in this manner consider the equation

$$v(x,y,z) - \psi[u(x,y,z)] = 0 \qquad (1\text{-}63)$$

Clearly, if in this relation x, y, z are replaced by $f(t)$, $g(t)$, $h(t)$, an identity in t results.   In particular this implies the equation is satisfied by $x = x_0$, $y = y_0$, $z = z_0$.   Observe that the left member of Eq. (1-63) is continuously differentiable in a neighborhood of (x_0,y_0,z_0).   Now it is asserted that

$$v_z - \psi'(u)u_z \neq 0 \qquad (1\text{-}64)$$

at (x_0,y_0,z_0).   To show this consider that

$$\psi'(c_1) = \frac{dc_2}{dc_1} = \frac{dc_2/dt}{dc_1/dt} = \frac{\nabla v \cdot d\mathbf{r}/dt}{\nabla u \cdot d\mathbf{r}/dt}$$

If $v_z - \psi'(u)u_z = 0$ at (x_0,y_0,z_0), then

$$v_z \, \nabla u \cdot \frac{d\mathbf{r}}{dt} - u_z \, \nabla v \cdot \frac{d\mathbf{r}}{dt} = 0$$

Recall the expression for the triple product $(\nabla u \times \nabla v) \times d\mathbf{r}/dt$.   Then the preceding equation takes the form

$$(\nabla u \times \nabla v) \times \frac{d\mathbf{r}}{dt} \cdot \mathbf{k} = 0$$

Hence

$$\mathbf{V} \times \frac{d\mathbf{r}}{dt} \cdot \mathbf{k} = 0$$

which contradicts hypothesis 5.   Thus (1-64) holds.   In turn (1-64) implies that Eq. (1-63) can be solved in a neighborhood of (x_0,y_0,z_0) to obtain a continuously differentiable function $z = \varphi(x,y)$ such that $z_0 = \varphi(x_0,y_0)$.   In fact there is a neighborhood of t_0 such that

$$h(t) = \varphi[f(t),g(t)]$$

hold identically in t. The proof that φ is a solution of Eq. (1-38) in a neighborhood of (x_0,y_0,z_0) follows from the discussion in the paragraph of Sec. 1-5 which precedes Example 1-12.

Under the hypothesis that Γ is noncharacteristic there can be at most one solution of the Cauchy problem in a neighborhood of (x_0,y_0,z_0). For if two distinct integral surfaces of Eq. (1-38) contain Γ near (x_0,y_0,z_0), then in a sufficiently small neighborhood of (x_0,y_0,z_0), Γ is the unique curve of intersection of these surfaces. But then Γ is a characteristic.

PROBLEMS

Sec. 1-2

1 Classify each of the following equations to the extent of the definitions given in Sec. 1-2.

(a) $u_{xy} + x^2 u_{yy} + (\cos y)u_x = (\tan xy)u + x^2 y^2$

(b) $\left(\dfrac{\partial \varphi}{\partial x}\right)^2 \dfrac{\partial^2 \varphi}{\partial x^2} + y^2 \dfrac{\partial^2 \varphi}{\partial y^2} + \varphi = 0$

(c) $x^2 r + yt - zp + (1 - z^3)q = x$ **(d)** $x^2 p - yz^2 q = 0$

(e) $u_y u_{xx} + u_x u_{yy} - u_z{}^2 + xy^2 = z$ **(f)** $pq = z$

(g) $\dfrac{\partial^2 u}{\partial x^2} + \dfrac{\partial^2 u}{\partial y^2} + \dfrac{\partial^2 u}{\partial z^2} = 0$ Laplace's equation

(h) $\dfrac{\partial u}{\partial x} = \dfrac{\partial v}{\partial y}$ $\dfrac{\partial u}{\partial y} = -\dfrac{\partial v}{\partial x}$ Cauchy-Riemann equations

Sec. 1-3

2 **(a)** Show that each sphere in the family $x^2 + y^2 + z^2 + c_1 z = 0$ intersects each sphere in the family $x^2 + c_1 x + y^2 + z^2 = 0$ orthogonally.

(b) Let G be a given differentiable function of the variables x, y, z such that $G(x,y,z) = c$ defines a surface for each value of the constant c in a certain range of values and so a family of surfaces. Let $F(x,y,z) = 0$ implicitly define a differentiable function $z = f(x,y)$ and so a surface S. If S intersects orthogonally each surface in the above family of surfaces, show that the quasilinear equation $pG_x + qG_y = G_z$ must be satisfied.

3 Eliminate the arbitrary functions which appear and obtain a differential equation of lowest order.

(a) $z = xf(y)$ **(b)** $u = e^{-x}F(x - 2y)$ **(c)** $G(x^2 + y^2 + z^2, z) = 0$

(d) $u = x^n F\left(\dfrac{y}{x}, \dfrac{z}{x}\right)$ n a positive integer

(e) $ze^{-x^2} + \psi(x^2 + y^2) = 0$

(f) $z = f(y) \cos x + g(y) \sin x$ **(g)** $u = f(x - ct) + g(x + ct)$

(h) $u = f_1(y - mx) + xf_2(y - mx) + x^2 f_3(y - mx)$

4   Consider the relation $\Phi(x^2 + y^2, y^2 + z^2, u^2 + xy) = 0$, where Φ is an arbitrary function. Let $F(x,y,z,u) = x^2 + y^2$, $G(x,y,z,u) = y^2 + z^2$, and $H(x,y,z,u) = u^2 + xy$. Let x, y, z be independent variables, and suppose for each choice of Φ the equation $\Phi(F,G,H) = 0$ implicitly defines u as a function of x, y, z. Follow the method of Example 1-4 and differentiate $\Phi = 0$ with respect to x, y, z so as to eliminate Φ and derive the first-order quasilinear equation

$$2yzuu_x - 2xzuu_y + 2xyuu_z = z(x^2 - y^2)$$

5   Consider the relation $\Phi(F_1, \ldots, F_n) = 0$, where Φ is arbitrary and $F_i(x_1, \ldots, x_n, u)$, $i = 1, \ldots, n$, are n differentiable functions. Let $x_1, \ldots, x_n$ be independent variables, and suppose for each choice of Φ the equation $\Phi = 0$ implicitly defines u as a function of $x_1, \ldots, x_n$. Differentiate $\Phi = 0$ successively with respect to $x_1, x_2, \ldots, x_n$ and so obtain the n equations

$$\sum_{i=1}^{n} \Phi_i(F_{ix_j} + F_{iu}u_{x_j}) = 0 \qquad j = 1, \ldots, n$$

where Φ_i means $\partial\Phi/\partial F_i$, F_{ix_j} means $\partial F_i/\partial x_j$, etc.   Eliminate Φ and show that a quasilinear equation

$$\sum_{k=1}^{n} A_k(x_1, \ldots, x_n, u)\frac{\partial u}{\partial x_k} = G(x_1, \ldots, x_n, u)$$

results.

6   Eliminate the arbitrary constants $a, b, c, \ldots$ which appear and obtain a partial differential equation of lowest order.

(a) $z = e^{ax+by}$

(b) $a(x^2 + y^2) + bz^2 = 1$

(c) $z = bx^ay^{1-a}$

(d) $z = ax^2 + 2bxy + cy^2$

(e) $u = Ae^{ax} \cos ay$

(f) $u = A \cos ax \cos at$

(g) $u = Ae^{-a^2t} \cos ax$

(h) $az + b = a^2x + y$

(i) $z = ax + by + a^2 + b^2$

7   Find a partial differential equation of lowest order satisfied by each surface in the given family surfaces.
(a)   All planes through the point $(1,0,0)$ not perpendicular to the xy plane.
(b)   All spheres of unit radius.
(c)   The family of all tangent planes to the surface $z = xy$.

Sec. 1-4

8   Hold one independent variable constant and integrate with respect to the remaining variable so as to obtain a solution involving an arbitrary function. Verify that your answer is correct by substituting into the differential equation.

(a) $q = x^2 + y^2$

(b) $p = \sin\dfrac{x}{y}$

(c) $z_x + xz = x^3 + 3xy$

9   Let $Ap + Bq = 0$ be a first-order equation, where A, B are constants. Assume a solution of the form $z = f(ax + by)$, f arbitrary and a, b constants. Substitute into the differential equation to determine suitable values of a, b.

10   Obtain the general solution.

(a) $3p - 4q = x^2$

(b) $p - 3q = \sin x + \cos y$

(c) $5p + 4q + z = x^3 + 1 + 2e^{3y}$

(d) $p + 2q - 5z = \cos x + y^3 + 1$

(e) $p - aq = e^{mx} \cos by$ a, m, b constants

11   Make the change of independent variables $\xi = \log x$, $\eta = \log y$ and reduce the differential equation to one with constant coefficients.   Obtain the general solution.

(a) $4xp - 2yq = 0$

(b) $2xp + 3yq = \log x$

(c) $xp - 7yq = x^2 y$

(d) $8xp - 5yq + 4z = x^2 \cos x$

(e) $axz_x + byz_y + cz = x^2 + y^2$

12   Follow the method of the text in Example 1-8 and obtain the general solution in the form (1-37).

(a) $xyp - x^2 q + yz = 0$

(b) $yp - xq = 0$

(c) $(x + a)p + (y + b)q + cz = 0$ a, b, c constants

13   Let $Lu = Au_x + Bu_y + Cu_z + Du = G$ be a linear first-order equation in three independent variables x, y, z and dependent variable u, where A, B, C, D are constants and $G(x,y,z)$ is a given function.   Assume $A \neq 0$.   To obtain a solution of the homogeneous equation $Lu = 0$ which involves an arbitrary function assume a solution of the form $u = e^{-Dx/A}f(ax + by + cz)$, substitute into $Lu = 0$, and show that the constants a, b, c must satisfy $Aa + Bb + Cc = 0$ if the assumed form satisfies $Lu = 0$ for arbitrary choice of f.   Conversely, if a, b, c are chosen so that the preceding equation holds, then $u = e^{-Dx/A}f(ax + by + cz)$ is a solution for arbitrary (differentiable) f.   If $A = 0$ but $B \neq 0$, one can proceed similarly with $u = e^{-Dx/B}f(ax + by + cz)$, etc.   Find a solution involving an arbitrary function for each of the following equations.

(a) $2u_x - u_y + 4u_z + u = 0$

(b) $u_x - 4u_z + 7u_z - u = x + y + z + 1$

Sec. 1-5

14   Obtain the general solution.

(a) $p + xq = z$

(b) $xp + yq = nz$ n constant

(c) $(x + z)p + (y + z)q = 0$

(d) $(y + x)p + (y - x)q = z$

(e) $zp + yq = x$

(f) $(x + y)(p - q) = z$

(g) $yp - xq = x^3 y + xy^3$

(h) $(mz - ny)p + (nx - lz)q = ly - mx$ l, m, n constants

(i) $x^2 p + y^2 q = axy$ $a \neq 0$, constant

(j) $(y^3 x - 2x^4)p + (2y^4 - x^3 y)q = 9z(x^3 - y^3)$

(k) $x(y - z)p + y(z - x)q = z(x - y)$

(l) $yp - xq + z + x^2 + y^2 - 1 = 0$

(m) $p - 2q = 3x^2 \sin (y + 2x)$

(n) $(x^2 - y^2 - z^2)p + 2xyq = 2xz$

(o) $\cos y \dfrac{\partial z}{\partial x} + \cos x \dfrac{\partial z}{\partial y} = \cos x \cos y$

(p) $(z + e^x)p + (z + e^y)q = z^2 - e^{x+y}$

(q) $xp + yq = 2xy(a^2 - z^2)^{1/2}$

15   Refer to Prob. 2b and find the general form of the equation of all surfaces orthogonal to the given family of surfaces.

(a) $x^2 + y^2 + z^2 = 2ax$ (b) $\dfrac{x^2}{a^2} + \dfrac{y^2}{b^2} + z^2 = c^2$

16   (a) Let $\Phi(x,y,z)$ be a continuously differentiable function and let c be a fixed constant such that $\Phi = c$ defines a smooth surface S.   Show that the family of all smooth space curves C orthogonal to S must satisfy the system of equations

$$\frac{dx}{\Phi_x} = \frac{dy}{\Phi_y} = \frac{dz}{\Phi_z}$$

(b) The velocity potential Φ of a stationary velocity field of fluid flow is $\Phi(x,y,z) = xy + xyz^2$.   The velocity field is $\mathbf{V} = \nabla\Phi$.   The *trajectories* of the field are the curves C such that the tangent to the curve at each point has the direction of $\mathbf{V}$ at the point, i.e., the field lines.   Find the trajectories.
(c) The potential Φ of an electrostatic field is $\Phi(x,y,z) = 1/x + 1/y + 1/z$.   The electric field is $\mathbf{E} = -\nabla\Phi$.   Find the field lines of the $\mathbf{E}$ field.

17   Find the general form of the equation of all surfaces such that the tangent plane to each point on the surface passes through the fixed point $(0,0,a)$.

18   Let $x_1, \ldots, x_n$ be independent variables in a region $\mathscr{R}$ of n-dimensional space, and let

$$\sum_{i=1}^{n} P_i \frac{\partial u}{\partial x_i} = 0 \tag{1}$$

be a linear equation with coefficients P_i which are continuously differentiable and such that they are not simultaneously zero in $\mathscr{R}$.   A function $u = \varphi(x_1, \ldots, x_n)$ is said to be a *solution of* (1) in $\mathscr{R}$ if φ is continuously differentiable and (1) holds identically in $\mathscr{R}$.   The system of ordinary differential equations

$$\frac{dx_1}{P_1} = \cdots = \frac{dx_n}{P_n} \tag{2}$$

is called the *subsidiary system* of (1).   These equations define an $(n-1)$-parameter family of curves in n space, called the *characteristic curves* of (1).   If x_n is chosen as the independent variable, then (2) can be written in the form

$$\frac{dx_1}{dx_n} = \frac{P_1}{P_n} \cdots \frac{dx_{n-1}}{dx_n} = \frac{P_{n-1}}{P_n} \tag{3}$$

The general solution of the system (3) is of the form

$$x_i = x_i(x_n, c_1, \ldots, c_{n-1}) \qquad i = 1, \ldots, n-1$$

where the c_i are arbitrary constants.   If these are solvable for the c_i's, the general solution of (2) can be written as $u_i(x_1, \ldots, x_n) = c_i$, $i = 1, \ldots, n-1$, where the $n-1$ functions u_i are functionally independent in $\mathscr{R}$.   Each relation $u_i = c_i$ is called an *integral* of the subsidiary equations (2).   For fixed c_i the equation $u_i(x_1, \ldots, x_n) = c_i$ defines a hypersurface (of dimension $n-1$) in n space.   For each fixed set of values $c_1, \ldots, c_n$ the $n-1$ hypersurfaces $u_i = c_i$, $i = 1, \ldots, n-1$, intersect in a characteristic curve C in n space. Through each point $(x_1^{(0)}, \ldots, x_n^{(0)})$ of $\mathscr{R}$ there passes one, and only one, characteristic C. Each function $u_i(x_1, \ldots, x_n)$ satisfies (1) in $\mathscr{R}$.   For given a point in $\mathscr{R}$ there exist constants

$c_1, \ldots, c_n$ such that a characteristic C passes through the point, and so equations (2) hold at the point. Differentiate the equation $u_j = c_j$; then

$$0 = du_j = \sum_{i=1}^{n} \frac{\partial u_j}{\partial x_i}\, dx_i = \sum_{i=1}^{n} \frac{\partial u_j}{\partial x_i}\left(\frac{P_i}{P_n}\, dx_n\right) = \frac{dx_n}{P_n} \sum_{i=1}^{n} P_i \frac{\partial u_j}{\partial x_i}$$

If $u_1, \ldots, u_{n-1}$ are functionally independent in $\mathscr{R}$, the *general solution* of (1) is $u = f(u_1, \ldots, u_{n-1})$, f arbitrary. For each choice of f this defines a solution of (1), since

$$\sum_{i=1}^{n} P_i \frac{\partial u}{\partial x_i} = \sum_{i=1}^{n} P_i \left(\sum_{j=1}^{n-1} \frac{\partial f}{\partial u_j} \frac{\partial u_j}{\partial x_i}\right) = \sum_{j=1}^{n-1} \frac{\partial f}{\partial u_j} \left(\sum_{i=1}^{n} P_i \frac{\partial u_j}{\partial x_i}\right) = 0$$

Conversely, let u be a solution of (1) in $\mathscr{R}$. Then the n equations

$$\sum_{i=1}^{n} P_i \frac{\partial u}{\partial x_i} = 0 \qquad \sum_{i=1}^{n} P_i \frac{\partial u_j}{\partial x_i} = 0$$

$j = 1, \ldots, n - 1$, hold simultaneously in $\mathscr{R}$. Since the P_i have values different from zero at each point of $\mathscr{R}$, it follows that the Jacobian

$$\frac{\partial(u, u_1, \ldots, u_{n-1})}{\partial(x_1, \ldots, x_n)} = 0$$

in $\mathscr{R}$. This implies that (locally at least) there exists a function f such that $u = f(u_1, \ldots, u_{n-1})$.

Let v be a particular solution of the linear equation

$$\sum_{i=1}^{n} P_i \frac{\partial u}{\partial x_i} + Ru = 0 \tag{4}$$

where $R(x_1, \ldots, x_n)$ is continuously differentiable, in $\mathscr{R}$. Let $u_1, \ldots, u_{n-1}$ be $n - 1$ functionally independent integrals of the subsidiary equations (2). Then the *general solution* of (4) is $u = vf(u_1, \ldots, u_{n-1})$, f arbitrary. For if f is a given function, then $u_{x_i} = v_{x_i} f + v f_{x_i}$. Hence

$$\sum_{i=1}^{n} P_i \frac{\partial u}{\partial x_i} = f \sum_{i=1}^{n} P_i \frac{\partial v}{\partial x_i} + v \sum_{i=1}^{n} P_i \frac{\partial f}{\partial x_i}$$

$$= f(-Rv) = -Ru$$

so u is a solution of (4). Conversely, if u is a solution of (4) and $w = u/v$, then $w_{x_i} = (vu_{x_i} - uv_{x_i})/v^2$, so that

$$\sum_{i=1}^{n} P_i \frac{\partial w}{\partial x_i} = \frac{1}{v} \sum_{i=1}^{n} P_i \frac{\partial u}{\partial x_i} - \frac{u}{v^2} \sum_{i=1}^{n} P_i \frac{\partial v}{\partial x_i}$$

$$= \frac{1}{v}(-Ru) - \frac{u}{v^2}(-Rv) = 0$$

Thus w is a solution of (1). It follows that $w = f(u_1, \ldots, u_{n-1})$ for some f, and so $u = vf(u_1, \ldots, u_{n-1})$. The *general solution* of the inhomogeneous equation

$$\sum_{i=1}^{n} P_i \frac{\partial u}{\partial x_i} + Ru = G \tag{5}$$

where G is a given continuously differentiable function, is $u = u_p + u_h$, where u_p is a particular solution of (5) and u_h is the general solution of (4). Find the general solution of each of the following.

(a) $P_1 u_x + P_2 u_y + P_3 u_z + Ru = 0$ P_1, P_2, P_3, R constants

(b) $(y + z)u_x + (z + x)u_y + (x + y)u_z = 0$

(c) $3u_x + 5u_y - u_z = \cos y - 2e^{-z}$

(d) $u_x - u_y + 7u_z + u = Ae^{ax+by+cz}$ A, a, b, c constants

(e) $(x + z)u_x + yu_y - 2u_z = 4e^z$

(f) $xzu_x + yzu_y - (x^2 + y^2)u_z = zy^2 \sin y$

(g) $(z - y)u_x + yu_y - zu_z = y(x + z) - y^2$

(h) $(\tan x)u_x + (\tan y)u_y + u_z = \sin z$

(i) $xu_x + yu_y + zu_z = nu$ n constant

(j) $\sqrt{x}u_x + \sqrt{y}u_y + \sqrt{z}u_z + u = x + y$

(k) $\dfrac{\partial u}{\partial x_1} + x_1 \dfrac{\partial u}{\partial x_2} + x_1 x_2 \dfrac{\partial u}{\partial x_3} + x_1 x_2 x_3 \dfrac{\partial u}{\partial x_4} = 0$

(l) $x_1 \dfrac{\partial u}{\partial x_1} + x_1 x_2 \dfrac{\partial u}{\partial x_2} + x_1 x_2 x_3 \dfrac{\partial u}{\partial x_3} + x_1 x_2 x_3 x_4 \dfrac{\partial u}{\partial x_4} + u = x_1^2 x_2^2$

19 A first-order quasilinear equation in n independent variables $x_1, \ldots, x_n$ and dependent variable u has the form

$$\sum_{i=1}^{n} P_i \frac{\partial u}{\partial x_i} = R \tag{1}$$

where the functions are assumed to be continuously differentiable functions of the x_i as well as u, and such that the P_i do not vanish in some region $\mathscr{T}$ of $(x_1, \ldots, x_n, u)$ space. Let $\mathscr{R}$ be the projection of $\mathscr{T}$ onto the hyperplane $u = 0$. A *solution* $u = \varphi(x_1, \ldots, x_n)$ of (1) on $\mathscr{R}$ is a continuously differentiable function of the x_i such that if φ and its derivatives are substituted into (1), an identity results. Assume $w = \psi(x_1, \ldots, x_n, u)$ is a solution of the linear homogeneous equation

$$\sum_{i=1}^{n} P_i \frac{\partial w}{\partial x_i} + R \frac{\partial w}{\partial u} = 0 \tag{2}$$

in the $n + 1$ independent variables $x_1, \ldots, x_n, u$. Assume also that $w_u \neq 0$ and that $w = 0$ implicitly defines a continuously differentiable function $u = \varphi(x_1, \ldots, x_n)$. Then $u_{x_j} = -w_{x_j}/w_u, j = 1, \ldots, n$. Substitute these expressions into the left-hand side of (1):

$$\sum_{i=1}^{n} P_i \frac{\partial u}{\partial x_i} = \sum_{i=1}^{n} P_i \left(\frac{-w_{x_j}}{w_u} \right) = -\left[\frac{\sum_{i=1}^{n} P_i(\partial w/\partial x_i)}{w_u} \right] = R$$

where (2) has been used. Thus u is a solution of (1). The problem of solving (1) is reduced to solving (2). From the results of Prob. 17 the general solution of (2) is $w = f(w_1, \ldots, w_n)$, where $w_1, \ldots, w_n$ are n functionally independent integrals of the subsidiary equations

$$\frac{dx_1}{P_1} = \cdots = \frac{dx_n}{P_n} = \frac{du}{R} \tag{3}$$

Then the *general solution* of (1) is $f(w_1, \ldots, w_n) = 0$, f arbitrary. Obtain the general solution of the following.

(a) $u_x + xu_y + xyu_z = xyzu$

(b) $xu_x + (z + u)u_y + (y + u)u_z = y + z$

(c) $xu_x + yu_y + zu_z = u + \dfrac{xy}{z}$

(d) $(s - x)u_x + (s - y)u_y + (s - z)u_z = s - u,\ s = x + y + z + u$

Sec. 1-6

20 Determine the integral surface which passes through the given curve.

(a) $p + q = z;\ z = \cos x,\ y = 0$

(b) $z(p - q) = y - x;\ x = 1,\ z = y^2$

(c) $xp - yq = 0;\ x = y = z = t$

(d) $(x + z)p + (y + z)q = 0;\ x = 1 - t,\ y = 1 + t,\ z = t$

(e) $x^2p + y^2q = z^2;\ x = t,\ y = 2t,\ z = 1$

(f) $xzp + yzq + xy = 0;\ xy = a^2,\ z = h$

(g) $2xzp + 2yzq = z^2 - x^2 - y^2;\ x + y + z = 0,\ x^2 + y^2 + z^2 = a^2$

(h) $(y - z)p + (z - x)q = x - y;\ x = t,\ y = 2t,\ z = 0$

(i) $x(x^2 + y^2)p + 2y^2(xp + yq - z) = 0;\ x^2 + y^2 = a^2,\ z = h$

(j) $(x^2 + y^2)p + 2xyq = xz;\ x = a,\ y^2 + z^2 = a^2$

(k) $x(y - z)p + y(z - x)q = z(x - y);\ x = y = z = t$

(l) $z(x + z)p - y(y + z)q = 0;\ x = 1,\ y = t,\ z = \sqrt{t}$

(m) $yp - xq = 2xyz;\ x = t,\ y = t,\ z = t$

(n) $(y^2 - x^2 + 2xz)p + 2y(z - x)q = 0;\ x = 0,\ y^2 + 4z^2 = 4a^2$

(o) $p \sec x + aq = \cot y;\ z(0,y) = \sin y$

21 Find a surface orthogonal to the sphere $x^2 + y^2 + z^2 - 2ax = 0$, $a > 0$, and passing through the line $y = x$, $z = h$, $0 < h < a$.

22 In Eq. (1-38) let the coefficients P, Q, R be analytic functions in $\mathcal{T}$, that is, differentiable any number of times and such that the Taylor's series expansion about a point of $\mathcal{T}$ converges. Let Γ be a curve described by Eq. (1-54), where f, g, h are analytic functions of t. Let Γ lie in $\mathcal{T}$, and let (x_0, y_0, z_0) be a point on Γ corresponding to $t = t_0$. We wish to determine a solution of the Cauchy problem in a neighborhood by means of a power-series expansion. The problem is whether or not the given data determine the coefficients in the expansion

$$z = z_0 + p_0(x - x_0) + q_0(y - y_0)$$

$$+ \frac{1}{2!} [\varphi_{xx}|_0 (x - x_0)^2 + \varphi_{xy}|_0 (x - x_0)(y - y_0) + \varphi_{yy}|_0 (y - y_0)^2] + \cdots$$

about (x_0, y_0, z_0). The condition that $z = \varphi(x,y)$ contain Γ near (x_0, y_0, z_0) is

$$h(t) = \varphi[f(t), g(t)] \tag{1}$$

Differentiate (1); then

$$pf'(t) + qg'(t) = h'(t) \tag{2}$$

Set $t = t_0$; then

$$p_0 f'(t_0) + q_0 g'(t_0) = h'(t_0) \tag{3}$$

From Eq. (1-38),

$$P_0 p_0 + Q_0 q_0 = R_0 \tag{4}$$

Accordingly, if the determinant

$$\Delta = Pg' - Qf' \tag{5}$$

is not zero at t_0, then (3) and (4) determine p_0 and q_0 uniquely. In this event successive differentiation of (2) and Eq. (1-38) determines the higher derivatives of φ uniquely also. Note that the condition $\Delta \neq 0$ at t_0 is hypothesis 5 of Theorem 1-1 and ensures that Γ is noncharacteristic at (x_0, y_0, x_0). If $\Delta = 0$ at $t = t_0$, either Eqs. (2) and (4) are dependent, or else they are incompatible. The first possibility occurs if, and only if, Γ is characteristic at (x_0, y_0, z_0). In this event there are infinitely many distinct integral surfaces containing Γ. If the second possibility occurs, no solution of the Cauchy problem exists. In each of the following Cauchy problems determine whether or not a solution exists. If a unique solution exists, find the power-series expansion to second-degree terms about the indicated point containing the given curve.

(a) $xp + yq = 0$; $(1,1,1)$; $x = 1$, $y = t^2$, $z = t^3$
(b) $yp - xq = 0$; $(1,0,1)$; $x = t$, $y = 0$, $z = t^2$
(c) $yp - xq = z(x^2 + y^2)$; $(1,0,0)$; $x = \cos t$, $y = \sin t$, $z = t$
(d) $yp - xq = (x^2 + y^2)$; $(0,1,0)$; $x = 0$, $y = t$, $z = 0$

23 With reference to Prob. 19, Eq. (1) in the case $n = 3$ becomes

$$P_1 u_x + P_2 u_y + P_3 u_z = R \tag{1}$$

where the P_i and R are functions of x, y, z, u. Let S be a given surface in xyz space defined parametrically by

$$x = f(s,t) \qquad y = g(s,t) \qquad z = h(s,t) \tag{2}$$

The Cauchy problem for (1) is to determine a solution of (1) such that at each point (x,y,z) on the surface S

$$u(x,y,z) = F(x,y,z) \tag{3}$$

where F is a given continuously differentiable function. A method of solution is as follows. Let w_1, w_2, w_3 be functionally independent integrals of the subsidiary equations (3) in Prob. 19. Eliminate the parameters s, t from the simultaneous equations

$$w_1\{f(s,t), g(s,t), h(s,t), F[f(s,t),(s,t),h(s,t)]\} = c_1$$

$$w_2 = c_2 \qquad w_3 = c_3$$

to obtain a functional relation $\Phi(c_1,c_2,c_3) = 0$. Then the solution is $\Phi(w_1,w_2,w_3) = 0$. Obtain a solution of each of the following Cauchy problems.

(a) $xu_x + yu_y + zu_z = nu$; $u = x + y + z$ on the surface $x = t$, $y = s$, $z = st$
(b) $u_x + xu_y + xyu_z = xyzu$; $u = x^2 + y^2$ on the surface $x = s$, $y = t$, $z = 0$

LINEAR SECOND-ORDER EQUATIONS

2-1 INTRODUCTION

The general solution of a linear first-order partial differential equation with suitably differentiable coefficients involves an arbitrary function, as shown in Chap. 1. It might be surmised that (under appropriate hypotheses on the coefficients) a linear second-order equation has a general solution containing two arbitrary functions. This is not true in general, even in the case of constant coefficients. In Secs. 2-2 and 2-3 some methods for deriving a solution which involves an arbitrary function are discussed for a few special types of linear second-order equations. Necessary and sufficient conditions for the existence of a solution which contains an arbitrary function are given in Sec. 2-2.

There are three principal classes of linear second-order partial differential equations: hyperbolic, parabolic, and elliptic. The classification arises in a natural way when a transformation of the independent variables is made so as to simplify the form of the differential equation. Section 2-3 is devoted to the reduction to normal form and the three types. As in the first-order case, the Cauchy problem is important to the theory of second-order equations. If there are two independent variables, the Cauchy problem is to determine an integral surface of the differential equation which passes through a given space curve and is such that the normal to the surface has a prescribed orientation along the curve. The Cauchy problem for a linear second-order equation in two independent variables is considered in Sec. 2-5 and that for a linear second-order equation in n independent variables in Sec. 2-6.

In Sec. 2-7 Green's formula, which relates a linear second-order partial differential operator to its adjoint, is derived. This is the generalization of the corresponding relation between a linear second-order ordinary differential operator and its adjoint, familiar in the study of linear ordinary differential equations and boundary-value problems. Green's formula is used in the discussion of the properties of eigenvalues and eigenfunctions of a self-adjoint elliptic-type linear differential operator in Chap. 3. It is also employed in Chap. 4 in connection with linear second-order hyperbolic operators.

The concept of an analytic function of a real variable, or of several real variables, is used in the sequel. In the case of two independent variables, let f be a real-valued function defined on an open set $\mathcal{S}$ in the xy plane, and let (x_0, y_0) be a point of $\mathcal{S}$. Then f is *analytic at* (x_0, y_0) if f has continuous partial derivatives of all orders with respect to x and y, and the Taylor's series of f about the point (x_0, y_0)

$$f(x_0, y_0) + \left.\frac{\partial f}{\partial x}\right|_0 (x - x_0) + \left.\frac{\partial f}{\partial y}\right|_0 (y - y_0)$$

$$+ \frac{1}{2}\left[\left.\frac{\partial^2 f}{\partial x^2}\right|_0 (x - x_0)^2 + 2 \left.\frac{\partial^2 f}{\partial x\,\partial y}\right|_0 (x - x_0)(y - y_0) + \left.\frac{\partial^2 f}{\partial y^2}\right|_0 (y - y_0)^2\right] + \cdots$$

[where the zero subscripts denote evaluation of f and its partial derivatives at (x_0, y_0)] converges to $f(x,y)$ for points (x,y) in some neighborhood of (x_0, y_0). The function f is said to be *analytic on* S if f is analytic at each point of S. Corresponding definitions are made when the number of independent variables is other than two.

2-2 LINEAR SECOND-ORDER EQUATIONS IN TWO INDEPENDENT VARIABLES

The general form of a linear second-order equation in two independent variables x, y is

$$Az_{xx} + 2Bz_{xy} + Cz_{yy} + Dz_x + Ez_y + Fz = G \tag{2-1}$$

It is assumed that the coefficients and the given function G are real-valued and twice continuously differentiable on a region $\mathscr{R}$ of the xy plane. The operational notation

$$D_x = \frac{\partial}{\partial x} \qquad D_y = \frac{\partial}{\partial y} \qquad D_x^2 = \frac{\partial^2}{\partial x^2} \qquad D_x D_y = \frac{\partial^2}{\partial x\,\partial y}$$

and so on, is useful here. Equation (2-1) can be written simply

$$Lz = G \tag{2-2}$$

where

$$L = AD_x^2 + 2BD_x D_y + CD_y^2 + DD_x + ED_y + F \tag{2-3}$$

is the general form of a linear second-order partial differential operator in two independent variables. Let z denote a function with continuous second partial derivatives in the region $\mathscr{R}$. Then L applied to z yields a function $w = Lz$ which is (at least) continuous in $\mathscr{R}$. To each such function z there corresponds a unique function $w = Lz$. Thus L induces a mapping, or transformation, of the class of all functions with continuous second derivatives on $\mathscr{R}$ into the class of all continuous functions on $\mathscr{R}$. The operator L is a *linear operator*, characterized by the property that

$$L(c_1 z_1 + c_2 z_2) = c_1 L z_1 + c_2 L z_2 \tag{2-4}$$

holds for every pair of constants c_1, c_2 and for every pair of functions z_1, z_2 with continuous second derivatives on $\mathcal{R}$.

A *solution of* Eq. (2-2) *on* $\mathcal{R}$ is a function $z = \varphi(x,y)$ with continuous second derivatives such that if φ and its derivatives are substituted into the left member of Eq. (2-2), an identity in the variables x, y on $\mathcal{R}$ results. When φ is real-valued, the surface in xyz space defined by $z = \varphi(x,y)$ is called an *integral surface* of Eq. (2-2). The *homogeneous equation* corresponding to Eq. (2-2) is

$$Lz = 0 \qquad\qquad (2\text{-}5)$$

From the linearity of the operator it follows that if $z_1, \ldots, z_q$ are solutions of Eq. (2-5), then for every choice of constants $c_1, \ldots, c_q$ the function $z = c_1 z_1 + \cdots + c_q z_q$ also constitutes a solution. Further, if z_p is a particular solution of Eq. (2-2), then

$$L(c_1 z_1 + \cdots + c_q z_q + z_p) = L(c_1 z_1 + \cdots + c_q z_q) + L z_p$$
$$= L z_p = G$$

Thus $z = c_1 z_1 + \cdots + c_q z_q + z_p$ is also a solution of Eq. (2-2) for every choice of constants $c_1, \ldots, c_q$.

Equations with Constant Coefficients

The simplest case is when the coefficients in the operator L in Eq. (2-2) are real constants. Assume also that the given function G is a real-valued analytic function in $\mathcal{R}$. Then Eq. (2-2) always has solutions, and indeed infinitely many distinct solutions. In some cases a relation involving two arbitrary functions can be obtained such that for each choice for the arbitrary functions (subject to differentiability requirements) a solution of Eq. (2-5) results. Such a relation is called the *general solution of the homogeneous equation*. If z_h denotes the general solution of Eq. (2-5) and z_p is any particular solution of Eq. (2-2), then $z = z_h + z_p$ is termed the *general solution of the inhomogeneous equation*. The following example illustrates these definitions.

EXAMPLE 2-1

$$\frac{\partial^2 z}{\partial x^2} - \frac{\partial^2 z}{\partial y^2} = 4x + 3 \cos 2y$$

The corresponding homogeneous equation is $(D_x{}^2 - D_y{}^2)z = 0$. The operator L in this example is factorable:

$$D_x{}^2 - D_y{}^2 = (D_x + D_y)(D_x - D_y) = (D_x - D_y)(D_x + D_y)$$

Suppose z is a solution of the first-order linear equation $D_x z + D_y z = 0$. Then z is a solution of the homogeneous equation, since

$$(D_x{}^2 - D_y{}^2)z = [(D_x - D_y)(D_x + D_y)]z = (D_x - D_y)(D_x z + D_y z) = 0$$

Similarly, if z is a solution of the first-order equation $D_x z - D_y z = 0$, then z is a solution of the homogeneous equation. Now the general solution of $D_x z + D_y z = 0$ is $z = f(x - y)$, f arbitrary, and the general solution of $D_x z - D_y z = 0$ is $z = g(x + y)$, g arbitrary. Hence the general solution of the homogeneous equation is

$$z_h = f(x - y) + g(x + y)$$

To find a particular solution corresponding to the term $4x$ hold y constant and integrate the equation $z_{xx} = 4x$. Similarly to find a particular solution corresponding to the term $3 \cos 2y$ hold x constant and integrate the equation $-z_{yy} = 3 \cos 2y$. Thus a particular solution of the given differential equation is $z_p = \frac{2}{3}x^3 - \frac{3}{4} \cos 2y$. The general solution of the inhomogeneous equation is

$$z = f(x - y) + g(x + y) + \frac{2x^3}{3} - \frac{3 \cos 2y}{4}$$

where f and g are arbitrary functions.

Factorable Operators

As in the preceding example, an operator L with constant coefficients may be factorable as the product of linear first-order operators. When this is possible, the general solution of Eq. (2-5) can be written down with the aid of the results of Sec. 1-4. Suppose L is such that

$$L = L_1 L_2 = (a_1 D_x + b_1 D_y + c_1)(a_2 D_x + b_2 D_y + c_2) \qquad (2\text{-}6)$$

Since the coefficients are constants and $D_x D_y = D_y D_x$, the operators L_1, L_2 commute: $L_1 L_2 = L_2 L_1$. Thus, if z_1 is a solution of the linear first-order equation $L_1 z = 0$, then

$$L z_1 = (L_1 L_2) z_1 = (L_2 L_1) z_1 = L_2(L_1 z_1) = 0$$

In the same way, if z_2 is a solution of $L_2 z = 0$, then z_2 is a solution of Eq. (2-5). Since L is a linear operator, $z = z_1 + z_2$ is also a solution. Accordingly, if $A = a_1 a_2 \neq 0$ and the factors L_1, L_2 are distinct, then the general solution of Eq. (2-5) is

$$z_h = e^{-c_1 x/a_1} f(b_1 x - a_1 y) + e^{-c_2 x/a_2} g(b_2 x - a_2 y) \qquad (2\text{-}7)$$

where f, g are arbitrary functions. If the factors of L are repeated, i.e., when $L_1 = L_2$, then the general solution is

$$z_h = e^{-c_1 x/a_1} [x f(b_1 x - a_1 y) + g(b_1 x - a_1 y)] \qquad (2\text{-}8)$$

If one of the coefficients a_1, a_2 is zero, the corresponding term in Eq. (2-7) is replaced by a term of the form in Eq. (1-26).

A special case where L is always factorable is when L is a *homogeneous operator:*

$$L = AD_x{}^2 + 2BD_xD_y + CD_y{}^2 \tag{2-9}$$

In this case the order of the derivative of z occurring in each term of Lz is the same, namely, two. If $A \neq 0$, let r_1, r_2 be the roots of the polynomial $Ar^2 + 2Br + C$. Then

$$L = A(D_x - r_1D_y)(D_x - r_2D_y)$$

If $A = 0$, then $L = D_y(2BD_x + CD_y)$. Note that the roots r_1, r_2 are real if, and only if, $B^2 - AC \geq 0$.

Several methods of finding particular solutions of Eq. (2-2) are outlined in Prob. 3 at the end of the chapter. These methods are analogous to those used to determine particular solutions of linear ordinary differential equations with constant coefficients. Consider the following example.

EXAMPLE 2-2 $r - t + 2p + z = y^2 + 2 \sin (2x + y) - x^2y$

Since $L = (D_x - D_y + 1)(D_x + D_y + 1)$, it follows from Eq. (2-7) that the general solution of the corresponding homogeneous equation is

$$z_h = e^{-x}f(x + y) + e^{-x}g(x - y)$$

To find a particular solution of the equation $Lz = y^2$ assume $z = z(y)$ and solve the ordinary differential equation $z'' - z = -y^2$. Then a particular solution of this equation is $z = y^2 + 2$. Refer now to Prob. 3. To find a particular solution of $Lz = 2 \sin (2x + y)$ write

$$\frac{2 \sin (2x + y)}{D_x{}^2 - D_y{}^2 + 2D_x + 1} = \frac{2 \sin (2x + y)}{-2^2 + 1^2 + 2D_x + 1} = \frac{\sin (2x + y)}{D_x - 1}$$

$$= \frac{(D_x + 1)[\sin (2x + y)]}{D_x{}^2 - 1} = \frac{(D_x + 1)[\sin (2x + y)]}{-2^2 - 1}$$

$$= -\frac{2 \cos (2x + y) + \sin (2x + y)}{5}$$

Also from Prob. 3, one obtains

$$\frac{-x^2y}{D_x{}^2 - D_y{}^2 + 2D_x + 1} = \frac{-x^2y}{(D_x + 1)^2\{1 - [D_y/(D_x + 1)]^2\}}$$

$$= \frac{1}{(D_x + 1)^2}\left[1 + \left(\frac{D_y}{D_x + 1}\right)^2 + \left(\frac{D_y}{D_x + 1}\right)^4 + \cdots\right](-x^2y)$$

$$= -\frac{x^2y}{(D_x + 1)^2} = -y(1 + D_x)^{-2}x^2$$

$$= -y(1 - 2D_x + 3D_x{}^2 - \cdots)x^2 = -y(x^2 - 4x + 6)$$

Thus the general solution of the given equation is

$$z = -\tfrac{1}{5}[2 \cos (2x + y) + \sin (2x + y)] - y(x^2 - 4x + 6)$$
$$+ e^{-x}[f(x + y) + g(x - y)] + y^2 + 2$$

2-3 LINEAR SECOND-ORDER EQUATIONS IN n INDEPENDENT VARIABLES

A linear second-order equation in n independent variables $x_1, \ldots, x_n$ has the form

$$\sum_{i=1}^{n} \sum_{j=1}^{n} A_{ij} \frac{\partial^2 u}{\partial x_i \, \partial x_j} + \sum_{i=1}^{n} B_i \frac{\partial u}{\partial x_i} + Cu = G \tag{2-10}$$

With the aid of the operational notation $D_{x_i} = \partial/\partial x_i$, $i = 1, \ldots, n$, the linear operator L appearing in Eq. (2-10) can be written

$$L = \sum_{i=1}^{n} \sum_{j=1}^{n} A_{ij} D_{x_i} D_{x_j} + \sum_{i=1}^{n} B_i D_{x_i} + C \tag{2-11}$$

The corresponding homogeneous equation is

$$Lu = 0 \tag{2-12}$$

Assume the coefficients A_{ij}, B_i, C in L are real constants, and $A_{ij} = A_{ji}$, $i, j = 1, \ldots, n$. When L is factorable

$$L = L_1 L_2 = (a_1 D_{x_1} + \cdots + a_n D_{x_n} + c)(b_1 D_{x_1} + \cdots + b_n D_{x_n} + d)$$

as a product of linear first-order operators, the ideas of the preceding paragraph extend to Eq. (2-10). From the results of Prob. 18, Chap. 1, the general solution of the first-order equations $L_1 u = 0$, $L_2 u = 0$ can be written down. Accordingly the *general solution* of Eq. (2-12) is

$$u_h = e^{-cx_1/a_1} f(a_2 x_1 - a_1 x_2, \ldots, a_n x_1 - a_1 x_n)$$
$$+ e^{-dx_1/b_1} g(b_2 x_1 - b_1 x_2, \ldots, b_n x_1 - b_1 x_n) \tag{2-13}$$

where f, g are arbitrary functions. The foregoing assumes that $a_1 \neq 0$ and $b_1 \neq 0$. If either a or b is zero, the form of the general solution is modified as in the case for $n = 2$ above. The *general solution* of Eq. (2-10) is

$$u = u_h + u_p \tag{2-14}$$

where u_p is a particular solution.

Functionally Invariant Pairs

A pair of functions ψ, φ twice continuously differentiable and such that

$$u = \psi f(\varphi) \tag{2-15}$$

constitutes a solution of Eq. (2-12) for arbitrary (twice differentiable) choice of f is called a *functionally invariant pair* of Eq. (2-12). It is assumed in the preceding definition that ψ is not identically zero and φ is not identically a constant. In Example 2-2 a functionally independent pair of the differential equation of that example is $\psi(x,y) = e^{-x}$, $\varphi(x,y) = x + y$. In order to determine necessary and sufficient conditions that ψ, φ form a functionally invariant pair of Eq. (2-12), assume Eq. (2-15) defines a solution for every choice of f. Differentiation yields

$$u_{x_i} = \psi_{x_i} f(\varphi) + \psi f'(\varphi)\varphi_{x_i} \qquad i = 1, \ldots, n$$

$$u_{x_i x_j} = \psi_{x_i x_j} f + \psi_{x_i} f' \varphi_{x_j} + \psi_{x_j} f' \varphi_{x_i} + \psi f'' \varphi_{x_j} \varphi_{x_i} + \psi f' \varphi_{x_i x_j}$$

$$i, j = 1, \ldots, n$$

If these expressions are substituted into Eq. (2-12) and the terms are rearranged, the equation

$$f'' \left(\psi \sum_{i=1}^{n} \sum_{j=1}^{n} A_{ij} \varphi_{x_i} \varphi_{x_j} \right)$$

$$+ f' \left[2 \sum_{i=1}^{n} \sum_{j=1}^{n} A_{ij} \psi_{x_i} \varphi_{x_j} + \psi \left(\sum_{i=1}^{n} \sum_{j=1}^{n} A_{ij} \varphi_{x_i x_j} + \sum_{i=1}^{n} B_i \varphi_{x_i} \right) \right] + f L \psi = 0$$

results. Here use has been made of $A_{ij} = A_{ji}$, $i, j = 1, \ldots, n$. Since f is arbitrary and ψ does not vanish identically, the following equations must hold:

$$\sum_{i=1}^{n} \sum_{i=1}^{n} A_{ij} \varphi_{x_i} \varphi_{x_j} = 0 \tag{2-16}$$

$$2 \sum_{i=1}^{n} \sum_{j=1}^{n} A_{ij} \psi_{x_i} \varphi_{x_j} + \psi \left(\sum_{i=1}^{n} \sum_{j=1}^{n} A_{ij} \varphi_{x_i x_j} + \sum_{i=1}^{n} B_i \varphi_{x_i} \right) = 0 \tag{2-17}$$

$$L \psi = 0 \tag{2-18}$$

Conversely, if ψ is not identically zero, φ is not identically constant, and Eqs. (2-16) to (2-18) hold, then ψ, φ constitute a functionally independent pair of Eq. (2-12). Equation (2-16) is called the *characteristic equation* of the operator L. It is of importance in connection with the Cauchy problem (Sec. 2-4) and also in the next section.

A number of the classical partial differential equations of mathematical physics are of the form

$$\sum_{i=1}^{n} A_i \frac{\partial^2 u}{\partial x_i{}^2} + \sum_{i=1}^{n} B_i \frac{\partial u}{\partial x_i} + Cu = G \tag{2-19}$$

where the coefficients are constants. This is a particular constant-coefficient case of Eq. (2-10), in which $A_{ij} = 0$ if $i \neq j$. Of particular interest are functionally independent pairs ψ, φ of the homogeneous equation corresponding to Eq. (2-19), where φ is of the form

$$\varphi = \sum_{i=1}^{n} k_i x_i \tag{2-20}$$

and $k_1, \ldots, k_n$ are real constants, not all zero. In this event the characteristic equation (2-16) implies that the constants k_i must satisfy

$$\sum_{i=1}^{n} A_i k_i^2 = 0 \qquad (2\text{-}21)$$

The following examples illustrate the ideas in terms of three important equations of mathematical physics in which $n = 3$.

EXAMPLE 2-3 $\dfrac{\partial^2 u}{\partial x^2} + \dfrac{\partial^2 u}{\partial y^2} = \dfrac{1}{c^2}\dfrac{\partial^2 u}{\partial t^2}$

Here c is a constant. The equation is called the *wave equation*, and the variable t generally has the significance of time. Comparison with Eq. (2-19) shows that $x_1 = x$, $x_2 = y$, $x_3 = t$, $A_1 = A_2 = 1$, $A_3 = -1/c^2$, $B_i = 0$, $i = 1, 2, 3$, and $C = 0$. The characteristic equation (2-16) is

$$\varphi_x{}^2 + \varphi_y{}^2 - \frac{\varphi_t{}^2}{c^2} = 0$$

Assume $\varphi = k_1 x + k_2 y + \omega t$, where the k_i and ω are constants. The characteristic equation becomes

$$k_1{}^2 + k_2{}^2 = \frac{\omega^2}{c^2}$$

Clearly $\psi = 1$ satisfies Eqs. (2-17) and (2-18). Accordingly 1, φ constitute a functionally independent pair of the wave equation, and

$$u = f(k_1 x + k_2 y + \omega t)$$

is a solution for arbitrary choice of f provided k_1, k_2, and ω satisfy the characteristic equation.

EXAMPLE 2-4 $\dfrac{\partial^2 u}{\partial x^2} + \dfrac{\partial^2 u}{\partial y^2} = \dfrac{1}{K}\dfrac{\partial u}{\partial t}$

This is the *diffusion* (or heat) equation, and K denotes a constant. Comparison with Eq. (2-19) shows that $A_1 = A_2 = 1$, $A_3 = 0$, $B_1 = B_2 = 0$, $B_3 = -1/K$, and $C = 0$. The characteristic equation is

$$\varphi_x{}^2 + \varphi_y{}^2 = 0$$

Hence $\varphi = \varphi(t)$. Equation (2-17) reduces to $\psi \varphi_t / K = 0$. This implies that φ is a constant. Hence the diffusion equation does not have a functionally invariant pair, and a solution which involves an arbitrary function of the real independent variables is not possible.

EXAMPLE 2-5 $\dfrac{\partial^2 u}{\partial x^2} + \dfrac{\partial^2 u}{\partial y^2} + \dfrac{\partial^2 u}{\partial z^2} = 0$

Here $x_1 = x$, $x_2 = y$, and $x_3 = z$. This classical equation is called *Laplace's equation*. The characteristic equation is

$$\varphi_x{}^2 + \varphi_y{}^2 + \varphi_z{}^2 = 0$$

The only real-valued solutions of this are those where φ is constant. Accordingly, under the restrictions of real-valuedness on ψ, φ, Laplace's equation does not possess functionally invariant pairs.

Exponential-type Solutions

In many problems involving linear partial differential equations (notably those arising in physics) exponential-type solutions, and superpositions of such solutions, are more useful than a form which contains an arbitrary function. Also, exponential-type solutions always exist for a homogeneous equation with constant coefficients. For example, consider Eq. (2-5), and assume the coefficients in L are constants. A solution of the form

$$z = e^{hx+my} \tag{2-22}$$

is attempted, where h, m are constants to be determined. Since $D_x z = hz$ and $D_y z = mz$, substitution of z into Eq. (2-5) gives

$$Ah^2 + 2Bhm + Cm^2 + Dh + Em + F = 0 \tag{2-23}$$

Conversely, if h and m are chosen such that Eq. (2-23) holds, the function z in Eq. (2-22) satisfies Eq. (2-5). Suppose that the relation (2-23) is solved for m so as to obtain a functional relationship $m = g(h)$. Then the function

$$z = f(h)e^{hx+g(h)y}$$

satisfies Eq. (2-5) for arbitrary choice of function f. More generally the superpositions

$$z = \sum_h f(h)e^{hx+g(h)y} \qquad z = \int f(h)e^{hx+g(h)y}\, dh$$

define solutions of Eq. (2-5) whenever they define twice continuously differentiable functions, and differentiation within the summation sign or within the integral sign is legitimate. The preceding ideas extend to Eq. (2-12) if the coefficients are constants.

EXAMPLE 2-6 $\dfrac{\partial^2 u}{\partial t^2} = c^2 \dfrac{\partial^2 u}{\partial x^2}$

This is the wave equation in one space variable x. Assume a solution

$$u = e^{hx+mt}$$

Then the relation corresponding to (2-23) is $c^2 h^2 - m^2 = 0$. Thus either $m = ch$ or $m = -ch$. For any value of h the functions

$$u = e^{hx+hct} \qquad u = e^{hx-hct}$$

are solutions of the wave equation. The wave equation occurs in many problems of vibration or wave motion. Often solutions which are periodic in time t are desired. Let $\omega > 0$ be a real constant, and choose $h = i\omega$, where $i = \sqrt{-1}$. Then

$$u = e^{i\omega(x+ct)} \qquad u = e^{i\omega(x-ct)}$$

are solutions which are periodic in time (and in distance x). Useful forms of solution of the wave equation are

$$u = \sum_n [A(\omega_n) \cos \omega_n x + B(\omega_n) \sin \omega_n x] e^{\pm i\omega_n ct}$$

$$u = \int_{-\infty}^{\infty} f(\omega) e^{i\omega(x \pm ct)} \, d\omega$$

EXAMPLE 2-7 $\dfrac{\partial u}{\partial t} = K \dfrac{\partial^2 u}{\partial x^2}$

This is the diffusion equation of Example 2-4 except that only one space variable is present. Recall that K is a positive constant. Assume a solution $u = e^{hx+mt}$. Substitution into the differential equation gives $Kh^2 - m = 0$. Thus $m = Kh^2$, and for any value h the function

$$u = e^{hx+Kh^2 t}$$

is a solution. So also are the superpositions

$$u = \sum_h f(h) e^{hx+Kh^2 t} \qquad u = \int_{h_1}^{h_2} f(h) e^{hx+Kh^2 t} \, dh$$

provided the expressions are meaningful in the sense that they define functions having sufficient differentiability for substitution into the differential equation to be possible. In physical problems the variable t often denotes time, and solutions which tend to zero with increasing time are desired. Choose $h = i\omega$, where $i = \sqrt{-1}$, and let ω be a real positive parameter. Then for any choice of f the function

$$u = f(\omega) e^{i\omega x - K\omega^2 t}$$

is a solution having the required property. Superpositions of the form

$$u = \sum_n f(\omega_n) e^{i\omega_n x - K\omega_n^2 t}$$

$$u = \sum_n (A_n \cos \omega_n x + B_n \sin \omega_n x) e^{-K\omega_n^2 t}$$

$$u = \int_{-\infty}^{\infty} f(\omega) e^{i\omega x - K\omega^2 t} \, d\omega$$

are applicable to problems of heat conduction.

2-4 NORMAL FORMS. HYPERBOLIC, PARABOLIC, AND ELLIPTIC EQUATIONS

In analytic geometry the study of curves whose equations are of the second degree in x and y is simplified by reducing the equation to a normal form. By means of a linear transformation of the variables x, y the equation is transformed into the standard form of a hyperbola or parabola or ellipse (or a degenerate form of one of these) in terms of new variables ξ, η. The $\xi\eta$ coordinate system is the coordinate system relative to which the curve has its simplest and most natural algebraic representation, and the properties of the original curve are just those of the curve in the standard form.   An analogous approach is used in the study of linear partial differential equations of order two.

Classification of Almost-linear Equations in Two Independent Variables

Let

$$Lz = Az_{xx} + 2Bz_{xy} + Cz_{yy} + M(x,y,z,z_x,z_y) = 0 \tag{2-24}$$

be an almost-linear equation in the real independent variables x, y.   Let the coefficients A, B, C be real-valued functions with continuous second derivatives on a region $\mathcal{R}$ of the xy plane, and assume that A, B, C do not vanish simultaneously.   The sum of terms

$$AD_x^2 + 2BD_xD_y + CD_y^2 \tag{2-25}$$

is called the *principal part* of the operator L in Eq. (2-24).   Primarily it is the principal part which determines the properties of solutions of Eq. (2-24).
   The function Δ defined on $\mathcal{R}$ by

$$\Delta(x,y) = B^2(x,y) - A(x,y)C(x,y) \tag{2-26}$$

is called the *discriminant* of L.   The operator L [as well as Eq. (2-24)] is said to be

1. *Hyperbolic* at a point (x,y) if $\Delta(x,y) > 0$
2. *Parabolic* at a point (x,y) if $\Delta(x,y) = 0$
3. *Elliptic* at a point (x,y) if $\Delta(x,y) < 0$

The operator L is called *hyperbolic in* $\mathcal{R}$ if L is hyperbolic at each point of $\mathcal{R}$. Corresponding definitions are made for the terms *parabolic in* $\mathcal{R}$ and *elliptic in* $\mathcal{R}$.   In general an operator may run the gamut of types in the domain of definition of its coefficients.   On the other hand, if the coefficients A, B, C are constants, then the operator is of one type in the entire plane.

EXAMPLE 2-8 $(1 - x^2)z_{xx} - 2xyz_{xy} + (1 - y^2)z_{yy} + xz_x + 3x^2yz_y - 2z = 0$
The discriminant is

$$\Delta = (-xy)^2 - (1 - x^2)(1 - y^2) = -1 + x^2 + y^2$$

Accordingly, the operator L, as well as the differential equation, is hyperbolic in the region where $x^2 + y^2 > 1$, parabolic at points on the circle $x^2 + y^2 = 1$, and elliptic in region where $x^2 + y^2 < 1$ (see Fig. 2-1).

   The property of L as regards type is an invariant under continuous one-to-one real transformations of the variables x, y.   Let

$$\xi = \xi(x,y) \qquad \eta = \eta(x,y) \tag{2-27}$$

be real-valued functions with continuous second derivatives on $\mathcal{R}$ such that the Jacobian $\partial(\xi,\eta)/\partial(x,y) \neq 0$ on $\mathcal{R}$. These functions map the region $\mathcal{R}$ onto a region $\mathcal{R}'$ of the $\xi\eta$ plane. In order to determine the representation of L in the new $\xi\eta$ variables, calculate

$$z_x = z_\xi \xi_x + z_\eta \eta_x \qquad z_y = z_\xi \xi_y + z_\eta \eta_y$$

$$z_{xx} = z_{\xi\xi}\xi_x^2 + 2z_{\xi\eta}\xi_x\eta_x + z_{\eta\eta}\eta_x^2 + z_\xi \xi_{xx} + z_\eta \eta_{xx}$$

and so on. If the expressions for the derivatives of z are substituted into Eq. (2-24), then

$$Lz = Q(\xi)z_{\xi\xi} + 2Q(\xi,\eta)z_{\xi\eta} + Q(\eta)z_{\eta\eta} + M'(\xi,\eta,z,z_\xi,z_\eta) = 0 \qquad (2\text{-}28)$$

where

$$Q(\xi) = A\xi_x^2 + 2B\xi_x\xi_y + C\xi_y^2$$
$$Q(\xi,\eta) = A\xi_x\eta_x + B(\xi_x\eta_y + \xi_y\eta_x) + C\xi_y\eta_y$$
$$Q(\eta) = A\eta_x^2 + 2B\eta_x\eta_y + C\eta_y^2$$

The principal part of L in its new representation is

$$Q(\xi)\frac{\partial^2}{\partial\xi^2} + 2Q(\xi,\eta)\frac{\partial^2}{\partial\xi\,\partial\eta} + Q(\eta)\frac{\partial^2}{\partial\eta^2} \qquad (2\text{-}29)$$

with discriminant $\Delta' = [Q(\xi,\eta)]^2 - Q(\xi)Q(\eta)$. A direct calculation yields

$$\Delta' = \left[\frac{\partial(\xi,\eta)}{\partial(x,y)}\right]^2 \Delta \qquad (2\text{-}30)$$

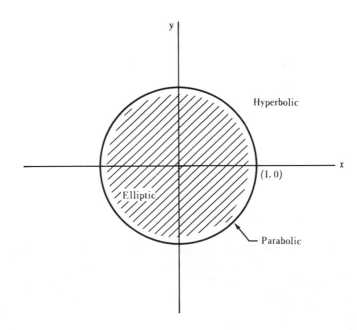

Figure 2-1

Accordingly L is hyperbolic (or parabolic, or elliptic) on $\mathcal{R}'$ if, and only if, L is hyperbolic (or parabolic, or elliptic) on $\mathcal{R}$.

Hyperbolic Equations. Characteristic Curves

Assume L is hyperbolic on $\mathcal{R}$ and that either A or C does not vanish in $\mathcal{R}$. Then there exists a transformation which is locally continuous, one to one, and such that Eq. (2-24) is transformed into

$$z_{\xi\eta} + G(\xi,\eta,z,z_\xi,z_\eta) = 0 \tag{2-31}$$

called the *normal* (or canonical) *form for hyperbolic equations in two indepen-dent variables*. From Eq. (2-28) it is clear that the normal form results if the transformation is such that $Q(\xi) = Q(\eta) = 0$. Hence consider the first-order partial differential equation

$$Q(\varphi) = A\varphi_x{}^2 + 2B\varphi_x\varphi_y + C\varphi_y{}^2 = 0 \tag{2-32}$$

called the *characteristic equation* of L. This is a nonlinear equation; however, $Q(\varphi)$ can be factored as follows. Assume first that $A \neq 0$ in $\mathcal{R}$. Since $\Delta > 0$, define the real-valued continuous functions m_1, m_2 by

$$m_1 = \frac{-B + \sqrt{\Delta}}{A} \qquad m_2 = \frac{-B - \sqrt{\Delta}}{A} \tag{2-33}$$

Then

$$Q(\varphi) = A(\varphi_x - m_1\varphi_y)(\varphi_x - m_2\varphi_y)$$

Accordingly if ξ, η are chosen such that

$$\xi_x - m_1\xi_y = 0 \qquad \eta_x - m_2\eta_y = 0 \tag{2-34}$$

then $Q(\xi) = Q(\eta) = 0$. From Sec. 1-5, the corresponding subsidiary equations are

$$\frac{dx}{1} = \frac{dy}{-m_1} = \frac{d\xi}{0} \qquad \frac{dx}{1} = \frac{dy}{-m_2} = \frac{d\eta}{0}$$

respectively. Often the ordinary differential equations

$$\frac{dy}{dx} = -m_1 \qquad \frac{dy}{dx} = -m_2 \tag{2-35}$$

are referred to as the characteristic equations of L. However, in this text the term characteristic equation is reserved for Eq. (2-32). Let the general solution of Eqs. (2-35) be

$$u(x,y) = c_1 \qquad v(x,y) = c_2 \tag{2-36}$$

respectively.   Recall from Sec. 1-5 that the function u satisfies the first partial differential equation in Eq. (2-34) and v satisfies the second.   Choose

$$\xi = u(x,y) \qquad \eta = v(x,y) \tag{2-37}$$

Then these functions have continuous second derivatives, and

$$\frac{\partial(\xi,\eta)}{\partial(x,y)} = \begin{vmatrix} \xi_x & \xi_y \\ \eta_x & \eta_y \end{vmatrix} = \begin{vmatrix} m_1\xi_y & \xi_y \\ m_2\eta_y & \eta_y \end{vmatrix} = \xi_y\eta_y(m_1 - m_2) \neq 0$$

The transformed equation is

$$2Q(\xi,\eta)z_{\xi\eta} + M'(\xi,\eta,z,z_\xi,z_\eta) = 0$$

Since

$$\Delta' = [Q(\xi,\eta)]^2 = \left[\frac{\partial(\xi,\eta)}{\partial(x,y)}\right]^2 \Delta \neq 0$$

the normal form (2-31) results on dividing the transformed equation by $Q(\xi,\eta)$.

The general solutions of the subsidiary differential equations (2-35) define a two-parameter family of curves in $\mathcal{R}$.   Through each point of $\mathcal{R}$ there passes exactly one curve $\xi = c_1$ and exactly one curve $\eta = c_2$.   The family of curves obtained in this way is called the family of *characteristic curves* (or simply characteristics) of the partial differential equation (2-24).   From this viewpoint Eq. (2-37) defines a curvilinear coordinate system in $\mathcal{R}$.   The ξ curves and the η curves are just the characteristics.   It is with respect to

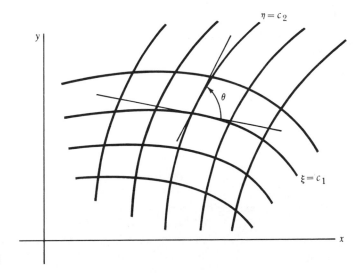

Figure 2-2

this particular set of curvilinear coordinates that the partial differential equation takes on its simplest, i.e., normal, form.   Let (x_0, y_0) be a point in $\mathfrak{R}$, and let $\xi = c_1$, $\eta = c_2$ be the characteristics through (x_0, y_0).   The slopes of these curves at (x_0, y_0) are $m_1(x_0, y_0)$ and $m_2(x_0, y_0)$, respectively.   If θ denotes the angle of intersection, then

$$\tan \theta = \frac{-m_1(x_0, y_0) + m_2(x_0, y_0)}{1 + m_1(x_0, y_0) m_2(x_0, y_0)} = -\frac{2\sqrt{\Delta}}{A(x_0, y_0) + C(x_0, y_0)}$$

provided $C(x_0, y_0) \neq -A(x_0, y_0)$.   If $C(x_0, y_0) = -A(x_0, y_0)$, then the characteristics intersect orthogonally at (x_0, y_0).   More generally, if $C = -A$ in Eq. (2-24), then the curvilinear coordinate system formed by the characteristic curves is an orthogonal system.

In the foregoing it was assumed that $A \neq 0$ in $\mathfrak{R}$.   If $A(x, y) = 0$ at some point of $\mathfrak{R}$, then $C(x, y) \neq 0$, and C remains different from zero in some neighborhood of (x, y).   In this event define the functions m_1, m_2 by Eq. (2-33) after A has been replaced by C and proceed in the same manner as above. Note that if $A = C = 0$ on $\mathfrak{R}$, then Eq. (2-24) is immediately reducible to the normal form.   This follows from the fact that in the hyperbolic case $\Delta > 0$, and so $B \neq 0$.

Parabolic Equations

If L is parabolic on $\mathfrak{R}$, then there exists a transformation (2-27) such that Eq. (2-24) is transformed into

$$z_{\eta\eta} + G(\xi, \eta, z, z_\xi, z_\eta) = 0 \tag{2-38}$$

called the *normal form for parabolic equations in two independent variables*. Since $\Delta = 0$, either $A \neq 0$ or $C \neq 0$.   Assume $A \neq 0$.   Now the roots of $Am^2 + 2Bm + C$ are real and equal.   Let

$$m = -\frac{B}{A}$$

Then $Q(\varphi) = A(\varphi_x - m\varphi_y)^2$.   Let $u(x, y) = c$ be the general solution of

$$\frac{dy}{dx} = -m$$

and choose $\xi = u(x, y)$.   Accordingly in the parabolic case there is a one-parameter family of characteristic curves $\xi = c$, and through each point of $\mathfrak{R}$ there passes exactly one characteristic.   From the manner of choice of ξ it follows that $Q(\xi) = 0$.   Now choose a simple function $\eta = v(x, y)$ with continuous second derivatives such that $\partial(\xi, \eta)/\partial(x, y) \neq 0$ in $\mathfrak{R}$.   Then $[Q(\xi, \eta)]^2 = \Delta' = 0$.   It follows from Eq. (2-28) that if the transformation is constructed in this manner, the transformed equation is

$$Q(\eta)z_{\eta\eta} + M'(\xi,\eta,z,z_\xi,z_\eta) = 0$$

Now $Q(\eta) \neq 0$. For if so, then

$$\frac{\partial(\xi,\eta)}{\partial(x,y)} = \begin{vmatrix} m\xi_y & \xi_y \\ m\eta_y & \eta_y \end{vmatrix} = 0$$

This contradicts the manner in which η was chosen.   Hence division of the transformed differential equation by $Q(\eta)$ is possible, and the normal form (2-38) results.   If $A(x,y) = 0$, then $C(x,y) \neq 0$, and a procedure similar to the one used above leads to the normal form.

Elliptic Equations

If $\Delta = B^2 - AC < 0$, then L is an elliptic operator in $\mathcal{R}$.   In this case the roots of $Am^2 + 2Bm + C$ are the complex-valued functions

$$m_1 = \frac{-B + i\sqrt{-\Delta}}{A} \qquad m_2 = \frac{-B - i\sqrt{-\Delta}}{A}$$

where $i = \sqrt{-1}$. The solutions of the differential equations (2-35) are complex-valued, and there exist no real characteristic curves for elliptic differential equations.   Suppose that the coefficients A, B, C are analytic functions of the variables x, y in $\mathcal{R}$.   Then there exists a transformation such that Eq. (2-24) is transformed into

$$z_{\xi\xi} + z_{\eta\eta} + G(\xi,\eta,z,z_\xi,z_\eta) = 0 \tag{2-39}$$

called the *normal form for elliptic equations in two independent variables.* Let $\varphi = u + iv$ be a solution of $\varphi_x - m_1\varphi_y = 0$ in $\mathcal{R}$ such that $u(x,y)$, $v(x,y)$ are real-valued functions and $u_y \neq 0$, $v_y \neq 0$.   Let $\xi = u(x,y)$, $\eta = v(x,y)$. Then

$$0 = \varphi_x - m_1\varphi_y = (\xi_x + i\eta_x) - m_1(\xi_y + i\eta_y)$$

$$= \xi_x + \frac{B}{A}\xi_y + \frac{\sqrt{-\Delta}}{A}\eta_y + i\left(\eta_x + \frac{B}{A}\eta_y - \frac{\sqrt{-\Delta}}{A}\xi_y\right)$$

It follows that

$$\xi_x = -\frac{B}{A}\xi_y - \frac{\sqrt{-\Delta}}{A}\eta_y$$

$$\eta_x = -\frac{B}{A}\eta_y + \frac{\sqrt{-\Delta}}{A}\xi_y$$

Hence

$$\frac{\partial(\xi,\eta)}{\partial(x,y)} = \xi_x \eta_y - \xi_y \eta_x = -\frac{\sqrt{-\Delta}}{A}(\xi_y{}^2 + \eta_y{}^2) \neq 0$$

Now

$$Q(\varphi) = A(\varphi_x - m_1\varphi_y)(\varphi_x - m_2\varphi_y) = 0$$

and

$$Q(\varphi) = Q(\xi + i\eta) = Q(\xi) + 2iQ(\xi,\eta) - Q(\eta)$$

Accordingly $Q(\xi) = Q(\eta)$ and $Q(\xi,\eta) = 0$. From Eq. (2-28) it follows that the transformed form is

$$Q(\xi)(z_{\xi\xi} + z_{\eta\eta}) + M'(\xi,\eta,z,z_\xi,z_\eta) = 0$$

If

$$Q(\xi) = 0$$

then

$$0 = Q^2(\xi,\eta) - Q(\xi)Q(\eta) = \left[\frac{\partial(\xi,\eta)}{\partial(x,y)}\right]^2 \Delta$$

This implies that $\Delta = 0$, contrary to the hypothesis that $\Delta < 0$. Thus $Q(\xi) \neq 0$, and the normal form (2-39) now follows.

EXAMPLE 2-9 $yz_{xx} + (x + y)z_{xy} + xz_{yy} = 0$

Here

$$\Delta = \frac{(x + y)^2}{4} - xy = \frac{(x - y)^2}{4} > 0$$

so the equation is hyperbolic everywhere except along the line $y = x$, where it is parabolic. The subsidiary differential equations (2-35) are $y' = 1$, $y' = x/y$. The characteristic curves are

$$\xi = u(x,y) = y - x = c_1 \qquad \eta = v(x,y) = y^2 - x^2 = c_2$$

These are straight lines with slope 1 and rectangular hyperbolas, respectively (see Fig. 2-3). Through each point of the region (consisting of the xy plane with the line $y = x$ deleted) there passes exactly one pair of characteristic curves $\xi = c_1$, $\eta = c_2$. Note that the Jacobian

$$\frac{\partial(\xi,\eta)}{\partial(x,y)} = 2(x - y) \neq 0$$

in $\mathscr{R}$.   To determine the transformed differential equation,

$$z_x = z_\xi\xi_x + z_\eta\eta_x = -z_\xi - 2xz_\eta \qquad z_y = z_\xi + 2yz_\eta$$
$$z_{xx} = -z_{\xi\xi}\xi_x - z_{\xi\eta}\eta_x - 2z_\eta - 2x(z_{\eta\xi}\xi_x + z_{\eta\eta}\eta_x)$$
$$= z_{\xi\xi} + 4xz_{\xi\eta} + 4x^2z_{\eta\eta} - 2z_\eta$$

and so on. Substitute the expressions for z_{xx}, z_{xy}, z_{yy} into the differential equation, and the result is

$$\xi \frac{\partial^2 z}{\partial \xi \, \partial \eta} + \frac{\partial z}{\partial \eta} = 0 \qquad \text{or} \qquad \frac{\partial}{\partial \xi}\left(\xi \frac{\partial z}{\partial \eta}\right) = 0$$

Integration gives

$$\xi \frac{\partial z}{\partial \eta} = h(\eta)$$

Hence

$$z = \frac{1}{\xi} \int h(\eta) \, d\eta + f(\xi)$$

$$= f(y - x) + \frac{g(y^2 - x^2)}{y - x}$$

is the general solution in regions where $y \neq x$.

Classification of Almost-linear Equations in n Independent Variables

An almost-linear second-order equation in n independent variables $x_1, \ldots, x_n$ is of the form

$$\sum_{i=1}^{n} \sum_{j=1}^{n} A_{ij} \frac{\partial^2 u}{\partial x_i \, \partial x_j} + M(x_1, \ldots, x_n, u, u_{x_1}, \ldots, u_{x_2}) = 0 \qquad (2\text{-}40)$$

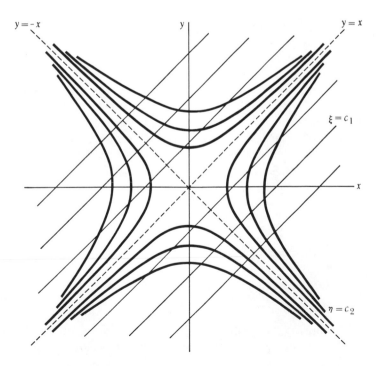

Figure 2-3

It is assumed that the coefficients A_{ij} are real-valued continuously differentiable functions of $x_1, \ldots, x_n$ and that $A_{ij} = A_{ji}, i,j = 1, \ldots, n$.   The linear operator

$$L_1 = \sum_{i=1}^{n} \sum_{j=1}^{n} A_{ij} D_{x_i} D_{x_j}$$

where $D_{x_i} = \partial/\partial x_i$, $i = 1, \ldots, n$, is called the *principal part* of the operator L appearing in Eq. (2-40).   The classification of L [and of Eq. (2-40)] at $x_1 = x_{10}, \ldots, x_n = x_{n0}$ is based on the classification of the *characteristic form*

$$Q(\xi) = \sum_{i=1}^{n} \sum_{j=1}^{n} A_{ij} \xi_i \xi_j$$

Here it is understood that the functions A_{ij} are evaluated at $x_1 = x_{10}, \ldots,$ $x_n = x_{n0}$, and $(\xi_1, \ldots, \xi_n)$ is a real n-tuple.   A well-known property of such a real quadratic form is that there exists a linear transformation $\xi_i = \sum_{j=1}^{n} S_{ij} \eta_j$, $i = 1, \ldots, n$, where $S = (S_{ij})$ is a nonsingular matrix, such that $Q(\xi)$ is reduced to the *canonical* (or normal) *form*

$$Q(\eta) = \eta_1{}^2 + \cdots + \eta_p{}^2 - \eta_{p+1}^2 - \cdots - \eta_{p+q}^2 \qquad (2\text{-}41)$$

where $\eta_i \neq 0, i = 1, \ldots, p + q$ (see Ref. 3).   The number p of positive terms appearing in Eq. (2-41) is called the *positive index*, the number q of negative terms is called the *negative index*, and the number $r = p + q$ is called the *rank* of the characteristic form Q at the point $x_1 = x_{10}, \ldots, x_n = x_{n0}$.   The rank $r \leq n$, and r is also the rank of the matrix $A = (A_{ij})$ at the point. The number $\nu = n - r$ is called the *nullity* (or defect) of the characteristic form, and $\nu \geq 0$.   Thus $\nu > 0$ if, and only if, the rank of the matrix A is less than n, that is, if, and only if, A is a singular matrix.   The important thing is that these numbers are *invariants* with respect to real nonsingular linear transformations of the variables $\xi_1, \ldots, \xi_n$; that is, they have the same value regardless of the mode of reduction of $Q(\xi)$ to the form given in Eq. (2-41).   At $x_1 = x_{10}, \ldots, x_n = x_{n0}$ the operator L [and Eq. (2-40)] is said to be

1. *Elliptic* if $\nu = 0$, and either $p = 0$ or $q = 0$
2. *Hyperbolic* if $\nu = 0$, and either $p = n - 1$ and $q = 1$, or $p = 1$ and $q = n - 1$
3. *Ultrahyperbolic* if $\nu = 0$, and $1 < q < n - 1$ (so $1 < p < n - 1$)
4. *Parabolic* if $\nu > 0$

The operator L is elliptic at the point if, and only if, the characteristic form is *definite*, being either *positive definite* or *negative definite*.   The form is positive definite if $Q(\xi) \geq 0$ holds for all real n-tuples $(\xi_1, \ldots, \xi_n)$ and $Q(\xi) = 0$ if, and only if, $\xi_1 = \cdots = \xi_n = 0$. The meaning of the term negative definite is obtained by reversing the inequality in the previous sentence. In the particular case $n = 2$

$$Q(\xi) = A_{11}\xi_1^2 + 2A_{12}\xi_1\xi_2 + A_{22}\xi_2^2$$

This form is definite if, and only if, $\Delta = A_{12}^2 - A_{11}A_{22} < 0$. Thus the criterion for ellipticity of L in the general case reduces in the particular case $n = 2$ to the criterion stated previously for the operator of Eq. (2-24). In the same way it follows that the criteria for hyperbolicity and parabolicity stated above reduce to $\Delta > 0$ and $\Delta = 0$, respectively, if $n = 2$. Also, when $n = 2$, types 2 and 3 are identical. If $n > 2$, the types are not the same. The preceding classification is a pointwise classification. The operator L in Eq. (2-26) is said to be *elliptic in* $\mathscr{R}$ if L is elliptic at each point of $\mathscr{R}$. Similar definitions are given for the phrases *hyperbolic in* $\mathscr{R}$, *parabolic in* $\mathscr{R}$, etc.

If $n > 2$, it is not possible in general to reduce Eq. (2-40) to normal form in a region. However, in the special case where the coefficients A_{ij} are constants it is possible to reduce the differential equation to normal form. A proof of this fact is given in Ref. 3. Only the results for the linear equation (2-10) are stated here. If L is elliptic, then by a linear transformation of the independent variables and a change in the dependent variable Eq. (2-10) is reduced to the normal form

$$\Delta v + cv = F(x_1, \ldots, x_n)$$

where c is a constant and

$$\Delta v = \frac{\partial^2 v}{\partial x_1^2} + \cdots + \frac{\partial^2 v}{\partial x_n^2}$$

(the n-dimensional laplacian of v). If L is hyperbolic, then Eq. (2-10) can be reduced to the normal form

$$\frac{\partial^2 v}{\partial x_1^2} + \cdots + \frac{\partial^2 v}{\partial x_{n-1}^2} - \frac{\partial^2 v}{\partial x_n^2} + cv = F(x_1, \ldots, x_n)$$

If L is ultrahyperbolic, the equation can be reduced to the normal form

$$\frac{\partial^2 v}{\partial x_1^2} + \cdots + \frac{\partial^2 v}{\partial x_p^2} - \frac{\partial^2 v}{\partial x_{p+1}^2} - \cdots - \frac{\partial^2 v}{\partial x_{p+q}^2} + cv = F(x_1, \ldots, x_n)$$

where $1 < p < n - 1$ and $p + q = n$. When L is parabolic, Eq. (2-10) can be reduced to the normal form

$$\frac{\partial^2 v}{\partial x_1^2} + \cdots + \frac{\partial^2 v}{\partial x_r^2} + B_{r+1}\frac{\partial v}{\partial x_{r+1}} + \cdots + B_n\frac{\partial x}{\partial x_n} + cv = F(x_1, \ldots, x_n)$$

where $0 < r < n$.

If $n = 2$, the normal form written above in the hyperbolic case is

$$\frac{\partial^2 v}{\partial x^2} - \frac{\partial^2 v}{\partial y^2} + cv = F(x,y) \qquad (2\text{-}42)$$

Under the nonsingular linear transformation

$$\xi = x + y \qquad \eta = x - y$$

Eq. (2-42) is transformed into Eq. (2-31).

2-5 CAUCHY PROBLEM FOR LINEAR SECOND-ORDER EQUATIONS IN TWO INDEPENDENT VARIABLES

For a first-order partial differential equation in two independent variables the Cauchy problem is to determine an integral surface which passes through a prescribed space curve.   Under suitable hypotheses there exists a unique solution.   In the Cauchy problem for a second-order equation in two independent variables there is an additional requirement: the normal to the integral surface is prescribed along the curve.

The Cauchy problem for a linear second-order equation

$$Lz = Az_{xx} + 2Bz_{xy} + Cz_{yy} + Dz_x + Ez_y + Fz = G \qquad (2\text{-}43)$$

is as follows.   Assume the coefficients and given function G are continuous on a region $\mathfrak{R}$ of the xy plane.   Let Γ_0 be a smooth curve lying in $\mathfrak{R}$ and defined by parametric equations

$$x = f(\tau) \qquad y = g(\tau) \qquad a < \tau < b \qquad (2\text{-}44)$$

where f and g are continuously differentiable and

$$[f'(\tau)]^2 + [g'(\tau)]^2 \neq 0 \qquad a < \tau < b \qquad (2\text{-}45)$$

Given continuously differentiable functions h, H, determine a solution $z = \varphi(x,y)$ of Eq. (2-43) such that

$$\varphi[f(\tau),g(\tau)] = h(\tau) \qquad \varphi_n[f(\tau),g(\tau)] = H(\tau) \qquad (2\text{-}46)$$

where $\varphi_n = \partial \varphi / \partial n$ denotes the derivative in the direction normal to Γ_0.

The functions f, g, h, and H constitute the *Cauchy*, or *initial, data* of the problem.   Often Γ_0 is called the *initial curve*.   The origin of the term initial used here lies in the physical nature of many Cauchy problems.   In such problems time t is one of the independent variables, and the differential equation involves partial derivatives with respect to t.   If u denotes the dependent variable, the values of u and $\partial u / \partial t$ are known at $t = 0$, that is, at initial time.   Thus, if there is but one other independent variable, say x, then the conditions are prescribed along the line $t = 0$ in the xt plane.   An example of this is the Cauchy problem for the wave equation discussed in Chap. 4.

Let Γ denote the smooth curve in xyz space defined by the parametric equations

$$x = f(\tau) \qquad y = g(\tau) \qquad z = h(\tau) \qquad a < \tau < b \tag{2-47}$$

Then the first condition in (2-46) together with Eq. (2-44) expresses the requirement that the integral surface contain Γ. The second condition prescribes the orientation of the normal to the integral surface along Γ. In order to see this, differentiate the first condition in (2-46) with respect to τ. Then

$$\frac{d\varphi}{d\tau} = \varphi_x \frac{dx}{d\tau} + \varphi_y \frac{dy}{d\tau}$$

implies the relation

$$\varphi_x f' + \varphi_y g' = h' \tag{2-48}$$

along Γ_0. Recall that the direction numbers of the normal $\mathbf{n}$ to Γ_0 are $-g'$, f'. Thus

$$-\varphi_x g' + \varphi_y f' = [(f')^2 + (g')^2]^{\frac{1}{2}} H \tag{2-49}$$

The two equations in φ_x, φ_y uniquely determine the partial derivatives along Γ_0 since the determinant is different from zero. But φ_x, φ_y, -1 are direction numbers of the normal to the integral surface.

In the Cauchy problem for second-order equations the values of the dependent variable and the values of the normal derivative are assigned *arbitrarily* (to within the smoothness requirements) on the initial curve. Accordingly the differential equation together with the initial curve should not imply a relationship between the data if the problem is to be solvable. In this connection it will be seen that the type of the differential equation and the characteristic curves are important.

Instead of prescribing φ and $\partial\varphi/\partial n$, the values of φ, φ_x, and φ_y may be prescribed along Γ_0. In this event Eq. (2-46) is replaced by

$$\varphi[f(\tau),g(\tau)] = h(\tau) \qquad a < \tau < b$$
$$\varphi_x[f(\tau),g(\tau)] = \rho(\tau) \qquad \varphi_y[f(\tau),g(\tau)] = \sigma(\tau) \qquad a < \tau < b \tag{2-50}$$

where h, ρ, and σ are given smooth functions. The data now consist of the five functions f, g, h, ρ, and σ. However they cannot be assigned arbitrarily if a solution is to exist. From Eq. (2-48) it follows that the relation

$$\rho(\tau)f'(\tau) + \sigma(\tau)g'(\tau) = h'(\tau) \tag{2-51}$$

must hold. Often Eq. (2-51) is called the *strip condition*. The normal derivative of φ along Γ_0 is now determined by the other data. It is given by

$$\varphi_n(\tau) = \frac{-\rho(\tau)g'(\tau) + \sigma(\tau)f'(\tau)}{\{[f'(\tau)]^2 + [g'(\tau)]^2\}^{\frac{1}{2}}} \tag{2-52}$$

Direction numbers of the tangent $\mathbf{T}$ to the curve Γ are f', g', h', and direction numbers of the normal $\mathbf{N}$ to the integral surface at a point on Γ are ρ, σ, -1. Hence Eq. (2-51) states that $\mathbf{N} \cdot \mathbf{T} = 0$, a relation which is geometrically evident (see Fig. 2-4).

EXAMPLE 2-10   Let Γ_0 be the segment of the x axis defined by the parametric equations

$$x = \tau \qquad y = 0 \qquad -a < \tau < a$$

Find a solution $z = \varphi(x,y)$ of

$$z_{xx} + 2z_{xy} - 3z_{yy} = 0$$

in a region containing Γ_0 such that

$$\varphi(\tau,0) = h(\tau) \qquad \varphi_x(\tau,0) = \rho(\tau) \qquad \varphi_y(\tau,0) = \sigma(\tau)$$

Here the strip condition is $\rho(\tau) = h'(\tau)$. If this relation does not hold, there is no solution of the problem. The derivative φ_y is in the direction normal to the initial curve and can be assigned arbitrarily. However φ_x is the derivative in the direction of the tangent to the initial curve and cannot be chosen arbitrarily. It is determined by the values of φ along Γ_0.

The general solution of the differential equation in a region is

$$z = F(3x - y) + G(x + y)$$

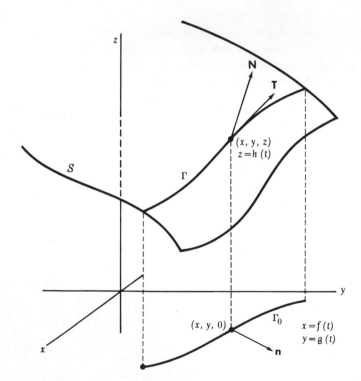

Figure 2-4

For a function of this form to satisfy the prescribed initial conditions it is necessary that

$$F(3\tau) + G(\tau) = h(\tau) \qquad -F'(3\tau) + G'(\tau) = \sigma(\tau)$$

Let β be a primitive of σ. Then the second relation implies

$$-\tfrac{1}{3}F(3\tau) + G(\tau) = \beta(\tau) + c$$

where c is a constant of integration. Hence

$$\tfrac{4}{3}F(3\tau) = h(\tau) - \beta(\tau) - c$$

$$F(\tau) = \frac{3}{4}\left[h\left(\frac{\tau}{3}\right) - \beta\left(\frac{\tau}{3}\right)\right] - \frac{3c}{4}$$

$$G(\tau) = \tfrac{1}{4}h(\tau) + \tfrac{3}{4}\beta(\tau) + \frac{3c}{4}$$

and so

$$z = \frac{3}{4}\left[h\left(\frac{3x - y}{3}\right) - \beta\left(\frac{3x - y}{3}\right)\right] + \tfrac{1}{4}h(x + y) + \tfrac{3}{4}\beta(x + y)$$

If h and β are twice continuously differentiable, then this function satisfies all the conditions of the problem. Moreover it is the unique solution.

As a particular case let

$$h(\tau) = 3\tau^2 \qquad \rho(\tau) = 6\tau \qquad \sigma(\tau) = 0$$

The space curve Γ is the parabola

$$z = 3x^2 \qquad y = 0$$

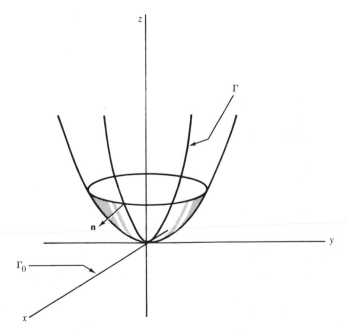

Figure 2-5

lying in the xz plane. The normal to the integral surface is required to be perpendicular to the y axis along Γ. The solution is

$$z = 3x^2 + y^2$$

(see Fig. 2-5).

The Cauchy problem formulated in Eqs. (2-43) to (2-46) is too general, and no solution need exist. Two reasons are (1) the hypotheses on the coefficients are too weak, and there may be no solutions of Eq. (2-43); (2) the type of the differential equation (hyperbolic, parabolic, or elliptic) must be taken into account, and it is important to know whether the initial curve Γ_0 is characteristic. The differential equation in Example 2-10 is hyperbolic, with analytic coefficients, and the data are analytic. Also, the initial curve is *noncharacteristic:* Γ_0 is nowhere tangent to a characteristic curve. There exists a unique analytic solution of that problem. The same differential equation is considered in Example 2-11, below. However the initial curve is characteristic. Now either there is no solution of the problem, or else there exist infinitely many distinct solutions.

EXAMPLE 2-11   Let Γ_0 be the line in the xy plane defined parametrically by

$$x = \tau \qquad y = 3\tau \qquad -a < \tau < a$$

Find a solution $z = \varphi(x,y)$ of

$$z_{xx} + 2z_{xy} - 3z_{yy} = 0$$

in a region containing Γ_0 such that

$$\varphi(\tau,3\tau) = h(\tau) \qquad \varphi_n(\tau,3\tau) = H(\tau)$$

where h, H are arbitrary continuously differentiable functions and $\varphi_n = \partial\varphi/\partial n$ is the derivative normal to Γ_0.

The characteristic curves of the differential equation are the lines

$$\xi = 3x - y = c_1 \qquad \eta = x + y = c_2$$

where c_1, c_2 are arbitrary real constants. Thus Γ_0 coincides with one of the two characteristics which pass through the origin. The unit tangent vector to Γ_0 is

$$\mathbf{T}_0 = \frac{\mathbf{i} + 3\mathbf{j}}{\sqrt{10}}$$

and the unit normal to Γ_0 is

$$\mathbf{n} = \frac{-3\mathbf{i} + \mathbf{j}}{\sqrt{10}}$$

The directional derivative of φ in the direction of $\mathbf{T}_0$ is

$$D_{T_0} = \frac{\varphi_x + 3\varphi_y}{\sqrt{10}}$$

More generally, if ψ is any continuously differentiable function of x and y defined in a region containing Γ_0, then the tangential derivative of ψ along Γ_0 is

$$D_{T_0}\psi = \frac{\psi_x + 3\psi_y}{\sqrt{10}}$$

Define the tangential operator D_{T_0} by

$$D_{T_0} = \frac{1}{\sqrt{10}}\left(\frac{\partial}{\partial x} + 3\frac{\partial}{\partial y}\right)$$

On Γ_0, ψ is a function of τ, and

$$\frac{d\psi}{d\tau} = \frac{\partial\psi}{\partial x}\frac{dx}{d\tau} + \frac{\partial\psi}{\partial y}\frac{dy}{d\tau} = \frac{\partial\psi}{\partial x} + 3\frac{\partial\psi}{\partial y}$$

Hence at points on Γ_0

$$D_{T_0} = \frac{1}{\sqrt{10}}\frac{d}{d\tau}$$

The operator in the direction of the normal **n** is

$$D_n = \frac{1}{\sqrt{10}}\left(-3\frac{\partial}{\partial x} + \frac{\partial}{\partial y}\right) = \frac{\partial}{\partial n}$$

Now, since

$$D_x = \frac{1}{\sqrt{10}}(D_{T_0} - 3D_n) \qquad D_y = \frac{1}{\sqrt{10}}(3D_{T_0} + D_n)$$

the operator L in the differential equation can be expressed in terms of the tangential and normal directional derivatives.

$$L = D_x^2 + 2D_xD_y - 3D_y^2 = (D_x + 3D_y)(D_x - D_y)$$

$$= -2D_{T_0}(D_{T_0} + 2D_n)$$

The differential equation $Lz = 0$ becomes, at points on Γ_0,

$$D_{T_0}^2 z = -2D_{T_0}D_n z$$

The left side of the differential equation involves only the tangential derivative, while the right side involves the normal and tangential derivatives. This is possible only because Γ_0 is characteristic. If φ is a solution, then

$$D_{T_0}\varphi = \frac{1}{\sqrt{10}}\frac{d\varphi}{d\tau} = \frac{h'(\tau)}{\sqrt{10}} \qquad D_{T_0}^2\varphi = \frac{h''(\tau)}{10}$$

$$D_n\varphi = \frac{\partial\varphi}{\partial n} = H(\tau) \qquad D_{T_0}D_n\varphi = \frac{H'(\tau)}{\sqrt{10}}$$

so that at points on Γ_0 the differential equation states

$$h''(\tau) = -2\sqrt{10}\,H'(\tau)$$

which is a relation between the data. But then the data cannot be assigned arbitrarily. The Cauchy problem is not properly solvable.

If the functions h, H are chosen such that

$$h'(\tau) = -2\sqrt{10}\, H(\tau)$$

then there exist infinitely many distinct solutions for the particular choice of data. Recall that the general solution of the differential equation is

$$z = F(3x - y) + G(x + y)$$

Impose the initial conditions on z; then

$$F(0) + G(4\tau) = h(\tau) \qquad -10F'(0) - 2G'(4\tau) = \sqrt{10}\, H(\tau)$$

Hence

$$G(\tau) = h\left(\frac{\tau}{4}\right) - F(0)$$

If the assumed relation between h' and H is used, the second relation between F and G yields

$$G'(\tau) = \frac{h'(\tau/4)}{4} - 5F'(0)$$

The last two conditions on G are consistent provided $F'(0) = 0$. Accordingly let F be a twice continuously differentiable function such that $F'(0) = 0$. Then

$$z = F(3x - y) + h\left(\frac{x + y}{4}\right) - F(0)$$

satisfies.

Cauchy-Kowalewski Theorem (Special Case)

Cauchy problems are formulated for nonlinear partial differential equations in n independent variables and, indeed, for systems of such equations. A fundamental existence and uniqueness theorem in the case where all the functions are analytic is the Cauchy-Kowalewski theorem. The following is the Cauchy-Kowalewski theorem for a particular case of the Cauchy problem formulated in the second paragraph of this section. A proof is given in Appendix 1. Note it is a local theorem: existence and uniqueness of solution are assured only in some neighborhood of a point.

THEOREM 2-1 Let the coefficients and G in Eq. (2-43) be analytic on a region $\mathcal{R}$ of the xy plane which contains the origin. Let $C(x,y) \neq 0$ on $\mathcal{R}$. Given arbitrary functions $h(x)$, $\sigma(x)$, analytic on the segment of the x axis contained in $\mathcal{R}$, there exists a neighborhood $\mathcal{N}$ of $(0,0)$ and a unique analytic solution $z = \varphi(x,y)$ of Eq. (2-43) on $\mathcal{N}$ such that

$$\varphi(x,0) = h(x) \qquad \varphi_y(x,0) = \sigma(x)$$

on the segment of the x axis contained in $\mathcal{N}$.

Significance of the Characteristic Curves

In the particular case considered in Theorem 2-1 the initial curve is a segment of the x axis. The hypothesis that C is different from zero on $\mathscr{R}$ ensures that the initial curve is noncharacteristic. Of course, if Eq. (2-43) is elliptic on $\mathscr{R}$, there are no characteristics, and a unique analytic solution exists. However in the hyperbolic and parabolic cases the following definitions are pertinent. A smooth curve is said to be *characteristic at a point* (x_0,y_0) if it is tangent there to a characteristic curve of Eq. (2-43). Otherwise the curve is *noncharacteristic at* (x_0,y_0). Suppose the equation of the curve is

$$\psi(x,y) = 0$$

where $\psi_y \neq 0$. Recall the characteristic equation (2-32). The curve is characteristic at (x_0,y_0) if, and only if,

$$Q(\psi) = A\psi_x{}^2 + 2B\psi_x\psi_y + C\psi_y{}^2 = 0 \tag{2-53}$$

at (x_0,y_0). This follows from the fact that the slope of the given curve is $m = -\psi_x/\psi_y$, and

$$Q(\psi) = \psi_y{}^2(Am^2 - 2Bm + C) = A\psi_y{}^2(m + m_1)(m + m_2)$$

where $-m_1, -m_2$ are the slopes of the characteristics [see Eq. (2-35)]. Now, in Theorem 2-1, the slope of the initial curve is $m = 0$. But clearly $m_1 \neq 0$, and $m_2 \neq 0$.

In the Cauchy problem formulated in Eqs. (2-43) to (2-47) the initial curve need not be a segment of the x axis. However, suppose that the coefficients, given function G, and the data are analytic in their arguments. If Γ_0 is noncharacteristic at a point (x_0,y_0), there exists a unique analytic solution of Eq. (2-43) which satisfies the prescribed conditions, at least in some neighborhood of (x_0,y_0). This follows from Theorem 2-1. First eliminate τ from the equations of Γ_0 to obtain the rectangular form $\Phi(x,y) = 0$. Let

$$\xi = x - x_0 \qquad \eta = \Phi(x,y) \tag{2-54}$$

The Jacobian of the transformation is

$$\xi_x\eta_y - \xi_y\eta_x = \Phi_y \neq 0$$

in some neighborhood of (x_0,y_0). Accordingly the mapping is one to one on a neighborhood of (x_0,y_0) onto a neighborhood of the origin in the $\xi\eta$ plane. The point (x_0,y_0) is mapped onto $(0,0)$, and the portion of Γ_0 lying in the neighborhood of (x_0,y_0) is mapped onto a segment of the ξ axis. Equation (2-43) is transformed into

$$Q(\xi)z_{\xi\xi} + 2Q(\xi,\eta)z_{\xi\eta} + Q(\eta)z_{\eta\eta} + D_1z_{\xi} + E_1z_{\eta} + F_1z = G_1 \tag{2-55}$$

The coefficients and G_1 in Eq. (2-55) are analytic in a neighborhood of (0,0). Observe that the coefficient of $z_{\eta\eta}$ is

$$Q(\eta) = A\Phi_x{}^2 + 2B\Phi_x\Phi_y + C\Phi_y{}^2 \neq 0$$

near (0,0), since Γ_0 is noncharacteristic at (x_0,y_0). The transformed initial conditions are

$$\varphi(\xi,0) = \hbar(\xi) \qquad \varphi_\eta(\xi,0) = \sigma(\xi)$$

where $\hbar$, σ are analytic functions of ξ in a neighborhood of $\xi = 0$. By Theorem 2-1 there exists a neighborhood of (0,0) and a unique analytic solution of Eq. (2-55) on the neighborhood which satisfies the transformed initial conditions on a segment of the ξ axis containing $\xi = 0$. Since the transformation (2-54) is invertible, it follows that there exists a neighborhood of (x_0,y_0) and a unique analytic solution of Eq. (2-43) which satisfies Eq. (2-46) on the portion of Γ_0 lying in the neighborhood.

Suppose Eq. (2-43) is elliptic and the coefficients and G are analytic functions on $\mathcal{R}$. Given analytic data, there exists a unique analytic solution of the Cauchy problem (at least locally) since Γ_0 cannot be characteristic at any point. However, there is a differentiability theorem for elliptic equations (see Ref. 4) which implies in the present case that each solution of Eq. (2-43) must be analytic in x and y. Hence a solution of the Cauchy problem must be analytic along the initial curve. Accordingly, if the prescribed data are not analytic, then there is no solution of the Cauchy problem.

Assume Eq. (2-43) is hyperbolic or parabolic and the coefficients, given function G, and data are analytic. If Γ_0 is a characteristic curve, then, in general, there is no solution of the Cauchy problem. To see why this is so, let $\alpha(\tau) = f'(\tau)/\{[f'(\tau)]^2 + [g'(\tau)]^2\}^{1/2}$ and $\beta(\tau) = g'(\tau)/\{[f'(\tau)]^2 + [g'(\tau)]^2\}^{1/2}$ and define the tangential operator D_{T_0} by

$$D_{T_0} = \alpha\frac{\partial}{\partial x} + \beta\frac{\partial}{\partial y}$$

Thus if $\psi(x,y)$ is an analytic function on $\mathcal{R}$, then on Γ_0

$$D_{T_0}\psi = \{[f'(\tau)]^2 + [g'(\tau)]^2\}^{-1/2}\frac{d\psi}{d\tau}$$

Define the normal operator D_n by

$$D_n = -\beta\frac{\partial}{\partial x} + \alpha\frac{\partial}{\partial y}$$

Then

$$D_x = \alpha D_{T_0} - \beta D_n \qquad D_y = \beta D_{T_0} + \alpha D_n$$

and the operator L in Eq. (2-43) can be written, for points on Γ_0,

$$L = (A\alpha^2 + 2B\alpha\beta + C\beta^2)D_{T_0}{}^2 + 2[(C - A)\alpha\beta + B(\alpha^2 - \beta^2)]D_{T_0}D_n$$
$$+ (A\beta^2 - 2B\alpha\beta + C\alpha^2)D_n{}^2 + (D\alpha + E\beta)D_{T_0} + (E\alpha - D\beta)D_n + F$$

Now suppose Γ_0 is characteristic at a point (x_0,y_0) which corresponds to $\tau = \tau_0$. Then

$$A\beta^2 - 2B\alpha\beta + C\alpha^2 = 0$$

at the point.   The differential equation (2-43) states that at (x_0,y_0), $\tau = \tau_0$,

$$(A\alpha^2 + 2B\alpha\beta + C\beta^2)h'' + (D\alpha + E\beta)\gamma h' + \gamma^2 Fh = \gamma^2 G - 2[(C - A)\alpha\beta$$
$$+ B(\alpha^2 - \beta^2)]\gamma H' - (E\alpha - D\beta)\gamma^2 H$$

where

$$\gamma = \{[f'(\tau_0)]^2 + [g'(\tau)]^2\}^{\frac{1}{2}}$$

Hence there is a relation between the data which must hold along a characteristic.   Since h and H cannot be assigned arbitrarily the problem is not properly solvable.

Some of the properties of solutions of the three types of equations have been mentioned in the preceding paragraphs.   If the differential equation is elliptic, solutions have as much continuity and differentiability as possessed by the coefficients.   The absence of characteristic curves allows the smoothness of solution.   This aspect is reflected in the types of problems of mathematical physics in which elliptic partial differential equations occur.   Elliptic equations describe equilibrium configurations and steady-state conditions which prevail after a sufficiently long time has elapsed.   On the other hand, hyperbolic and parabolic equations appear in problems where time t is an important independent variable.   The state is known at $t = 0$, and it is desired to determine the state for $t > 0$.   Hyperbolic equations describe dynamic configurations, often are a consequence of equations of motion, and occur in problems of fluid mechanics, acoustics, and electromagnetic theory. In these problems an essential feature is a propagation of change of state as t increases.

In further contrast with the elliptic type, a hyperbolic equation with analytic coefficients may have nonanalytic solutions.   It is possible to construct a solution of such an equation which is analytic in the regions adjacent to a given initial curve Γ_0, which takes on prescribed continuous data along the curve, but which has discontinuous derivatives at points of Γ_0.   Such a function is called a *discontinuous solution* (even though the function itself is continuous).   For example, there are problems in gas dynamics in which such solutions are important.   It can be shown (see Ref. 4) that the curves along which such discontinuities in derivatives may occur are the characteristic curves of the differential equation and no others.

EXAMPLE 2-12 $z_{xt} = 0$

The differential equation is hyperbolic everywhere in the xt plane, and the general solution is

$$z = f(x) + g(t)$$

Recall that the definition of the term solution given in the text requires continuous second derivatives. However, here it is clear that if f and g have continuous first derivatives, this suffices to ensure $z_{xt} = 0$. For example, let x_0 be a fixed value and define f by

$$f(x) = \begin{cases} (x - x_0)^2 & x \geq x_0 \\ 0 & x < x_0 \end{cases}$$

Then f has a continuous first derivative for all values of x, and $f'(x_0) = 0$. If $h > 0$,

$$\lim_{h \to 0} \frac{f'(x_0 + h) - f'(x_0)}{h} = \lim_{h \to 0} \frac{2h - 0}{h} = 2$$

$$\lim_{h \to 0} \frac{f'(x_0 - h) - f'(x_0)}{h} = \lim_{h \to 0} \frac{0 - 0}{h} = 0$$

Thus $f''(x_0)$ does not exist. However $z = f(x)$ satisfies the differential equation, and z_{xt}, z_{tt} are continuous everywhere. Note that the line $x = x_0$ in the xt plane is a characteristic curve of the differential equation, and the function $z = f(x)$ is such that z_{xx} fails to exist along this line. Similarly one can construct a function $z = g(t)$ which satisfies the differential equation everywhere but is such that z_{tt} fails to exist along a characteristic $t = t_0$. The function z defined by

$$z(x,t) = \begin{cases} (x - x_0)^3 & x \geq x_0 \\ 0 & x < x_0 \end{cases}$$

has continuous second derivatives and satisfies $z_{xt} = 0$ everywhere and hence is a solution of the differential equation in the sense of the definition of the text. But z is not analytic along the characteristic $x = x_0$.

   Suppose initial conditions

$$z = f(x) \qquad z_t = g(x)$$

are prescribed along the characteristic $t = 0$. If z is a solution of the problem, then z has continuous second derivatives and satisfies $z_{xt} = 0$. But

$$z_{xt} = z_{tx} = g'(x)$$

along $t = 0$. Hence $g'(x) = 0$. Accordingly no solution exists unless the prescribed function g is a constant. In this event there exist infinitely many solutions of the problem. In fact, if $g(x) = c$, then

$$z = f(x) + ct + G(t) - G(0)$$

is a solution of the problem whenever G is a function with continuous second derivatives such that $G'(0) = 0$.

EXAMPLE 2-13 $x^2 u_{xx} - t^2 u_{tt} = xt \qquad x > 0; t > 0$

$$u(x,1) = f(x) \qquad u_t(x,1) = g(x) \qquad x > 0$$

The initial curve is the line $t = 1$ in the xt plane. Since $\Delta = x^2 t^2 > 0$ in the region of interest, the equation is hyperbolic. The characteristic curves are the hyperbolas $\xi = xt = c_1$ and the straight lines $\eta = x/t = c_2$. Thus Γ_0 is nowhere characteristic. Assume the functions f, g are analytic. The general solution of the differential equation is

$$u = F(xt) + \sqrt{xt}\, G\left(\frac{x}{t}\right) + \frac{xt}{2}\log\frac{x}{t}$$

If the initial conditions are imposed, the equations

$$F(x) + \sqrt{x}\, G(x) = f_1(x)$$

$$xF'(x) + \frac{\sqrt{x}}{2} G(x) - x\sqrt{x}G'(x) = g_1(x)$$

result, where

$$f_1(x) = f(x) - \frac{x}{2}\log x \qquad g_1(x) = g(x) - \frac{x}{2}\log x + \frac{x}{2}$$

If the pair of equations is solved for F', then

$$2xF' - F = g_1 - f_1 + xf_1'$$

Integration gives

$$F(x) = \tfrac{1}{2}f_1(x) - \frac{\sqrt{x}}{4}\int^x \frac{f_1(s)}{s^{3/2}}\,ds + \frac{\sqrt{x}}{2}\int^x \frac{g_1(s)}{s^{3/2}}\,ds$$

and so

$$G(x) = \frac{1}{\sqrt{x}}\left[\tfrac{1}{2}f_1(x) - \frac{\sqrt{x}}{2}\int^x \frac{g_1(s)}{s^{3/2}}\,ds + \frac{\sqrt{x}}{4}\int^x \frac{f_1(s)}{s^{3/2}}\,ds\right]$$

Accordingly the solution of the Cauchy problem is

$$u(x,t) = \frac{1}{2}\left[f_1(xt) + tf_1\left(\frac{x}{t}\right)\right] + \frac{\sqrt{xt}}{4}\int_{x/t}^{xt} \frac{2g_1(s) - f_1(s)}{s^{3/2}}\,ds$$

2-6 CAUCHY PROBLEM FOR LINEAR SECOND-ORDER EQUATIONS IN n INDEPENDENT VARIABLES

Let n be a fixed positive integer. A *point* in real n-dimensional space is an ordered n-tuple

$$\mathbf{x} = (x_1, \ldots, x_n)$$

of real numbers x_i. The term *vector* is synonymous with point. Let $\mathbf{x} = (x_1, \ldots, x_n)$ and $\mathbf{y} = (y_1, \ldots, y_n)$ be points. Then *addition* of $\mathbf{x}$ and $\mathbf{y}$ is defined by

$$\mathbf{x} + \mathbf{y} = (x_1 + y_1, \ldots, x_n + y_n)$$

A scalar is a real number. Multiplication of a vector $\mathbf{x}$ by a scalar c is defined by

$$c\mathbf{x} = (cx_1, \ldots, cx_n)$$

The set of all vectors $\mathbf{x}$ together with the algebraic rules and properties constitutes *real cartesian vector space* $\mathscr{E}_n$. The *scalar* (or dot) product of vectors in $\mathscr{E}_n$ is defined by

$$\mathbf{x} \cdot \mathbf{y} = x_1 y_1 + \cdots + x_n y_n$$

The length of a vector $\mathbf{x}$ is defined as

$$|\mathbf{x}| = (x_1{}^2 + \cdots + x_n{}^2)^{\frac{1}{2}} = (\mathbf{x} \cdot \mathbf{x})^{\frac{1}{2}}$$

Let $\mathbf{x}_0$ be a given point of $\mathscr{E}_n$ and let r be a real positive number. The set of all $\mathbf{x}$ in $\mathscr{E}_n$ such that

$$|\mathbf{x} - \mathbf{x}_0| = r$$

is called the *sphere of radius r about* $\mathbf{x}_0$. The set of all $\mathbf{x}$ such that

$$|\mathbf{x} - \mathbf{x}_0| < r$$

is called the *interior of the sphere*, or, synonymously, the *r neighborhood of* $\mathbf{x}_0$. This set of points is symbolized by $N_r(\mathbf{x}_0)$. Let $\mathcal{S}$ be a set of points in $\mathscr{E}_n$. A point $\mathbf{x}_0$ is called an *interior point of* $\mathcal{S}$ if there exists a neighborhood $N_r(\mathbf{x}_0)$ which consists entirely of points of $\mathcal{S}$. A set $\mathcal{S}$ is called *open* if every point of $\mathcal{S}$ is an interior point of $\mathcal{S}$. An *arc* in $\mathscr{E}_n$ is a set of points defined by n functions

$$x_i = g_i(s) \qquad i = 1, \ldots, n; \, a < s < b$$

where s is a real parameter and the functions g_i are continuous on (a,b). A *smooth arc* is an arc in which the functions g_i are continuously differentiable on (a,b). If the functions g_i are only required to be piecewise continuously differentiable, the arc is called piecewise smooth. A *region* in $\mathscr{E}_n$ is an open set having the property that given a pair of points $\mathbf{x}, \mathbf{y}$ in the set, there exists a piecewise smooth arc which lies in the set and joins the points.

A *surface* in $\mathscr{E}_n$ is an $(n-1)$-dimensional manifold of points. A sphere is an example of a surface. Often a surface is defined by an equation

$$F(x_1, \ldots, x_n) = 0 \tag{2-56}$$

where F is a real-valued continuously differentiable function of the variables $x_1, \ldots, x_n$ such that

$$\sum_{i=1}^{n} \left(\frac{\partial F}{\partial x_i} \right)^2 \neq 0$$

Such a surface is called *smooth*. An example of a smooth surface in $\mathscr{E}_n$

is the set of all points **x** such that

$$\mathbf{x} = (x_1, \ldots, x_{n-1}, 0)$$

This is called the *plane* $x_n = 0$.   If $n = 2$, it is the x axis in the xy plane, while if $n = 3$, it is the xy plane in xyz space.   A smooth surface in $\mathscr{E}_n$ may also be defined by n equations

$$x_i = \psi_i(s_1, \ldots, s_{n-1}) \qquad i = 1, \ldots, n \tag{2-57}$$

in $n - 1$ real parameters $s_1, \ldots, s_{n-1}$.   It is assumed the functions ψ_i are continuously differentiable, and the Jacobians

$$J_i = \frac{\partial(\psi_1, \ldots, \psi_{i-1}, \psi_{i+1}, \ldots, \psi_n)}{\partial(s_1, \ldots, s_{n-1})} \qquad i = 1, \ldots, n \tag{2-58}$$

do not vanish simultaneously.

If $u = \varphi(x_1, \ldots, x_n)$ is a differentiable function of the n variables $x_1, \ldots, x_n$, then the *gradient* of u is the vector

$$\nabla u = \left(\frac{\partial u}{\partial x_1}, \ldots, \frac{\partial u}{\partial x_n} \right) \tag{2-59}$$

Let S be a smooth surface defined by Eq. (2-56).   At a point $\mathbf{x}_0$ on S the *normal* is the vector

$$\gamma = \frac{\nabla F}{|\nabla F|} \tag{2-60}$$

where it is understood that ∇F and $|\nabla F|$ are evaluated at $\mathbf{x} = \mathbf{x}_0$.   The *tangent plane* to S at $\mathbf{x}_0$ is the smooth surface consisting of all points **x** satisfying

$$(\mathbf{x} - \mathbf{x}_0) \cdot \nabla F = 0 \tag{2-61}$$

where again ∇F is evaluated at $\mathbf{x}_0$.   Suppose instead S is defined parametrically by Eq. (2-57).   Let

$$\gamma_i = (-1)^{n-i} J_i \qquad i = 1, \ldots, n \tag{2-62}$$

where $J_1, \ldots, J_n$ are the Jacobians defined in Eq. (2-58).   Then, if $\mathbf{x}_0$ is a point on S corresponding to the values $s_{10}, \ldots, s_{n-1,0}$ of the parameters, the normal at $\mathbf{x}_0$ is the vector

$$\gamma = \frac{(\gamma_1, \ldots, \gamma_n)}{(\gamma_1^2 + \cdots + \gamma_n^2)^{\frac{1}{2}}} \tag{2-63}$$

where the γ_i are evaluated at $s_{10}, \ldots, s_{n-1,0}$.   The derivative of a function u in the direction of the normal γ to a surface is defined by

$$\frac{\partial u}{\partial \gamma} = \nabla u \cdot \gamma \tag{2-64}$$

This is referred to briefly as the *normal derivative*.

Let

$$Lu = \sum_{i=1}^{n} \sum_{j=1}^{n} A_{ij} \frac{\partial^2 u}{\partial x_i \, \partial x_j} + \sum_{i=1}^{n} B_i \frac{\partial u}{\partial x_i} + Cu = G \tag{2-65}$$

be a linear second-order equation. Assume the coefficients and given function G are continuous on a region $\mathscr{R}$. Then the Cauchy problem for Eq. (2-65) is as follows. Let S_0 be a smooth surface in $\mathscr{R}$ defined by Eq. (2-57). Given arbitrary continuously differentiable functions h, H of the parameters $s_1, \ldots, s_{n-1}$, determine a solution u of Eq. (2-65) such that

$$u = h(s_1, \ldots, s_{n-1}) \qquad \frac{\partial u}{\partial \gamma} = H(s_1, \ldots, s_{n-1}) \tag{2-66}$$

on S, where $\partial u/\partial \gamma$ is the normal derivative. The functions ψ_i, $i = 1, \ldots, n$, h, and H constitute the *data*, and S_0 is called the *initial surface* of the problem.

If, instead of the normal derivative, the n first partial derivatives of u are prescribed on S, then Eq. (2-66) is replaced by

$$u = h(s_1, \ldots, s_{n-1})$$

$$\frac{\partial u}{\partial x_i} = \rho_i(s_1, \ldots, s_{n-1}) \qquad i = 1, \ldots, n \tag{2-67}$$

where $\rho_1, \ldots, \rho_n$ are continuously differentiable functions. In this case the normal derivative is also known by virtue of Eqs. (2-59) and (2-64). However, now the data cannot be prescribed in an arbitrary manner, since

$$\frac{\partial u}{\partial s_k} = \sum_{i=1}^{n} \frac{\partial u}{\partial x_i} \frac{\partial x_i}{\partial s_k} \qquad k = 1, \ldots, n-1$$

implies that the $n - 1$ relations

$$\frac{\partial h}{\partial s_k} = \sum_{i=1}^{n} \rho_i \frac{\partial \psi_i}{\partial s_k} \qquad k = 1, \ldots, n-1 \tag{2-68}$$

must hold between the data.

Cauchy-Kowalewski Theorem (Special Case, n Independent Variables)

In the *analytic case* the coefficients and G in Eq. (2-65) are analytic functions of the variables $x_1, \ldots, x_n$ in $\mathscr{R}$, and the data are analytic in the parameters $s_1, \ldots, s_{n-1}$. Henceforth only the analytic case is considered. In the particular case where the initial surface is the plane $x_n = 0$ there is the following form of the Cauchy-Kowalewski theorem. It is stated below without proof.

THEOREM 2-2 Let the coefficients and G in Eq. (2-65) be analytic in a region $\mathscr{R}$ which contains the origin $(0, \ldots, 0)$. Assume $A_{nn} \neq 0$ in $\mathscr{R}$.

Let $\mathscr{R}_0$ be the projection of $\mathscr{R}$ onto the plane $x_n = 0$. Given arbitrary functions h, H of the variables $x_1, \ldots, x_{n-1}$, analytic on $\mathscr{R}_0$, there exists a neighborhood $N_r(0, \ldots, 0)$ and a unique analytic solution u of Eq. (2-65) in N_r such that

$$u(x_1, \ldots, x_{n-1}, 0) = h(x_1, \ldots, x_{n-1}) \qquad \frac{\partial u}{\partial x_n}\bigg|_{x_n=0} = H(x_1, \ldots, x_{n-1}) \quad (2\text{-}69)$$

on the projection of N_r onto $x_n = 0$.

In the problem formulated in Eqs. (2-65) and (2-66) the initial surface S_0 may not coincide with the hyperplane $x_n = 0$. To reduce the problem to one to which Theorem 2-2 is applicable a transformation which maps S_0 onto a portion of the hyperplane is constructed. Let

$$\Phi(x_1, \ldots, x_n) = 0$$

be the equation of S_0 in rectangular form obtained by eliminating the parameters s_i from Eqs. (2-57). Let $\mathbf{x}_0$ be a point on S_0 at which $\partial\Phi/\partial x_n \neq 0$. Consider the change of independent variables

$$\xi_i = x_i - x_{i0} \qquad i = 1, \ldots, n-1 \qquad \xi_n = \Phi(x_1, \ldots, x_n)$$

from x space to ξ space. The Jacobian of the transformation is different from zero at $\mathbf{x}_0$. Accordingly there is a neighborhood of $\mathbf{x}_0$ in x space which is mapped in a one-to-one fashion onto a neighborhood of the origin $(0, \ldots, 0)$ in ξ space, and in the process a portion of the surface S_0 containing $\mathbf{x}_0$ is mapped onto a portion of the plane $\xi_n = 0$ containing the origin $(0, \ldots, 0)$ in ξ space.

Theorem 2-2 is applicable to the transformed problem provided the coefficient of $\partial^2 u/\partial\xi_n^2$ in the transformed differential equation has a nonzero value at the point $(0, \ldots, 0)$ in ξ space. Now

$$\frac{\partial u}{\partial x_i} = \sum_{k=1}^{n} \frac{\partial u}{\partial\xi_k}\frac{\partial\xi_k}{\partial x_i}$$

$$\frac{\partial^2 u}{\partial x_i\,\partial x_j} = \frac{\partial}{\partial x_j}\left(\sum_{k=1}^{n} \frac{\partial u}{\partial\xi_k}\frac{\partial\xi_k}{\partial x_i}\right) = \sum_{k=1}^{n}\frac{\partial}{\partial x_j}\left(\frac{\partial u}{\partial\xi_k}\frac{\partial\xi_k}{\partial x_i}\right)$$

$$= \sum_{k=1}^{n}\sum_{q=1}^{n}\frac{\partial^2 u}{\partial\xi_k\,\partial\xi_q}\frac{\partial\xi_q}{\partial x_j}\frac{\partial\xi_k}{\partial x_i} + \sum_{k=1}^{n}\frac{\partial u}{\partial\xi_k}\frac{\partial^2\xi_k}{\partial x_i\,\partial x_j}$$

Hence

$$Lu = \sum_{k=1}^{n}\sum_{k=1}^{n} A_{ij}\frac{\partial^2 u}{\partial x_i\,\partial x_j} + \cdots$$

$$= \sum_{k=1}^{n}\sum_{q=1}^{n} P_{kq}\frac{\partial^2 u}{\partial\xi_k\,\partial\xi_q} + \cdots$$

where

$$P_{kq} = \sum_{i=1}^{n} \sum_{j=1}^{n} A_{ij} \frac{\partial \xi_k}{\partial x_i} \frac{\partial \xi_q}{\partial x_j} \qquad \begin{matrix} k = 1, \ldots, n \\ q = 1, \ldots, n \end{matrix}$$

In particular

$$P_{nn} = \sum_{i=1}^{n} \sum_{j=1}^{n} A_{ij} \frac{\partial \xi_n}{\partial x_i} \frac{\partial \xi_n}{\partial x_j} = \sum_{i=1}^{n} \sum_{j=1}^{n} A_{ij} \frac{\partial \Phi}{\partial x_i} \frac{\partial \Phi}{\partial x_j}$$

Define

$$Q(\Phi) = \sum_{i=1}^{n} \sum_{j=1}^{n} A_{ij} \frac{\partial \Phi}{\partial x_i} \frac{\partial \Phi}{\partial x_j} \tag{2-70}$$

If $Q(\Phi) \neq 0$ at $\mathbf{x}_0$, the coefficient of $\partial^2 u / \partial \xi_n{}^2$ in the transformed differential equation has a value different from zero at $(0, \ldots, 0)$, and hence the transformed problem has a unique analytic solution in a neighborhood of the origin in ξ space.   Since the transformation is invertible in this neighborhood, this implies the existence and uniqueness of an analytic solution of the original problem in some neighborhood of the point $\mathbf{x}_0$.

Characteristic Surfaces

A smooth surface defined by an equation $\Psi(x_1, \ldots, x_n) = 0$ is said to be *characteristic at* $\mathbf{x}_0$ if

$$Q(\Psi) = 0 \tag{2-71}$$

at the point.   Otherwise the surface is called *noncharacteristic* at the point. A surface is called a *characteristic surface* of Eq. (2-65) if Eq. (2-71) holds at each point on the surface.   Often Eq. (2-71) is referred to as the *characteristic equation* of the differential equation (2-65).   From the preceding discussion it follows that if the initial surface S_0 is noncharacteristic, there exists a unique analytic solution of the Cauchy problem (at least locally).

Recall the classification as to type of Eq. (2-65) given in Sec. 2-3.   If the differential equation is elliptic, then $Q(\Psi) = 0$ if, and only if,

$$\frac{\partial \Psi}{\partial x_1} = \cdots = \frac{\partial \Psi}{\partial x_n} = 0$$

Hence in the elliptic case no smooth surface is characteristic, and there always exists a unique analytic solution of the analytic Cauchy problem at least locally.   However, if Eq. (2-65) is hyperbolic or parabolic in a region, then characteristic surfaces exist and play an important role in the theory as well as the applications.

If $n = 2$, Eq. (2-71) reduces to Eq. (2-53), and the characteristic surfaces are just the characteristic curves.   The following examples illustrate the case $n = 3$.

EXAMPLE 2-14 $\dfrac{\partial^2 u}{\partial x^2} + \dfrac{\partial^2 u}{\partial y^2} - \dfrac{1}{c^2}\dfrac{\partial^2 u}{\partial t^2} = 0$

The differential equation is the wave equation of Example 2-3. This equation is hyperbolic everywhere. Take $x_1 = x$, $x_2 = y$, and $x_3 = t$ in Eq. (2-71). Then the characteristic equation is

$$Q(\Psi) = \left(\frac{\partial \Psi}{\partial x}\right)^2 + \left(\frac{\partial \Psi}{\partial y}\right)^2 - \frac{1}{c^2}\left(\frac{\partial \Psi}{\partial t}\right)^2 = 0$$

Let S be a surface in xyt space defined by an equation of the form

$$\Psi(x,y,t) = 0$$

If γ_1, γ_2, γ_3 denote the direction cosines of the normal to S, then

$$\gamma_1 = \lambda\frac{\partial \Psi}{\partial x} \qquad \gamma_2 = \lambda\frac{\partial \Psi}{\partial y} \qquad \gamma_3 = \lambda\frac{\partial \Psi}{\partial t}$$

where λ is a proportionality factor. Thus S is a characteristic surface if, and only if,

$$\gamma_1^2 + \gamma_2^2 - \frac{\gamma_3^2}{c^2} = 0$$

Since $\gamma_1^2 + \gamma_2^2 + \gamma_3^2 = 1$, this condition is equivalent to

$$\gamma_3^2 = \frac{c^2}{1 + c^2}$$

Accordingly S is a characteristic surface if the normal makes the constant angle $\theta = \tan^{-1}(1/c)$ with the t axis. If (x_0, y_0, t_0) is a fixed point in the xyt space, then the planes

$$\gamma_1(x - x_0) + \gamma_2(y - y_0) + \gamma_3(t - t_0) = 0$$

through (x_0, y_0, t_0) are characteristic surfaces provided

$$\gamma_3^2 = \frac{c^2}{1 + c^2} \qquad \gamma_1^2 + \gamma_2^2 = \frac{1}{1 + c^2}$$

There is a one-parameter family of such planes, and these envelop a cone with vertex at (x_0, y_0, t_0). This cone is called the *characteristic cone* at (x_0, y_0, t_0). The significance of the characteristic cone is discussed further in Chap. 4.

If the initial surface S_0 in the Cauchy problem for the wave equation is defined by an equation

$$t = \psi(x,y)$$

then the initial conditions take the form

$$u = h(x,y) \qquad \frac{\partial u}{\partial \gamma} = H(x,y) \qquad \text{on } S_0$$

where γ is the normal to S_0. The initial surface is noncharacteristic if

$$\psi_x^2 + \psi_y^2 \neq \frac{1}{c^2}$$

since ψ_x, ψ_y, -1 are direction numbers for γ. In this case if the data are analytic, there exists a unique analytic solution of the Cauchy problem. Of considerable interest in the applications is the case where the initial surface is the plane $t = 0$. Since the xy plane is noncharacteristic, there exists a unique analytic solution of the wave equation such that

$$u(x,y,0) = h(x,y) \qquad u_t(x,y,0) = H(x,y)$$

provided h and H are analytic functions of x and y. This is called the initial-value problem for the wave equation.

Assume the initial surface S_0 is defined by $t = \psi(x,y)$, where ψ is continuously differentiable. Recall that a vector $\boldsymbol{\xi} = (\xi_1,\xi_2,\xi_3)$ lies in the tangent plane to S_0 at a point if

$$\boldsymbol{\xi} \cdot \boldsymbol{\gamma} = \xi_1\gamma_1 + \xi_2\gamma_2 + \xi_3\gamma_3 = 0$$

where γ_1, γ_2, γ_3 are the direction cosines of the normal to S_0 at the point. An operator

$$L_\xi = \xi_1 D_x + \xi_2 D_y + \xi_3 D_t$$

is called *tangential* at a point on S_0 if $\boldsymbol{\xi}$ lies in the tangent plane to S_0 at the point. An important property of a tangential operator L_ξ is that if $u(x,y,t)$ is continuously differentiable and

$$u = h(x,y) \qquad \text{on } S_0$$

then the value $L_\xi u$ is determined at the point of tangency by h. Observe that on S_0

$$u(x,y,t) = u[x,y,\psi(x,y)] = h(x,y)$$

so that

$$u_x + u_t\psi_x = h_x \qquad u_x = h_x - u_t\psi_x$$

Similarly

$$u_y = h_y - u_t\psi_y$$

Since L_ξ is tangential,

$$\xi_1\psi_x + \xi_2\psi_y - \xi_3 = 0$$

at the point, and so

$$L_\xi u = \xi_1(h_x - u_t\psi_x) + \xi_2(h_y - u_t\psi_y) + \xi_3 u_t$$
$$= \xi_1 h_x + \xi_2 h_y - u_t(\xi_1\psi_x + \xi_2\psi_y - \xi_3) = \xi_1 h_x + \xi_2 h_y$$

Thus $L_\xi u$ is determined by h at the point of tangency.

Let D_γ denote the normal derivative on S_0. Each directional derivative can be expressed in terms of a tangential operator and D_γ. This is analogous to writing an arbitrary vector as the sum of a vector lying in the tangent plane to S_0 and a vector along the normal. In particular D_x and D_y can be so expressed. Note that

$$L_{\xi_x} = D_x - \gamma_1 D_\gamma$$

is a tangential operator since

$$D_x - \gamma_1 D_\gamma = D_x - \gamma_1(\gamma_1 D_x + \gamma_2 D_y + \gamma_3 D_t)$$
$$= (1 - \gamma_1^2)D_x - \gamma_1\gamma_2 D_y - \gamma_1\gamma_3 D_t$$

and

$$(1 - \gamma_1^2)\gamma_1 - (\gamma_1\gamma_2)\gamma_2 - (\gamma_1\gamma_3)\gamma_3 = 0$$

Hence

$$D_x = L_{\xi_x} + \gamma_1 D_\gamma$$

Similarly

$$D_y = L_{\xi_y} + \gamma_2 D_\gamma \qquad D_t = L_{\xi_t} + \gamma_3 D_\gamma$$

where L_{ξ_y}, L_{ξ_t} are tangential operators. Now the operator L in the wave equation can be written in terms of tangential and normal operators.

$$L = D_x^2 + D_y^2 - \frac{1}{c^2} D_t^2$$

$$= \left(L_{\xi_x}^2 + L_{\xi_y}^2 - \frac{1}{c^2} L_{\xi_t}^2 \right) + 2D_\gamma \left(\gamma_1 L_{\xi_x} + \gamma_2 L_{\xi_y} - \frac{\gamma_3}{c^2} L_{\xi_t} \right) + \left(\gamma_1^2 + \gamma_2^2 - \frac{\gamma_3^2}{c^2} \right) D_\gamma^2$$

Suppose S_0 is a characteristic surface. Then on S_0 the wave equation states

$$\left(L_{\xi_x}^2 + L_{\xi_y}^2 - \frac{1}{c^2} L_{\xi_t}^2 \right) u = 2 \left(\frac{\gamma_3}{c^2} L_{\xi_t} - \gamma_1 L_{\xi_x} - \gamma_2 L_{\xi_y} \right) \frac{\partial u}{\partial \gamma}$$

The operator on the left is tangential, and so the values involve h. The values of the right side are determined by H. Hence a relation between the data must hold. In general there is no solution of the Cauchy problem if the initial surface is characteristic since the data cannot be arbitrarily prescribed.

EXAMPLE 2-15 $\dfrac{\partial^2 u}{\partial x^2} + \dfrac{\partial^2 u}{\partial y^2} - \dfrac{1}{K} \dfrac{\partial u}{\partial t} = 0$

This is the diffusion (or heat) equation of Example 2-4. It is everywhere parabolic. The characteristic equation is

$$Q(\Psi) = \Psi_x^2 + \Psi_y^2 = 0$$

Hence $\Psi = \Psi(t)$, and the characteristic surfaces are planes, $t = $ constant, in xyt space. An important physical case occurs when the initial surface is the plane $t = 0$. The initial conditions are

$$u(x,y,0) = h(x,y) \qquad u_t(x,y,0) = H(x,y)$$

Since the xy plane is a characteristic surface, there is no solution of this Cauchy problem. Observe that the differential equation itself implies the relation

$$K(h_{xx} + h_{yy}) = H$$

must hold between the data.

EXAMPLE 2-16 $\dfrac{\partial^2 u}{\partial x^2} + \dfrac{\partial^2 u}{\partial y^2} + \dfrac{\partial^2 u}{\partial z^2} = 0$

This elliptic equation is Laplace's equation (recall Example 2-5). The characteristic equation is

$$Q(\Psi) = \Psi_x^2 + \Psi_y^2 + \Psi_z^2 = 0$$

Accordingly Laplace's equation has no real characteristic surfaces.

Significance of the Characteristic Surfaces

Assume Eq. (2-65) is hyperbolic or parabolic and the coefficients and G are analytic in $\mathscr{R}$.   Let the data be analytic also.   If the initial surface S_0 is characteristic, then, in general, there is no solution of the Cauchy problem. This is due to the fact that on a characteristic surface the differential equation implies a relationship between the data, and hence these cannot be assigned arbitrarily.   In order to show this, use is made of the concept of a tangential operator.   Let γ be the normal to S_0, and recall that a vector $\xi = (\xi_1, \ldots, \xi_n)$ lies in the tangent plane to S_0 at a point if

$$\xi \cdot \gamma = \xi_1 \gamma_1 + \cdots + \xi_n \gamma_n = 0$$

A linear operator

$$L_\xi = \xi_1 D_{x_1} + \cdots + \xi_n D_{x_n} \qquad D_{x_i} = \frac{\partial}{\partial x_i}$$

is called *tangential* at a point on S_0 if ξ lies in the tangent plane to S_0 at the point.   The important property of tangential operator here is that if u is a twice continuously differentiable function and L_ξ is tangential at a point $\mathbf{x}_0$ on S_0, then the value of $L_\xi u$ at $\mathbf{x}_0$ is determined in terms of the values of u on S_0.   Suppose the equation of S_0 is

$$x_n = \psi(x_1, \ldots, x_{n-1})$$

and suppose

$$u = u[x_1, \ldots, x_{n-1}, \psi(x_1, \ldots, x_{n-1})] = h(x_1, \ldots, x_{n-1}) \qquad \text{on } S_0$$

Then

$$u_{x_i} + u_{x_n}\psi_{x_i} = h_{x_i} \qquad u_{x_i} = h_{x_i} - u_{x_n}\psi_{x_i} \qquad i = 1, \ldots, n-1$$

Since L_ξ is tangential at $\mathbf{x}_0$,

$$\xi_1 \psi_{x_1} + \cdots + \xi_{n-1}\psi_{x_{n-1}} - \xi_n = 0$$

where the partial derivatives ψ_{x_i} are evaluated at $\mathbf{x}_0$.   Thus

$$
\begin{aligned}
L_\xi u &= \xi_1(h_{x_1} - u_{x_n}\psi_{x_1}) + \cdots + \xi_{n-1}(h_{x_{n-1}} - u_{x_n}\psi_{x_{n-1}}) + \xi_n u_{x_n} \\
&= \xi_1 h_{x_1} + \cdots + \xi_{n-1}h_{x_{n-1}} - u_{x_n}(\xi_1 \psi_{x_1} + \cdots + \xi_{n-1}\psi_{x_{n-1}} - \xi_n) \\
&= \xi_1 h_{x_1} + \cdots + \xi_{n-1}h_{x_{n-1}}
\end{aligned}
$$

Each directional derivative can be expressed in terms of a tangential operator and the normal operator D_γ, defined in Eq. (2-64).   In particular the operators D_{x_i} can be so expressed.   For each fixed i, $i = 1, \ldots, n$, let

$$L_{\xi_i} = D_{x_i} - \gamma_i D_\gamma$$

Then L_{ξ_i} is tangential since

$$L_{\xi_i} = D_{x_i} - \gamma_i(\gamma_1 D_{x_1} + \cdots + \gamma_n D_{x_n})$$
$$= -\gamma_i\gamma_1 D_{x_1} - \cdots - \gamma_i\gamma_{i-1} D_{x_{i-1}} + (1 - \gamma_i^2)D_{x_i}$$
$$\qquad\qquad - \gamma_i\gamma_{i+1}D_{x_{i+1}} - \cdots - \gamma_1\gamma_n D_{x_n}$$

and

$$-\gamma_i\gamma_1^2 - \cdots - \gamma_i\gamma_{i+1}^2 + (1 - \gamma_i^2)\gamma_i - \gamma_i\gamma_{i+1}^2 - \cdots - \gamma_i\gamma_n^2 = 0$$

Hence

$$D_{x_i} = L_{\xi_i} + \gamma_i D_\gamma \qquad i = 1, \ldots, n$$

The operator L in Eq. (2-65) can be written in terms of the tangential operators L_{ξ_i} and D_γ.

$$L = \sum_{i=1}^{n}\sum_{j=1}^{n} A_{ij}D_{x_i}D_{x_j} + \sum_{i=1}^{n} B_i D_{x_i} + C$$
$$= \sum_{i=1}^{n}\sum_{j=1}^{n} A_{ij}L_{\xi_i}L_{\xi_j} + 2\left(\sum_{i=1}^{n}\sum_{j=1}^{n} A_{ij}\gamma_i L_{\xi_j}\right)D_\gamma$$
$$+ \left(\sum_{i=1}^{n}\sum_{j=1}^{n} A_{ij}\gamma_i\gamma_j\right)D_\gamma^2 + \sum_{i=1}^{n} B_i L_{\xi_i} + \left(\sum_{i=1}^{n} B_i\gamma_i\right)D_\gamma + C$$

Assume now that S_0 is a characteristic surface and u is a solution of the Cauchy problem with initial conditions

$$u = h(x_1, \ldots, x_{n-1}) \qquad \frac{\partial u}{\partial \gamma} = H(x_1, \ldots, x_{n-1}) \qquad \text{on } S_0$$

Then

$$\sum_{i=1}^{n}\sum_{j=1}^{n} A_{ij}\gamma_i\gamma_j = 0$$

and the differential equation (2-65) becomes

$$\left(\sum_{i=1}^{n}\sum_{j=1}^{n} A_{ij}L_{\xi_i}L_{\xi_j}\right)u + \sum_{i=1}^{n} B_i L_{\xi_i}u + Cu = -\left(2\sum_{i=1}^{n}\sum_{j=1}^{n} A_{ij}\gamma_i L_{\xi_j} + \sum_{i=1}^{n} B_i\gamma_i\right)D_\gamma u$$

on S_0. The left side of the preceding equation involves only h, while the right side involves H. Hence a relation between the data must hold.

2-7 ADJOINT OPERATOR. GREEN'S FORMULA. SELF-ADJOINT DIFFERENTIAL OPERATOR

In this section some useful relations for a second-order linear operator of the form

$$L = \sum_{i=1}^{n}\sum_{j=1}^{n} A_{ij}\frac{\partial^2}{\partial x_i\,\partial x_j} + \sum_{i=1}^{n} B_i\frac{\partial}{\partial x_i} + C \qquad\qquad (2\text{-}72)$$

are derived. Assume that the coefficients in L have continuous second derivatives with respect to the independent variables $x_1, \ldots, x_n$ in some region $\mathscr{R}$ of n-dimensional space. Assume also that $A_{ij} = A_{ji}$, $i, j = 1, \ldots, n$, in $\mathscr{R}$. Then for any pair of functions u, v with continuous second derivatives in $\mathscr{R}$ the following identities hold:

$$vA_{ij}\frac{\partial^2 u}{\partial x_i\, \partial x_j} = \frac{\partial}{\partial x_i}\left(vA_{ij}\frac{\partial u}{\partial x_j}\right) - \frac{\partial u}{\partial x_j}\frac{\partial}{\partial x_i}(A_{ij}v)$$

$$= \frac{\partial}{\partial x_i}\left(A_{ij}v\frac{\partial u}{\partial x_j}\right) - \frac{\partial}{\partial x_j}\left[u\frac{\partial}{\partial x_i}(A_{ij}v)\right] + u\frac{\partial^2}{\partial x_i\, \partial x_j}(A_{ij}v)$$

for each fixed pair of integers i, j in the range $1, \ldots, n$. Also

$$vB_i\frac{\partial u}{\partial x_i} = \frac{\partial}{\partial x_i}(B_i uv) - u\frac{\partial}{\partial x_i}(B_i v)$$

for each fixed integer $i, i = 1, \ldots, n$. Hence, since $A_{ij} = A_{ji}$,

$$vLu = u\left[\sum_{i=1}^{n}\sum_{j=1}^{n}\frac{\partial^2}{\partial x_i\, \partial x_j}(A_{ij}v) - \sum_{i=1}^{n}\frac{\partial}{\partial x_i}(B_i v) + Cv\right]$$

$$+ \sum_{i=1}^{n}\frac{\partial}{\partial x_i}\left[v\sum_{j=1}^{n}A_{ij}\frac{\partial u}{\partial x_j} - u\sum_{j=1}^{n}\frac{\partial}{\partial x_j}(A_{ij}v) + B_i uv\right] \qquad (2\text{-}73)$$

The linear second-order operator L^* defined by

$$L^*v = \sum_{i=1}^{n}\sum_{j=1}^{n}\frac{\partial^2}{\partial x_i\, \partial x_j}(A_{ij}v) - \sum_{j=1}^{n}\frac{\partial}{\partial x_i}(B_i v) + Cv \qquad (2\text{-}74)$$

is called the *adjoint* of L. For any pair of twice continuously differentiable functions u, v on $\mathscr{R}$ the identity

$$vLu - uL^*v = \sum_{i=1}^{n}\frac{\partial}{\partial x_i}\left[\sum_{j=1}^{n}A_{ij}\left(v\frac{\partial u}{\partial x_j} - u\frac{\partial v}{\partial x_j}\right) + uv\left(B_i - \sum_{i=1}^{n}\frac{\partial A_{ij}}{\partial x_j}\right)\right] \qquad (2\text{-}75)$$

holds. Identity (2-75) is called *Lagrange's identity*.

Recall a vector (or point) in real n-dimensional space is an ordered n-tuple $(a_1, \ldots, a_n)$ of real numbers. The symbol $\mathbf{x}$ is used for a variable point $(x_1, \ldots, x_n)$. A *vector field* $\mathbf{V}(\mathbf{x})$ is an ordered n-tuple of real-valued functions $V_1(\mathbf{x}), \ldots, V_n(\mathbf{x})$. The divergence of a vector field $\mathbf{V}(\mathbf{x})$ is the scalar function

$$\nabla \cdot \mathbf{V} = \sum_{i=1}^{n}\frac{\partial V_i}{\partial x_i}$$

Let $\mathscr{R}$ be a bounded region in n-dimensional space and let $b(\mathscr{R})$ denote the boundary of $\mathscr{R}$. Then $\mathscr{R}$ is called a *regular region* if

$$\int_{\mathscr{R}}\nabla \cdot \mathbf{V}\, d\tau = \int_{b(\mathscr{R})}\left(\sum_{i=1}^{n}V_i\gamma_i\right)d\sigma \qquad (2\text{-}76)$$

holds for every vector field $\mathbf{V}$ whose components V_i are continuously differentiable on $\mathscr{R}$ and the boundary $b(\mathscr{R})$. Here $d\tau = dx_1 \cdots dx_n$ is the volume element in $\mathscr{R}$, $d\sigma$ is the surface element on the boundary $b(\mathscr{R})$, and $\boldsymbol{\gamma} = (\gamma_1, \ldots, \gamma_n)$ is the unit exterior normal to $b(\mathscr{R})$. The statement embodied in Eq. (2-76) is called the *divergence theorem*. A sufficient condition for $\mathscr{R}$ to be a regular region is that the boundary $b(\mathscr{R})$ consists of a finite number of contiguous surfaces, each piece representable parametrically as in Eq. (2-57), where each function ψ_i has continuous first derivatives with respect to the parameters $s_1, \ldots, s_{n-1}$. In this event the normal $\boldsymbol{\gamma}$ is a piecewise continuous function on the bounding surface $b(\mathscr{R})$. For example, if $n = 3$, a region is regular whenever the boundary consists of a closed surface on which the normal is a continuous function of position except possibly at a finite number of points or along a finite number of ridges of the surface.

Let $\mathscr{R}$ be a regular region and u, v a pair of functions twice continuously differentiable in $\mathscr{R}$ and also on the boundary $b(\mathscr{R})$. Assume that each coefficient in L has this property as well. Define the vector field $\mathbf{P} = \{P_1(x), \ldots, P_n(x)\}$ by

$$P_i(x) = v \sum_{j=1}^{n} A_{ij} \frac{\partial u}{\partial x_j} - u \sum_{j=1}^{n} A_{ij} \frac{\partial v}{\partial x_j} + uv \left(B_i - \sum_{j=1}^{n} \frac{\partial A_{ij}}{\partial x_j} \right) \qquad i = 1, \ldots, n$$

Integrate Lagrange's identity over $\mathscr{R}$; then

$$\int_{\mathscr{R}} (vLu - uL^*v) \, d\tau = \int_{\mathscr{R}} \nabla \cdot \mathbf{P} \, d\tau$$

Apply the divergence theorem to the right-hand member of this equation. Then the relation

$$\int_{\mathscr{R}} (vLu - uL^*v) \, d\tau = \int_{b(\mathscr{R})} \left(\sum_{i=1}^{n} P_i \gamma_i \right) d\sigma \tag{2-77}$$

results. Equation (2-77) is called *Green's formula* for the operator L.

It is necessary here to discuss briefly the concept of equality for linear second-order operators. Let L be the operator in Eq. (2-72), and let M be the operator

$$M = \sum_{i=1}^{n} \sum_{j=1}^{n} E_{ij} \frac{\partial^2}{\partial x_i \, \partial x_j} + \sum_{i=1}^{n} F_i \frac{\partial}{\partial x_i} + G \tag{2-78}$$

where the coefficients are twice continuously differentiable on $\mathscr{R}$. The domain of the operator L is the class of all functions u such that Lu is a well-defined function on $\mathscr{R}$. This is also the domain of M. Then $M = L$ means

$$Mu = Lu$$

is an identity in $\mathbf{x}$ on $\mathscr{R}$ for every u in the domain. A necessary and sufficient condition that $M = L$ is that

$$E_{ij}(\mathbf{x}) = A_{ij}(\mathbf{x}) \qquad F_i(\mathbf{x}) = B_i(\mathbf{x}) \qquad G(\mathbf{x}) = C(\mathbf{x}) \qquad \begin{array}{l} \text{on } \mathscr{R} \\ i, j = 1, \ldots, n \end{array}$$

that is, the coefficients are identical. Clearly the foregoing is sufficient to ensure that $M = L$. Conversely, suppose $M = L$. Then $Mu = Lu$ holds for every polynomial in the variables $x_1, \ldots, x_n$. Choose $u = 1$. Then $G(\mathbf{x}) = C(\mathbf{x})$. Choose $u = x_1$. Then $G = C$ and $Mu = Lu$ implies $F_1(\mathbf{x}) = B_1(\mathbf{x})$. In this manner the identity of the coefficients in M and L can be established.

The operator L in Eq. (2-72) is called *self-adjoint* if $L^* = L$. It is clear from the foregoing that a necessary and sufficient condition for L to be self-adjoint is that the equations

$$\sum_{j=1}^{n} \frac{\partial A_{ij}}{\partial x_j} = B_i \qquad i = 1, \ldots, n \tag{2-79}$$

hold on $\mathscr{R}$. Accordingly every self-adjoint linear second-order operator has the form

$$L = \sum_{i=1}^{n} \sum_{j=1}^{n} \frac{\partial}{\partial x_i} \left(A_{ij} \frac{\partial}{\partial x_j} \right) + C \tag{2-80}$$

If L is self-adjoint, then Green's formula states that for each pair of functions u, v with the continuity properties assumed above,

$$\int_{\mathscr{R}} (vLu - uLv) \, d\tau = \int_{b(\mathscr{R})} \left(\sum_{i=1}^{n} P_i \gamma_i \right) d\sigma \tag{2-81}$$

where

$$P_i(x) = \sum_{j=1}^{n} A_{ij} \left(v \frac{\partial u}{\partial x_j} - u \frac{\partial v}{\partial x_j} \right) \qquad i = 1, \ldots, n \tag{2-82}$$

In the case where the coefficients in the operator L are constants, it follows from Eq. (2-79) that L is self-adjoint if, and only if, $B_i = 0$, $i = 1, \ldots, n$. Some classical examples of self-adjoint operators when $n = 3$ appear in the equations of Examples 2-14 to 2-16.

PROBLEMS

Sec. 2-2

1 (a) Show that the general solution of the wave equation

$$\frac{\partial^2 u}{\partial t^2} = c^2 \frac{\partial^2 u}{\partial x^2}$$

is

$$u = f(x - ct) + g(x + ct)$$

(b) Find the general solution of the inhomogeneous wave equation $u_{tt} - c^2 u_{xx} = x^2 + xt - \sin \omega t$, $\omega > 0$ a constant.

2 Find the general solution of the equation of spherical waves

$$\frac{1}{c^2}\frac{\partial^2 u}{\partial t^2} = \frac{1}{r^2}\frac{\partial}{\partial r}\left(r^2 \frac{\partial u}{\partial r}\right)$$

Hint: Make the change of dependent variable $v = ru$.

3 Symbolic methods of obtaining particular solutions of Eq. (2-1) exist for the constant-coefficient case.   Let L be the linear operator in Eq. (2-3), where the coefficients are real constants.   Then $L = P(D_x, D_y)$ is a polynomial in the operators D_x, D_y.   If $f(x,y)$ is a function, it is understood in the following that the symbolic equation

$$\frac{f(x,y)}{P(D_x,D_y)} = \varphi(x,y)$$

means φ is a function such that

$$P(D_x,D_y)[\varphi(x,y)] = f(x,y)$$

Derive each of the following symbolic equations.

(a) $e^{ax+by}/P(D_x,D_y) = e^{ax+by}/P(a,b)$ provided the constants a, b are such that $P(a,b) \neq 0$.

(b) $\sin (ax + by)/P(D_x^2, D_y^2) = \sin (ax + by)/P(-a^2, -b^2)$ provided $P(-a^2, -b^2) \neq 0$.   A similar equation holds for $\cos (ax + by)$.

(c) $\cos (ax + by)/P(D_x,D_y) = \mathrm{Re}\,[e^{i(ax+by)}/P(ia,ib)]$ where $i = \sqrt{-1}$ and $\mathrm{Re}\,[\ ]$ means "the real part of [ ]," provided a, b are real constants such that $P(ia,ib) \neq 0$.

(d) $\sin (ax + by)/P(D_x,D_y) = \mathrm{Im}\,[e^{i(ax+by)}/P(ia,ib)]$ where $i = \sqrt{-1}$ and $\mathrm{Im}\,[\ ]$ means "the imaginary part of [ ]," provided a, b are as stated in **c**.

(e) $e^{mx}\cos (ax + by)/P(D_x,D_y) = \mathrm{Re}\,[e^{mx+i(ax+by)}/P(m + ia, ib)]$ provided m is a real constant and a, b are as stated in **c**.

(f) $x^n/(D_x + a)^m = (a^{-m} - ma^{-(m+1)}D_x + [m(m + 1)/2!]a^{-(m+2)}D_x^2 - \cdots)x^n$ provided $a \neq 0$ and n, m are positive integers.   A similar result holds for $y^n/(D_y + a)^m$.

4 Obtain the general solution.

(a) $r - 10s + 9t = 0$

(b) $4z_{xy} + z_{yy} = \cos y + 1$

(c) $z_{xx} + z_x + x + y + 1 = 1$

(d) $r = xy$

(e) $z_{yy} + z = e^{x+y}$

(f) $r - t + p + q + x + y + 1 = 0$

(g) $2s + 3t - q = 6\cos (2x - 3y) - 30\sin (2x - 3y)$

(h) $s + ap + bq + abz = e^{mx+ny}$ a, b, m, n constants

(i) $r - 4t = 12x^2 + \cos y + 4$

(j) $r - t - 3p + 3q = xy + e^{x+2y}$

(k) $r - 2s + t = 4e^{3y} + \cos x$

5 If the operator L with constant coefficients is factorable as in Eq. (2-6), a particular solution of Eq. (2-1) may sometimes be obtained by the following procedure.   With reference to Eq. (2-6) let $L_2 z = v$.   Now obtain a particular solution v_p of the first-order linear equation $L_1 v = G$.   In turn derive a particular solution z_p of the first-order equation $L_2 z = v_p$.   Then $L z_p = L_1(L_2 z_p) = L_1 v_p = G$.   Use this method to obtain a particular solution of each of the following equations.   Also find the general solution.

(a) $r + 5s + 6t = \log(y - 2x)$ (b) $r - s - 2t = (y - 1)e^x$

(c) $r - 4t = \dfrac{4x}{y^2} - \dfrac{y}{x^2}$

6 Use simple integrations, reduction to an ordinary differential equation, etc., to obtain a solution involving two arbitrary functions.

(a) $t = x^3 + y^3$ (b) $z_{xy} = \dfrac{x}{y} + a$ $a = \text{const}$

(c) $t - xq = x^2$ (d) $xr = c^2p + x^2y^2$ $c = \text{const}$

(e) $yz_{xy} + z_x = \cos(x + y) - y\sin(x + y)$

(f) $y^2t + 2yq = 1$

(g) $(x - y)z_{xy} - z_x + z_y = 0$ Euler-Poisson-Darboux equation

Hint: Let $u = (x - y)z$.

7 Show that the equation

$$ax^2z_{xx} + 2bxyz_{xy} + cy^2z_{yy} + dxz_x + eyz_y + fz = G(x,y)$$

where a, b, c, d, e, f are constants, is transformed into an equation with constant coefficients under the transformation of independent variables $\xi = \log x$, $\eta = \log y$.

8 Use the method of Prob. 7 to obtain the general solution of each of the following equations.

(a) $x^2r - y^2t - 2xp + 2yq = 0$ (b) $x^2r - xys - xp = 1$

(c) $xyz_{xy} - y^2z_{yy} - 2xz_x + 2yz_y - 2z = 0$

(d) $x^2r - 2xys - 3y^2t + xp - 3yq = 2\log xy + 4x$

(e) $x^2r - y^2t = xy$

(f) $x^2z_{xx} + 2xyz_{xy} + y^2z_{yy} - nxz_x - nyz_y + nz = x + y$ $n = \text{const} \neq 0$

(g) $xp + yq - xys = z$

(h) $(axD_x + byD_y)^2z + \lambda^2z = 0$ a, b, λ constants

9 (a) Let L be the linear operator in Eq. (2-3).   Define a functionally invariant pair of Eq. (2-5) as done in the text for Eq. (2-12).   Refer to the method in the text and show that ψ, φ is a functionally invariant pair of Eq. (2-5) if, and only if, ψ, φ satisfy the system of partial differential equations

$$A\varphi_x^2 + 2B\varphi_x\varphi_y + C\varphi_y^2 = 0 \tag{1}$$

$$2[A\psi_x\varphi_x + B(\psi_x\varphi_y + \psi_y\varphi_x) + C\psi_y\varphi_y] + \psi(A\varphi_{xx} + 2B\varphi_{xy} + C\varphi_{yy} + D\varphi_x + E\varphi_y) = 0 \tag{2}$$

$$L\psi = 0 \tag{3}$$

(b) Let the coefficients in L be real constants.   Then the characteristic equation (1) has real-valued solutions if, and only if, $\Delta = B^2 - AC \geq 0$.   Assume $\Delta > 0$.   If $A \neq 0$, let r_1, r_2 be the distinct roots of $Ar^2 + 2Br + C$, and let $\varphi_1 = r_1x + y$, $\varphi_2 = r_2x + y$.   If $A = 0$, let $\varphi_1 = x$, $\varphi_2 = Cx - 2By$.   Then φ_1, φ_2 are functionally independent solutions of (1).   If $\varphi = r_1x + y$, show that (2) becomes

$$(Ar_1 + B)\psi_x + (Br_1 + C)\psi_y + \dfrac{Dr_1 + E}{2}\psi = 0 \tag{4}$$

Let $\alpha = Ar_1 + B$, $\beta = Br_1 + C$, $\gamma = (Dr + E)/2$. Then the general solution of (4) is $\psi = e^{-\gamma x/\alpha}h(\beta x - \alpha y)$. Let $t = \beta x - \alpha y$, substitute this form of ψ into (3), and show that $h(t)$ must satisfy a first-order equation

$$ah'(t) + bh(t) = 0 \tag{5}$$

and so

$$h(t) = e^{-bt/a}$$

(c) Let the coefficients in L be real constants, $\Delta > 0$,

$$AE^2 - 2BDE + CD^2 + 4\Delta F = 0 \tag{6}$$

Show that Eq. (2-5) has the general solution

$$z - e^{(cx+dy)/2\Delta}[f(r_1x \mid y) \mid g(r_2x + y)]$$

$c = CD - BE$, $d = AE - BD$, r_1, r_2 as described in **b** above.

10 With reference to Prob. 9a, a functionally independent pair of Eq. (2-5) when the coefficients are not constants can often be constructed as follows. Factor (1) as

$$A(\varphi_x - m_1\varphi_y)(\varphi_x - m_2\varphi_y) = 0$$

and obtain particular solutions of the linear first-order equations $\varphi_x - m_1\varphi_y = 0$, $i = 1, 2$. Now choose a simple but nontrivial solution of $L\psi = 0$ such that Eq. (2) in Prob. 9a holds. Obtain the general solution of each of the following equations.

(a) $e^{2y}(r - p) = e^{2x}(t - q)$ (b) $(y^2 - x^2)(r - t) + 4(xp + yq - z) = 0$

(c) $xr + (y - x)s - yt = q - p$

Sec. 2-3

11 Obtain the general solution.

(a) $u_{xx} + 2u_{xy} + u_{yy} - 2u_{yz} - 2u_{xz} + u_{zz} = x\cos x + y\cos y + z\cos z$

(b) $2u_{xx} + u_{xy} + 2u_{xz} - u_{yy} - u_{yz} = x^2 + y^2 + z^2$

(c) $u_{xx} + 2u_{xy} + u_{yy} - u_{zz} - 2u_z - u = e^{2x} + 2\cos y + z + 2$

(d) $x^2u_{xx} + 2xyu_{xy} + y^2u_{yy} + 2yzu_{yz} + 2xzu_{xz} + z^2u_{zz} = \ln xyz$

(e) $u_{tt} = Au_{xx} + 2Bu_{xy} + Cu_{yy}$, A, B, C positive constants such that $B^2 - AC = 0$

12 Construct exponential-type solutions. Also find a particular solution if the equation is inhomogeneous.

(a) $z_{xx} - 2z_{xy} + z_y - z = 0$ (b) $z_{xx} + 4z_{xy} + z_{yy} + z_x + z_y = 3e^{2x} + xy$

(c) $z_{xx} - 2z_x - z_y = x + y$ (d) $z_{xx} + z_{yy} = x + ye^y$

(e) $u_{xx} + u_{yy} = 0$ (f) $u_{xx} + u_{yy} + u_{zz} = 0$

(g) $u_{xx} + u_{yy} + k^2u = 0$ $k = $ const   (h) $u_{xx} + u_{yy} = ku_t$ $k = $ const

(i) $c^2(u_{xx} + u_{yy}) = u_{tt}$ $c = $ const

13 Find the equation which the constants l, m, n must satisfy if the function u defined by

$$u(x,y,t) = \int_{\alpha}^{\beta} f(lx + my + bt, \xi)\, d\xi$$

satisfies the two-dimensional wave equation in Example 2-3.   Assume that it is legitimate to differentiate within the integral sign.

14   Assume that it is legitimate to differentiate within the integral sign.   Verify that the function u defined by

$$u(x,y,z,t) = \int_{-\pi}^{\pi} \int_{-\pi}^{\pi} f(x \sin \xi \cos \eta + y \sin \xi \sin \eta + z \cos \xi + ct, \xi, \eta) \, d\xi \, d\eta$$

(Whittaker's solution) satisfies the three-dimensional wave equation

$$u_{tt} = c^2(u_{xx} + u_{yy} + u_{zz})$$

15   Assume that it is legitimate to differentiate within the integral sign.   Verify that the function u defined by

$$u(x,y,z) = \int_{0}^{2\pi} f(x \cos \xi + y \sin \xi + iz, \xi) \, d\xi \qquad i = \sqrt{-1}$$

satisfies Laplace's equation in Example 2-5.

Sec. 2-4

16   Show that Eq. (2-30) holds.

17   Let L be the linear operator (2-3) in which the coefficients may be variable.   Let Eq. (2-27) define a transformation of the independent variables where the functions have continuous second derivatives.   Carry out the calculation of the derivatives of z, substitute into Eq. (2-1), and show that the transformed equation is

$$Q(\xi)z_{\xi\xi} + 2Q(\xi,\eta)z_{\xi\eta} + Q(\eta)z_{\eta\eta} + (L_1\xi + D\xi_x + E\xi_y)z_\xi$$
$$+ (L_1\eta + D\eta_x + E\eta_y)z_\eta + Fz = G$$

where $Q(\xi)$, $Q(\xi,\eta)$, and $Q(\eta)$ are given in the text [following Eq. (2-28)] and

$$L_1\varphi = A\varphi_{xx} + 2B\varphi_{xy} + C\varphi_{yy}$$

Hence show that the normal forms of Eq. (2-1) in the three types are

$$z_{\xi\eta} + \alpha z_\xi + \beta z_\eta + \gamma z = G'(\xi,\eta) \qquad \text{hyperbolic}$$
$$z_{\eta\eta} + \alpha z_\xi + \beta z_\eta + \gamma z = G'(\xi,\eta) \qquad \text{parabolic}$$
$$z_{\xi\xi} + z_{\eta\eta} + \alpha z_\xi + \beta z_\eta + \gamma z = G'(\xi,\eta) \qquad \text{elliptic}$$

where α, β, γ are (in general) functions of ξ, η.

18   **(a)** In the hyperbolic equation

$$z_{xy} + Dz_x + Ez_y + Fz = G$$

suppose that $\partial D/\partial x = \partial E/\partial y$ holds.   Then $E \, dx + D \, dy$ is an exact differential.   Choose a function $a(x,y)$ such that $da = E \, dx + D \, dy$.   Then $E = \partial a/\partial x$, $D = \partial a/\partial y$.   The equation now takes the form

$$z_{xy} + a_y z_x + a_x z_y + Fz = G$$

Make the change of dependent variable $z = ue^{-a(x,y)}$ and show that the equation is transformed into

$$u_{xy} + bu = Ge^a$$

$$b(x,y) = F - a_x a_y - a_{xy} = F - DE - \frac{\partial D}{\partial x}$$

In particular the transformation is possible whenever $D = D(y)$, $E = E(x)$.

(b) Consider the parabolic equation

$$z_{xx} + Dz_x + Ez_y + Fz = G$$

Define the function a by

$$a(x,y) = \int^x D(x,y)\, dx$$

y held fast.   Make the change of dependent variable $z = ue^{-a/2}$ and show that the equation is transformed into

$$u_{xx} + Eu_y + bu = Ge^{a/2}$$

$$b(x,y) = F - \tfrac{1}{4}a_x^2 - \tfrac{1}{2}a_{xx} - \tfrac{1}{2}Ea_y$$

(c) In the elliptic equation

$$z_{xx} + z_{yy} + Dz_x + Ez_y + Fz = G$$

suppose that $\partial D/\partial y = \partial E/\partial x$ holds.   Then $D\,dx + E\,dy$ is an exact differential.   Choose a function $a(x,y)$ such that $da = D\,dx + E\,dy$.   Then $D = \partial a/\partial x$, $E = \partial a/\partial y$.   The equation now takes the form

$$z_{xx} + z_{yy} + a_x z_x + a_y z_y + Fz = G$$

Make the change of dependent variable $z = ue^{-a/2}$ and show that the equation is transformed into

$$u_{xx} + u_{yy} + bu = Ge^{a/2}$$

$$b(x,y) = F - \tfrac{1}{4}a_x^2 - \tfrac{1}{4}a_y^2 - \tfrac{1}{2}a_{xx} - \tfrac{1}{2}a_{yy}$$

In particular the transformation is possible whenever

$$D = D(x) \qquad E = E(y)$$

19   In the hyperbolic equation

$$z_{xy} + Dz_x + Ez_y + Fz = 0$$

the functions $h = \partial D/\partial x + DE - F$, $k = \partial E/\partial y + DE - F$ are called *invariants* of the differential equation.   For transformations of the independent variables of the type $\xi = \xi(x)$, $\eta = \eta(y)$, or a transformation of the dependent variable $z = \psi(x,y)u$, the differential equation is transformed into one of the same form with invariants

$$h_1 = \mu h \qquad k_1 = \mu k$$

Show that if either $h = 0$ or $k = 0$, the general solution can be obtained by quadratures.

Hint: Suppose $h = 0$. Then

$$\frac{\partial^2 z}{\partial x \, \partial y} + D \frac{\partial z}{\partial x} + E \frac{\partial z}{\partial y} + Fz = \frac{\partial}{\partial x}\left(\frac{\partial z}{\partial y} + Dz\right) + E\left(\frac{\partial z}{\partial y} + Dz\right)$$

Let $u = \partial z/\partial y + Dz$ and solve the first-order linear equation $u_x + Eu = G$ to obtain an intermediate integral.

20 Obtain the general solution:

(a) $z_{xy} + z_x + z_y + z = 0$

(b) $z_{xy} + xyz_x + yz = 0$

(c) $z_{xy} - \dfrac{z_x}{x-y} + \dfrac{z_y}{x-y} = \dfrac{1}{x-y}$

(d) $z_{xy} - \dfrac{2z_x}{x-y} + \dfrac{3z_y}{x-y} - \dfrac{3z}{(x-y)^2} = 0$

21 Let $L = AD_x^2 + 2BD_xD_y + CD_y^2 + DD_x + ED_y + F$ be a second-order linear operator with real constant coefficients.

(a) Assume L is hyperbolic, $A \neq 0$. Show that the characteristics are the two families of straight lines

$$\xi = y + m_1 x = c_1 \qquad \eta = y + m_2 x = c_2$$

where $m_{1,2}$ are defined by Eq. (2-33) and Δ is defined by Eq. (2-26). Use the equations of the characteristics to obtain a transformation of independent variables such that the equation $Lz = G$ is transformed into the normal form

$$z_{\xi\eta} + dz_\xi + pz_\eta + fz = H(\xi,\eta)$$

where d, p, f are constants. Thus in the special case where $D = E = F = 0$ show that the general solution of the homogeneous equation $Lz = 0$ is $z = f(y + m_1 x) + g(y + m_2 x)$. If one of the coefficients D, E, F is different from zero, show that the change of dependent variable $z = e^{-(d\eta + p\xi)}u$ further transforms the equation into

$$u_{\xi\eta} + \beta u = H(\xi,\eta)e^{(d\eta + p\xi)}$$

where $\beta = f - pd$. Discuss fully the case when $A = 0$.

(b) Assume that L is parabolic, $A \neq 0$. Show that the characteristic curves are the straight lines $\xi = Bx - Ay = C_1$. Choose as a second independent function $\eta = x$ and show directly that the equation $Lz = G$ is transformed into the normal form

$$z_{\eta\eta} + dz_\xi + pz_\eta + fz = H(\xi,\eta)$$

where d, p, f are constants. Thus in the special case where $D = E = F = 0$ show that the general solution of the homogeneous equation $Lz = 0$ is $z = xf(Bx - Ay) + g(Bx - Ay)$. If one of the coefficients D, E, F is different from zero, show that the change of dependent variable $z = ue^{-p\eta/2}$ further transforms the equation into $u_{\eta\eta} + du_\xi + \beta u = H(\xi,\eta)e^{p\eta/2}$, where $\beta = f - \frac{1}{4}p^2$. Discuss fully the case when $A = 0$.

(c) Assume L is elliptic, and show that this implies $A \neq 0$. Show that the characteristic differential equations are

$$y' = -m_1 \qquad y' = -\overline{m}_1$$

where $m_1 = (-B + i\sqrt{-\Delta})/A$, $i = \sqrt{-1}$, and $\overline{m}_1$ denotes the complex conjugate of m_1. The general solutions of these are $\varphi(x,y) = y + m_1 x = c_1$ and $\psi(x,y) = y + \overline{m}_1 x = c_2$ [so that $\psi(x,y) = \overline{\varphi(x,y)}$]. Write $\varphi = \xi + i\eta$ and determine the expressions for $\xi(x,y)$, $\eta(x,y)$ (these are real-valued functions). Under the one-to-one transformation $\xi = \xi(x,y)$,

$\eta = \eta(x,y)$ show directly that the equation $Lz = G$ is transformed into the normal form

$$z_{\xi\xi} + z_{\eta\eta} + dz_\xi + pz_\eta + fz = H(\xi,\eta)$$

Thus, in the special case where $D = E = F = 0$ show that the general solution of $Lz = 0$ is $z = f(y + m_1x) + g(y + \bar{m}_1x)$. If one of the coefficients D, E, F is different from zero, show that the change of dependent variable $z = ue^{-(d\xi+p\eta)/2}$ further transforms the equation into

$$u_{\xi\xi} + u_{\eta\eta} + \beta u = H(\xi,\eta)e^{(d\xi+p\eta)/2}$$

where $\beta = f - \frac14 d^2 - \frac14 p^2$.

22 Classify each equation. Reduce to normal form and obtain the general solution.

(a) $4z_{aa} - 8z_{xy} + 4z_{yy} = 1$ (b) $4z_{xx} - 4z_{xy} + 5z_{yy} = 0$

(c) $r - 2s + t + a(p - q) + cz = (x + 2y)^2$ a, c nonzero constants

(d) $r - t + p + q + x + y + 1 = 0$ (e) $x^2r + 2xys + y^2t = 4x^2$

(f) $xr - (x + y)s + yt = \dfrac{x + y}{x - y}(p - q)$

(g) $x^2r - y^2t = xy$ (h) $\dfrac{\partial}{\partial x}\left(x^2\dfrac{\partial z}{\partial x}\right) = x^2\dfrac{\partial^2 z}{\partial y^2}$

(i) $r - (2\sin x)s - (\cos^2 x)t - (\cos x)q = 0$

(j) $(\sec^4 y)r - t + (2\tan y)q = 0$

(k) $xy(t - r) + (x^2 - y^2)s = py - qx - 2(x^2 - y^2)$

(l) $y^2z_{xx} - 2yz_{xy} + z_{yy} = z_x + 6y$

(m) $\dfrac{\partial}{\partial x}\left[\left(1 - \dfrac{x}{h}\right)^2 z_x\right] = \dfrac{1}{a^2}\left(1 - \dfrac{x}{h}\right)^2\dfrac{\partial^2 z}{\partial t^2}$ a, h real positive constants

23 Classify each of the following equations.

(a) $c^2(u_{xx} + u_{yy} + u_{zz}) = u_{tt}$ $c > 0$ a real constant

(b) $\dfrac{\partial u}{\partial t} = k\left(\dfrac{\partial^2 u}{\partial x^2} + \dfrac{\partial^2 u}{\partial y^2}\right)$ $k > 0$ a real constant

(c) $u_{xx} + u_{yy} + u_{zz} = 0$

(d) $u_{xx} + u_{yy} - u_{tt} = 0$

(e) $u_{xx} + u_{yy} + u_{zz} + \lambda u = 0$ λ a constant

(f) $u_{xx} + u_{yy} - u_{zz} - u_{tt} = 0$

Sec. 2-5

24 In each of the following Cauchy problems determine whether (i) there exists a unique solution, (ii) there exist infinitely many distinct solutions, (iii) no solution exists. If i, find the solution.

(a) $u_{xx} - u_{tt} = 0$; $u(x,0) = b$, b a constant, $u_t(x,0) = \sin x$

(b) $u_{xx} - u_{tt} = 0$; $u(x,0) = \sin x$, $u_t(x,0) = b$, b a constant

(c) $u_{xx} - u_{tt} = 0$; $u = 0$, $u_x = 1$ on C, C is the line $t = x$

(d) $u_{xx} - u_{tt} = 0$; $u = 0$, $u_x = x$ on C, C is the line $t = x$

(e) $u_{xx} - 10u_{xt} + 9u_{tt} = 0$; $u(x,0) = x^2$, $u_t(x,0) = 1$

(f) $z_{xy} = 0$; $z = \cos x$, $z_x = 1$ on C, C is the line $y = x$

(g) $z_{xy} = 0$; $z(x,0) = 1$, $z_y(x,0) = x^2$

(h) $z_{xy} = xy$; $z = \cos x$, $z_y = x^3$ on C, C is the line $y = x$

(i) $z_{xy} + z_x = x$; $z = x^2$, $z_x = 0$ on C, C is the line $y = x$

(j) $z_{xy} = 0$; $z(x,0) = x$, $z_y(x,0) = 1$

(k) $z_{xy} - 4z_{yy} = 16y + 4e^{-y}$; $z(x,0) = -1$, $z_y(x,0) = 4x + 1$

(l) $z_{xx} - 2z_{xy} + z_{yy} = 4e^{3v}$; $z(x,0) = x^2$, $z_y(x,0) = \frac{1}{3}$

(m) $z_{xx} = x^2 + y^2$; $z(0,y) = 1$, $z_x(0,y) = 0$

25 Solve each of the following Cauchy problems.

(a) $x^2 u_{xx} - 2xy u_{xy} - 3y^2 u_{yy} = 0$; $u(x,1) = x$, $u_y(x,1) = 1$

(b) $y u_{xx} + (x + y)u_{xy} + x u_{yy} = xy$; $u(0,y) = y$, $u_x(0,y) = 1$, $y > 0$

(c) $2u_{xx} - 2y u_{yy} - u_y = 0$; $u(x,1) = \dfrac{x^2}{2}$, $u_y(x,1) = 2$

(d) $u_{xx} + (2 \cos x)u_{xy} - (\sin^2 x)u_{yy} - (\sin x)u_y = 0$; $u = x^2$, $u_y = 1$, on C, C is the curve $y = \sin x$

(e) $x^2 u_{xx} - 2x u_{xy} + u_{yy} + u_y = 0$; $u(1,y) = y^2$, $u_x(1, y) = e^y$

Sec. 2-7

26 Let

$$L = A \frac{\partial^2}{\partial x^2} + 2B \frac{\partial^2}{\partial x\, \partial y} + C \frac{\partial^2}{\partial y^2} + D \frac{\partial}{\partial x} + E \frac{\partial}{\partial y} + F$$

where the coefficients are functions with continuous second derivatives in a region $\mathscr{R}$ of the xy plane.

(a) Write out in full the expression for the adjoint L^* of L.

(b) Show from Eq. (2-75) that for any pair of functions u, v with continuous second derivatives in $\mathscr{R}$ the identity

$$vLu - uL^*v = \frac{\partial Q}{\partial x} - \frac{\partial P}{\partial y}$$

holds, where

$$Q = A\left(v \frac{\partial u}{\partial x} - u \frac{\partial v}{\partial x}\right) + B\left(v \frac{\partial u}{\partial y} - u \frac{\partial v}{\partial y}\right) + \left(D - \frac{\partial A}{\partial x} - \frac{\partial B}{\partial y}\right)uv$$

$$P = B\left(u \frac{\partial v}{\partial x} - v \frac{\partial u}{\partial x}\right) + C\left(u \frac{\partial v}{\partial y} - v \frac{\partial u}{\partial y}\right) + \left(\frac{\partial B}{\partial x} + \frac{\partial C}{\partial y} - E\right)uv$$

(c) Let Γ be a simple closed curve lying in $\mathscr{R}$, $\mathscr{R}_1$ the region enclosed by Γ. Use Green's formula (2-77) in the case $n = 2$ and show that

$$\int_{\mathscr{R}_1} (vLu - uL^*v)\, dx\, dy = \int_{\Gamma} P\, dx + Q\, dy$$

the line integral being taken in the positive sense around Γ.

(d) Show directly from the expression for the adjoint of L derived in **a** that the adjoint of the adjoint of L is L: $L^{**} = L$.

(e) Show that L is self-adjoint if, and only if, the equations

$$\frac{\partial A}{\partial x} + \frac{\partial B}{\partial y} = D \qquad \frac{\partial B}{\partial x} + \frac{\partial C}{\partial y} = E$$

hold in $\mathcal{R}$.   Thus every second-order self-adjoint operator, in the case $n = 2$, has the form

$$Lu = \frac{\partial}{\partial x}\left(A\frac{\partial u}{\partial x} + B\frac{\partial u}{\partial y} \right) + \frac{\partial}{\partial y}\left(B\frac{\partial u}{\partial x} + C\frac{\partial u}{\partial y} \right) + Fu$$

(f) A special case is the laplacian operator

$$L = \frac{\partial^2}{\partial x^2} + \frac{\partial^2}{\partial y^2}$$

Use Green's formula derived in **c** to obtain the relation

$$\int_{\mathcal{R}_1} (v\,\nabla^2 u - u\,\nabla^2 v)\,dx\,dy = \int_{\Gamma} \left(v\frac{\partial u}{\partial n} - u\frac{\partial v}{\partial n} \right) ds$$

where $\partial u/\partial n$ denotes the directional derivative of u in the direction of the exterior **normal** **n** on Γ.   The line integral is taken in the positive sense around Γ.   This relation is **called** the symmetric form of Green's theorem in the plane.

CHAPTER 3

ELLIPTIC DIFFERENTIAL EQUATIONS

3-1 INTRODUCTION

A problem in a class of boundary-value problems of interest in the applications is described as follows. Let $\mathscr{R}$ be a bounded region with boundary S, and let $\bar{\mathscr{R}} = \mathscr{R} + S$, the union of $\mathscr{R}$ with its boundary S, that is, the *closure* of $\mathscr{R}$. Let L be a linear second-order self-adjoint partial differential operator which is elliptic on $\bar{\mathscr{R}}$. A solution of the differential equation

$$Lu = f \quad \text{in } \mathscr{R} \tag{3-1}$$

is desired such that u is continuous on $\bar{\mathscr{R}}$ and

$$u = g \quad \text{on } S \tag{3-2}$$

Here f is a given continuous function on $\bar{\mathscr{R}}$, and g is a given continuous function on the boundary S. This problem is called the *Dirichlet problem* for the region $\mathscr{R}$. Equation (3-2) is termed the *Dirichlet boundary condition*. A problem of a somewhat different type is to determine a solution of Eq. (3-1) which has continuous first derivatives on $\bar{\mathscr{R}}$ and satisfies

$$\frac{\partial u}{\partial n} = g \quad \text{on } S \tag{3-3}$$

where $\partial u/\partial n$ denotes the derivative in the direction of the exterior normal on S. This problem is called the *Neumann problem*, and Eq. (3-3) is termed the *Neumann boundary condition*. A boundary condition of the form

$$a\frac{\partial u}{\partial n} + bu = g \quad \text{on } S \tag{3-4}$$

is a *mixed boundary condition*. It is assumed that the given functions a, b, and g are continuous on S, and a and b do not vanish simultaneously. The problem of finding a solution of Eq. (3-1) such that the solution has continuous first derivatives on $\bar{\mathscr{R}}$ and satisfies Eq. (3-4) on S is called the *mixed* (or *Robin's*) *problem*. Often the terms first, second, and third boundary-value problem are applied to the Dirichlet, Neumann, and mixed boundary-value problem, respectively. In some problems combinations of the above boundary conditions occur. For example, a Dirichlet condition may be prescribed over part of S and a Neumann condition over the remainder.

If the region $\mathscr{R}$ in which a solution of the differential equation (3-1) is sought is the unbounded region exterior to a closed surface (or the region in the plane exterior to a closed curve if the problem is two-dimensional), the problem is called an *exterior problem*. In this case an additional condition is often imposed, termed a *condition at infinity*, which governs the behavior of the solution at large distances from the origin.

Each of the boundary-value problems described above is a *linear boundary-value* problem; i.e., the differential equation and the boundary conditions are linear relations in the unknown function. Several consequences of the linearity should be noted. If u_1, u_2 are solutions of the Dirichlet problems

$$Lu_1 = f_1 \quad \text{in } \mathscr{R} \qquad u_1 = g_1 \quad \text{on } S$$
$$Lu_2 = f_2 \quad \text{in } \mathscr{R} \qquad u_2 = g_2 \quad \text{on } S$$

and c_1, c_2 are constants, then the function $u = c_1 u_1 + c_2 u_2$ is a solution of the Dirichlet problem

$$Lu = c_1 f_1 + c_2 f_2 \quad \text{in } \mathscr{R} \qquad u = c_1 g_1 + c_2 g_2 \quad \text{on } S$$

This important property is called the *superposition principle*. Consider a Dirichlet problem in which it is desired to solve Eq. (3-1) subject to the condition (3-2). If the function f is not identically zero, it may be more convenient to transform the problem into one involving the homogeneous differential equation $Lu = 0$. Choose a simple function v which is continuous on $\mathscr{R}$ and such that $Lv = f$ in $\mathscr{R}$. Let g_1 denote the function v restricted to points on the boundary S. If w satisfies $Lw = 0$ in $\mathscr{R}$ and $w = g - g_1$ on S, the function $u = v + w$ is a solution of the original problem. In another instance suppose the boundary S consists of two parts S_1 and S_2. Let the problem be

$$Lu = 0 \quad \text{in } \mathscr{R} \qquad u = g_1 \quad \text{on } S_1 \qquad u = g_2 \quad \text{on } S_2$$

If u_1, u_2 are solutions of $Lu = 0$ in $\mathscr{R}$, and if $u_1 = g_1$ on S_1, $u_1 = 0$ on S_2, $u_2 = 0$ on S_1, $u_2 = g_2$ on S_2, then $u = u_1 + u_2$ satisfies the original problem. Analogous procedures can be used for Neumann and mixed boundary-value problems. If inhomogeneous boundary conditions are inconvenient, the problem can be reduced to one involving homogeneous boundary conditions and an inhomogeneous differential equation. For example, in the mixed problem a solution of Eq. (3-1) which satisfies Eq. (3-4) on S is desired. If g is not identically zero, choose a simple function v with continuous second derivatives such that v satisfies Eq. (3-4) on S. Let $f_1 = Lv$. If w satisfies $Lw = f - f_1$ in $\mathscr{R}$ and $\partial w/\partial n + bw = 0$ on S, then $u = v + w$ is a solution of the original problem.

The first several sections of this chapter are concerned with properties of a very important elliptic equation in mathematical physics, namely, Laplace's equation. The method of separation of variables is applied to boundary-value problems involving Laplace's equation in several coordinate systems. The concluding section contains an outline of some general properties of linear second-order elliptic type operators.

3-2   LAPLACE'S EQUATION AND POISSON'S EQUATION. PROPERTIES OF HARMONIC FUNCTIONS

The prototype of the class of linear second-order homogeneous partial differential equations of elliptic type in two independent variables with analytic coefficients is

$$\frac{\partial^2 u}{\partial x^2} + \frac{\partial^2 u}{\partial y^2} = 0 \tag{3-5}$$

called *Laplace's equation in the plane* (or in two dimensions). The corresponding equation in three independent variables

$$\frac{\partial^2 u}{\partial x^2} + \frac{\partial^2 u}{\partial y^2} + \frac{\partial^2 u}{\partial z^2} = 0 \tag{3-6}$$

is called *Laplace's equation in space* (or in three dimensions). The laplacian of u, whether in two or three dimensions, is usually denoted by Δu, and this convention is followed in the text. From the context it will be clear how many independent variables are involved. The inhomogeneous equation corresponding to Laplace's equation (3-5) is

$$\Delta u = f(x,y) \tag{3-7}$$

called *Poisson's equation in the plane*. *Poisson's equation in space* is

$$\Delta u = f(x,y,z) \tag{3-8}$$

These are important classical equations, and they (as well as their extension to n independent variables) have been studied extensively. A branch of mathematical analysis which includes the study of the properties of solutions of these equations is called *potential theory*. In this section several characteristic properties of solutions of Laplace's equations are derived.

Harmonic Functions

Recall (see Chap. 2) that a solution of a linear second-order partial differential equation in a region is a function with continuous second derivatives which satisfies the differential equation in the region. Solutions of Laplace's equation are termed *harmonic functions*. More specifically, a function u is *harmonic at a point P* in three-dimensional space if there exists a sphere S with center at P such that u has continuous second derivatives and satisfies Eq. (3-6) at each point inside S. A function u is *harmonic in a region $\mathcal{R}$* if u is harmonic at each point of $\mathcal{R}$. Corresponding definitions are made for harmonic functions in two dimensions, with circle replacing sphere. Solutions of Laplace's equation are also called *potential functions*, because of the physical significance of the solution.

Let S be a closed surface in xyz space, and let $\mathscr{R}$ denote the region enclosed by S.   Recall that the symbol $\bar{\mathscr{R}}$ denotes the set of points consisting of all points of $\mathscr{R}$ together with the points on S, that is, the closure of $\mathscr{R}$.   It is assumed that the surface S is such that the divergence theorem of Sec. 2-6 is applicable to $\mathscr{R}$; that is, $\mathscr{R}$ is a regular region.   For example, let the closed surface S consist of a finite number of contiguous smooth surface elements such that at points where adjacent elements are joined the normal has an ordinary jump discontinuity.

Spherical Means

In the derivation of properties of solutions of partial differential equations the concept of the spherical mean of a function is very useful.   Let $P(x,y,z)$ be a given point $\mathscr{R}$, and let $S(P,r)$ denote the sphere with center at P and with radius r.   It is assumed henceforth that r is chosen such that $S(P,r)$ lies within the region $\mathscr{R}$.   If φ is a continuous function on $\mathscr{R}$, the *spherical mean* of φ is the function $\bar{\varphi}$ defined by

$$\bar{\varphi}(r) = \frac{1}{4\pi r^2} \iint\limits_{S(P,r)} \varphi(Q)\, dS \tag{3-9}$$

Here Q denotes a variable point on $S(P,r)$, and dS is the surface element of integration.   In Eq. (3-9) it is assumed that the center $P(x,y,z)$ is held fixed, otherwise a more appropriate notation for the value of the spherical mean is $\bar{\varphi}(x,y,z;r)$.   For a fixed radius r the value $\bar{\varphi}(r)$ is the average of the values of φ taken over the sphere $S(P,r)$; hence the name spherical mean.   Introduce spherical coordinates r, θ, φ with origin at P, and let

$$\xi = x + r \sin\theta \cos\varphi \qquad \eta = y + r\sin\theta\sin\varphi \qquad \zeta = z + r\cos\theta$$

for $0 \le \theta \le 2\pi$, $0 \le \varphi \le \pi$ (see Fig. 3-1).   Then (ξ,η,ζ) are the coordinates

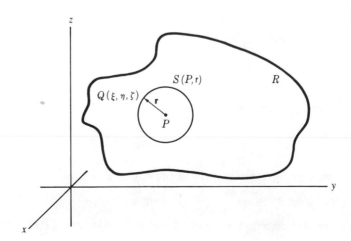

Figure 3-1

of the variable point Q on $S(P,r)$, and the spherical mean can be rewritten

$$\bar{\varphi}(r) = \frac{1}{4\pi} \int_0^{2\pi} \int_0^\pi \varphi(x + r\sin\theta\cos\varphi, y + r\sin\theta\sin\varphi, z + r\cos\varphi)\sin\theta \, d\theta \, d\varphi$$

(3-10)

since $dS = r^2 \sin\theta \, d\theta \, d\varphi$. It is clear that $\bar{\varphi}$ is a continuous function of r on some interval $0 < r \leq R$. Also

$$\bar{\varphi}(r) = \frac{\varphi(\bar{Q})}{4\pi} \int_0^{2\pi} \int_0^\pi \sin\theta \, d\theta \, d\varphi = \varphi(\bar{Q})$$

by the mean-value theorem for integrals, where $\bar{Q}$ is some point on $S(P,r)$. Hence

$$\lim_{r\to 0} \bar{\varphi}(r) = \varphi(x,y,z)$$

(3-11)

since $\bar{Q} \to P$ as $r \to 0$. Now if $\bar{\varphi}$ is defined at $r = 0$ by $\bar{\varphi}(0) = \varphi(x,y,z)$, then $\bar{\varphi}$ is continuous on the closed interval $[0,R]$. It is also clear that if φ is continuously differentiable on $\mathcal{R}$, then $\bar{\varphi}$ has a continuous first derivative for $0 < r < R$.

Properties of Harmonic Functions

The first property of harmonic functions demonstrated here is embodied in Theorem 3-1, often called Gauss' mean-value theorem. This theorem states that for a harmonic function the spherical mean has a constant value over spheres $S(P,r)$ about P, and indeed this constant mean value is the value of the function at the center P.

THEOREM 3-1 Let u be harmonic in $\mathcal{R}$. If P is a given point of $\mathcal{R}$ and $S(P,r)$ is a sphere with center at P such that $S(P,r)$ is contained in $\mathcal{R}$, then

$$u(P) = \bar{u}(r) = \frac{1}{4\pi r^2} \iint_{S(P,r)} u(Q) \, dS$$

(3-12)

Proof The proof consists in showing that $\bar{u}(r) = \text{const}, 0 < r < R$. Then the conclusion follows by the continuity of $\bar{u}$ on $[0,R]$ and Eq. (3-11). Replace φ by u in Eq. (3-10) and differentiate with respect to r. Then

$$\frac{d\bar{u}}{dr} = \frac{1}{4\pi} \int_0^{2\pi} \int_0^\pi (u_\xi n_1 + u_\eta n_2 + u_\zeta n_3) \sin\theta \, d\theta \, d\varphi$$

where ξ, η, ζ are as defined previously and

$$n_1 = \sin\theta\cos\varphi \qquad n_2 = \sin\theta\sin\varphi \qquad n_3 = \cos\theta$$

(3-13)

are the direction cosines of the normal $\mathbf{n}$ on $S(P,r)$. Let ∇u denote the gradient vector

$$\mathbf{i}\,\frac{\partial u}{\partial \xi} + \mathbf{j}\,\frac{\partial u}{\partial \eta} + \mathbf{k}\,\frac{\partial u}{\partial \zeta} \tag{3-14}$$

and let $V(P,r)$ denote the spherical region (ball) whose boundary is $S(P,r)$. Then

$$\frac{1}{4\pi} \int_0^{2\pi} \int_0^{\pi} (u_\xi n_1 + u_\eta n_2 + u_\zeta n_3) \sin\theta \; d\theta \; d\varphi$$

$$= \frac{1}{4\pi r^2} \iint_{S(P,r)} \nabla u \cdot \mathbf{n}\, dS = \frac{1}{4\pi r^2} \iiint_{V(P,r)} \Delta u \; d\tau = 0$$

by the divergence theorem. Hence $d\bar{u}/dr = 0$, and so $\bar{u}$ has a constant value for $0 < r < R$. The continuity of $\bar{u}$ at $r = 0$ now implies

$$\bar{u}(r) = \bar{u}(0) = u(x,y,z)$$

The converse of Gauss' theorem is also true. That is, if u is continuous and has the mean-value property expressed in the conclusion of Theorem 3-1 at each point of $\mathscr{R}$, then u must be harmonic in $\mathscr{R}$. Accordingly the mean-value property over spheres characterizes harmonic functions. Indeed, it can be proved that this property implies that u is an analytic function in $\mathscr{R}$ (see Ref. 9). The mean-value property can also be used to prove the powerful maximum-minimum principle. A well-known theorem of calculus asserts that a function u which is continuous on a closed and bounded set must be bounded there, and moreover it attains its maximum and minimum in the set. For harmonic functions considerably more can be said.

THEOREM 3-2 (*Maximum-minimum principle for harmonic functions*) Let $\mathscr{R}$ be a bounded region with boundary S. Let u be a function which is continuous on $\bar{\mathscr{R}}$ and harmonic in $\mathscr{R}$. If u is not everywhere constant in value on $\bar{\mathscr{R}}$, the maximum M and minimum m of u on $\mathscr{R}$ must be attained on the boundary S, and the inequality $m < u(P) < M$ holds for each point P in $\mathscr{R}$.

Proof We assume that the maximum value M is attained at a point P in $\mathscr{R}$ and show that u has the constant value M everywhere in $\bar{\mathscr{R}}$. Let R be any other point in $\mathscr{R}$, and join P to R by a polygonal line lying entirely within $\mathscr{R}$. Let d be the distance from this line to the boundary S, and choose r_0 such that $0 < r_0 < d$. Let T_0 denote the sphere about P of radius r_0. Then T_0 and its interior lie within $\mathscr{R}$. By Eq. (3-12) and the assumption that $u(P) = M$, it follows that

$$\frac{1}{4\pi r_0^2} \iint_{T_0} [M - u(Q)]\, dS = \frac{M}{4\pi r_0^2}\, 4\pi r_0^2 - \frac{1}{4\pi r_0^2} \iint_{T_0} u(Q)\, dS = 0$$

The function $M - u$ is continuous, and $M - u(Q) \geq 0$ holds everywhere, in particular for all points Q on T_0. The vanishing of the integral implies that u has the constant value M on T_0. In fact u must be constant everywhere inside T_0 as well. To show this, choose a radius $r < r_0$ and apply the same argument used for T_0 to the sphere T_r about P of radius r. Then u has the constant value M on T_r. Thus u is identically equal to M everywhere inside and on T_0. Now let Q_1 be the point of intersection of the polygonal line with the sphere T_0. Then $u(Q_1) = M$. Let T_1 be the sphere of radius r_0 about Q_1. Then T_1 and its interior lie within $\mathcal{R}$, and by a repetition of the previous argument it follows that u has the constant value M everywhere inside and on the sphere T_1. Let Q_2 be the point of intersection of the polygonal line with the sphere T_1; then $u(Q_2) = M$, etc. After a finite number of steps the point R is included within or on the surface of a sphere T_n such that u has the constant value M everywhere inside and on T_n. Hence $u(R) = M$. This proves that u has the constant value M everywhere in $\mathcal{R}$. By continuity u must have the constant value M everywhere on $\bar{\mathcal{R}}$. Accordingly if u is not constant in value on $\mathcal{R}$, then $u(P) < M$ must hold in $\mathcal{R}$. The statement regarding the minimum value m now follows on noting that the function $-u$ is continuous on $\bar{\mathcal{R}}$ and harmonic in $\mathcal{R}$, and the maximum of $-u$ is $-m$.

COROLLARY 3-3 Let $\mathcal{R}$ be a bounded region with boundary S. Let u be harmonic in $\mathcal{R}$ and continuous on $\bar{\mathcal{R}}$. Let M be a constant such that $|u(P)| \leq M$ holds for all points P on the boundary S. Then $|u(P)| \leq M$ holds everywhere on $\bar{\mathcal{R}}$, including all points in $\mathcal{R}$. If strict inequality holds on S, strict inequality must hold in $\mathcal{R}$ as well. In particular if $u = 0$ at all points of S, then $u = 0$ everywhere on $\bar{\mathcal{R}}$.

Proof Let M_1 be the maximum value of u on S. Since u is continuous on S, $M_1 \leq M$. By the maximum-minimum principle, $u(P) \leq M_1$ must hold for every point P in $\mathcal{R}$. The proof that $u(P) \geq -M$ in $\mathcal{R}$ is accomplished in the same way. Hence $|u(P)| \leq M$.

It should be noted that Theorem 3-2 (and its consequences) holds for any bounded region and not just those to which the divergence theorem is applicable. The theorems below regarding the Dirichlet problem are obtained immediately from the preceding properties of harmonic functions. The first theorem is a consequence of Corollary 3-3 on noting that the difference of two solutions of the problem is harmonic in $\mathcal{R}$, continuous on $\bar{\mathcal{R}}$, and vanishes everywhere on S. The second theorem pertains to the stability of the solution of the Dirichlet problem. A small change in the boundary values produces only a small change in the values of the solution at interior points.

THEOREM 3-4 (*Uniqueness*) Let f be a given continuous function on $\mathcal{R}$, and let g be a given continuous function defined on the boundary S. Then there is at most one solution of the Dirichlet problem

$$\Delta u = f \quad \text{in } \mathscr{R} \qquad\qquad u = g \quad \text{on } S$$

THEOREM 3-5   (*Continuous dependence on boundary values*)   Let f be a given continuous function on $\mathscr{R}$, and let g_1, g_2 be given continuous functions on the boundary S such that $|g_1(P) - g_2(P)| < \epsilon$ holds everywhere on S. If u_1, u_2 are solutions of the Dirichlet problems

$$\Delta u_1 = f \quad \text{in } \mathscr{R} \qquad\qquad u_1 = g_1 \quad \text{on } S$$
$$\Delta u_2 = f \quad \text{in } \mathscr{R} \qquad\qquad u_2 = g_2 \quad \text{on } S$$

then $|u_1(P) - u_2(P)| < \epsilon$ holds at each point P in $\mathscr{R}$.

Proof   Let $v = u_1 - u_2$. Then v satisfies the hypotheses of Corollary 3-3 with ϵ replacing M. Accordingly, $|v(P)| < \epsilon$ holds at each point P in $\mathscr{R}$.

From Theorem 3-4 it follows that the unique solution of the problem $\Delta u = 0$ in $\mathscr{R}$, $u = c$ on S, c a constant, is $u = c$ on $\bar{\mathscr{R}}$. In particular, the only solution of the homogeneous problem $\Delta u = 0$ in $\mathscr{R}$, $u = 0$ on S, is the trivial solution. It should be emphasized at this point that although uniqueness is guaranteed by Theorem 3-4, the hypotheses of the theorem are too weak to ensure the existence of a solution to the Dirichlet problem. The fact is that there exist bounded regions $\mathscr{R}$ such that the Dirichlet problem has no solution regardless of the choice of continuous boundary function g (see Ref. 4).

In some applications the function g prescribed on the boundary in the Dirichlet problem has points of discontinuity. Of course it is then impossible to find a solution u under the requirement that u be continuous on $\bar{\mathscr{R}}$. The Dirichlet problem with piecewise continuous boundary values is described as follows. Let g be a given function defined and piecewise continuous on the boundary S. A solution of $\Delta u = f$ in $\mathscr{R}$ is described such that (1) $u = g$ at each point of S where g is continuous and (2) u is bounded on $\bar{\mathscr{R}}$. It can be proved that there is at most one solution of this problem.

For the Neumann problem

$$\Delta u = f \quad \text{in } \mathscr{R} \qquad \frac{\partial u}{\partial n} = g \quad \text{on } S$$

at most uniqueness to within an additive constant can be expected. For if u is a solution, so is $u + c$, where c is a constant. Also the functions f and g cannot be prescribed in a completely arbitrary manner if solutions are to exist. To see this, suppose u is a solution of the problem. By the divergence theorem

$$\iiint_{\mathscr{R}} f \, d\tau = \iiint_{\mathscr{R}} \Delta u \, d\tau = \iint_S \frac{\partial u}{\partial n} \, dS$$

Hence the relation

$$\iiint_{\mathscr{R}} f \, d\tau = \iint_S g \, dS \tag{3-15}$$

must hold.   Thus the Neumann problem is not solvable for arbitrary choices of f and g.   In particular a necessary condition that there exists a solution of the problem

$$\Delta u = 0 \quad \text{in } \mathscr{R} \qquad \frac{\partial u}{\partial n} = g \qquad \text{on } S$$

is that g satisfy

$$\iint_S g \, dS = 0 \tag{3-16}$$

Of importance in many applications is the *exterior Dirichlet problem*.   For Laplace's equation it may be stated as follows.   Let S be a simple closed surface of the type described above, and let $\mathscr{R}$ be the unbounded region exterior to S.   Let g be a given continuous function defined on S.   One seeks a function u which is harmonic in $\mathscr{R}$, continuous on $\bar{\mathscr{R}}$, and such that $u = g$ on S.   The preceding theorems cannot be immediately applied to this problem since they were derived under the assumption that $\mathscr{R}$ was a bounded region. Also, simple examples show that the theorems are false for unbounded regions. As a counterexample for Theorem 3-4, let S be the unit sphere about the origin, and let the prescribed function be the function $g(P) = 1$.   Consider the functions $u(P) = 1, v(P) = 1/r$, where $r^2 = x^2 + y^2 + z^2$.   Each function is harmonic outside the unit sphere, and $u(P) = v(P) = 1$ on the sphere. Note also that the function $w = u - v$ is harmonic in the region exterior to the sphere, has the value zero everywhere on S, but does not vanish identically outside of S.   It is clear that an additional requirement must be imposed before uniqueness of solution for the exterior Dirichlet problem can be obtained.   Let φ be a function which is defined at all points exterior to some sphere about the origin.   As before, let $r = (x^2 + y^2 + z^2)^{\frac{1}{2}}$ be the distance of the variable point $P(x,y,z)$ from the origin.   Then φ *vanishes uniformly at infinity* if, given $\epsilon > 0$, there exists a number $r_\epsilon > 0$ such that if $r \geq r_\epsilon$, then $|\varphi(P)| < \epsilon$.   Geometrically visualize the sphere S of radius r_ϵ about the origin.   Then $|\varphi(P)| < \epsilon$ holds for every point P external to S.   To symbolize this property write

$$\lim_{r \to \infty} \varphi(P) = 0 \text{ uniformly}$$

With the additional requirement that the solution vanish uniformly at infinity the uniqueness property holds for the exterior Dirichlet problem.

THEOREM 3-6   Let S be a simple closed surface, $\mathscr{R}$ the region exterior to S, and let g be a given function continuous on S.   There can be at most one function u which is continuous on $\bar{\mathscr{R}}$ and harmonic in $\mathscr{R}$ such that $u = g$ on S and $\lim\limits_{r \to \infty} u(P) = 0$ uniformly.

Proof   Suppose u_1, u_2 are functions with the properties described in the theorem.   Then the function $v = u_1 - u_2$ is continuous on $\bar{\mathscr{R}}$, harmonic in $\mathscr{R}$, vanishes on S, and vanishes uniformly at infinity.   Now it is shown that v must be identically zero everywhere in $\mathscr{R}$.   Let P be a fixed point in $\mathscr{R}$, and let $\epsilon > 0$.   Choose a sphere S_r about the origin with radius r sufficiently large so that (1) S is contained within S_r, (2) P is contained within S_r, and (3) $|v(Q)| < \epsilon$ holds for all points Q on S_r.   Let $\mathscr{R}_r$ denote the annular region bounded by the surfaces S and S_r.   Since $v = 0$ on S, $|v(Q)| < \epsilon$ holds on the complete boundary $S + S_r$ of $\mathscr{R}_r$.   It follows from Corollary 3-3 that $|v(P)| < \epsilon$.   The only way in which this inequality can hold for every choice of $\epsilon > 0$ is when $v(P) = 0$.

Analogous theorems hold for harmonic functions of two real variables. Let C be a simple closed piecewise smooth curve in the xy plane, and let $\mathscr{R}$ be the region interior to C.   If $P(x,y)$ is a given point of $\mathscr{R}$, denote by $C(P,r)$ the circle of radius r about P as center.   Assume $C(P,r)$ is contained in $\mathscr{R}$. Given a continuous function φ, the *circular mean* of φ is the function $\bar{\varphi}$ defined by

$$\bar{\varphi}(r) = \frac{1}{2\pi r} \int_{C(P,r)} \varphi(Q)\, ds \tag{3-17}$$

where Q is a variable point on $C(P,r)$ and ds is the arc element.   Let r, θ be polar coordinates with origin at P.   Then the circular mean can be rewritten

$$\bar{\varphi}(r) = \frac{1}{2\pi} \int_0^{2\pi} \varphi(x + r \cos \theta, y + r \sin \theta)\, d\theta \tag{3-18}$$

since $ds = r\, d\theta$.   By exactly similar arguments Gauss' mean-value theorem can now be shown to hold for a harmonic function u on $\mathscr{R}$:

$$u(P) = \bar{u}(r) = \frac{1}{2\pi r} \int_{C(P,r)} u(Q)\, dS \tag{3-19}$$

Then the maximum-minimum principle in two dimensions states that except in the trivial case of a constant, a harmonic function in a bounded plane region $\mathscr{R}$ cannot attain its maximum and minimum in $\mathscr{R}$ but must attain these values on the boundary of $\mathscr{R}$.   The uniqueness theorem for the Dirichlet problem and the stability property embodied in Theorem 3-5 follow from the mean-value theorem in the same manner as before.

3-3   SEPARATION OF VARIABLES IN LAPLACE'S EQUATION

Even when the general solution of a linear partial differential equation can be written down, it is more than difficult (except in special instances) to fit the arbitrary functions which appear to prescribed boundary conditions. Generally other techniques must be used. One consideration in the choice of method and the coordinate system used is the shape of the boundary. A method applicable to a number of the classical linear homogeneous equations of mathematical physics, in various coordinate systems, is called the *method of separation of variables* (also Bernoulli's method, or the Fourier method). Instead of starting with the general solution the procedure is to derive a sequence of particular solutions, of a form called *separable*, in such a way that superposition yields a solution which satisfies the boundary conditions. In this section the method is applied to several boundary-value problems of potential theory. Its use in boundary- and initial-value problems involving hyperbolic and parabolic equations is illustrated in Chaps. 4 and 5.

Plane Polar Coordinates

If the boundary is rectangular, rectangular coordinates are suggested. Separation of variables for Laplace's equation in rectangular coordinates is carried out in Probs. 16 and 24. The first case discussed here is Laplace's equation in plane polar coordinates r, θ. This is the appropriate coordinate system if the boundary is a circle or a part of a circle. Laplace's equation in polar coordinates r, θ is

$$\Delta u = \frac{\partial^2 u}{\partial r^2} + \frac{1}{r} \frac{\partial u}{\partial r} + \frac{1}{r^2} \frac{\partial^2 u}{\partial \theta^2} = 0 \tag{3-20}$$

A solution, Eq. (3-20), of the form

$$u(r,\theta) = R(r)\Theta(\theta) \tag{3-21}$$

is called a *separable solution* of Laplace's equation in plane polar coordinates. In an effort to find separable solutions a solution of this form is assumed, and then the equations which the functions R, Θ must satisfy are deduced. Substitution of u from Eq. (3-21) into Eq. (3-20) yields

$$r^2 R''\Theta + rR'\Theta + R\Theta'' = 0$$

or

$$\frac{r^2 R'' + rR'}{R} = -\frac{\Theta''}{\Theta} \tag{3-22}$$

where

$$R' = \frac{dR}{dr} \qquad \Theta' = \frac{d\Theta}{d\theta}$$

The left side of Eq. (3-22) involves only r, and the right side involves only θ. Thus the variables in Eq. (3-20) have been *separated*.   Now it is argued that if Eq. (3-22) holds, then since this is an identity in the independent variables r, θ, it follows that each side is identically equal to a constant:

$$\frac{r^2R'' + rR'}{R} = \lambda = \frac{-\Theta''}{\Theta}$$

The constant λ is called a *separation constant*.   Accordingly, if $u = R\Theta$ is a solution of Eq. (3-20), the functions R, Θ must satisfy the ordinary differential equations

$$r^2R'' + rR' - \lambda R = 0 \tag{3-23}$$
$$\Theta'' + \lambda\Theta = 0 \tag{3-24}$$

Equation (3-23) is of the Euler type and can be solved by assuming a solution of the form $R = r^\alpha$, where α is a constant to be determined.   The general solution is

$$R = \begin{cases} C_0 + D_0 \log r & \text{if } \lambda = 0 \\ C_1 r^\mu + D_1 r^{-\mu} & \text{if } \mu = \sqrt{\lambda} \neq 0 \end{cases}$$

The general solution of Eq. (3-24) is

$$\Theta = \begin{cases} A_0 + B_0\theta & \text{if } \lambda = 0 \\ A_1 \cos \mu\theta + B_1 \sin \mu\theta & \text{if } \mu = \sqrt{\lambda} \neq 0 \end{cases}$$

Hence separable solutions of Eq. (3-20) are of the form

$$u = (C_0 + D_0 \log r)(A_0 + B_0\theta) + (C_1 r^\mu + D_1 r^{-\mu})(A_1 \cos \mu\theta + B_1 \sin \mu\theta) \tag{3-25}$$

where the capital letters denote constants and μ is a constant.   Conversely, a function of this form is a solution of Eq. (3-20) in some region.

Dirichlet Problem for a Circle

To illustrate how the prescribed conditions in a problem serve to determine certain values of the separation constant as well as the arbitrary constants which appear in Eq. (3-25), the Dirichlet problem for a circle is solved.   Let C be a circle of radius a about the origin in the xy plane, and let $\mathcal{R}$ denote the region interior to C.   Let $f(\theta)$ be a given continuous function on C.   The problem is to find a solution u of Eq. (3-20) in $\mathcal{R}$ such that u is single-valued and continuous on $\mathcal{R} + C$ and such that

$$u(a,\theta) = f(\theta) \qquad 0 \leq \theta \leq 2\pi \tag{3-26}$$

The requirement of single-valuedness on $\mathcal{R} + C$ implies the *periodicity condition*

$$u(r, \theta + 2\pi) = u(r,\theta) \qquad 0 \le r \le a; -\infty < \theta < \infty \tag{3-27}$$

in addition to the boundary condition (3-26). This condition is satisfied by the separable solution (3-25) provided Θ satisfies

$$\Theta(\theta + 2\pi) = \Theta(\theta) \qquad -\infty < \theta < \infty \tag{3-28}$$

that is, Θ is periodic, of period 2π. Equation (3-28) implies that

$$\Theta(-\pi) - \Theta(\pi) = 0 \qquad \Theta'(-\pi) - \Theta'(\pi) = 0 \tag{3-29}$$

Equations (3-24) and (3-29) constitute the regular self-adjoint Sturm-Liouville problem considered in Sec. 2 of Appendix 2. Hence the values of λ are restricted to

$$\lambda = \lambda_n = n^2 \qquad n = 0, 1, \ldots$$

Also $B_0 = 0$, $D_0 = 0$ are necessary in Eq. (3-25). The functions

$$u_n(r,\theta) = (C_n r^n + D_n r^{-n})(A_n \cos n\theta + B_n \sin n\theta) \tag{3-30}$$

are called *circular harmonics*. In the present problem the region of interest includes $r = 0$, and so continuity demands that the coefficient $D_n = 0$. The circular harmonics

$$u_n(r,\theta) = r^n(A_n \cos n\theta + B_n \sin n\theta) \qquad n = 0, 1, \ldots$$

are appropriate here. The superposition

$$u(r,\theta) = \sum_{n=0}^{\infty} r^n(A_n \cos n\theta + B_n \sin n\theta)$$

will also be a solution provided the series is suitably convergent. The boundary condition (3-26) remains to be satisfied. This requires

$$\sum_{n=0}^{\infty} a^n(A_n \cos n\theta + B_n \sin n\theta) = f(\theta) \qquad 0 \le \theta \le 2\pi \tag{3-31}$$

Assume now that f is periodic, of period 2π, and continuous with a piecewise continuous first derivative for all θ. From Sec. 2 of Appendix 2 the Fourier coefficients of f are

$$a_n = \frac{1}{\pi} \int_0^{2\pi} f(t) \cos nt \, dt \qquad b_n = \frac{1}{\pi} \int_0^{2\pi} f(t) \sin nt \, dt \qquad n = 0, 1, 2, \ldots$$

If the constants A_n, B_n are chosen $A_0 = a_0/2$, $A_n = a_n/a^n$, $B_n = b_n/a^n$, $n = 1, 2, \ldots$, then Eq. (3-31) holds. The series solution obtained by separation of variables is

$$u(r,\theta) = \frac{a_0}{2} + \sum_{n=0}^{\infty} \left(\frac{r}{a}\right)^n (a_n \cos n\theta + b_n \sin n\theta) \tag{3-32}$$

where the coefficients a_0, a_n, b_n are the Fourier coefficients of f. Note that

$$u(0,\theta) = \frac{a_0}{2} = \frac{1}{2\pi} \int_0^{2\pi} f(\theta)\, d\theta$$

which expresses the mean-value property of the harmonic function u over the circle C.

With the aid of the maximum-minimum principle one can prove that the formal series solution (3-32) converges and actually yields the solution of the Dirichlet problem for the circle. Recall that f is assumed continuous, with a piecewise continuous first derivative, and periodic, with period 2π. Hence the Fourier series

$$\frac{a_0}{2} + \sum_{n=1}^{\infty} (a_n \cos n\theta + b_n \sin n\theta)$$

converges uniformly to f on $[0,2\pi]$. Now the Cauchy criterion states that given $\epsilon > 0$, there exists an integer $N_\epsilon > 0$ such that if n, m are integers with $n \geq m \geq N_\epsilon$, then

$$\left| \frac{a_0}{2} + \sum_{k=m}^{n} (a_k \cos k\theta + b_k \sin k\theta) \right| < \epsilon \qquad 0 \leq \theta \leq 2\pi$$

Fix integers n, m with $n \geq m \geq N_\epsilon$. Define the function v on the closed disk $\bar{\mathscr{R}} = \mathscr{R} + C$ by

$$v(r,\theta) = \frac{a_0}{2} + \sum_{k=m}^{n} \left(\frac{r}{a}\right)^k (a_k \cos k\theta + b_k \sin k\theta)$$

Observe that v is continuous on $\bar{\mathscr{R}}$, harmonic in $\mathscr{R}$, and on the boundary C

$$|v(a,\theta)| < \epsilon \qquad 0 \leq \theta \leq 2\pi$$

Accordingly, by Corollary 3-3,

$$|v(r,\theta)| < \epsilon \qquad \text{on } \bar{\mathscr{R}}$$

What has been shown is that given $\epsilon > 0$, there exists an integer $N_\epsilon > 0$ such that whenever n, m are integers with $n \geq m \geq N_\epsilon$, then

$$\left| \frac{a_0}{2} + \sum_{k=m}^{n} \left(\frac{r}{a}\right)^k (a_k \cos k\theta + b_k \sin k\theta) \right| < \epsilon$$

for all points (r,θ) belonging to $\bar{\mathscr{R}}$. But this is just the Cauchy criterion for the uniform convergence of the series (3-32) on $\bar{\mathscr{R}}$. Thus the series defines a continuous function u on $\bar{\mathscr{R}}$. Moreover this function u takes on the correct boundary values in the following continuous sense:

$$\lim_{(r,\theta) \to (a,\theta_0)} u(r,\theta) = f(\theta_0) \qquad 0 \leq \theta_0 \leq 2\pi \tag{3-33}$$

where it is understood that the approach is taken only over points of $\bar{\mathscr{R}}$. The relation (3-33) follows immediately from the fact that u is continuous on $\bar{\mathscr{R}}$, and

$$u(a,\theta_0) = \frac{a_0}{2} + \sum_{k=1}^{\infty} (a_k \cos k\theta_0 + b_k \sin k\theta_0) = f(\theta_0)$$

for $0 \leq \theta_0 \leq 2\pi$.

It remains to show that u is harmonic in $\mathscr{R}$. Since the Fourier series of f converges, there exists a constant $M > 0$ such that

$$|a_0| < M \qquad |a_n| < M \qquad |b_n| < M \qquad n = 1, 2, \ldots$$

Now define the sequence of functions $\{u_n\}$ by

$$u_n(r,\theta) = \left(\frac{r}{a}\right)^n (a_n \cos n\theta + b_n \sin n\theta) \qquad n = 1, 2, \ldots$$

and $u_0(r,\theta) = a_0/2$. Observe that each u_n is continuous on $\bar{\mathscr{R}}$ and harmonic in $\mathscr{R}$. Choose a value r_0 such that $0 < r_0 < a$. Then

$$\left|\frac{\partial u_n}{\partial r}\right| = \left|\left(\frac{n}{a}\right)\left(\frac{r}{a}\right)^{n-1} (a_n \cos n\theta + b_n \sin n\theta)\right|$$

$$\leq \left(\frac{n}{a}\right)\left(\frac{r}{a}\right)^n (|a_n| + |b_n|) < 2M\left(\frac{n}{a}\right)\left(\frac{r_0}{a}\right)^n \qquad n = 1, 2, \ldots$$

for $0 \leq r \leq r_0$. The series of positive constants

$$\sum_{n=0}^{\infty} 2M\left(\frac{n}{a}\right)\left(\frac{r_0}{a}\right)^n$$

converges and dominates, term by term, the series

$$\sum_{n=0}^{\infty} \frac{\partial u_n}{\partial r} \qquad 0 \leq r \leq r_0; 0 \leq \theta \leq 2\pi$$

Thus the series obtained by differentiating the series (3-32), term by term with respect to r, converges uniformly on the closed disk of radius r_0 about the origin. In turn this implies that the function u has a continuous first derivative with respect to r, and $\partial u/\partial r$ can be calculated by term-by-term differentiation of the series for u with respect to r whenever $0 \leq r < r_0$. Repetition of this type of argument shows that the second derivatives $\partial^2 u/\partial r^2$, $\partial^2 u/\partial \theta^2$ exist and are continuous functions and can be obtained by termwise differentiation of the series for u for points within the circle of radius r_0 about the origin. Accordingly

$$\Delta u = \sum_{n=0}^{\infty} \Delta u_n = 0 \qquad 0 \leq r < r_0; 0 \leq \theta \leq 2\pi$$

Recall now that r_0 was arbitrarily chosen to be within the requirement $0 < r_0 < a$. Hence the function u is harmonic in $\mathscr{R}$.

The solution (3-32) can be interpreted physically in several ways. In Chap. 5 it is shown that in a homogeneous thermally conducting solid in which no sources of heat are present the steady-state temperature u satisfies Laplace's equation. Suppose such a solid occupies a region $\mathscr{R}$ whose boundary is S. If the temperature on the boundary is a prescribed function f (independent of time), the problem of determining the steady temperature distribution in $\mathscr{R}$ due to the surface temperature is the Dirichlet problem $\Delta u = 0$ in $\mathscr{R}$, $u = f$ on S. Consider a thin circular disk of conducting material of radius a lying in the xy plane with center at the origin. Assume the faces of the disk are insulated. If the temperature on the rim $r = a$ is held at $u = f(\theta)$, Eq. (3-32) gives the steady temperature distribution in the disk. Another interpretation is the steady temperature in an infinitely long (i.e., length so great that end effects are neglected) solid homogeneous cylinder of radius a with axis the z axis.

Poisson Integral Formula for a Circle

An important and useful integral representation of the solution of the Dirichlet problem can be derived from Eq. (3-32). Substitution of the expressions for the Fourier coefficients a_n, b_n of f into Eq. (3-32) gives

$$u(r,\theta) = \frac{1}{2\pi} \int_0^{2\pi} f(\varphi)\, d\varphi$$
$$+ \frac{1}{\pi} \sum_{n=1}^{\infty} \left(\frac{r}{a}\right)^n \left[\int_0^{2\pi} f(\varphi)(\cos n\varphi \cos n\theta + \sin n\varphi \sin n\theta)\, d\varphi \right]$$

$$= \frac{1}{2\pi} \int_0^{2\pi} f(\varphi)\, d\varphi + \frac{1}{\pi} \sum_{n=1}^{\infty} \left(\frac{r}{a}\right)^n \left[\int_0^{2\pi} f(\varphi) \cos n(\varphi - \theta)\, d\varphi \right]$$

$$= \frac{1}{2\pi} \int_0^{2\pi} f(\varphi)\, d\varphi + \frac{1}{\pi} \int_0^{2\pi} f(\varphi) \left[\sum_{n=1}^{\infty} \left(\frac{r}{a}\right)^n \cos n(\varphi - \theta) \right] d\varphi$$

$$= \frac{1}{2\pi} \int_0^{2\pi} f(\varphi) \left[1 + 2 \sum_{n=1}^{\infty} \left(\frac{r}{a}\right)^n \cos n(\varphi - \theta) \right] d\varphi$$

Note that the interchange of summation and integration is valid in view of the uniform convergence of the series. Now if x is real, $|x| < 1$, then

$$1 + 2 \sum_{n=1}^{\infty} x^n \cos n\psi = \frac{1 - x^2}{1 - 2x \cos \psi + x^2} \tag{3-34}$$

To show this, let $z = xe^{i\psi}$, $i = \sqrt{-1}$. Recall the expansion

$$\frac{1+z}{1-z} = -1 + \frac{2}{1-z} = -1 + 2 + 2 \sum_{n=1}^{\infty} z^n = 1 + 2 \sum_{n=1}^{\infty} z^n$$

Let Re ζ denote the real part of a complex number ζ.   Then

$$\text{Re} \; \frac{1+z}{1-z} = \frac{1 - |z|^2}{1 - 2\,\text{Re}\,(z) + |z|^2} = \frac{1 - x^2}{1 - 2x \cos \psi + x^2}$$

and

$$\text{Re}\!\left(1 + 2\sum_{n=1}^{\infty} z^n\right) = \text{Re}\left[1 + 2\sum_{n=1}^{\infty} x^n(\cos n\psi + i \sin n\psi)\right]$$

$$= 1 + 2\sum_{n=1}^{\infty} x^n \cos n\psi$$

In Eq. (3-34) set $x = r/a$, $\psi = \varphi - \theta$, and substitute the result into the preceding equation.   Then

$$u(r,\theta) = \frac{1}{2\pi} \int_0^{2\pi} \frac{a^2 - r^2}{a^2 - 2ar \cos (\varphi - \theta) + r^2} \, f(\varphi) \, d\varphi \qquad (3\text{-}35)$$

Equation (3-35) is called *Poisson's integral for the circle*.   The assumptions made are that $0 \leq r < a$ and that f is continuous with a piecewise continuous derivative.   However the integral exists and defines a differentiable function if only continuity is assumed for f.   In fact the integrand involves r, θ as parameters, and the integrand has continuous derivatives of all orders with respect to r and θ, provided $0 \leq r < a$.   This implies that the function u defined by the integral has continuous derivatives of all orders with respect to r and θ for $0 \leq r < a$, and the derivatives of u can be calculated by differentiating within the integral sign.   A calculation of the derivatives involved shows that the function u defined by Eq. (3-35) is harmonic inside the circle.   Moreover it can be shown that the function u so defined takes on the prescribed values $f(\theta)$ for $0 \leq \theta \leq 2\pi$ when $r = a$ (see Ref. 9).   Thus Poisson's integral furnishes the unique solution of the Dirichlet problem.

EXAMPLE 3-1 $\Delta u = 0$ $0 \leq r < a$ $u(a,\theta) = f(\theta)$ where $f(\theta) = 1$, $0 < \theta < \pi$; $f(\theta) = 0$, $\pi < \theta < 2\pi$.

To obtain the series solution (3-32) of the problem a calculation of the Fourier coefficients of f yields $a_0 = 1$ and

$$a_n = 0 \qquad n = 1, 2, \ldots \qquad\qquad b_n = \frac{1 - \cos n\pi}{n\pi} \qquad n = 1, 2, \ldots$$

Accordingly the series solution is

$$u(r,\theta) = \frac{1}{2} + \frac{2}{\pi} \sum_{n=1}^{\infty} \left(\frac{r}{a}\right)^{2n-1} \frac{\sin (2n - 1)\theta}{2n - 1}$$

From Poisson's integral (3-35) the solution must also be given by

$$u(r,\theta) = \frac{1}{2\pi} \int_0^{\pi} \frac{a^2 - r^2}{a^2 - 2ar \cos (\varphi - \theta) + r^2} \, d\varphi$$

Integral tables furnish the formula, valid if $c^2 < d^2$,

$$\int \frac{dx}{c + d \cos x} = \frac{2}{\sqrt{c^2 - d^2}} \tan^{-1} \frac{\sqrt{c^2 - d^2} \tan (x/2)}{c + d}$$

Let $f(x) = (c + d \cos x)^{-1}$, and let $F(x)$ denote the right-hand member in the integration formula above. Differentiation shows that $F'(x) = f(x)$ if $x \neq \pm(2n - 1)\pi$, $n = 1, 2, \ldots$. However $F'(x)$ does not exist at odd multiples of π. Thus in this case

$$\int_a^b f(x)\, dx \neq F(b) - F(a)$$

for an arbitrary choice of interval $[a,b]$. Suppose $0 < x < \pi$. Then $F'(x) = f(x)$, and

$$\int_0^\pi f(x)\, dx = \lim_{\epsilon \to 0} \int_0^{\pi - \epsilon} f(x)\, dx = \lim_{\epsilon \to 0} \left\{ \frac{2}{\sqrt{c^2 - d^2}} \tan^{-1} \frac{\sqrt{c^2 - d^2} \tan [(\pi - \epsilon)/2]}{c + d} \right\}$$

$$= \frac{\pi}{\sqrt{c^2 - d^2}}$$

In the expression for the solution given by Poisson's integral set $c = a^2 + r^2$, $d = -2ar$, $x = \varphi - \theta$. Now the integration interval is $0 \le \varphi \le \pi$, while the variable θ is such that $0 \le \theta \le 2\pi$. Hence if $0 < \theta < \pi$, then $-\pi < x < \pi$. Accordingly since we are not integrating across an odd multiple of π, it follows that Poisson's integral gives for $0 \le r < a$, $0 < \theta < \pi$,

$$u(r,\theta) = \lim_{\epsilon \to 0} \left(\frac{a^2 - r^2}{2\pi} \frac{2}{a^2 - r^2} \left\{ \tan^{-1} \frac{(a^2 - r^2) \tan [(\varphi - \theta)/2]}{(a - r)^2} \right\}\Big|_\epsilon^{\pi - \epsilon} \right)$$

$$= \frac{1}{\pi} \tan^{-1} \left(\frac{a + r}{a - r} \cot \frac{\theta}{2} \right) + \frac{1}{\pi} \tan^{-1} \left(\frac{a + r}{a - r} \tan \frac{\theta}{2} \right)$$

If $\pi < \theta < 2\pi$, then $-2\pi < x < 0$, and this includes $x = -\pi$, where the function $F(x)$ is discontinuous. To evaluate Poisson's integral for $0 \le r \le a$, $\pi < \theta < 2\pi$, we have

$$u(r,\theta) = \frac{1}{\pi} \lim_{\epsilon \to 0} \left[\tan^{-1} \left(\frac{a + r}{a - r} \tan \frac{\varphi - \theta}{2} \right) \right]\Big|_\epsilon^{\theta - \pi - \epsilon}$$

$$+ \frac{1}{\pi} \lim_{\epsilon \to 0} \left[\tan^{-1} \left(\frac{a + r}{a - r} \tan \frac{\varphi - \theta}{2} \right) \right]\Big|_{\theta - \pi + \epsilon}^{\pi - \epsilon}$$

$$= \frac{1}{\pi} \frac{\pi}{2} + \frac{1}{\pi} \tan^{-1} \left(\frac{a + r}{a - r} \tan \frac{\theta}{2} \right) + \frac{1}{\pi} \tan^{-1} \left(\frac{a + r}{a - r} \cot \frac{\theta}{2} \right) + \frac{1}{\pi} \frac{\pi}{2}$$

$$= 1 + \frac{1}{\pi} \tan^{-1} \left(\frac{a + r}{a - r} \tan \frac{\theta}{2} \right) + \frac{1}{\pi} \tan^{-1} \left(\frac{a + r}{a - r} \cot \frac{\theta}{2} \right)$$

From these expressions it can be verified that

$$\lim_{r \to a} u(r,\theta) = 1 \qquad 0 < \theta < \pi \qquad\qquad \lim_{r \to a} u(r,\theta) = 0 \qquad \pi < \theta < 2\pi$$

Cylindrical Coordinates

Cylindrical coordinates r, θ, z are related to rectangular coordinates by the equations

$$x = r \cos \theta \qquad y = r \sin \theta \qquad z = z$$

(see Fig. 3-2).   In cylindrical coordinates Laplace's equation has the form

$$\Delta u = \frac{\partial^2 u}{\partial r^2} + \frac{1}{r}\frac{\partial u}{\partial r} + \frac{1}{r^2}\frac{\partial^2 u}{\partial \theta^2} + \frac{\partial^2 u}{\partial z^2} = 0 \tag{3-36}$$

Note that Eq. (3-36) reduces to Eq. (3-20) in the case of functions which are independent of z.   A separable solution of the form

$$u = \varphi(r,\theta)Z(z)$$

is sought first.   Substitution into Eq. (3-36) yields

$$\varphi_{rr}Z + \frac{1}{r}\varphi_r Z + \frac{1}{r^2}\varphi_{\theta\theta}Z + \varphi Z'' = 0$$

Now division by φZ gives

$$\frac{\Delta\varphi}{\varphi} = \frac{\varphi_{rr} + (1/r)\varphi_r + (1/r^2)\varphi_{\theta\theta}}{\varphi} = -\frac{Z''}{Z}$$

The right side of this equation involves only z, while the left side is independent of z.   Hence there must exist a constant λ such that

$$\frac{\Delta\varphi}{\varphi} = -\lambda = -\frac{Z''}{Z}$$

where Δ is the laplacian in plane polar coordinates r, θ.   The factors φ, Z must satisfy

$$\Delta\varphi + \lambda\varphi = \varphi_{rr} + \frac{1}{r}\varphi_r + \frac{1}{r^2}\varphi_{\theta\theta} + \lambda\varphi = 0 \tag{3-37}$$

$$Z'' - \lambda Z = 0 \tag{3-38}$$

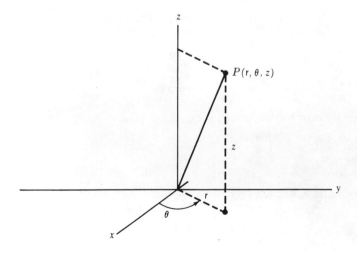

Figure 3-2

The variables r, θ are separated in Eq. (3-37) by assuming

$$\varphi = R(r)\Theta(\theta)$$

Substitution into Eq. (3-37) yields

$$R''\Theta + \frac{1}{r} R'\Theta + \frac{1}{r^2} R\Theta'' + \lambda R\Theta = 0$$

where primes denote derivatives with respect to the appropriate independent variables. If the equation is multiplied through by $r^2/R\Theta$, then one obtains the relation

$$\frac{r^2 R'' + rR'}{R} + \lambda r^2 = \mu = -\frac{\Theta''}{\Theta}$$

where μ is a second independent separation constant. Accordingly the factors R, Θ must satisfy the respective ordinary differential equations

$$r^2 R'' + rR' + (\lambda r^2 - \mu)R = 0 \tag{3-39}$$

$$\Theta'' + \mu\Theta = 0 \tag{3-40}$$

Up to this point the separation constants λ, μ are completely arbitrary. If assumptions are made for these constants, various useful separable harmonic functions are obtained. Suppose λ is real and positive, and let

$$\alpha = \sqrt{\lambda}$$

Make the change of independent variable $\xi = \alpha r$ in Eq. (3-39). Then the transformed equation is

$$\xi^2 \frac{d^2 R}{d\xi^2} + \xi \frac{dR}{d\xi} + (\xi^2 - \mu)R = 0 \tag{3-41}$$

which is Bessel's equation (see Sec. 4, Appendix 2). Assume μ is real and nonnegative, and let

$$\nu = \sqrt{\mu}$$

Then the general solution of Eq. (3-41) is

$$R = AJ_\nu(\xi) + BY_\nu(\xi)$$

where A, B are arbitrary constants and J_ν, Y_ν are the Bessel functions of the first and second kind, respectively, of order ν. Hence the general solution of Eq. (3-39) is

$$R = AJ_\nu(\alpha r) + BY_\nu(\alpha r)$$

The general solutions of Eqs. (3-38) and (3-40) respectively are

$$Z = Ee^{\alpha z} + Fe^{-\alpha z}$$

$$\Theta = C \cos \nu\theta + D \sin \nu\theta$$

where E, F, C, and D are arbitrary constants. Accordingly there are the separable harmonic functions

$$u = [AJ_\nu(\alpha r) + BY_\nu(\alpha r)](C \cos \nu\theta + D \sin \nu\theta)(Ee^{\alpha z} + Fe^{-\alpha z}) \qquad (3\text{-}42)$$

in cylindrical coordinates, where α, ν are arbitrary real nonnegative constants. Here it is understood that if $\nu = 0$, the trigonometric θ dependence is replaced by a linear function [recall Eq. (3-40)]. Thus there are the separable solutions

$$u = [AJ_0(\alpha r) + BY_\nu(\alpha r)](C + D\theta)(Ee^{\alpha z} + Fe^{-\alpha z}) \qquad (3\text{-}43)$$

If $\lambda = 0$, Eq. (3-37) reduces to Laplace's equations (3-20) in plane polar coordinates. Thus there are the separable solutions

$$u = (Ar^\nu + Br^{-\nu})(C \cos \nu\theta + D \sin \nu\theta)(E + Fz) \qquad (3\text{-}44)$$

of Eq. (3-36) where ν is arbitrary. If u is to be z-independent, $F = 0$ is chosen, and these solutions are just the solutions of Eq. (3-20) obtained in the previous paragraphs.

If the region under consideration allows the variable θ to vary over the interval $[0,2\pi]$, then, as in the polar-coordinate case, the separation constant ν is quantized to integral values. In this case $\lambda = n^2$, $n = 0, 1, \ldots$, and the functions

$$u_n = [AJ_n(\alpha r) + BY_n(\alpha r)](C \cos n\theta + D \sin n\theta)(Ee^{\alpha z} + Fe^{-\alpha z}) \qquad (3\text{-}45)$$

are solutions of Eq. (3-36) which are single-valued and continuous for $r > 0$, $0 \le \theta \le 2\pi$, $-\infty < z < \infty$. If the region includes the z axis and continuity of the solution demanded there, it is necessary to choose $B = 0$ in Eqs. (3-42) to (3-45), since the Bessel function of the second kind Y_ν is unbounded near $r = 0$.

The following example illustrates the use of separable solutions of Laplace's equation in cylindrical coordinates.

EXAMPLE 3-2   Find the potential φ inside the cylinder

$$0 \le r \le a \qquad 0 \le \theta \le 2\pi \qquad 0 \le z \le h$$

if the potential on the top $z = h$ and on the lateral surface $r = a$ is held at zero, while on the base $z = 0$ the potential

$$\varphi(r,\theta,0) = f(r,\theta)$$

where f is a given continuous function such that $f(a,\theta) = 0$, $0 \le \theta \le 2\pi$. Consider the particular case $f(r,\theta) = V_0(1 - r^2/a^2)$, where V_0 is a constant.

The potential must be a single-valued and continuous solution of Laplace's equation inside the cylinder. Accordingly the separable forms of solution

$$u_n(r,\theta,z) = J_n(\alpha r)(A_n \cos n\theta + B_n \sin n\theta)e^{\pm \alpha z} \qquad n = 0, 1, \ldots$$

obtained from Eq. (3-45) are chosen. The z-dependent factor can be rewritten

$$Z = E_1 \cosh \alpha z + F_1 \sinh \alpha z$$

If the condition $Z(h) = 0$ is imposed, then a z dependence of the form $Z = C \sinh [\alpha(h - z)]$, where C is an arbitrary constant, results. The solutions

$$u_n(r,\theta,z) = J_n(\alpha r)(A_n \cos n\theta + B_n \sin n\theta) \sinh [\alpha(h - z)]$$

satisfy $u_n(r,\theta,h) = 0$, for $n = 0, 1, \ldots$. The condition on the lateral surface $r = a$ is

$$u_n(a,\theta,z) = 0 \qquad 0 \le \theta \le 2\pi; 0 \le z \le h$$

This implies the condition $J_n(\alpha a) = 0$ if one is to avoid the trivial solution. For each fixed n the equation $J_n(\xi) = 0$ has infinitely many positive roots, and these can be ordered

$$0 < \xi_{n1} < \xi_{n2} < \cdots < \xi_{nk} < \xi_{n,k+1} < \cdots$$

For each fixed n choose $\alpha = \alpha_{nk} = \xi_{nk}/a$, $k = 1, 2, \ldots$, and obtain the ∞^2 harmonic functions

$$u_{nk}(r,\theta,z) = J_n\left(\frac{\xi_{nk}r}{a}\right)(A_{nk} \cos n\theta + B_{nk} \sin n\theta) \sinh \frac{\xi_{nk}(h - z)}{a}$$

$$n = 0, 1, 2, \ldots; k = 1, 2, \ldots$$

Each u_{nk} satisfies the boundary conditions on the top and sides. It remains to satisfy the condition on the base. For this purpose consider a superposition

$$\varphi(r,\theta,z) = \sum_{n=0}^{\infty} \sum_{k=1}^{\infty} J_n\left(\frac{\xi_{nk}r}{a}\right)(A_{nk} \cos n\theta + B_{nk} \sin n\theta) \sinh \frac{\xi_{nk}(h - z)}{a} \tag{3-46}$$

of the separable solutions. On the base $z = 0$ it is desired that

$$\sum_{n=0}^{\infty} \sum_{k=1}^{\infty} J_n\left(\frac{\xi_{nk}r}{a}\right)(A_{nk} \cos n\theta + B_{nk} \sin n\theta) \sinh \frac{\xi_{nk}h}{a} = f(r,\theta) \qquad 0 \le r \le z; 0 \le \theta \le 2\pi$$

As shown in Sec. 4 of Appendix 2, for each fixed n the functions $\sqrt{r}\, J_n(\xi_{nk}r/a)$, $k = 1, 2, \ldots$, form an orthogonal sequence on the interval $0 \le r \le a$. Hence, by the orthogonality of the trigonometric functions on $0 \le \theta \le 2\pi$, the functions

$$\varphi_{nk}^{(e)}(r,\theta) = J_n\left(\frac{\xi_{nk}r}{a}\right)\cos n\theta \qquad \varphi_{nk}^{(0)}(r,\theta) = J_n\left(\frac{\xi_{nk}r}{a}\right)\sin n\theta$$

possess the following orthogonality properties:

$$\int_0^a \int_0^{2\pi} \varphi_{nk}^{(e)}\varphi_{ml}^{(e)}r\, dr\, d\theta = 0 \qquad (m,l) \ne (n,k)$$

and a similar equation involving $\varphi_{nk}^{(0)}$, $\varphi_{ml}^{(0)}$; also

$$\int_0^a \int_0^{2\pi} \varphi_{nk}^{(e)}\varphi_{ml}^{(0)}r\, dr\, d\theta = 0$$

for all pairs (n,k), (m,l) in the admissible ranges of these indices. It remains to determine the coefficients A_{nk}, B_{nk}, such that

$$\sum_{n=0}^{\infty} \sum_{k=1}^{\infty} \left[A_{nk} \sinh\left(\frac{\xi_{nk}h}{a}\right) \varphi_{nk}^{(e)} + B_{nk} \sinh\left(\frac{\xi_{nk}h}{a}\right) \varphi_{nk}^{(0)} \right] = f(r,\theta) \qquad (3\text{-}47)$$

Let (p,q) be a fixed, but otherwise arbitrary, pair of positive integers. Multiply both sides of Eq. (3-47) by $r\varphi_{pq}^{(e)}$, integrate over $0 \leq r \leq a$, $0 \leq \theta \leq 2\pi$, and then interchange the order of summation and integration. The result is

$$\sum_{n=0}^{\infty} \sum_{k=1}^{\infty} \left(A_{nk} \sinh\frac{\xi_{nk}h}{a} \int_0^a \int_0^{2\pi} \varphi_{nk}^{(e)} \varphi_{pq}^{(e)} r \, dr \, d\theta \right.$$

$$\left. + B_{nk} \sinh\frac{\xi_{nk}h}{a} \int_0^a \int_0^{2\pi} \varphi_{nk}^{(0)} \varphi_{pq}^{(e)} r \, dr \, d\theta \right) = \int_0^a \int_0^{2\pi} f(r,\theta)\varphi_{pq}^{(e)}(r,\theta) r \, dr \, d\theta$$

By the orthogonality properties stated above the foregoing equation reduces to

$$A_{pq} \sinh\frac{\xi_{pq}h}{a} \int_0^a \int_0^{2\pi} |\varphi_{pq}^{(e)}|^2 \, r \, dr \, d\theta = \int_0^a \int_0^{2\pi} f(r,\theta)\varphi_{pq}^{(e)}(r,\theta) r \, dr \, d\theta$$

This determines the coefficient A_{pq}. Now from Eq. (97) in Appendix 2 and the properties of $\cos p\theta$ it follows that

$$\int_0^a \int_0^{2\pi} |\varphi_{pq}^{(e)}|^2 \, r \, dr \, d\theta = \frac{a^2}{2} [J_{p+1}(\xi_{pq})]^2 \pi \qquad p = 1, 2, \ldots\,; q = 1, 2, \ldots$$

$$\int_0^a \int_0^{2\pi} |\varphi_{0q}|^2 \, r \, dr \, d\theta = \frac{a^2}{2} [J_1(\xi_{0q})]^2 2\pi \qquad q = 1, 2, \ldots$$

Hence

$$A_{pq} = \frac{2}{\pi a^2 \sinh (\xi_{pq}h/a)[J_{p+1}(\xi_{pq})]^2} \int_0^a \int_0^{2\pi} f(r,\theta)\varphi_{pq}^{(e)}(r,\theta) r \, dr \, d\theta$$

$$p = 1, 2, \ldots\,; q = 1, 2, \ldots \qquad (3\text{-}48)$$

and

$$A_{0q} = \frac{1}{\pi a^2 \sinh (\xi_{0q}h/a)[J_1(\xi_{0q})]^2} \int_0^a \int_0^{2\pi} f(r,\theta)\varphi_{0q}^{(e)}(r,\theta) r \, dr \, d\theta \qquad q = 1, 2, \ldots \qquad (3\text{-}49)$$

In the same way,

$$B_{pq} = \frac{2}{\pi a^2 \sinh (\xi_{pq}h/a)[J_{p+1}(\xi_{pq})]^2} \int_0^a \int_0^{2\pi} f(r,\theta)\varphi_{pq}^{(0)}(r,\theta) r \, dr \, d\theta$$

$$p = 1, 2, \ldots\,; q = 1, 2, \ldots \qquad (3\text{-}50)$$

With the coefficients determined in this way, the solution is given by the series (3-46).

In the particular case where $f(r,\theta) = V_0(1 - r^2/a^2)$, the potential is symmetric about the z axis. This is shown directly by the fact that Eq. (3-48) implies $A_{pq} = 0$, and from Eq. (3-50) it follows that $B_{pq} = 0$, for $p = 1, 2, \ldots\,; q = 1, 2, \ldots$. Accordingly the potential is given by

$$\varphi(r,z) = \sum_{k=1}^{\infty} A_k J_0\left(\frac{\xi_k r}{a}\right) \sinh\frac{\xi_k(h-z)}{a} \qquad (3\text{-}51)$$

where $\{\xi_k\}$ is the sequence of positive roots of $J_0(\xi)$, and

$$A_k = \frac{2V_0}{a^2 \sinh\,(\xi_k h/a)[J_1(\xi_k)]^2} \int_0^a \frac{1-r^2}{a^2} J_0\left(\frac{\xi_k r}{a}\right)\, r\, dr \qquad k = 1, 2, \ldots$$

To evaluate the integral, let $x = \xi_k r/a$ and integrate by parts twice.   Thus

$$\int_0^a \frac{1-r^2}{a^2} r J_0\left(\frac{\xi_k r}{a}\right) dr = \frac{a^2}{\xi_k^2}\int_0^{\xi_k} \frac{1-x^2}{\xi_k^2} x J_0(x)\, dx$$

$$= \frac{a^2}{\xi_k^2}\left[\frac{1-x^2}{\xi_k^2} x J_1(x)\,\Big|_0^{\xi_k} + \frac{2}{\xi_k^2}\int_0^{\xi_k} x^2 J_1(x)\, dx\right]$$

$$= \frac{2a^2}{\xi_k^4}\,[x^2 J_2(x)]\,\Big|_0^{\xi_k} = \frac{2a^2 J_2(\xi_k)}{\xi_k^2}$$

By the recursion formula given in Sec. 4 of Appendix 2, it follows that

$$J_2(\xi_n) = \frac{2J_1(\xi_n)}{\xi_n} - J_0(\xi_n) = \frac{2J_1(\xi_n)}{\xi_n}$$

Hence

$$A_k = \frac{8V_0}{\xi_k^3 \sinh\,(\xi_k h/a) J_1(\xi_k)} \qquad k = 1, 2, \ldots$$

and the potential inside the cylinder is given by

$$\varphi(r,z) = 8V_0 \sum_{k=1}^{\infty} \frac{J_0(\xi_k r/a)\,\sinh\,[\xi_k(h-z)/a]}{\xi_k^3 J_1(\xi_k)\,\sinh\,(\xi_k h/a)}$$

3-4   SPHERICAL HARMONICS

Spherical coordinates r, θ, φ are related to rectangular coordinates by the equations

$$x = r \sin\theta \cos\varphi \qquad y = r \sin\theta \sin\varphi \qquad z = r \cos\theta \tag{3-52}$$

(see Fig. 3-3).   In spherical coordinates Laplace's equation has the form

$$\Delta u = \frac{\partial^2 u}{\partial r^2} + \frac{2}{r}\frac{\partial u}{\partial r} + \frac{1}{r^2}\frac{\partial^2 u}{\partial \theta^2} + \frac{\cot\theta}{r^2}\frac{\partial u}{\partial \theta} + \frac{1}{r^2 \sin^2\theta}\frac{\partial^2 u}{\partial \varphi^2} = 0 \tag{3-53}$$

Equation (3-53) can be written

$$\frac{\partial}{\partial r}\left(r^2 \frac{\partial u}{\partial r}\right) + \frac{1}{\sin\theta}\frac{\partial}{\partial \theta}\left(\sin\theta \frac{\partial u}{\partial \theta}\right) + \frac{1}{\sin^2\theta}\frac{\partial^2 u}{\partial \varphi^2} = 0 \tag{3-54}$$

To separate the radial dependence from the angular coordinates θ, φ assume a solution of the form

$$u = R(r)\,Y(\theta,\varphi) \tag{3-55}$$

After substitution of u into Eq. (3-54) it follows by the usual argument that the functions R, Y must satisfy the equations

$$(r^2 R')' - \mu R = 0 \tag{3-56}$$

$$\frac{1}{\sin \theta} \left[\frac{\partial}{\partial \theta} \left(\sin \theta \, \frac{\partial Y}{\partial \theta} \right) + \frac{1}{\sin \theta} \frac{\partial^2 Y}{\partial \varphi^2} \right] + \mu Y = 0 \tag{3-57}$$

respectively, where μ is a separation constant. Equation (3-56) is of the Euler type. If a solution of the form $R = r^p$ is assumed, then p must satisfy $p^2 + p - \mu = 0$. One root of the quadratic is $\alpha = (-1 + \sqrt{1 + 4\mu})/2$, and the other is $-(1 + \alpha)$. Accordingly, the general solution of Eq. (3-56) is

$$R = b_1 r^\alpha + b_2 r^{-(1+\alpha)} \tag{3-58}$$

It is convenient to introduce the operator

$$\Delta^* = \frac{1}{\sin \theta} \left[\frac{\partial}{\partial \theta} \left(\sin \theta \, \frac{\partial}{\partial \theta} \right) + \frac{1}{\sin \theta} \frac{\partial^2}{\partial \varphi^2} \right] \tag{3-59}$$

called the *surface laplacian*. Since $\mu = \alpha(\alpha + 1)$, it follows that Eq. (3-57) can be rewritten as

$$\Delta^* Y + \alpha(\alpha + 1) Y = 0 \tag{3-60}$$

In order to separate the variables θ, φ in Eq. (3-60) assume $Y = \Theta(\theta)\Phi(\varphi)$. The resulting equation can be put into the form

$$\frac{\sin \theta}{\Theta} \frac{d}{d\theta} \left(\sin \theta \, \frac{d\Theta}{d\theta} \right) + \alpha(\alpha + 1) \sin^2 \theta = - \frac{\Phi''}{\Phi}$$

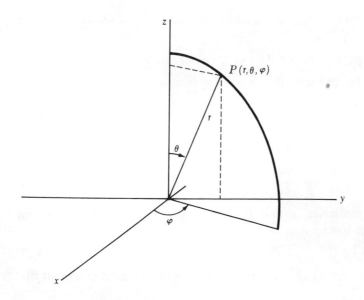

Figure 3-3

Again, both sides must be identically equal to a constant. Let ν^2 denote the separation constant. Then Θ, Φ satisfy the ordinary differential equations

$$\sin\theta\,\frac{d}{d\theta}\left(\sin\theta\,\frac{d\Theta}{d\theta}\right) + [\alpha(\alpha+1)\sin^2\theta - \nu^2]\Theta = 0 \tag{3-61}$$

$$\Phi'' + \nu^2\Phi = 0 \tag{3-62}$$

respectively. The general solution of Eq. (3-62) is

$$\Phi = c_1\cos\nu\varphi + c_2\sin\nu\varphi \tag{3-63}$$

provided $\nu \neq 0$. If $\nu = 0$, the general solution is

$$\Phi = c_1 + c_2\varphi$$

In Eq. (3-61) make the change of independent variable $\xi = \cos\theta$. Since $d/d\theta = -\sin\theta\,d/d\xi$, this equation becomes

$$\frac{d}{d\xi}\left[(1-\xi^2)\frac{dy}{d\xi}\right] + \left[\alpha(\alpha+1) - \frac{\nu^2}{1-\xi^2}\right]y = 0 \tag{3-64}$$

where y replaces Θ. Equation (3-64) is Legendre's associated equation. At this point the separation constants α, μ, ν are arbitrary (real or complex) constants. With α, ν unrestricted, the two independent solutions of Eq. (3-64) are the Legendre associated functions $P_\alpha^\nu(\xi)$ and $Q_\alpha^\nu(\xi)$. These functions are expressible in terms of hypergeometric functions. Thus separable forms of solution for Laplace's equation in spherical coordinates are

$$u(r,\theta,\varphi) = (b_1 r^\alpha + b_1 r^{-(1+\alpha)})(c_1\cos\nu\varphi + c_2\sin\nu\varphi)$$

$$[A_1 P_\alpha^\nu(\cos\theta) + A_2 Q_\alpha^\nu(\cos\theta)]$$

where α, ν are arbitrary constants, $\nu \neq 0$. If $\nu = 0$, the trigonometric expression in φ is replaced by $c_1 + c_2\varphi$.

Surface Harmonics

In many problems separable solutions of Laplace's equation which are single-valued and continuous over a sphere $r = a$ are desired. In a problem of this type a separable solution $Y(\theta,\varphi) = \Theta(\theta)\Phi(\varphi)$ of Eq. (3-60) is required such that Y is continuous for $0 \leq \theta \leq \pi$, $0 \leq \varphi \leq 2\pi$, and periodic (of period 2π) in φ. As in the case of polar coordinates in the plane, the periodicity requirement implies that the separation constant $\nu = \nu_m = m$, $m = 0, 1, \ldots$. If $m = 0$ in Eq. (3-62), the corresponding solution is $\Phi = \text{constant}$. The requirement that Y be continuous at the poles $\theta = 0$, $\theta = \pi$, that is, continuity from the right and left respectively, implies that the solution of Legendre's associated equation (3-64) must be continuous at $\xi = \pm 1$. From Sec. 3 of Appendix 2 it follows that the separation constant $\alpha = n$, $n = 0$,

1, ... Now if $\alpha = n$, $\nu = m$, where n, m are nonnegative integers, in Eq. (3-64), the corresponding solution continuous for $-1 \leq \xi \leq 1$ is the associated Legendre function $P_n{}^m(\xi)$. Accordingly a function of the form

$$Y_{nm}(\theta,\varphi) = P_n{}^m(\cos\theta)(A_{nm} \cos m\varphi + B_{nm} \sin m\varphi) \qquad (3\text{-}65)$$

is a single-valued and continuous solution of Eq. (3-60) with α replaced by n.

The associated Legendre function $P_n{}^m$ is related to the Legendre polynomial P_n by the equation

$$P_n{}^m(\xi) = (1 - \xi^2)^{m/2} \frac{d^m P_n(\xi)}{d\xi^m} \qquad (3\text{-}66)$$

Thus $P_n{}^m(\xi)$ is identically zero if $m > n$. Hence in Eq. (3-65) it can be assumed that m is one of the integers $0, 1, \ldots, n$. By the linearity of the differential equation a function of the form

$$Y_n(\theta,\varphi) = A_{n0} P_n(\cos\theta) + \sum_{m=1}^{n} P_n{}^m(\cos\theta)(A_{nm} \cos m\varphi + B_{nm} \sin m\varphi) \qquad (3\text{-}67)$$

where A_{n0}, A_{ni}, B_{ni}, $i = 1, \ldots, n$ are constants, constitutes a solution of Eq. (3-60). Moreover Y_n is single-valued and continuous on the sphere $r = a$. Such a function is called a *surface harmonic* of degree n. Particular surface harmonics of degree n are

$$Y_{nm}^{(e)}(\theta,\varphi) = P_n{}^m(\cos\theta) \cos m\varphi \qquad Y_{nm}^{(0)}(\theta,\varphi) = P_n{}^m(\cos\theta) \sin m\varphi \qquad (3\text{-}68)$$

$m = 0, 1, \ldots, n$. For $m \geq 1$ these are even and odd functions, respectively, of φ. If $m = 0$, then $Y_{n0} = P_n$. If $m < n$, the surface harmonics $Y_{nm}^{(e)}$, $Y_{nm}^{(0)}$ are called *tesseral harmonics*, while if $m = n$, they are called *sectoral harmonics*. The Legendre polynomial is often termed a *zonal harmonic*. Along each great circle (meridian) $\varphi = (2k - 1)\pi/2m$, $k = 1, 2, \ldots, m$, on the sphere $r = a$ the surface harmonic $Y_{nm}^{(e)}$ vanishes, while along each great circle $\varphi = k\pi/m$, $k = 0, 1, \ldots, m - 1$, the surface harmonic $Y_{nm}^{(0)}$ vanishes. Since $P_n{}^m(\xi)$ has exactly $n - m$ zeros in the open interval $-1 < \xi < 1$, it follows that $P_n{}^m(\cos\theta)$ has exactly $n - m$ zeros in the interval $0 < \theta < \pi$. If $\theta_1, \ldots, \theta_{n-m}$ are these zeros, then along each circle of latitude $\theta = \theta_i$ the surface harmonics $Y_{nm}^{(e)}$, $Y_{nm}^{(0)}$ vanish. If $m < n$, the curves (meridians and circles of latitude) on which the harmonics $Y_{nm}^{(e)}$, $Y_{nm}^{(0)}$ vanish subdivide the surface of the sphere into regions called tesserae, and within each such region the tesseral harmonic is of one sign. If $m = n$, then from Eq. (3-66) it follows that $P_n{}^n(\cos\theta) = c \sin^n \theta$, c a constant. Hence the curves on which the sectoral harmonics vanish subdivide the sphere into sectors. Within each sector the sectoral harmonic is of one sign. The curves on which the zonal harmonic Y_{n0} vanish are the circles of latitude $\theta = \theta_i$, $i = 1, \ldots, n$, and these subdivide the sphere into zones, within which Y_n is of one sign.

The surface harmonics $Y_{nm}(\theta,\varphi)$ have many interesting and useful properties, of which only a few can be discussed in this text. One important property is orthogonality on a sphere S of radius $r = a$. Let (k,l) and (p,q) be distinct pairs of nonnegative integers, $l \leq k$ and $q \leq p$. Then

$$\iint_S Y_{kl} Y_{pq} \, dS = a^2 \int_0^{2\pi} \int_0^{2\pi} Y_{kl} Y_{pq} \sin \theta \, d\theta \, d\varphi = 0 \tag{3-69}$$

Recall Green's formula (2-77). In terms of the laplacian operator Green's formula states that if u, v are twice continuously differentiable functions on $\bar{\mathcal{V}} = \mathcal{V} + S$, where $\mathcal{V}$ is the interior of the sphere, then

$$\iiint_{\mathcal{V}} (v \, \Delta u - u \, \Delta v) \, d\tau = \iint_S \left(v \, \frac{\partial u}{\partial n} - u \, \frac{\partial v}{\partial n} \right) dS \tag{3-70}$$

Hence, if u, v are harmonic on $\mathcal{V}$, then

$$\iint_S v \, \frac{\partial u}{\partial n} \, dS = \iint_S u \, \frac{\partial v}{\partial n} \, dS$$

Now to establish relation (3-69) choose $u = u_{kl} = r^k Y_{kl}$, $v = u_{pq} = r^p Y_{pq}$. On S

$$\frac{\partial u}{\partial n} = \frac{\partial u_{kl}}{\partial r} \bigg|_{r=a} = ka^{k-1} Y_{kl} \qquad dS = a^2 \sin \theta \, d\theta \, d\varphi$$

and similarly for $\partial v / \partial n$. Thus

$$(p - k)a^{k+p+1} \int_0^{2\pi} \int_0^\pi Y_{kl} Y_{pq} \sin \theta \, d\theta \, d\varphi = 0$$

If $p \neq k$, then (3-69) follows immediately. If $p = k$, the orthogonality properties of the trigonometric functions $\cos l\varphi$, $\sin l\varphi$, $\cos q\varphi$, $\sin q\varphi$ on $[0,2\pi]$ imply (3-69). Note that

$$\iint_S Y_{kl}^{(e)} Y_{pq}^{(0)} \, dS = 0 \qquad \text{all } (k,l), (p,q), \, l \leq k; q \leq p$$

Also there are the useful relations

$$\int_0^{2\pi} \int_0^\pi (Y_{kl}^{(i)})^2 \sin \theta \, d\theta \, d\varphi = \frac{2\pi}{2k+1} \frac{(k+l)!}{(k-l)!} \qquad k = 1, 2, \ldots ; l = 1, 2, \ldots, k \tag{3-71}$$

where the superscript i means either e or 0, and

$$\int_0^{2\pi} \int_0^\pi (Y_{k0}^{(e)})^2 \sin \theta \, d\theta \, d\varphi = \frac{4\pi}{2k+1} \qquad k = 1, 2, \ldots \tag{3-72}$$

The orthogonality properties imply that for each nonnegative integer n there are exactly $2n + 1$ linearly independent surface harmonics of degree n. The surface harmonics are the eigenfunctions of the problem

$$\Delta^* Y + \lambda Y = 0 \qquad \text{on } S$$

the prescribed condition being that Y be continuous on S and periodic, of period 2π, in φ. The eigenvalues are $\lambda_n = n(n + 1), n = 0, 1, \ldots$, and corresponding to the nth eigenvalue λ_n there are exactly $2n + 1$ independent eigenfunctions Y_{nm}. It can be shown that the set of surface harmonics is complete in the space of all functions which are continuous and single-valued on the sphere S (see Ref. 3).

Spherical Harmonics

Separable solutions of Laplace's equation in spherical coordinates which are continuous for $r > 0$ and all θ, φ, and periodic in φ, are of the form

$$u_n(r,\theta,\varphi) = (C_n r^n + D_n r^{-(n+1)}) Y_n(\theta,\varphi) \tag{3-73}$$

where Y_n is a surface harmonic of degree n.   A function of the form

$$u_n(r,\theta,\varphi) = r^n Y_n(\theta,\varphi) \tag{3-74}$$

is called a *spherical harmonic* of degree n. These are continuous solutions for $r \geq 0$. If a solution of Laplace's equation in spherical coordinates is desired in the unbounded region exterior to the sphere $r = a$, such that

$$\lim_{r \to \infty} u = 0 \qquad \text{uniformly}$$

then solutions of the form

$$u_n(r,\theta,\varphi) = r^{-(n+1)} Y_n(\theta,\varphi) \qquad n = 0, 1, \ldots \tag{3-75}$$

are useful. The following example illustrates the application of spherical harmonics to a potential problem.

EXAMPLE 3-3 The sphere S has the equation $r = a$ in spherical coordinates. The potential ψ is prescribed on S:

$$\psi(a,\theta,\varphi) = f(\theta,\varphi) \qquad 0 \leq \theta \leq \pi; 0 \leq \varphi \leq 2\pi$$

Here f is a given continuous function. Find the potential inside S. Consider the special case where $f(\theta,\varphi) = 100 \cos^2 \theta$.

The potential ψ is a harmonic function inside S. One starts with the spherical harmonics defined by Eq. (3-74). These are single-valued continuous solutions of Laplace's equation inside and on S. A superposition of spherical harmonics

$$\psi(r,\theta,\varphi) = \sum_{n=0}^{\infty} r^n Y_n(\theta,\varphi) = \sum_{n=0}^{\infty} \sum_{m=0}^{n} r^n [A_{nm} Y_{nm}^{(e)}(\theta,\varphi) + B_{nm} Y_{nm}^{(0)}(\theta,\varphi)] \tag{3-76}$$

will also be a single-valued continuous solution provided the series is suitably convergent.

In order to satisfy the prescribed condition on S it is necessary that

$$A_{00} + \sum_{n=1}^{\infty} a^n \left\{ A_{n0}\, Y_{n0}^{(e)}(\theta,\varphi) + \sum_{m=1}^{n} [A_{nm} Y_{nm}^{(e)}(\theta,\varphi) + B_{nm} Y_{nm}^{(0)}(\theta,\varphi)] \right\}$$

$$= f(\theta,\varphi) \qquad 0 \le \theta \le \pi; 0 \le \varphi \le 2\pi \quad (3\text{-}77)$$

Let (p,q) be a fixed pair of integers, with $p \ge 1$ and q one of the integers $1, 2, \ldots, p$. Multiply both sides of Eq. (3-77) by $Y_{pq}^{(e)}$, integrate over S, and interchange the order of integration and summation. By the orthogonality properties stated in the text, the resulting equation reduces to

$$a^p A_{pq} \int_0^{2\pi} \int_0^{\pi} (Y_{pq}^{(e)})^2 \sin\theta \; d\theta \; d\varphi = \int_0^{2\pi} \int_0^{\pi} f(\theta,\varphi) Y_{pq}^{(e)}(\theta,\varphi) \sin\theta \; d\theta \; d\varphi$$

Hence by Eq. (3-71)

$$A_{pq} = \frac{2p+1}{2\pi a^p} \frac{(p-q)!}{(p+q)!} \int_0^{2\pi} \int_0^{\pi} f(\theta,\varphi) Y_{pq}^{(e)}(\theta,\varphi) \sin\theta \; d\theta \; d\varphi$$

$$p = 1, 2, \ldots; q = 1, 2, \ldots, p \quad (3\text{-}78)$$

In the same way

$$B_{pq} = \frac{2p+1}{2\pi a^p} \frac{(p-q)!}{(p+q)!} \int_0^{2\pi} \int_0^{\pi} f(\theta,\varphi) Y_{pq}^{(0)}(\theta,\varphi) \sin\theta \; d\theta \; d\varphi$$

$$p = 1, 2, \ldots; q = 1, 2, \ldots, p \quad (3\text{-}79)$$

and

$$A_{p0} = \frac{2p+1}{4\pi a^p} \int_0^{2\pi} \int_0^{\pi} f(\theta,\varphi) P_p(\cos\theta) \sin\theta \; d\theta \; d\varphi \qquad p = 0, 1, \ldots \quad (3\text{-}80)$$

A potential function is called *axially symmetric* if it is independent of the coordinate φ. From Eqs. (3-69), (3-78), and (3-79) it follows that if the prescribed function f is independent of φ, then

$$A_{pq} = B_{pq} = 0 \qquad p = 1, 2, \ldots; q = 1, 2, \ldots, p$$

By Eq. (3-76) an axially symmetric potential has the series representation

$$\psi(r,\theta) = \sum_{n=0}^{\infty} r^n Y_{n0} = \sum_{n=0}^{\infty} r^n A_n P_n(\cos\theta) \quad (3\text{-}81)$$

where [from Eq. (3-80)] the coefficient

$$A_n = \frac{2n+1}{2a^n} \int_0^{\pi} f(\theta) P_n(\cos\theta) \sin\theta \; d\theta \quad (3\text{-}82)$$

Thus if the prescribed function f is independent of φ, the potential is axially symmetric. In the particular case where $f(\theta) = 100\cos^2\theta$, the coefficient

$$A_n = \frac{50(2n+1)}{a^n} \int_0^{\pi} \cos^2\theta \; P_n(\cos\theta) \sin\theta \; d\theta$$

Now $\cos^2\theta = \frac{2}{3}P_2(\cos\theta) + \frac{1}{3}P_0(\cos\theta)$. By the orthogonality properties of the Legendre polynomials (Sec. 3 of Appendix 2), it follows that $A_n = 0$ if $n \ne 0$ and $n \ne 2$. Also

$A_0 = \frac{100}{3}$, and $A_2 = 200/3a^2$. Accordingly the potential

$$\psi(r,\theta) = \frac{100}{3} + \frac{200}{3a^2} r^2 P_2(\cos\theta) \tag{3-83}$$

The mathematics of this example also describes the steady temperature u in a homogeneous thermally conducting solid bounded by the sphere $r = a$, provided no heat sources are present within the sphere and the surface temperature $u(a,\theta,\varphi)$ is held at $f(\theta,\varphi)$. In the particular case discussed above, the surface temperature is held at $100 \cos^2\theta$ degrees for $0 \le \theta \le \pi$, and the steady temperature u in the sphere is given by the right-hand member of Eq. (3-83). Note that the temperature at the center is $u(0,0) = \frac{100}{3}$, and this is the average value of u over the surface of the sphere.

EXAMPLE 3-4 Find the potential at points exterior to the sphere S of Example 3-3. Consider the special case where the prescribed function $f(\theta,\varphi) = V_0$, V_0 a constant.

In the region exterior to S the potential must satisfy $\lim_{r \to \infty} \psi = 0$ uniformly in θ, φ (recall the discussion of properties of the potential in Sec. 3-2). A superposition

$$\psi(r,\theta,\varphi) = \sum_{n=0}^{\infty} r^{-(n+1)} \sum_{m=0}^{n} [C_{nm} Y_{nm}^{(e)}(\theta,\varphi) + D_{nm} Y_{nm}^{(0)}(\theta,\varphi)]$$

defines a harmonic function in the region exterior to S and vanishes at infinity in the manner required, provided the series is suitably convergent. The coefficients C_{nm}, D_{nm} in the series are determined in the same manner as demonstrated in Example 3-3. In fact the expressions for C_{pq}, D_{pq} are given by the right-hand members of Eqs. (3-78) to (3-80), except that the factor a^p appears in the numerator rather than the denominator. When f is independent of the angular coordinate φ, the potential ψ is axially symmetric and is of the form

$$\psi(r,\theta) = \sum_{n=0}^{\infty} A_n r^{-(n+1)} P_n(\cos\theta)$$

where

$$A_n = \frac{(2n + 1)a^n}{2} \int_0^{\pi} f(\theta) P_n(\cos\theta) \sin\theta \, d\theta$$

In the special case $f(\theta) = V_0$, where V_0 is a constant, the coefficients $A_n = 0$, $n > 0$, and the potential is given by

$$\psi(r) = \frac{aV_0}{r} \qquad r \ge a$$

Also, in this special case, the potential inside S is

$$\psi(r) = V_0 \qquad 0 \le r \le a$$

The potential is continuous across the sphere S. The normal derivative of ψ is discontinuous across S, however, and the magnitude of the jump discontinuity in the normal derivative is

$$\frac{\partial \psi}{\partial r}\bigg|_{a+} - \frac{\partial \psi}{\partial r}\bigg|_{a-} = -4\pi\omega$$

where ω is the density of the surface-charge distribution on S. Since

$$\frac{\partial \psi}{\partial r}\Big|_{a+} = -\frac{V_0}{a} \qquad \frac{\partial \psi}{\partial r}\Big|_{a-} = 0$$

it follows that $\omega = V_0/4\pi a$, a constant. The total charge on S is $Q = 4\pi a^2\omega = aV_0$. Thus the potential can be rewritten $\psi(r) = Q/r$ for $r \geq a$. The result shows that the potential at external points is the same as the potential due to a point charge of strength Q located at the origin.

3-5 SELF-ADJOINT ELLIPTIC-TYPE BOUNDARY-VALUE PROBLEMS

A number of important boundary-value problems in two dimensions involve a linear operator of the form

$$L = \frac{\partial}{\partial x}\left(p_1 \frac{\partial}{\partial x}\right) + \frac{\partial}{\partial y}\left(p_2 \frac{\partial}{\partial y}\right) + q \tag{3-84}$$

where p_1, p_2, q are real-valued functions of the independent variables x, y. For example, if $p_1(x,y) = p_2(x,y) = 1$, $q(x,y) = 0$, then $L = \Delta$, the two-dimensional laplacian operator. An operator of the form in Eq. (3-84) is self-adjoint (recall Sec. 2-7). If

$$p_1(x,y) > 0 \qquad p_2(x,y) > 0 \tag{3-85}$$

then in addition the operator is elliptic. Also of frequent occurrence in three-dimensional problems of mathematical physics are linear self-adjoint elliptic operators of the form

$$L = \frac{\partial}{\partial x}\left(p_1 \frac{\partial}{\partial x}\right) + \frac{\partial}{\partial y}\left(p_2 \frac{\partial}{\partial y}\right) + \frac{\partial}{\partial z}\left(p_3 \frac{\partial}{\partial z}\right) + q \tag{3-86}$$

where

$$p_1(x,y,z) > 0 \qquad p_2(x,y,z) > 0 \qquad p_3(x,y,z) > 0$$

The operators in Eqs. (3-84) and (3-86) are generalizations to two and three dimensions, respectively, of the one-dimensional linear self-adjoint operator in Eq. (2) of Appendix 2. Thus it is not unexpected that many of the theorems on boundary-value problems for ordinary differential equations can be extended to include these elliptic operators. For convenience the theorems are stated in terms of the two-dimensional operator (3-84). However, the corresponding statements for the three-dimensional operator (3-85) are made obvious by the procedure; indeed they can be easily generalized to n independent variables.

Let C be a simple closed piecewise smooth curve which bounds a region $\mathscr{R}$ of the xy plane, and let $\bar{\mathscr{R}}$ denote $\mathscr{R}$ together with its boundary C. Henceforth in this section it is assumed that L is a linear operator of the form in

Eq. (3-84), where the coefficients p_1, p_2 are real-valued continuously differenti-able and such that the inequalities (3-85) hold for all points in $\bar{\mathscr{R}}$.   In addition q is a real-valued continuous function on $\bar{\mathscr{R}}$.   Let $B(u)$ be a linear form in u and the first derivatives u_x, u_y with real coefficients.   Examples are

$$B(u) = xu - yu_x \qquad B(u) = 4u_x - 5u_y + e^y u$$

The boundary-value problem

$$Lu = f \quad \text{in } \mathscr{R} \qquad\qquad B(u) = g \qquad \text{on } C \tag{3-87}$$

is said to be of *elliptic type* since L is an elliptic operator.   Here f is a given real-valued continuous function on $\mathscr{R}$, and g is a given real-valued function which is continuous on the boundary curve C.   Types of boundary conditions which arise often in the applications are

$$B(u) = u = g \qquad\qquad \text{Dirichlet} \tag{3-88}$$

$$B(u) = \frac{\partial u}{\partial n} = g \qquad\qquad \text{Neumann} \tag{3-89}$$

$$B(u) = a\frac{\partial u}{\partial n} + bu = g \qquad \text{mixed} \tag{3-90}$$

In Eqs. (3-89) and (3-90) $\partial u/\partial n$ denotes the derivative in the direction of the exterior normal **n** to C.   It is assumed that $a^2 + b^2 \neq 0$ in Eq. (3-90).

Associated with problem (3-87) is the *homogeneous problem*

$$Lu = 0 \quad \text{in } \mathscr{R} \qquad\qquad B(u) = 0 \qquad \text{on } C \tag{3-91}$$

The linearity of the operator L and the form $B(u)$ imply that a linear com-bination of solutions of the homogeneous problem is also a solution of the homogeneous problem.   Also, if u is a nontrivial solution of the homogene-ous problem and v is a particular solution of the inhomogeneous problem (3-87), then $\varphi = v + cu$ is a solution of (3-87) for arbitrary choice of constant c.   Hence, in this case, there are infinitely many distinct solutions of the boundary-value problem.   On the other hand, it is true that if the only solution of the homogeneous problem is the trivial solution (the function which is identically zero on $\bar{\mathscr{R}}$), and if a solution of problem (3-87) exists, the solution is unique.   To show this let φ, ψ be solutions of problem (3-87), and let $w = \varphi - \psi$.   Then

$$Lw = L(\varphi - \psi) = L\varphi - L\psi = 0 \qquad\qquad \text{in } \mathscr{R}$$

$$B(w) = B(\varphi - \psi) = B(\varphi) - B(\psi) = 0 \qquad \text{on } C$$

and w is a solution of the homogeneous problem.   Thus w is identically zero, and $\varphi = \psi$ on $\bar{\mathscr{R}}$.

Uniqueness of Solution

It is reasonable to expect uniqueness of solution for the boundary-value problems of mathematical physics. Sufficient conditions for uniqueness of solution for problem (3-87) are stated in Theorem 3-8. First, however, Theorem 3-7 is proved. An identity which holds for arbitrary choice of twice continuously differentiable function u is

$$\iint\limits_{\mathscr{R}} uLu\, dA = -\iint\limits_{\mathscr{R}} (p_1 u_x{}^2 + p_2 u_y{}^2 - qu^2)\, dA$$

$$+ \int_C u[p_1 u_x \cos(n,x) + p_2 u_y \cos(n,y)]\, ds \quad (3\text{-}92)$$

where dA is the area element in $\mathscr{R}$, ds is the arc element on C, and $\cos(n,x)$, $\cos(n,y)$ are the direction cosines of the exterior normal along C. To derive Eq. (3-92) consider the identity

$$u\frac{\partial}{\partial x}\left(p_1 \frac{\partial u}{\partial x}\right) = \frac{\partial}{\partial x}\left(up_1 \frac{\partial u}{\partial x}\right) - p_1 \left(\frac{\partial u}{\partial x}\right)^2$$

and a similar identity with p_2 replacing p_1 and y replacing x. Then

$$\iint\limits_{\mathscr{R}} uLu\, dA = \iint\limits_{\mathscr{R}} \left[\frac{\partial}{\partial x}\left(up_1 \frac{\partial u}{\partial x}\right) + \frac{\partial}{\partial y}\left(up_2 \frac{\partial u}{\partial y}\right)\right] dA$$

$$- \iint\limits_{\mathscr{R}} \left[p_1 \left(\frac{\partial u}{\partial x}\right)^2 + p_2 \left(\frac{\partial u}{\partial y}\right)^2 - qu^2\right] dA$$

Now the first integral on the right side of the preceding equation is transformed into an integral around the boundary curve C by means of the divergence theorem. Equation (3-92) results immediately.

THEOREM 3-7   In problem (3-87) let $q(x,y) \leq 0$ on $\mathscr{R}$. If the linear form $B(u)$ is such that

$$\int_C u[p_1 u_x \cos(n,x) + p_2 u_y \cos(n,y)]\, ds \leq 0 \quad\quad\quad (3\text{-}93)$$

whenever u is a real-valued twice continuously differentiable function which satisfies the homogeneous boundary condition $B(u) = 0$ on C, then

$$\iint\limits_{\mathscr{R}} uLu\, dA \leq 0 \quad\quad\quad (3\text{-}94)$$

holds for every such function u. If q does not vanish identically on $\mathscr{R}$, equality holds in Eq. (3-94) if, and only if, u vanishes identically on $\bar{\mathscr{R}}$. If q is identically zero on $\mathscr{R}$, equality holds in Eq. (3-94) if, and only if, u is identically constant on $\bar{\mathscr{R}}$.

Proof   The inequality (3-94) follows immediately from Eq. (3-92), inequality (3-93), and the fact that

$$p_1 u_x{}^2 + p_2 u_y{}^2 - qu^2 \geq 0 \qquad (3\text{-}95)$$

on $\mathscr{R}$.   Suppose equality holds in Eq. (3-94).   Then each term on the right side of Eq. (3-92) must be zero, since each is nonpositive.   The fact that the first integral on the right in Eq. (3-92) has the value zero together with inequality (3-95) implies that the integrand of the first integral vanishes everywhere in $\mathscr{R}$.   Since p_1, p_2 are positive and $-q(x,y) \geq 0$, it follows that

$$u_x = u_y = 0 \qquad \text{in } \mathscr{R}$$

Hence $u = c$, a constant, in $\mathscr{R}$.   By continuity $u = c$ on $\bar{\mathscr{R}}$.   Now

$$0 = \iint\limits_{\mathscr{R}} qu^2 \, dA = c^2 \iint\limits_{\mathscr{R}} q \, dA$$

Suppose q does not vanish identically on $\mathscr{R}$.   Then the integral of q over $\mathscr{R}$ has a value different from zero.   Hence $c = 0$; that is, $u = 0$ everywhere in $\bar{\mathscr{R}}$.   If q is identically zero in $\mathscr{R}$, only $u = c$ can be concluded.

THEOREM 3-8   In problem (3-87) let $q(x,y) \leq 0$ on $\mathscr{R}$.   If the boundary condition is the Dirichlet condition (3-88), the only solution of the corresponding homogeneous problem (3-91) which has continuous second derivatives on $\mathscr{R}$ is the trivial solution.   If $p_1 = p_2$ and q is not identically zero on $\mathscr{R}$, the same statement holds for the Neumann condition (3-89).   If $p_1 = p_2$ and q is identically zero on $\mathscr{R}$, the only solution of the homogeneous problem with the Neumann boundary condition $\partial u / \partial n = 0$ is $u = \text{const.}$ If $p_1 = p_2$ and the boundary condition of problem (3-87) is the mixed condition (3-90) with $a \neq 0$ and $a/b > 0$, the only solution of the corresponding homogeneous problem (3-86) is the trivial solution.

Proof   If the boundary condition is the Dirichlet condition, the homogeneous boundary condition is $u = 0$ on C.   Hence the inequality in Eq. (3-93) becomes equality whenever u satisfies the homogeneous boundary condition. If, in addition, u satisfies $Lu = 0$ in $\mathscr{R}$, equality holds in Eq. (3-94).   The conclusion of Theorem 3-7 now asserts $u = c$ on $\bar{\mathscr{R}}$.   But $u = 0$ on C. Accordingly $u = 0$ everywhere in $\bar{\mathscr{R}}$.   Assume now that $p_1 = p_2 = p$ on $\mathscr{R}$. Then

$$p_1 u_x \cos (n,x) + p_2 u_y \cos (n,y) = p[u_x \cos (n,x) + u_y \cos (n,y)] = p \frac{\partial u}{\partial n}$$

If u satisfies $\partial u / \partial n = 0$ on C, equality holds in Eq. (3-93).   Suppose also $Lu = 0$ in $\mathscr{R}$.   Then (again by Theorem 3-7) $u = c$ on $\bar{\mathscr{R}}$.   Moreover if q does not vanish everywhere, then $u = 0$ on $\bar{\mathscr{R}}$.   The proof for the mixed boundary condition is made in the same manner.

Associated Eigenvalue Problem

Also of importance in connection with the boundary-value problem (3-87) is the *eigenvalue problem*

$$L\varphi + \lambda\rho\varphi = 0 \quad \text{in } \mathscr{R} \qquad B(\varphi) = 0 \quad \text{on } C \qquad (3\text{-}96)$$

Here λ is a parameter, and ρ is a given real-valued continuous function, $\rho(x,y) > 0$, on $\bar{\mathscr{R}}$. Problem (3-96) is a homogeneous problem, so that, in general, the only solution is the trivial solution. However for certain values of λ there may exist a nontrivial solution. An *eigenvalue* is a value λ such that there exists a nontrivial function φ_λ with continuous second derivatives which satisfies (3-96). The corresponding solution φ_λ is called an *eigenfunction* of the operator L with boundary condition $B(\varphi) = 0$. The terms *characteristic value* and *characteristic function* are synonymous with eigenvalue and eigenfunction, respectively. The multiplicity (or index) of an eigenvalue is the number of linearly independent eigenfunctions corresponding to that eigenvalue.

Green's formula (2-81) expressed in terms of the operator (3-86) is

$$\iint\limits_{\mathscr{R}} (vLu - uLv)\, dA$$

$$= \int_C [p_1(vu_x - uv_x) \cos(n,x) + p_2(vu_y - uv_y) \cos(n,y)]\, ds \qquad (3\text{-}97)$$

Equation (3-97) holds for every pair of real-valued twice continuously differentiable function u, v on $\bar{\mathscr{R}}$. Of particular interest are those boundary-value problems in which the boundary condition is such that the right side of Green's formula has the value zero if $B(u) = 0$, $B(v) = 0$. The boundary-value problem is called *self-adjoint* if

$$\int_C [p_1(vu_x - uv_x) \cos(n,x) + p_2(vu_y - uv_y) \cos(n,y)]\, ds = 0 \qquad (3\text{-}98)$$

holds for every pair of functions u, v which satisfy the homogeneous boundary conditions $B(u) = 0$, $B(v) = 0$, on C. Note that u and v are not required to satisfy the differential equation, only the homogeneous boundary condition on C. Now it follows from Green's formula (3-97) that if the problem is self-adjoint, then

$$\iint\limits_{\mathscr{R}} vLu\, dA = \iint\limits_{\mathscr{R}} uLv\, dA \qquad (3\text{-}99)$$

for every pair of real-valued twice continuously differentiable functions u, v which satisfy the homogeneous boundary conditions. Relation (3-99) expresses an important symmetry property of the operator L on the set of all real-valued twice continuously differentiable functions which satisfy the homogeneous boundary condition on C. Since the coefficients in L and $B(u)$ are real-valued, it is an immediate consequence of the foregoing that if the problem is self-adjoint, then

$$\iint_{\mathcal{R}} v\overline{Lu}\, dA = \iint_{\mathcal{R}} \bar{u}Lv\, dx \tag{3-100}$$

holds for every pair of (possibly complex-valued) twice continuously differentiable functions u, v which satisfy the homogeneous boundary conditions. In Eq. (3-100) the bars denote the complex conjugate.

THEOREM 3-9   If the boundary-value problem is self-adjoint, the eigenvalues of (3-96) are real whenever they exist.   If $q(x,y) \leq 0$ on $\mathcal{R}$ and the boundary condition is such that inequality (3-93) holds for every real-valued twice continuously differentiable function u which satisfies the homogeneous boundary condition, the eigenvalues are nonnegative.   If the boundary condition is the Dirichlet condition, the eigenvalues are positive.   The same is true for the case of the mixed condition provided $a \neq 0$, $a/b > 0$, and the coefficients $p_1 = p_2$ in L.   In the case of the Neumann condition the eigenvalues are positive provided q is not zero everywhere on $\mathcal{R}$ and $p_1 = p_2$.   If $p_1 = p_2$ and q is everywhere zero, then $\lambda = 0$ is the smallest eigenvalue, and $\varphi = 1$ is a corresponding eigenfunction.   In any case real-valued eigenfunctions of the problem have the property

$$\iint_{\mathcal{R}} \rho\varphi_\lambda\varphi_\mu\, dA = 0 \qquad \lambda \neq \mu \tag{3-101}$$

whenever φ_λ, φ_μ correspond to distinct eigenvalues λ, μ.

Proof   Suppose λ is an eigenvalue and φ is a corresponding eigenfunction. Then

$$L\varphi + \lambda\rho\varphi = 0 \qquad \text{in } \mathcal{R}$$

implies

$$L\bar{\varphi} + \bar{\lambda}\rho\bar{\varphi} = 0 \qquad \text{in } \mathcal{R}$$

and

$$B(\varphi) = 0 \qquad\qquad \text{on } C$$

implies

$$B(\bar{\varphi}) = 0 \qquad\qquad \text{on } C$$

since the coefficients involved are real-valued.   Set $u = v = \varphi$ in Eq. (3-100).

Then

$$(\lambda - \bar{\lambda}) \iint\limits_{\mathscr{R}} \rho \, |\varphi|^2 \, dA = 0$$

But

$$\iint\limits_{\mathscr{R}} \rho \, |\varphi|^2 \, dA > 0$$

Hence $\lambda = \bar{\lambda}$; that is, λ is a real number.   Note that if φ is a complex-valued eigenfunction corresponding to an eigenvalue λ and

$$\varphi = u + iv \qquad i = \sqrt{-1}; \ u, v \text{ real-valued on } \bar{\mathscr{R}}$$

then u and v are eigenfunctions corresponding to λ provided these are nontrivial functions.   This is seen from the relation $u = (\varphi + \bar{\varphi})/2$ and

$$Lu = L\left(\frac{\varphi + \bar{\varphi}}{2}\right) = \frac{L\varphi + L\bar{\varphi}}{2} = -\frac{\lambda\rho\varphi + \lambda\rho\bar{\varphi}}{2} = -\lambda\rho u$$

$$B(u) = B\left(\frac{\varphi + \bar{\varphi}}{2}\right) = \frac{B(\varphi) + B(\bar{\varphi})}{2} = 0$$

Similarly for v.   Thus if λ is an eigenvalue, there must always exist a real-valued eigenfunction of (3-96) corresponding to λ.   Now let λ, μ be distinct eigenvalues, and let φ_λ, φ_μ be real-valued eigenfunctions corresponding to λ, μ, respectively.   Set $u = \varphi_\lambda$, $v = \varphi_\mu$ in Eq. (3-99).   Then

$$(\lambda - \mu) \iint\limits_{\mathscr{R}} \rho\varphi_\lambda\varphi_\mu \, dA = 0$$

Since $\lambda \neq \mu$, the relation (3-101) follows.

To establish the nonnegativeness of the eigenvalues let λ be an eigenvalue and φ a corresponding real-valued eigenfunction.   Set $u = \varphi$ in Eq. (3-92). The result is

$$\lambda = \frac{\displaystyle\iint\limits_{\mathscr{R}} (p_1\varphi_x{}^2 + p_2\varphi_y{}^2 - q\varphi^2) \, dA - \int_C \varphi[p_1\varphi_x \cos(n,x) + p_2\varphi_y \cos(n,y)] \, ds}{\|\varphi\|^2}$$

$$\tag{3-102}$$

where

$$\|\varphi\|^2 = \iint\limits_{\mathscr{R}} \rho\varphi^2 \, dA \tag{3-103}$$

The right-hand member in Eq. (3-102) is termed the *Rayleigh quotient*.   It is clear that if the hypotheses stated in the theorem hold, the numerator, and so also the quotient, is nonnegative. Also, $\lambda = 0$ is an eigenvalue if, and only if, the homogeneous problem (3-91) has a nontrivial solution. But if the boundary condition is the Dirichlet condition, then (by Theorem 5-8) the only solution of the homogeneous problem is the trivial solution. Accordingly in this case all the eigenvalues are positive. The remaining conclusions of the theorem are proved in a similar manner.

The proof of the existence of eigenvalues of problem (3-96) is not given in this text (see Ref. 3). It can be shown that the eigenvalues constitute a real sequence $\{\lambda_n\}$ such that

$$\lambda_1 \leq \lambda_2 \leq \cdots \leq \lambda_n \leq \lambda_{n+1} \leq \cdots \qquad \lim_{n=\infty} \lambda_n = +\infty$$

Each eigenvalue has finite multiplicity and in the listing is repeated the same number of times as its multiplicity. If φ_n, φ_m are real-valued eigenfunctions corresponding to distinct eigenvalues λ_n, λ_m, then φ_n, φ_m are *orthogonal on $\mathscr{R}$ with weight ρ*:

$$\iint\limits_{\mathscr{R}} \rho\varphi_n\varphi_m \, dA = 0 \qquad\qquad (3\text{-}104)$$

Gram-Schmidt Orthogonalization Method

If λ_p is an eigenvalue of multiplicity $q > 1$, the q linearly independent eigenfunctions $\varphi_p, \ldots, \varphi_{p+q-1}$ corresponding to

$$\lambda_p = \lambda_{p+1} = \cdots = \lambda_{p+q-1}$$

can be assumed to be real-valued and also mutually orthogonal on $\mathscr{R}$ with weight function ρ. Thus whenever eigenfunctions have distinct indices, the orthogonality relation (3-104) holds. This follows from the *Gram-Schmidt orthogonalization process.* Suppose

$$f_1, \ldots, f_r$$

is a linearly independent set of continuous functions on $\mathscr{R}$. Then one can construct an orthogonal set

$$g_1, \ldots, g_r$$

(with weight ρ) on $\mathscr{R}$ from the first set as follows. Let $g_1 = f_1$ and let

$$g_2 = f_2 - c_1 g_1$$

where c_1 is the constant determined by the equation

$$0 = \iint\limits_{\mathscr{R}} \rho g_1 g_2 \, dA = \iint\limits_{\mathscr{R}} \rho g_1 f_2 \, dA - c_1 \iint\limits_{\mathscr{R}} \rho g_1{}^2 \, dA$$

Observe that

$$\|g_1\|^2 = \iint_\mathscr{R} \rho g_1{}^2 \, dA \neq 0$$

otherwise g_1 is the trivial function, and so the original set of functions is not linearly independent. Suppose now that functions $g_1, \ldots, g_s$, $s < r$, have been constructed such that g_i, g_j, $i \neq j$, $i = 1, \ldots, s$, $j = 1, \ldots, s$, are orthogonal on $\mathscr{R}$ with weight ρ. Let

$$g_{s+1} = f_{s+1} - c_1 g_1 - \cdots - c_s g_s$$

where the constants c_i are determined by the equations

$$0 = \iint_\mathscr{R} \rho g_i g_{s+1} \, dA = \iint_\mathscr{R} \rho g_i f_{s+1} \, dA - c_i \iint_\mathscr{R} \rho g_i{}^2 \, dA \qquad i = 1, \ldots, s$$

Then the set $g_1, \ldots, g_{s+1}$ is an orthogonal set on $\mathscr{R}$ with weight ρ. Now if this process is applied to the set $\psi_p, \ldots, \psi_{p+q-1}$ of real-valued eigenfunctions which correspond to the repeated eigenvalue λ_p, a set $\varphi_p, \ldots, \varphi_{p+q-1}$ results which is orthogonal on $\mathscr{R}$ with weight function ρ. It is easily verified that each φ_k is an eigenfunction corresponding to λ_p.

Expansion in Eigenfunctions

Recall that if **A**, **B**, **C**, are three mutually perpendicular vectors in three-dimensional euclidian space, it is impossible to find a nonzero vector **X** in the space which is perpendicular to all three vectors, i.e., such that

$$\mathbf{A} \cdot \mathbf{X} = \mathbf{B} \cdot \mathbf{X} = \mathbf{C} \cdot \mathbf{X} = 0 \qquad \mathbf{X} \neq 0$$

The eigenfunctions $\{\varphi_n\}$ have an analogous property. In the space of all functions which are continuous on $\bar{\mathscr{R}}$ the set $\{\varphi_n\}$ is *complete*: if φ is continuous on $\bar{\mathscr{R}}$, and

$$\iint_\mathscr{R} \rho \varphi \varphi_n \, dA = 0 \qquad \text{all } n \tag{3-105}$$

then φ is identically zero on $\bar{\mathscr{R}}$. For a proof of this fact see Ref. 3. It should be mentioned that some authors use the term *closed* for what here is called complete.

Often it is desired to *expand a function in the eigenfunctions* $\{\varphi_n\}$. Let f be an arbitrary function on $\bar{\mathscr{R}}$; then the series representation

$$f(x,y) = \sum_{n=1}^{\infty} a_n \varphi_n(x,y) \tag{3-106}$$

is sought. Assume the series converges uniformly to f on $\bar{\mathscr{R}}$. Then the necessary form of the coefficients can be derived as follows. Let m be a fixed positive integer.   Multiply both sides of Eq. (3-106) by $\rho\varphi_m$, integrate the result over $\mathscr{R}$, and interchange the order of summation and integration:

$$\iint_{\mathscr{R}} \rho f \varphi_m \, dA = \sum_{n=1}^{\infty} a_n \iint_{\mathscr{R}} \rho \varphi_n \varphi_m \, dA$$

By the orthogonality property expressed in Eq. (3-104) it follows that

$$a_m = \frac{1}{\|\varphi_m\|^2} \iint_{\mathscr{R}} \rho f \varphi_m \, dA \qquad (3\text{-}107)$$

where $\|\varphi_m\|^2$ is defined by Eq. (3-103).   The constants a_m so determined are called the *Fourier coefficients of f relative to the* $\{\varphi_n\}$, with weight ρ.   The series on the right in Eq. (3-106) is then called the *Fourier series of f* relative to the $\{\varphi_n\}$, with weight ρ.   It is clear that if the series converges to f on $\bar{\mathscr{R}}$, then necessarily $B(f) = 0$ on C since each eigenfunction φ_n satisfies the homogeneous boundary condition.   It can be proved that if f is continuously differentiable on $\bar{\mathscr{R}}$ and satisfies the homogeneous boundary condition on C, the Fourier series of f converges uniformly to f on $\bar{\mathscr{R}}$ (see Ref. 3).

Solution of the Boundary-value Problem by the Method of Eigenfunctions

Consider the problem

$$Lu = f \quad \text{in } \mathscr{R} \qquad\qquad B(u) = 0 \qquad \text{on } C \qquad (3\text{-}108)$$

where f is the given nontrivial real-valued function on $\mathscr{R}$.   Assume the only solution of the homogeneous problem (3-91) is the trivial solution.   If a twice continuously differentiable solution of problem (3-108) exists, it must have a representation as a Fourier series in the real-valued eigenfunctions $\{\varphi_n\}$ of problem (3-96)

$$u(x,y) = \sum_{n=1}^{\infty} c_n \varphi_n(x,y)$$

The coefficients in the expansion can be obtained as follows.   First, the c_n must be the Fourier coefficients of u

$$c_n = \frac{1}{\|\varphi_n\|^2} \iint_{\mathscr{R}} \rho u \varphi_n \, dA \qquad n = 1, 2, \ldots$$

Now for each fixed positive integer m,

$$\iint_{\mathscr{R}} \rho u \varphi_m \, dA = \frac{1}{\lambda_m} \iint_{\mathscr{R}} u(\lambda_m \rho \varphi_m) \, dA = -\frac{1}{\lambda_m} \iint_{\mathscr{R}} u L \varphi_m \, dA$$

$$= -\frac{1}{\lambda_m} \iint_{\mathscr{R}} \varphi_m L u \, dA = -\frac{1}{\lambda_m} \iint_{\mathscr{R}} f \varphi_m \, dA$$

Thus

$$c_m = \frac{-b_m}{\lambda_m \|\varphi_m\|^2} \qquad b_m = \iint_{\mathscr{R}} f\varphi_m \, dA \tag{3-109}$$

The coefficients in the series representation of the solution are determined by the given function f and the eigenfunctions. Thus the formal series solution of problem (3-106) is

$$u(x,y) = -\sum_{n=1}^{\infty} \frac{b_n}{\lambda_n \|\varphi_n\|^2} \varphi_n(x,y) \tag{3-110}$$

In the preceding paragraph the existence of a classical, i.e., twice continuously differentiable, solution of problem (3-108) was assumed. However a solution need not exist without further hypotheses on f and the region $\mathscr{R}$. One approach to proving the existence of a classical solution is to show that the series (3-110) converges in a suitable fashion. This may be done by estimating the order of magnitude of the eigenvalues λ_n for large n and then constructing a dominating series. In the particular case of a rectangular region the eigenvalues can be explicitly exhibited, and in Example 3-5 (page 137) the proof of the existence of a solution is made following this method. The general case is not considered in this text (see Ref. 3).

Green's Function

If the expressions for the coefficients b_n are inserted into Eq. (3-110) and the operations of summation and integration are formally interchanged, then

$$u(x,y) = -\sum_{n=1}^{\infty} \left[\iint_{\mathscr{R}} f(\xi,\eta)\varphi_n(\xi,\eta) \, d\xi \, d\eta \right] \frac{\varphi_n(x,y)}{\lambda_n \|\varphi_n\|^2}$$

$$= \iint_{\mathscr{R}} \left[-\sum_{n=1}^{\infty} \frac{\varphi_n(x,y)\varphi_n(\xi,\eta)}{\lambda_n \|\varphi_n\|^2} \right] f(\xi,\eta) \, d\xi \, d\eta$$

Define the function $G(x,y;\xi,\eta)$ by

$$G(x,y;\xi,\eta) = -\sum_{n=1}^{\infty} \frac{\varphi_n(x,y)\varphi_n(\xi,\eta)}{\lambda_n \|\varphi_n\|^2} \tag{3-111}$$

Then the solution can be rewritten

$$u(x,y) = \iint_{\mathscr{R}} G(x,y;\xi,\eta) f(\xi,\eta) \, d\xi \, d\eta \tag{3-112}$$

The function G is called the *Green's function* of problem (3-87). It is an important and useful tool in both the theory and the applications. The series (3-111) is called the *bilinear series*. It can be shown to be convergent for all (x,y) and (ξ,η) in $\bar{\mathcal{R}}$ (see Ref. 3). Note that G satisfies the boundary condition on C. Also G is a symmetric function of the points (x,y), (ξ,η):

$$G(\xi,\eta;x,y) = G(x,y;\xi,\eta)$$

In many applications the idea of a *unit impulse function*, or δ *function*, is useful. Imagine a fictitious function δ having the symbolic properties

$$\delta(x) = 0 \qquad x \neq 0 \qquad \delta(0) = +\infty$$

$$\int_{-\infty}^{\infty} \delta(x)\,dx = 1 = \int_{-a}^{a} \delta(x)\,dx \qquad a > 0$$

$$\int_{-a}^{a} f(x)\delta(x)\,dx = f(0) \qquad a > 0$$

for every function f continuous on an interval containing the origin. The last property listed is called the *reproducing property*. Although there does not exist a function in the classical sense having these properties, consider the function δ_ϵ defined by

$$\delta_\epsilon(x) = \begin{cases} \dfrac{3(\epsilon^2 - x^2)}{4\epsilon^3} & |x| \leq \epsilon \\ 0 & |x| > \epsilon \end{cases}$$

The graph of δ_ϵ is shown in Fig. 3-4. For small $\epsilon > 0$ the function δ_ϵ is

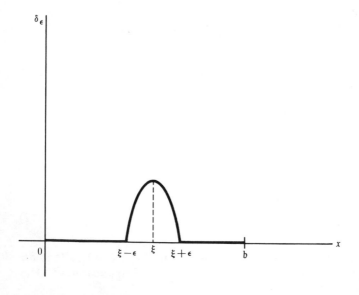

Figure 3-4

visualized as an approximation to the δ function. Observe that

$$\lim_{\epsilon \to 0} \delta_\epsilon(x) = 0 \qquad x \neq 0 \qquad \lim_{\epsilon \to 0} \delta_\epsilon(0) = +\infty$$

$$\int_{-\infty}^{\infty} \delta_\epsilon(x) \, dx = 1 = \int_{-a}^{a} \delta_\epsilon(x) \, dx \qquad a \geq \epsilon$$

$$\lim_{\epsilon \to 0} \int_{-\infty}^{\infty} \delta_\epsilon(x) f(x) \, dx = f(0) = \lim_{\epsilon \to 0} \int_{-a}^{a} \delta_\epsilon(x) f(x) \, dx \qquad a > 0$$

Thus one can proceed by limiting processes using the continuous function δ_ϵ and obtain the symbolic properties of the δ function. Other approximating functions can easily be constructed. A rigorous definition and derivation of the properties of the δ function can be given with the aid of concepts from functional analysis, particularly the notion of a continuous linear functional. In this setting the δ function is an example of a *distribution*, or *generalized function* (see Ref. 5).

Let (ξ, η) be a fixed but otherwise arbitrarily chosen point of $\mathscr{R}$. Then the two-dimensional δ function has the reproducing property

$$\iint_{\mathscr{R}} \varphi(x,y) \delta(x - \xi, y - \eta) \, dx \, dy = \varphi(\xi, \eta)$$

for every continuous function φ on $\mathscr{R}$. The properties of the two-dimensional δ function may be defined by limiting processes as indicated in the one-dimensional case. Symbolically write

$$\delta(x - \xi, y - \eta) = \delta(x - \xi)\delta(y - \eta)$$

In problem (3-108) choose $f(x,y) = \delta(x - \xi, y - \eta)$; that is, the "input" to the problem is the unit impulse applied at (ξ, η). From Eq. (3-109) it follows that

$$b_m = \iint_{\mathscr{R}} \varphi_m(x,y) \delta(x - \xi, y - \eta) \, dx \, dy = \varphi_m(\xi, \eta)$$

Thus

$$c_m = \frac{-\varphi_m(\xi, \eta)}{\lambda_m \|\varphi_m\|^2}$$

and the corresponding formal solution is, by Eq. (3-110),

$$u(x,y) = -\sum_{n=1}^{\infty} \frac{\varphi_n(x,y)\varphi_n(\xi, \eta)}{\lambda_n \|\varphi_n\|^2} = G(x,y;\xi,\eta)$$

Hence there is the interpretation of the Green's function of problem (3-87) as the "response" of the "system" with homogeneous boundary conditions due to a δ-function input:

$$LG = \delta(x - \xi, y - \eta) \quad \text{in } \mathscr{R} \qquad B(G) = 0 \qquad \text{on } C \qquad (3\text{-}113)$$

The Green's function of problem (3-87) has the following characteristic properties:

1. G is continuous in the four variables x, y, ξ, η for (x,y) and (ξ,η) in $\mathscr{R}$; also G is a symmetric function of points (x,y), (ξ,η)

$$G(\xi,\eta;x,y) = G(x,y;\xi,\eta)$$

2. G satisfies the homogeneous boundary conditions on C.

3. Let (ξ,η) be a fixed but otherwise arbitrarily chosen point of $\mathscr{R}$. Then, as a function of x and y, G satisfies the homogeneous differential equation $LG = 0$ everywhere in $\mathscr{R}$ except at the point (ξ,η).

4. G has discontinuous first partial derivatives at (ξ,η), and if C_ϵ is a circle of radius $\epsilon > 0$ about (ξ,η), then

$$\lim_{\epsilon \to 0} \int_{C_\epsilon} [p_1 G_x \cos(n,x) + p_2 G_y \cos(n,y)] \, ds = 1$$

To establish property 4 let $\mathscr{R}_\epsilon$ denote the region interior to C_ϵ. Integrate both sides of the differential equation satisfied by G over $\mathscr{R}_\epsilon$. Then

$$\iint_{\mathscr{R}_\epsilon} [(p_1 G_x)_x + (p_2 G_y)_y] \, dx \, dy + \iint_{\mathscr{R}_\epsilon} qG \, dx \, dy = 1$$

The first integral can be transformed into a line integral around C_ϵ by means of the divergence theorem. Now let $\epsilon \to 0$. Since q and G are continuous in $\mathscr{R}_\epsilon$,

$$\lim_{\epsilon \to 0} \iint_{\mathscr{R}_\epsilon} qG \, dx \, dy = 0$$

and the relation follows.

Boundary-value Problems for Poisson's Equation

In the particular case where the operator in problem (3-87) is $L = \Delta$ (the two-dimensional laplacian), the differential equation in the eigenvalue problem (3-96) is

$$\Delta\varphi + \lambda\rho\varphi = 0$$

called the *scalar Helmholtz equation*. It is of importance in Chaps. 4 and 5, where it arises when harmonic time dependence is assumed for solutions of the wave equation and diffusion equation, respectively. The following example illustrates the method of eigenfunctions for Poisson's equation (3-7) for the case of a rectangular boundary.

EXAMPLE 3-5   Determine the eigenfunction expansion of the solution of the problem

$$\Delta u = f(x,y) \quad \text{in } \mathscr{R} \qquad u = 0 \quad \text{on } C$$

where C is the boundary of the rectangle

$$0 \leq x \leq a \qquad 0 \leq y \leq b$$

Consider the eigenvalue problem

$$\Delta \varphi + \lambda \varphi = 0 \quad \text{in } \mathscr{R} \qquad \varphi = 0 \quad \text{on } C$$

The eigenfunctions can be obtained explicitly by separation of variables. Assume a solution of the differential equation of the form $u(x,y) = X(x)Y(y)$. Substitution into the differential equation yields

$$\frac{X''}{X} = -\nu = -\left(\frac{Y''}{Y} + \lambda\right)$$

where ν is the separation constant. Since $u = 0$ on the boundary, the factor X must satisfy

$$X'' + \nu X = 0 \qquad X(0) = X(a) = 0$$

Real-valued eigenfunctions of this Sturm-Liouville problem are

$$X_n = \sin \frac{n\pi x}{a} \qquad n = 1, 2, \ldots$$

corresponding to the eigenvalues $\nu_n = n^2\pi^2/a^2$, $n = 1, 2, \ldots$. The factor Y must satisfy

$$Y'' + (\lambda - \nu_n)Y = 0 \qquad Y(0) = Y(b) = 0$$

Real-valued eigenfunctions of this problem are

$$Y_m = \sin \frac{m\pi y}{b} \qquad m = 1, 2, \ldots$$

corresponding to eigenvalues $\lambda - \nu_n = m^2\pi^2/b^2$, $m = 1, 2, \ldots$. Accordingly the ∞^2 functions

$$\varphi_{nm}(x,y) = \sin \frac{n\pi x}{a} \sin \frac{m\pi y}{b} \qquad n = 1, 2, \ldots; \, m = 1, 2, \ldots$$

are eigenfunctions of the two-dimensional problem corresponding to the ∞^2 eigenvalues

$$\lambda_{nm} = \frac{n^2\pi^2}{a^2} + \frac{m^2\pi^2}{b^2} \qquad n = 1, 2, \ldots; \, m = 1, 2, \ldots$$

The set of eigenfunctions $\{\varphi_{nm}\}$ has the orthogonality property

$$\int_0^a \int_0^b \varphi_{nm}(x,y)\varphi_{pq}(x,y)\, dx\, dy = 0 \qquad (n,m) \neq (p,q)$$

Also

$$\|\varphi_{nm}\|^2 = \int_0^a \int_0^b \sin^2 \frac{n\pi x}{a} \sin^2 \frac{m\pi y}{b}\, dx\, dy = \frac{ab}{4}$$

The set of eigenfunctions $\{\varphi_{nm}\}$ just derived is complete in the space of all functions which are continuous on $\overline{\mathscr{R}}$. To prove this fact use is made of the completeness properties of the individual sequences

$$\left\{\sin\frac{n\pi x}{a}\right\}\qquad\left\{\sin\frac{m\pi y}{b}\right\}$$

Let ψ be a function which is continuous on $\overline{\mathscr{R}}$ and orthogonal to all the φ_{nm}:

$$\int_0^a\int_0^b\psi(x,y)\sin\frac{n\pi x}{a}\sin\frac{m\pi y}{b}\,dx\,dy=0\qquad n=1,2,\ldots;\,m=1,2,\ldots$$

Choose a positive integer p and define the function h_p by

$$h_p(y)=\int_0^a\psi(x,y)\sin\frac{p\pi x}{a}\,dx\qquad 0\le y\le b$$

Then h_p is orthogonal to the sequence $\{\sin(m\pi y/b)\}$

$$\int_0^b h_p(y)\sin\frac{m\pi y}{b}\,dy=0\qquad m=1,2,\ldots$$

Since the sequence $\{\sin(m\pi y/b)\}$ is complete in the space $C[0,b]$ of all functions $g(y)$ which are continuous on the interval $[0,b]$, it follows that h_p must vanish identically on $[0,b]$

$$\int_0^a\psi(x,y)\sin\frac{p\pi x}{a}\,dx=0\qquad 0\le y\le b$$

Evidently this is true for $p=1,2,\ldots$. Fix $y=y_0$; then the function $\psi(x,y_0)$ is continuous on $0\le x\le a$ and orthogonal to the sequence $\{\sin(n\pi x/a)\}$. But this sequence is complete in the space $C[0,a]$ of all functions $g(x)$ which are continuous on $[0,a]$. Hence

$$\psi(x,y_0)=0\qquad 0\le x\le a$$

Since y_0 was chosen arbitrarily in $[0,b]$, it follows that

$$\psi(x,y)=0\qquad 0\le x\le a;\,0\le y\le b$$

The importance of the completeness property just shown is that it implies that (1) all the eigenvalues are contained in the set $\{\lambda_{nm}\}$; that is, the separation-of-variables technique obtains all the eigenvalues; and (2) any eigenfunction must be a constant multiple of some φ_{pq}. To show (1) suppose λ is an eigenvalue not included in the set $\{\lambda_{nm}\}$. Let ψ be an eigenfunction corresponding to λ. Then ψ is orthogonal to every φ_{nm}. But then ψ is identically zero on $\overline{\mathscr{R}}$, and λ is not an eigenvalue. To establish (2) let ψ be an eigenfunction of the problem. Then ψ corresponds to some eigenvalue λ_{pq} in the set $\{\lambda_{nm}\}$. Assume first that ψ is orthogonal to the eigenfunction φ_{pq}. Then ψ is orthogonal to all the φ_{nm}, and again a contradiction arises. Hence ψ is not orthogonal to φ_{pq}. Accordingly a nonzero constant c exists such that the function

$$\psi_1=\psi-c\varphi_{pq}$$

is orthogonal to φ_{pq}. By the linearity and homogeneity of the eigenvalue problem it follows that ψ_1 is a solution of the differential equation with $\lambda=\lambda_{pq}$, and ψ_1 satisfies the boundary condition on C. Hence ψ_1 is orthogonal to all φ_{nm} and so vanishes identically on $\overline{\mathscr{R}}$; that is,

$$\psi=c\varphi_{pq}$$

The series expansion of f in the eigenfunctions φ_{nm} is

$$f(x,y) = \sum_{n=1}^{\infty} \sum_{m=1}^{\infty} A_{nm} \sin \frac{n\pi x}{a} \sin \frac{m\pi y}{b}$$

often called a *double Fourier sine series*.   The Fourier coefficients are

$$A_{nm} = \frac{4}{ab} \int_{0}^{a} \int_{0}^{b} f(x,y) \sin \frac{n\pi x}{a} \sin \frac{m\pi y}{b} \, dx \, dy \qquad n = 1, 2, \ldots ; m = 1, 2, \ldots$$

It is evident that continuity of f on $\overline{\mathscr{R}}$ suffices to ensure the existence of the Fourier coefficients of f.   However the Fourier series of f need not be convergent.   From Eq. (3-110) the formal series solution of Poisson's equation which satisfies the boundary condition on C is

$$u(x,y) = -\sum_{n=1}^{\infty} \sum_{m=1}^{\infty} \frac{A_{nm}}{\lambda_{nm}} \sin \frac{n\pi x}{a} \sin \frac{m\pi y}{b}$$

A condition sufficient to ensure that the formal series defines a function on $\overline{\mathscr{R}}$ which satisfies all the conditions of the boundary-value problem is that f be twice continuously differentiable on $\overline{\mathscr{R}}$ and satisfies the boundary condition; that is, $f = 0$ on C.   To show this use is made of the relation

$$\int_{0}^{a} \int_{0}^{b} u \, \Delta v \, dx \, dy = \int_{0}^{a} \int_{0}^{b} v \, \Delta u \, dx \, dy$$

which holds for every pair twice continuously differentiable functions on $\overline{\mathscr{R}}$ which vanish on C.   Now

$$A_{nm} = \frac{4}{ab\lambda_{nm}} \int_{0}^{a} \int_{0}^{b} f(x,y)\lambda_{nm}\varphi_{nm}(x,y) \, dx \, dy = \frac{-4}{ab\lambda_{nm}} \int_{0}^{a} \int_{0}^{b} f(x,y) \, \Delta\varphi_{nm} \, dx \, dy$$

$$= \frac{-4}{ab\lambda_{nm}} \int_{0}^{a} \int_{0}^{b} \varphi_{nm}(x,y) \, \Delta f \, dx \, dy$$

Since f is twice continuously differentiable on $\overline{\mathscr{R}}$, there exists a constant $M > 0$ such that

$$|\Delta f| \leq M \qquad 0 \leq x \leq a \qquad 0 \leq y \leq b$$

Hence $|A_{nm}| < 4M/\lambda_{nm}$, and

$$\left| \frac{A_{nm} \sin (n\pi x/a) \sin (m\pi y/b)}{\lambda_{nm}} \right| < \frac{4M}{\lambda_{nm}^{2}}$$

Thus a dominant convergent series of positive constants for the series is

$$\sum_{n=1}^{\infty} \sum_{m=1}^{\infty} \frac{4M}{\lambda_{nm}^{2}}$$

It follows that the series converges uniformly on $\overline{\mathscr{R}}$ and defines a continuous function u there.   Clearly u satisfies the boundary condition on C, since each φ_{nm} does.   The series obtained by applying the operator Δ termwise is also uniformly convergent on $\overline{\mathscr{R}}$.   Hence

u is twice continuously differentiable on $\overline{\mathscr{R}}$, and

$$\Delta u = -\sum_{n=1}^{\infty}\sum_{m=1}^{\infty}\frac{A_{nm}}{\lambda_{nm}}\Delta\varphi_{nm} = \sum_{n=1}^{\infty}\sum_{m=1}^{\infty}A_{nm}\varphi_{nm}(x,y) = f(x,y)$$

Thus u is the (unique) solution of the problem.

For example, in the problem

$$\Delta u = xy \qquad 0 \le x \le a; 0 \le y \le b \qquad\qquad \varphi = 0 \qquad \text{on } C$$

the Fourier coefficients of $f(x,y) = xy$ are

$$A_{nm} = \frac{4}{ab}\left(\int_0^a x \sin\frac{n\pi x}{a}\,dx\right)\left(\int_0^b y\sin\frac{m\pi y}{b}\,dy\right) = \frac{4ab}{nm\pi^2}(-1)^{n+m}$$

The solution is

$$u(x,y) = \frac{4ab}{\pi^2}\sum_{n=1}^{\infty}\sum_{m=1}^{\infty}\frac{(-1)^{n+m+1}}{nm\lambda_{nm}}\sin\frac{n\pi x}{a}\sin\frac{m\pi y}{b}$$

From Eq. (3-111) the Green's function is

$$G(x,y;\xi,\eta) = -\frac{4}{ab}\sum_{n=1}^{\infty}\sum_{m=1}^{\infty}\frac{1}{\lambda_{nm}}\sin\frac{n\pi x}{a}\sin\frac{n\pi\xi}{a}\sin\frac{m\pi y}{b}\sin\frac{m\pi\eta}{b}$$

Since the bilinear series on the right is dominated by the convergent series

$$\frac{4}{ab}\sum_{n=1}^{\infty}\sum_{m=1}^{\infty}\frac{1}{\lambda_{nm}}$$

it follows that the series converges uniformly on

$$0 \le x \le a \qquad 0 \le y \le b \qquad 0 \le \xi \le a \qquad 0 \le \eta \le b$$

Hence G is continuous on its domain of definition.

Integral Representation of the Solution

As has been indicated above, the Green's function, when it exists, enables one to *invert* the boundary-value problem, i.e., write down an expression for the solution which involves only the given data and G. In the elliptic problem (3-87) let the boundary condition be one of the conditions (3-88) to (3-90) and assume also that the only solution of the corresponding homogenous problem is the trivial solution. Then a Green's function exists and satisfies (3-113). Now interchange (x,y) and (ξ,η) in Eq. (3-113); i.e., let (x,y) be a fixed point in $\mathscr{R}$, and let (ξ,η) be variable. Then the operator L is written in terms of the independent variables ξ and η. Similarly in the equations of problem (3-87) interchange the roles of (x,y) and (ξ,η), and let $u(\xi,\eta)$ be the value of the solution of the problem at (ξ,η). In Green's formula (3-97) replace x by ξ and y by η, and choose

$$v(\xi,\eta) = G(\xi,\eta;x,y) = G(x,y;\xi,\eta)$$

Substitution into Green's formula yields

$$\iint\limits_{\mathscr{R}} G(x,y;\xi,\eta)f(\xi,\eta)\, d\xi\, d\eta - \iint\limits_{\mathscr{R}} u(\xi,\eta)\delta(\xi - x)\delta(\eta - y)\, d\xi\, d\eta$$

$$= \int_C [p_1(Gu_\xi - uG_\xi)\cos(n,\xi) + p_2(Gu_\eta - uG_\eta)\cos(n,\eta)]\, ds$$

Hence

$$u(x,y) = \iint G(x,y;\xi,\eta)f(\xi,\eta)\, d\xi\, d\eta + \int_C u[p_1 G_\xi \cos(n,\xi) + p_2 G_\eta \cos(n,\eta)]$$

$$- G[p_1 u_\xi \cos(n,\xi) + p_2 u_\eta \cos(n,\eta)]\, ds \quad (3\text{-}114)$$

Suppose the boundary condition is the Dirichlet condition (3-88). Then $G = 0$ on C, and the solution has the representation

$$u(x,y) = \iint\limits_{\mathscr{R}} G(x,y;\xi,\eta)f(\xi,\eta)\, d\xi\, d\eta$$

$$+ \int_C g[p_1 G_\xi \cos(n,\xi) + p_2 G_\eta \cos(n,\eta)]\, ds \quad (3\text{-}115)$$

in terms of the given functions f and g. In the case where $p_1 = p_2 = p$ on $\bar{\mathscr{R}}$, the solution of the Dirichlet problem is given by

$$u(x,y) = \iint\limits_{\mathscr{R}} G(x,y;\xi,\eta)f(\xi,\eta)\, d\xi\, d\eta + \int_C pg\frac{\partial G}{\partial n}\, ds \quad (3\text{-}116)$$

where $\partial G/\partial n$ is the derivative in the direction of the exterior normal along C. Again if $p_1 = p_2 = p$ and the boundary condition is the Neumann condition (3-89), the solution is represented by

$$u(x,y) = \iint\limits_{\mathscr{R}} G(x,y;\xi,\eta)f(\xi,\eta)\, d\xi\, d\eta - \int_C pgG\, ds \quad (3\text{-}117)$$

If the boundary condition is the mixed condition (3-90), the solution has the representation (3-117).

The derivations above are formal. However, if the data are sufficiently smooth, the integrals define a function u which is continuously differentiable and satisfy the differential equation in $\mathscr{R}$ and the boundary conditions on C. Thus existence theorems can be obtained by means of the Green's function. It is clear that there are alternative and dual modes of representing the solution of a boundary-value problem. One is a series representation in terms of the complete set of eigenfunctions of the associated eigenvalue problem; the other is an integral representation involving the Green's function. This duality is present throughout the theory of linear boundary-value problems in one dimension or several.

Properly Posed Problems

A boundary-value problem is said to be *properly posed* (also well posed, reasonable, or correctly set) if (1) a solution exists for each choice of data (in a suitable class of functions); (2) the solution is uniquely determined by the data; and (3) the solution depends continuously on the data. Property 3 means that if perturbations are made in the given data, the resulting perturbations in the solution are small provided the perturbations in the data are made sufficiently small. Hence the solution is *stable*. These are just the properties which one expects if the equations of the mathematical problem correctly describe the physical problem at hand. The class of self-adjoint elliptic-type problems discussed above for which Green's functions exist are well posed. The existence and stability are deduced from the integral representations, and the uniqueness of solution is given by Theorem 3-8. Examples of such well-posed problems are the Dirichlet problem for the circle and sphere solved in Secs. 3-3 and 3-4 and the boundary-value problem for Poisson's equation in Example 3-5.

EXAMPLE 3-6   Determine the eigenfunction expansion of the solution of the problem

$$\Delta u = f(r,\theta) \quad \text{in } \mathscr{R} \qquad u = 0 \quad \text{on } C$$

where C is the circle of radius a about the origin, $\mathscr{R}$ is the interior of C, and r, θ are polar coordinates.

Written in polar coordinates, the corresponding eigenvalue problem is

$$\Delta\varphi + \lambda\varphi = \varphi_{rr} + \frac{\varphi_r}{r} + \frac{\varphi_{\theta\theta}}{r^2} + \lambda\varphi = 0 \qquad \text{in } \mathscr{R}$$

$$\varphi(a,\theta) = 0 \qquad\qquad\qquad\qquad 0 \leq \theta \leq 2\pi$$

Since the boundary condition is the Dirichlet condition, it is known that the eigenvalues are positive. If a separable form of solution

$$\varphi = R(r)\Theta(\theta)$$

is assumed, the real-valued eigenfunctions

$$\varphi_{nk}^{(e)} = J_n\!\left(\frac{\xi_{nk}r}{a}\right)\cos n\theta \qquad \varphi_{nk}^{(0)} = J_n\!\left(\frac{\xi_{nk}r}{a}\right)\sin n\theta$$

corresponding to the eigenvalues

$$\lambda_{nk} = \frac{\xi_{nk}^2}{a^2} \qquad n = 0, 1, \ldots; k = 1, 2, \ldots$$

are obtained. Here $\{\xi_{nk}\}$ denotes, for each fixed n, the positive zeros of $J_n(\xi)$ arranged in order of increasing magnitude. These are identical to the functions derived in Example 3-2 in obtaining the solution of a potential problem in cylindrical coordinates. This is not accidental, since the equation which results from splitting off the z dependence in Laplace's equation (written in cylindrical coordinates) is just the scalar Helmholtz equation written in plane polar coordinates. Also note that in Example 3-2 the potential is required to vanish on the lateral surface $r = a$ of the cylinder, and this corresponds in the present problem to the boundary condition $\varphi = 0$ on the circle C. The orthogonality properties of the function $\varphi_{nk}^{(e)}$, $\varphi_{nk}^{(0)}$ are stated in Example 3-2.

The completeness property for the set $\{\varphi_{nk}^{(e)}\}$, $\{\varphi_{nk}^{(0)}\}$ in the space of all functions which are continuous on $\mathscr{R} + C$ can be proved by an argument similar to the one given in Example 3-5. Such a proof is based on the completeness properties of the individual sequences

$$\left\{rJ_n\left(\frac{\xi_{nk}r}{a}\right)\right\} \qquad \{\cos n\theta, \sin n\theta\}$$

Let n be a fixed integer, $n \geq 0$, and let $\psi(r)$ be continuous for $0 \leq r \leq a$. If

$$\int_0^a r\psi(r)J_n\left(\frac{\xi_{nk}r}{a}\right) dr = 0 \qquad k = 1, 2, \ldots$$

then $\psi(r) = 0$, $0 \leq r \leq a$. Similarly, if $h(\theta)$ is continuous on $[0, 2\pi]$ and

$$\int_0^{2\pi} h(\theta) \cos n\theta \, d\theta = 0 \qquad \int_0^{2\pi} h(\theta) \sin n\theta \, d\theta = 0 \qquad n = 0, 1, \ldots$$

then $h(\theta) - 0$, $0 \leq \theta \leq 2\pi$. Assuming these properties, it follows that if $\psi(r,\theta)$ is continuous on $\mathscr{R} + C$ and

$$\iint_{\mathscr{R}} \varphi_{nk}^{(e)}\varphi \, dA = 0 \qquad \iint_{\mathscr{R}} \varphi_{nk}^{(0)}\varphi \, dA = 0 \qquad n = 0, 1, \ldots ; k = 1, 2, \ldots$$

then $\varphi(r,\theta) = 0$, $0 \leq r \leq a$, $0 \leq \theta \leq 2\pi$.

Suppose φ is an eigenfunction corresponding to an eigenvalue μ. Then $\mu = \lambda_{pq}$, some $p, q, p \geq 0, q \geq 1$; that is, all the eigenvalues of the problem are included in the sequence $\{\lambda_{nk}\}$. If not, then $\mu \neq \lambda_{nk}$, all $n \geq 0$, $k \geq 1$, and so φ is orthogonal to all the $\varphi_{nk}^{(e)}$, $\varphi_{nk}^{(0)}$. But then the completeness properties discussed in the previous paragraph imply that φ is identically zero on $\overline{\mathscr{R}}$. This is a contradiction. Hence $\mu = \lambda_{pq}$, some p, q. Now if φ is orthogonal to both $\varphi_{pq}^{(e)}$ and $\varphi_{pq}^{(0)}$, then φ is orthogonal to all $\varphi_{nk}^{(e)}$, $\varphi_{nk}^{(0)}$, and once again a contradiction arises. Thus it is impossible for φ to be orthogonal to both $\varphi_{pq}^{(e)}$ and $\varphi_{pq}^{(0)}$. Hence there exist constants c_1, c_2, not both zero, such that

$$\psi = \varphi - c_1\varphi_{pq}^{(e)} - c_2\varphi_{pq}^{(0)}$$

is orthogonal to both $\varphi_{pq}^{(e)}$ and $\varphi_{pq}^{(0)}$. By the linearity and homogeneity of the eigenvalue problem it follows that ψ is an eigenfunction corresponding to λ_{pq}. This implies that ψ is orthogonal to all $\varphi_{nk}^{(e)}$ and $\varphi_{nk}^{(0)}$. Accordingly ψ vanishes identically on $\overline{\mathscr{R}}$, and

$$\varphi = c_1\varphi_{pq}^{(e)} + c_2\varphi_{pq}^{(0)}$$

Thus every eigenfunction is a linear combination of some pair $\varphi_{nk}^{(e)}$, $\varphi_{nk}^{(0)}$.

From Eq. (3-111) and a trigonometric identity one obtains the eigenfunction expansion

$$G(r,\theta;r',\theta') = -\frac{1}{\pi}\sum_{n=0}^{\infty}\sum_{k=1}^{\infty}\frac{J_n(\xi_{nk}r/a)J_n(\xi_{nk}r'/a)}{\epsilon_n\xi_{nk}^2 J_{n+1}^2(\xi_{nk})}\cos[n(\theta - \theta')]$$

where $\epsilon_0 = 1$, $\epsilon_n = \frac{1}{2}$ if $n \geq 1$. Insert this expression for G into Eq. (3-116) and interchange the order of summation and integration. The result is

$$u(r,\theta) = -\sum_{n=0}^{\infty}\sum_{k=1}^{\infty}\left(\frac{a}{\xi_{nk}}\right)^2 J_n\left(\frac{\xi_{nk}r}{a}\right)(A_{nk}\cos n\theta + B_{nk}\sin n\theta)$$

where

$$A_{nk} = 2\frac{\int_0^a \int_0^{2\pi} r'f(r',\theta')J_n(\xi_{nk}r'/a)\cos n\theta'\, dr'\, d\theta'}{\pi a^2 J_{n+1}^2(\xi_{nk})}$$

$$B_{nk} = 2\frac{\int_0^a \int_0^{2\pi} r'f(r',\theta')J_n(\xi_{nk}r'/a)\sin n\theta'\, dr'\, d\theta'}{\pi a^2 J_{n+1}^2(\xi_{nk})}$$

$n = 1, 2, \ldots; \qquad k = 1, 2, \ldots$

and

$$A_{0k} = \frac{\int_0^a \int_0^{2\pi} r'f(r',\theta)J_0(\xi_{0k}r'/a)\, dr'\, d\theta'}{\pi a^2 J_1^2(\xi_{0k})}$$

PROBLEMS

Sec. 3-2

1 Determine conditions under which the following functions are harmonic in three dimensions.

(a) $u = ax^2 + by^2 + cz^2$ $\qquad$ a, b, c, d constants

(b) $u = f(ax + by + cz + d)$ $\qquad$ a, b, c, d constants

(c) $u = a_{11}x^2 + a_{22}y^2 + a_{33}z^2 + 2a_{12}xy + 2a_{13}xz + 2a_{23}yz$

where the a_{ij} are constants.

(d) $u = Ae^{ax+by+cz}$ $\qquad$ A, a, b, c constants

(e) $u = A\sin ax \cosh by + B\cos ax \sinh by$

2 **(a)** Let φ be a harmonic function, not identically a constant. When is $u = f(\varphi)$ a harmonic function?

(b) Let φ, ψ be harmonic functions. When is $u = \varphi\psi$ harmonic? Interpret the condition geometrically.

3 A harmonic function in three-dimensional space which depends only on the distance from a fixed point is called a purely radially dependent potential. Let $r = (x^2 + y^2 + z^2)^{1/2}$. Determine all twice differentiable functions f such that $u = f(r)$ is a potential. Do this also in two dimensions.

4 Let u be a harmonic function in a region. Show that any derivative of u is also harmonic in the region (assume the property that a harmonic function has derivatives of all orders in the region).

5 Let $i = \sqrt{-1}$. It is known that each function f of the complex variable $z = x + iy$, x, y real variables, can be written

$$f(z) = u(x,y) + iv(x,y)$$

where u, v are real-valued functions. It is shown in the theory of functions of a complex variable that if f is an analytic function of z, then u and v are harmonic functions. For each of the following cases find the functions u, v and verify that they are harmonic.

(a) $f(z) = z^2$

(b) $f(z) = 4z^3 - 2z^2 + 1$

(c) $f(z) = e^z$

(d) $f(z) = \cos z$, where $\cos z = \dfrac{e^{iz} + e^{-iz}}{2}$

(e) $f(z) = \dfrac{1}{z} \qquad z \neq 0$

6 (a) A linear transformation

$$x = a_{11}\xi + a_{12}\eta \qquad y = a_{21}\xi + a_{22}\eta$$

is called an orthogonal transformation if the coefficients (assumed real) satisfy

$$a_{11}{}^2 + a_{21}{}^2 = a_{12}{}^2 + a_{22}{}^2 = 1 \qquad a_{11}a_{21} + a_{12}a_{22} = 0$$

In this case the transformation represents a rotation of axes.   Let $u(x,y)$ be a twice differentiable function of the variables x, y, and let $v(\xi,\eta) = u[x(\xi,\eta),y(\xi,\eta)]$.   Show that

$$v_{\xi\xi} + v_{\eta\eta} = u_{xx} + u_{yy}$$

(b) Let the xy coordinate system be related to the coordinate system by the translation $x = \xi + a$, $y = \eta + b$.   Let u, v be as described in **a**.   Show that the same relation holds. It follows (by combining the results of **a** and **b**) that the laplacian is invariant under the group of rigid motions in the plane; that is, u is a scalar invariant.   In particular, if u is a harmonic function, so is v.

7 Let u, v be harmonic functions in a region $\mathscr{R}$.   Let S_1 be a closed surface such that (1) S_1 and its interior lie within $\mathscr{R}$, (2) the divergence theorem is applicable to S_1 and the region bounded by S_1.   If $\mathbf{n}$ is the outward normal on S_1, show that

$$\int_{S_1} v\,\frac{\partial u}{\partial n}\,dS = \int_{S_1} u\,\frac{\partial v}{\partial n}\,dS$$

8 (a) Let $\bar{\mathscr{R}}$ be the set of all points in the plane such that $x^2 + y^2 \leq 1$.   Let $u = x^3 - 3xy^2$. Find the maximum and minimum values of u on $\bar{\mathscr{R}}$.

(b) Let C be the circle $x^2 + y^2 = a^2$ about the origin.   Verify Gauss' mean-value theorem in the plane for the function u of part **a**.

9 Let $\mathscr{R}$ be a bounded region with boundary S.   Let u be harmonic in $\mathscr{R}$, continuous on $\bar{\mathscr{R}}$, and such that $u > 0$ on S.   Prove that $u > 0$ in $\mathscr{R}$.

10 Let u, v, w be harmonic in $\mathscr{R}$, continuous on $\bar{\mathscr{R}}$, and such that $v(Q) \leq u(Q) \leq w(Q)$ holds for all points Q on S.   Prove that this inequality must hold everywhere in $\mathscr{R}$.

11 Let u be harmonic in a region $\mathscr{R}$ of xyz space.   Let P be a point in $\mathscr{R}$, S_p a sphere of radius r_0 with P as center such that S_p and its interior lie within $\mathscr{R}$.   Prove the average-value property

$$u(P) = \frac{1}{V}\int_{\mathscr{R}_p} u\,d\tau$$

where $V = 4\pi r_0{}^3/3$ is the volume of the spherical region $\mathscr{R}_p$ whose boundary is S_p, and

$\mathcal{R}_p = \mathcal{R}_p + S_p$.   *Hint:* Multiply both sides of Eq. (3-12) by $4\pi r^2 \, dr$ and integrate from $r = 0$ to $r = r_0$.

12 **(a)** Let $\mathcal{R}$ be a region in space with boundary S such that the divergence theorem

$$\int_{\mathcal{R}} \nabla \cdot \mathbf{A} \, d\tau = \int_{S} \mathbf{A} \cdot \mathbf{n} \, dS$$

is applicable to $\bar{\mathcal{R}} = \mathcal{R} + S$.   Set $\mathbf{A} = v \, \nabla u$, and obtain the equation

$$\int_{\mathcal{R}} v \, \Delta u \, d\tau + \int_{\mathcal{R}} \nabla u \cdot \nabla v \, d\tau = \int_{S} v \, \frac{\partial u}{\partial n} \, dS$$

where $\mathbf{n}$ is the exterior normal on S (note that this assumes u has continuous second derivatives on $\bar{\mathcal{R}}$, v has continuous first derivatives on $\bar{\mathcal{R}}$).   If u has continuous second derivatives on $\bar{\mathcal{R}}$, show that

$$\int_{\mathcal{R}} u \, \Delta u \, d\tau + \int_{\mathcal{R}} |\nabla u|^2 \, d\tau = \int_{S} u \, \frac{\partial u}{\partial n} \, dS$$

(b) Prove that if u is harmonic in $\mathcal{R}$ and has continuous second derivatives on $\bar{\mathcal{R}}$ and $\partial u / \partial n = 0$ on S, then u is constant on $\bar{\mathcal{R}}$.

(c) Prove that any pair of solutions of the Neumann problem $\Delta u = f$ in $\mathcal{R}$, $\partial u / \partial n = g$ on S having continuous second derivatives on $\bar{\mathcal{R}}$, differ by a constant.

13   Prove that there is at most one solution having continuous second derivatives on $\bar{\mathcal{R}}$ of the mixed problem $\Delta u = f$ in $\mathcal{R}$, $\partial u / \partial n + h u = g$ on S, where h is a continuous non-negative function, not identically zero, on S.

14 **(a)** Derive Eq. (3-17).

(b) Derive Eq. (3-19).

(c) State and give a complete proof of the maximum-minimum principle for plane harmonic functions.

(d) Let $\mathcal{R}$ be a region in the plane, and let u be harmonic in $\mathcal{R}$.   Let P be a point in $\mathcal{R}$ and C a circle of radius r_0 such that C, and its interior, lies within $\mathcal{R}$.   Prove that

$$u(P) = \frac{1}{\pi r_0^2} \int_{\bar{\mathcal{R}}} u \, dx \, dy$$

where $\bar{\mathcal{R}}_1$ is the circular disk bounded by C.

15 **(a)** Let S be a simple closed surface, $\mathcal{R}$ the unbounded region exterior to S.   Let u be harmonic in $\mathcal{R}$, continuous on $\mathcal{R} + S$, and vanishing uniformly at infinity.   Suppose u is not identically zero and $M = \max |u(P)|$ for P on S.   Prove that $|u(P)| < M$ holds in $\mathcal{R}$.

(b) Suppose that u is harmonic everywhere and vanishes uniformly at infinity.   Prove that u must be identically zero.   *Hints:* If $u = 0$ on S, then u is identically zero everywhere (see proof of Theorem 3-6).   Hence assume $M \neq 0$.   Let P be a point of $\mathcal{R}$.   Choose a sphere S_r of radius r about the origin sufficiently large such that (1) P lies within S_r, (2) $|u(Q)| < M/2$ holds for all points Q on and outside S_r.   What does the maximum-minimum principle imply for the region bounded by S and S_r?   In b let $\epsilon > 0$, and choose a sphere S_r about 0 sufficiently large such that (1) P lies within S_r, and (2) $|u(Q)| < \epsilon$ holds for all points Q on and outside S_r.

Sec. 3-3

16 Laplace's equation in rectangular coordinates in two dimensions is $u_{xx} + u_{yy} = 0$. Assume a separable solution $u = X(x) Y(y)$. Substitute into the differential equation, and follow the method of the text used for the polar-coordinate case to show that

$$\frac{X''}{X} = -\lambda = \frac{Y''}{Y}$$

where λ is a separation constant (real or complex), and

$$X'' = \frac{d^2 X}{dx^2} \qquad Y'' = \frac{d^2 Y}{dy^2}$$

Let $\mu = \sqrt{\lambda}$, and show that separable solutions must be of the form

$$u = (ax + b)(cy + d) + (Ae^{i\mu x} + Be^{-i\mu x})(Ce^{\mu y} + De^{-\mu y})$$

17 (a) A thin rectangular homogeneous thermally conducting plate lies in the xy plane and occupies the rectangle $0 \le x \le a, 0 \le y \le b$. The faces of the plate are insulated, and no internal sources or sinks are present. The edge $y = 0$ is held at $100°$, while the remaining edges are held at $0°$. Find the steady temperature $u(x,y)$ in the plate.

(b) Find $u(x,y)$ if the edge $y = 0$ is held at temperature $T \sin (\pi x/a)$, $0 \le x \le a$, where T is a constant.

(c) Find $u(x,y)$ if the edge $y = 0$ is held at temperature $Tx(x - a)$, where T is a constant. *Hint:* Since no heat sources are present in the plate, the steady-state temperature u must satisfy

$$\Delta u = 0 \qquad 0 < x < a; 0 < y < b$$

The boundary conditions in **a** are

$$u(0,y) = u(a,y) = 0 \qquad 0 < y < b \qquad\qquad u(x,b) = 0 \qquad 0 \le x \le a$$
$$u(x,0) = 100 \qquad 0 \le x \le a$$

As in Prob. 16, assume a separable solution $u = XY$ of Laplace's equation in rectangular coordinates, and obtain the ordinary differential equations

$$X'' + \lambda X = 0 \qquad Y'' - \lambda Y = 0$$

Now the boundary conditions $u(0,y) = u(a,y) = 0, 0 < y < b$, are satisfied by the separable solution if the factor X satisfies

$$X(0) = X(a) = 0$$

Show that this leads to the values $\lambda_n = n^2 \pi^2/a^2, n = 1, 2, \ldots$, as the only possible values for the separation constant. The corresponding eigenfunctions are $X_n = \sin (n\pi x/a)$, $n = 1, 2, \ldots$. Thus the corresponding solutions for Y are

$$Y_n(y) = c_n \cosh \frac{n\pi y}{a} + d_n \sinh \frac{n\pi y}{a}$$

The separable solution $u_n = X_n(x) Y_n(y)$ will satisfy the boundary condition $u_n(x,b) = 0$ if

$$c_n \cosh \frac{n\pi b}{a} + d_n \sinh \frac{n\pi b}{a} = 0$$

Thus

$$Y_n(y) = c_n \left[\cosh \frac{n\pi y}{a} - \frac{\cosh (n\pi b/a) \sinh (n\pi y/a)}{\sinh (n\pi b/a)} \right]$$

$$= c_n' \sinh \frac{n\pi(b-y)}{a}$$

The separable solutions

$$u_n(x,y) = \sin \frac{n\pi x}{a} \sinh \frac{n\pi(b-y)}{a} \qquad n = 1, 2, \ldots$$

satisfy all the boundary conditions save one. Consider the superposition

$$u(x,y) = \sum_{n=1}^{\infty} B_n \sin \frac{n\pi x}{a} \sinh \frac{n\pi(b-y)}{a}$$

In order to satisfy the condition along $y = 0$ it is necessary that

$$\sum_{n=1}^{\infty} B_n \sinh \frac{n\pi b}{a} \sin \frac{n\pi x}{a} = 100 \qquad 0 \le x \le a$$

This will be so if the coefficients

$$B_n = \frac{2}{a \sinh (n\pi b/a)} \int_0^a 100 \sin \frac{n\pi x}{a} \, dx \qquad n = 1, 2, \ldots$$

18 Find the steady temperature in the plate of Prob. 17 if the edge $x = a$ is held at temperature 100° while the remaining edges are held at 0°.

19 Derive the steady temperature in the plate of Prob. 17 if the temperature is held at T_1 along the edge $y = 0$, at T_2 along the edge $x = a$, at T_3 along the edge $y = b$, and at T_4 along the edge $x = 0$, where T_1, T_2, T_3, T_4 are constants.

20 Let the prescribed conditions on the edges of the plate in Prob. 17 be as follows:

$$u(x,0) = T_1 \sin \frac{\pi x}{2a} \qquad u(x,b) = 0; 0 \le x \le a$$

$$u(0,y) = T_2 y(b-y) \qquad u_x(a,y) = 0; 0 \le y \le b$$

where T_1 and T_2 are constants. Derive the expression for the temperature in the plate. *Hint:* Construct the superposition $u = v + w$ where v and w are harmonic functions in the rectangle which satisfy the boundary conditions

$$v(x,0) = T_1 \sin \frac{\pi x}{2a} \qquad v(x,b) = 0$$

$$v(0,y) = 0 \qquad v_x(a,y) = 0$$
$$w(x,0) = 0 \qquad w(x,b) = 0$$
$$w(0,y) = T_2 y(b-y) \qquad w_x(a,y) = 0$$

21 **(a)** A thin homogeneous plate occupies the region

$$0 \le x \le a \qquad y \ge 0$$

in the xy plane.   There are no heat sources in the plate, and the faces are insulated.   The temperature is held at $0°$ along the edges $x = 0$, $x = a$, while $u = T$ along the edge $y = 0$ (T constant).   Also $\lim\limits_{y \to \infty} u(x,y) = 0$ uniformly for $0 \leq x \leq a$.   Derive the series representation for the temperature in the plate.   Obtain a closed-form expression for the temperature with the aid of the equation

$$\sum_{k=1}^{\infty} \frac{\sin [(2k - 1)x]e^{-(2k-1)y}}{2k - 1} = \frac{\tan^{-1} (\sin x/\sinh y)}{2}$$

(b) Derive the series expression for the temperature in the plate described in **a** if instead the boundary condition along the edge $x = a$ is

$$\frac{\partial u}{\partial x} + hu = 0$$

where h is a positive constant, the remaining boundary conditions being the same.

22   In the problem of the torsion of a beam in the theory of elasticity there occurs the *stress function* Ψ.   If the stress function is known, the tangential stresses and the torsion moment can be determined.   It is shown in the theory that if the axis of the beam coincides with the z axis and the beam is of uniform cross section, then Ψ must satisfy Poisson's equation $\Delta \Psi = -2$ in $\mathscr{R}$, where $\mathscr{R}$ is the generating cross section in the xy plane, and the boundary condition $\Psi = 0$ on C, where C is the simple closed curve which bounds $\mathscr{R}$. Show that if the beam has the rectangular cross section $0 \leq x \leq a$, $0 \leq y \leq b$, the stress function is

$$\Psi(x,y) = ax - x^2$$

$$- \frac{8a^3}{\pi^3} \sum_{k=1}^{\infty} \frac{\sin [(2k - 1)\pi x/a]}{(2k - 1)^3} \left\{ \frac{\sinh [(2k - 1)\pi(b - y)/a] + \sinh [(2k - 1)\pi y/a]}{\sinh [(2k - 1)\pi b/a]} \right\}$$

Hint: Recall the discussion in Sec. 3-1.   The function $v(x,y) = ax - x^2$ satisfies $\Delta v = -2$ in $\mathscr{R}$ and is zero along $x = 0$ and along $x = a$.   Now construct a function w which is harmonic in $\mathscr{R}$ and such that

$$w(0,y) = w(a,y) = 0 \qquad 0 \leq y \leq b$$
$$w(x,0) = w(x,b) = x^2 - ax \qquad 0 \leq x \leq a$$

23   Solve the boundary-value problem

$$\Delta u = cx + dy \qquad \text{in } \mathscr{R} \qquad u = 0 \qquad \text{on the boundary}$$

where c and d are constants and $\mathscr{R}$ denotes the interior of the rectangle $0 \leq x \leq a$, $0 \leq y \leq b$.

24   In rectangular coordinates in three dimensions Laplace's equation is

$$\Delta u = u_{xx} + u_{yy} + u_{zz}$$

Assume a separable solution $u = X(x)Y(y)Z(z)$.   Obtain

$$\frac{X''}{X} + \frac{Y''}{Y} + \frac{Z''}{Z} = 0$$

Argue as before that

$$\frac{X''}{X} = -\alpha^2 = -\left(\frac{Y''}{Y} + \frac{Z''}{Z}\right)$$

where α is a separation constant.   Further

$$\frac{Y''}{Y} = \alpha^2 - \frac{Z''}{Z} = -\beta^2$$

Hence the factors must satisfy

$$X'' + \alpha^2 X = 0 \qquad Y'' + \beta^2 Y = 0 \qquad Z'' - (\alpha^2 + \beta^2)Z = 0$$

Accordingly separable solutions are of the form

$$u = e^{\pm i\alpha x}e^{\pm i\beta y}e^{\pm \gamma z}$$

where $\alpha \neq 0$, $\beta \neq 0$, and $\gamma^2 = \alpha^2 + \beta^2$.   If $\alpha = 0$, $\beta \neq 0$, separable solutions are

$$u = (ax + b)e^{\pm i\beta y}e^{\pm \beta z}$$

If $\alpha = \beta = 0$, separable solutions are

$$u = (ax + b)(cy + d)(ez + f)$$

where a, b, c, d, e, f are constants.   It is clear from symmetry that other forms of solution can be written down from the above forms.   For example,

$$u = e^{\pm \alpha x}e^{\pm i\beta y}e^{\pm i\gamma z}$$

where $\alpha^2 = \beta^2 + \gamma^2$ and $\beta \neq 0$, $\gamma \neq 0$, is a separable solution in rectangular coordinates.

25   A homogeneous solid bar occupies the region $0 \leq x \leq a$, $0 \leq y \leq b$, $0 \leq z \leq c$. There are no heat sources within the bar.   The base $z = 0$ is held at constant temperature T, while the remaining sides are held at $0°$.   Show that the steady-temperature distribution in the bar is given by

$$u(x,y,z) = \frac{16T}{\pi^2} \sum_{n=1}^{\infty} \sum_{m=1}^{\infty} \frac{\sin [(2n-1)\pi x/a] \sin [(2m-1)\pi y/b]}{(2n-1)(2m-1) \sinh \omega_{nm}c} \sinh [\omega_{nm}(c-z)]$$

where

$$\omega_{nm} = \left[\frac{(2n-1)^2\pi^2}{a^2} + \frac{(2m-1)^2\pi^2}{b^2}\right]^{1/2} \qquad n, m = 1, 2, \ldots$$

What is the temperature at the center of the bar?   *Hint:* The steady temperature u must satisfy Laplace's equation inside the bar.   The boundary conditions are

$$u(x,0,z) = u(x,b,z) = u(0,y,z) = u(a,y,z) = 0$$

and

$$u(x,y,0) = T \qquad u(x,y,c) = 0$$

As in Prob. 24 assume a separable solution $u = XYZ$.   Show that X must satisfy

$$X'' + \alpha^2 X = 0 \qquad 0 \leq x \leq a; X(0) = X(a) = 0$$

and hence the separation constant $\alpha = \alpha_n = n\pi/a$, $n = 1, 2, \ldots$.   Similarly show that

$$Y'' + \beta^2 Y = 0 \qquad 0 \leq y \leq b; Y(0) = Y(b) = 0$$

implies $\beta = \beta_m = m\pi/b$, $m = 1, 2, \ldots$.   The z-dependent factor must satisfy

$$Z'' - (\alpha_n{}^2 + \beta_m{}^2)Z = 0 \qquad Z(c) = 0$$

Let $\gamma_{nm} = (\alpha_n{}^2 + \beta_m{}^2)^{1/2}$, and derive the ∞^2 separable solutions

$$u_{nm}(x,y,z) = \sin \frac{n\pi x}{a} \sin \frac{m\pi y}{b} \sinh [\gamma_{nm}(c - z)] \qquad n = 1, 2, \ldots ; \; m = 1, 2, \ldots$$

which satisfy all the boundary conditions save one.   Determine the coefficients B_{nm} in the superposition

$$u(x,y,z) = \sum_{n=1}^{\infty} \sum_{m=1}^{\infty} B_{nm} \sin \frac{n\pi x}{a} \sin \frac{m\pi y}{b} \sinh [\gamma_{nm}(c - z)]$$

so as to satisfy the remaining condition.   Use the orthogonality properties

$$\int_0^b \int_0^a \varphi_{pq}(x,y)\varphi_{nm}(x,y) \, dx \, dy = 0 \qquad (p,q) \neq (n,m)$$

where

$$\varphi_{jk}(x,y) = \sin \frac{j\pi x}{a} \sin \frac{k\pi y}{b} \qquad j = 1, 2, \ldots ; \; k = 1, 2, \ldots$$

26   Solve Prob. 25 if instead the faces $x = 0$, $x = a$, $y = 0$, $y = b$ of the solid are insulated, while conditions on the top and bottom remain the same as stated there.   An insulated face means that the derivative of the temperature u in the direction of the normal to the face is zero.

27   An infinitely long bar of homogeneous material occupies the region $0 \leq x \leq a$, $0 \leq y \leq b$, $0 \leq z < \infty$.   There are no heat sources within the bar.   The base $z = 0$ is held at the temperature $Txy(x - a)(y - b)$, where T is a constant, while the sides are held at $0°$.   Also the temperature satisfies the condition $\lim_{z \to \infty} u(x,y,z) = 0$ uniformly in x, y for $0 \leq x \leq a$, $0 \leq y \leq b$.   Find the steady temperature in the solid.

28   (a) A thin thermally conducting homogeneous disk with insulated faces occupies the region $0 \leq r \leq a$ in the xy plane (where r, θ are polar coordinates).   The rim is held at $100°$.   What is the steady temperature in the disk?
(b) Solve the preceding problem if instead the temperature on the rim is held at $100\theta(1 - \theta/2\pi)$, $0 \leq \theta \leq 2\pi$.   What is the temperature at the center of the disk?

29   The disk of Prob. 28a has the following prescribed temperature on the rim $r = a$: $u = c$, c a constant, $0 < \theta < \alpha$, $u = 0$, $\alpha < \theta < 2\pi$, where α is a given angle, $0 < \alpha < 2\pi$.   Find the series expression for temperature at interior points of the disk.   In particular consider the case where $c = 100$ and $\alpha = \pi/2$.
    For this case use Poisson's integral (3-35) to derive a closed-form expression for the temperature inside the disk.   Use the closed-form expression to show that

$$\lim_{r \to a} u(r,\theta) = 100 \qquad 0 < \theta < \frac{\pi}{2}$$

$$\lim_{r \to a} u(r,\theta) = 0 \qquad \frac{\pi}{2} < \theta < 2\pi$$

What is the temperature at the center of the disk?

30   A thin homogeneous metal sheet with insulated faces occupies the region $0 \leq r \leq a$, $0 \leq \theta \leq \alpha$, in the xy plane.   Here r, θ are polar coordinates and α is given angle, $0 < \alpha < 2\pi$.   The temperature along the edges $\theta = 0$, $\theta = \alpha$ is held at zero.   On $r = a$, $0 < \theta < \alpha$, the temperature is given by $f(\theta)$, where f is a continuous function. Derive the series representation

$$u(r,\theta) = \sum_{n=1}^{\infty} B_n \left(\frac{r}{a}\right)^{n\pi/\alpha} \sin \frac{n\pi\theta}{\alpha}$$

for the temperature in the sheet, where

$$B_n = \frac{2}{\alpha} \int_0^{\alpha} f(\theta) \sin \frac{n\pi\theta}{\alpha} \, d\theta \qquad n = 1, 2, \ldots$$

Consider the particular case where $f(\theta) = 100$ and $\alpha = \pi/2$. Use the formula for the sum of the series given in Prob. 21 to derive the closed-form expression

$$u(r,\theta) = \frac{200 \tan^{-1} [2a^2 r^2 \sin 2\theta/(a^4 - r^4)]}{\pi}$$

for the temperature. Observe that $u(r,0) = u(r,\pi/2) = 0$, and verify that

$$\lim_{\substack{r \to a \\ r < a}} u(r,\theta) = 100 \qquad 0 < \theta < \frac{\pi}{2}$$

31 Find a function u harmonic in the region $0 \leq r < a$, $0 < \theta < \pi/2$ such that $u = 1$ on the edge $\theta = 0$, $0 < r < a$, $u = 0$ on the edge $\theta = \pi/2$, $0 < r < a$, and $u = 0$ on the rim $r = a$, $0 < \theta < \pi/2$. *Hint:* Consider that the harmonic function $v = 1 - 2\theta/\pi$ satisfies the boundary conditions along $\theta = 0$, $\theta = \pi/2$. Now construct a harmonic function w in the region such that the superposition $u = v + w$ satisfies the conditions of the problem.

32 A thin annulus occupies the region $0 < a \leq r \leq b$, $0 \leq \theta \leq 2\pi$, where $b > a$. The faces are insulated, and along the inner edge the temperature is maintained at $0°$, while along the outer edge the temperature is held at $100°$. Show that the temperature in the annulus is given by

$$u = 100 \frac{\log (r/a)}{\log (b/a)}$$

33 Determine the temperature distribution in the annulus of Prob. 32 if instead the temperature on the outer rim $r = b$ is held at $u = T \cos (\theta/2)$, $0 < \theta < 2\pi$, where T is a constant.

34 After division by r, and replacement of R by y and λ by $-\lambda$, show that Eq. (3-23) can be rewritten as

$$Ly + \lambda \rho y = D(rDy) + \frac{\lambda y}{r} = 0 \qquad 0 < r$$

Here $D = d/dr$, and the differential operator is self-adjoint (see Sec. 1, Appendix 2). Let $0 < a < b$, a, b fixed, and consider the self-adjoint Sturm-Liouville problem

$$Ly + \lambda \rho y = 0 \qquad a \leq r \leq b; y(a) = 0; y(b) = 0$$

(a) Review Sec. 1 of Appendix 2, and show, directly from the differential equation and the boundary conditions, that eigenfunctions y_n, y_m corresponding to distinct eigenvalues λ_n, λ_m, respectively, are orthogonal on $[a,b]$ with weight function $\rho = 1/r$:

$$\int_b^a \frac{y_n y_m}{r} \, dr = 0 \qquad n \neq m$$

and also that the eigenvalues are real and positive.

(b) Find the linearly independent solutions

$$y = \exp{(\pm i\omega \log r)} \qquad \omega = \sqrt{\lambda}; \; i = \sqrt{-1}$$

and hence the general solution

$$y = A \cos{(\omega \log r)} + B \sin{(\omega \log r)}$$

(c) Show that the eigenvalues are $\lambda_n = n^2\pi^2/[\log{(b/a)}]^2$, $n = 1, 2, \dots$, and that corresponding real-valued eigenfunctions are

$$y_n = \sin{\left(\omega_n \log{\frac{r}{a}}\right)} \qquad \omega_n = \sqrt{\lambda_n}; \; n = 1, 2, \dots$$

35 A thin thermally conducting sheet occupies the region $0 < a \leq r \leq b$, $0 \leq \theta < \alpha$, in the xy plane, where r, θ are polar coordinates, a and b are given numbers such that $a < b$, and α is a given angle, $0 < \alpha < 2\pi$. The edges $r = a$, $r = b$ are held at $0°$, as is also the edge $\theta = 0$. The edge $\theta = \alpha$ is held at $100°$. Find the steady temperature in the sheet.

36 Let C be the circle $r = a$ in the xy plane and $\mathscr{R}$ the region interior to C. Derive the solution of the boundary-value problem

$$\Delta u = x^2 - y^2 \qquad \text{in } \mathscr{R} \qquad\qquad u = 0 \qquad \text{on } C$$

in the form

$$u(r,\theta) = \frac{(r^4 - a^2 r^2) \cos{2\theta}}{12}$$

Hint: Examine the form of the function on the right in the differential equation, and derive the particular solution

$$v = \frac{x^4 - y^4}{12} = \frac{r^4 \cos{2\theta}}{12}$$

of the Poisson equation. Now derive the solution of the Dirichlet problem

$$\Delta w = 0 \qquad \text{in } \mathscr{R} \qquad\qquad w(a,\theta) = -v(a,\theta) \qquad \text{on } C$$

37 (a) A homogeneous thermally conducting cylinder occupies the region $0 \leq r \leq a$, $0 \leq \theta \leq 2\pi$, $0 \leq z \leq h$, where r, θ, z are cylindrical coordinates. There are no sources of heat within the cylinder. The top $z = h$ and the lateral surface $r = a$ are held at $0°$, while the base $z = 0$ is held at $100°$. Find the steady-temperature distribution within the cylinder.

(b) Solve the problem in a if the top is held at $100°$ instead of $0°$, the remaining conditions being the same.

38 The cylinder of Prob. 37 has its base held at $100°$. The lateral surface and the top radiate into an infinite medium, which is at temperature $0°$. Thus there are the boundary conditions

$$\left(\frac{\partial u}{\partial r} + \gamma u\right)\bigg|_{r=a} = 0 \qquad \left(\frac{\partial u}{\partial z} + \gamma u\right)\bigg|_{z=h} = 0$$

where $\gamma > 0$ is a constant.

Derive the solution

$$u = 400\gamma a \sum_{n=1}^{\infty} \frac{J_0(\xi_n r/a)}{(\xi_n^2 + \gamma^2 a^2) J_0(\xi_n)} \left[\xi_n \cosh \frac{\xi_n(h-z)}{a} + \gamma a \sinh \frac{\mu_n(h-z)}{a} \right]$$

where $0 < \xi_1 < \xi_2 < \cdots$ are the positive roots of the equation

$$\xi J_0'(\xi) + \gamma a J_0(\xi) = 0$$

39   A wedge-shaped solid occupies the region described by the inequalities $0 \le r \le a$, $0 \le \theta \le \beta$, $0 \le z \le h$, where β is a given angle, $0 < \beta < 2\pi$. The top $z = h$, the lateral surface $r = a$, and the faces $\theta = 0$ and $\theta = \beta$ are insulated. The base $z = 0$ is held at temperature $f(r,\theta)$. Derive the expression for the steady temperature in the solid if there are no sources of heat within. Consider the special case $f(r,\theta) = 100$.

Sec. 3-4

40   In potential theory it is shown that the gravitational potential ψ due to matter distributed in space satisfies Laplace's equation in regions free of matter, and in a region containing matter of density p satisfies Poisson's equation

$$\Delta \psi = -4\pi\rho$$

If S is a simple closed surface which bounds a region $\mathscr{R}$ of space containing matter of density ρ, and if the region exterior to S is free of matter, the potential and its first partial derivatives are continuous across S. Also, the potential in this case must vanish uniformly at infinity. Let S be a sphere of radius a, and suppose the interior of S contains matter of constant density ρ. Derive the expression for the potential **(a)** at points inside S, and **(b)** at points outside S. *Hint:* Use spherical coordinates with origin at the center of the sphere. Then ψ is spherically symmetric and satisfies

$$\psi_{rr} + \frac{2\psi_r}{r} = \begin{cases} -4\pi\rho & 0 \le r \le a \\ 0 & r > a \end{cases}$$

Integrate the equations directly and impose the requirements which the potential must satisfy.

41   Let $0 < a < b$, a, b fixed numbers. The spherical annulus in space which is bounded by the spheres $r = a$, $r = b$ is filled with matter. The density varies according to the formula $\rho = 1/r$, where r is the distance from the origin. Derive the expressions in spherical coordinates for the potential in the regions $0 \le r \le a$, $0 \le r \le b$, $r \ge b$.

42   Determine the electrostatic potential φ in the annular region bounded by the concentric spheres $r = a$, $r = b$, $0 < a < b$, if the inner sphere $r = a$ is held at constant potential V_0 and the outer sphere $r = b$ is held at constant potential V_1, $V_1 \ne V_0$.

43   A homogeneous thermally conducting solid is bounded by the concentric spheres $r = a$, $r = b$, $0 < a < b$. There are no heat sources within the solid. The inner surface $r = a$ is held at constant temperature u_1, and at the outer surface there is radiation into the medium $r > b$, which is at constant temperature u_2. Determine the steady temperature u in the solid. *Hint:* The steady temperature is spherically symmetric and satisfies Laplace's equation in regions where there are no sources. At $r = b$ the boundary condition is

$$\frac{\partial u}{\partial r} + h(u - u_2) = 0 \qquad h \text{ a positive constant}$$

44 Heat is generated at a constant rate Q within a homogeneous solid ball of radius a. The surface $r = a$ is held at the constant temperature T. Show that the steady temperature inside the ball is given by

$$u = \frac{Q(a^2 - r^2)}{6K} + T$$

What is the net flux of heat out through the surface $r = a$? *Hint:* The temperature must be finite, spherically symmetric, and satisfy Poisson's equation

$$\Delta u = -\frac{Q}{K}$$

inside the sphere.

45 Solve Prob. 44 if instead there is radiation of heat out into the region $r > a$ and the external medium has constant temperature zero. In this case the boundary condition at $r = a$ is

$$\frac{\partial u}{\partial r} + hu = 0$$

46 (a) Let (r, θ, φ) be spherical coordinates, as shown in Fig. 3-3. A homogeneous solid ball of radius a contains no heat sources. The portion of the surface defined by $r = a$, $0 \leq \theta < \pi/2$, is held at a constant temperature T, while the remainder is at temperature zero. Show that the temperature inside the ball is given by

$$u = \frac{T}{2} + \frac{T}{2} \sum_{k=0}^{\infty} \left(\frac{r}{a}\right)^{2k+1} [P_{2k}(0) - P_{2k+2}(0)]P_{2k+1}(\cos \theta)$$

(b) Solve the problem in **a** if instead the bottom hemisphere is held at temperature $-T$. *Hint:* In **b** let u denote the solution of the original problem in **a**. What properties do the functions $v = 2u$, $w = -T$ possess?

47 Determine the temperature in the ball of Prob. 46 if instead the surface temperature is $u = T(1 + 2 \sin^2 \theta)$, $0 \leq \theta \leq \pi$, where T is a constant.

48 The solid described in Prob. 43 has the inner surface $r = a$ held at the temperature $u = f_1(\theta)$, and the outer surface $r = b$ is held at the temperature $u = f_2(\theta)$, where f_1, f_2 are given functions of θ. Show that the steady temperature in the solid is given by

$$u(r, \theta) = \sum_{n=0}^{\infty} \left(C_n r^n + \frac{D_n}{r^{n+1}}\right) P_n(\cos \theta)$$

where

$$C_n = \frac{a^{n+1}A_n - b^{n+1}B_n}{a^{2n+1} - b^{2n+1}} \qquad D_n = \frac{a^{-n}A_n + b^{-n}B_n}{a^{-(2n+1)} - b^{-(2n+1)}}$$

$$A_n = \frac{2n+1}{2} \int_0^{\pi} f_1(\theta) P_n(\cos \theta) \sin \theta \, d\theta$$

$$B_n = \frac{2n+1}{2} \int_0^{\pi} f_2(\theta) P_n(\cos \theta) \sin \theta \, d\theta$$

$n = 0, 1, 2, \ldots$.   Consider the particular case where $f_1(\theta) = T_1$, $f_2(\theta) = T_2(1 - \cos \theta)$,

T_1, T_2 constants.   *Hint:* The temperature is axially symmetric.   Assume a solution

$$u = \sum_{n=0}^{\infty} \left(C_n r^n + \frac{D_n}{r^{n+1}} \right) P_n(\cos \theta)$$

The boundary conditions imply

$$\sum_{n=0}^{\infty} \left(C_n a^n + \frac{D_n}{a^{n+1}} \right) P_n(\cos \theta) = f_1(\theta)$$

$$\sum_{n=0}^{\infty} \left(C_n b^n + \frac{D_n}{b^{n+1}} \right) P_n(\cos \theta) = f_2(\theta)$$

Let m be a fixed nonnegative integer. Multiply both sides of each equation by $P_m(\cos \theta) \sin \theta$ and integrate over $0 \le \theta \le \pi$. Use the orthogonality properties of the Legendre polynomials to obtain two equations in the unknowns C_m, D_m.

49 A homogeneous conducting solid hemisphere is bounded by the xy plane and the surface $r = a$, $0 \le \theta \le \pi/2$.   The curved surface is held at the temperature $u = T(1 - \cos \theta)$, T a constant.   The base is insulated.   Find the steady temperature in the solid.   *Hint:* Insulated base means $\dfrac{1}{r} \dfrac{\partial u}{\partial \theta}\Big|_{\theta = \pi/2} = 0.$

Sec. 3-5

50 Let $\bar{\mathscr{R}}$ be the closed rectangle $0 \le x \le a$, $0 \le y \le b$. Show that the eigenfunction expansion of the solution of the boundary-value problem

$$\Delta \Psi = -2 \quad \text{in } \mathscr{R} \qquad \Psi = 0 \quad \text{on } C$$

where C is the boundary of the rectangle and $\mathscr{R}$ is its interior, is

$$\Psi(x,y) = \frac{32}{\pi^4} \sum_{n=1}^{\infty} \sum_{m=1}^{\infty} \varphi_{nm}(x,y)$$

where

$$\varphi_{nm} = \frac{\sin [(2n-1)\pi x/a] \sin [(2m-1)\pi y/b]}{(2n-1)(2m-1)[(2n-1)^2/a^2 + (2m-1)^2/b^2]}$$

51 Find the eigenfunction expansion of the solution of the boundary-value problem in Prob. 23.

52 Let $\bar{\mathscr{R}}$ be the closed rectangle $0 \le x \le a$, $0 \le y \le b$.   Find the eigenfunction expansion of the solution of the boundary-value problem

$$D \Delta \varphi - c\varphi = F \quad \text{in } \mathscr{R} \qquad \frac{\partial \varphi}{\partial n} = 0 \quad \text{on } C$$

where D, c, and F are positive constants, C is the boundary of the rectangle, and $\mathbf{n}$ is the exterior normal on C.

53 (a) Let Δ be the three-dimensional laplacian.   Let $\bar{\mathscr{R}}$ be the closed parallelepiped in xyz space defined by $0 \le x \le a$, $0 \le y \le b$, $0 \le z \le c$, $\mathscr{R}$ the interior of $\bar{\mathscr{R}}$, and S the bounding surface of the parallelepiped.   Derive the complete set of eigenfunctions of the eigenvalue problem

$$\Delta\varphi + \lambda\varphi = 0 \quad \text{in } \mathscr{R} \qquad \varphi = 0 \quad \text{on } S$$

$$\varphi_{nmp} = \sin\frac{n\pi x}{a} \sin\frac{m\pi y}{b} \sin\frac{p\pi z}{c} \qquad n = 1, 2, \ldots; m = 1, 2, \ldots; p = 1, 2, \ldots$$

and show that the eigenvalues are

$$\lambda_{nmp} = \pi^2\left(\frac{n_2}{a^2} + \frac{m^2}{b^2} + \frac{p^2}{c^2}\right)$$

(b) Let $\mathscr{R}$ and S be as described in **a** above.   Find the eigenfunction expansion of the solution of the boundary-value problem

$$\Delta u = xyz \quad \text{in } \mathscr{R} \qquad u = 0 \quad \text{on } S$$

54   Find the eigenfunction expansion of the solution of the boundary-value problem

$$\Delta u = a^2 - r^2 \quad \text{in } \mathscr{R} \qquad u = 0 \quad \text{on } C$$

where C is the circle of radius a about the origin, $\mathscr{R}$ denotes the region interior to C, and r and θ are polar coordinates.

55   Solve the boundary-value problem in Prob. 50 by means of an eigenfunction expansion if the region $\mathscr{R}$ is the region interior to the circle $r = a$ in the xy plane.

56   Let $\mathscr{R}$ and C be as described in Prob. 54.   Show that the eigenfunction expansion of the solution of the boundary-value problem

$$\Delta\varphi = x^2 - y^2 \quad \text{in } \mathscr{R} \qquad \varphi = 0 \quad \text{on } C$$

is

$$\varphi(r,\theta) = -\left[2a^4 \sum_{m=1}^{\infty} \frac{J_2(\xi_{2m}r/a)}{\xi_{2m}{}^3 J_3(\xi_{2m})}\right] \cos 2\theta$$

57   In the theory of elasticity it is shown that if a thin elastic plate (of uniform thickness) lies with its midplane in the xy plane and a surface force density f (force/unit area) acts in the vertical z direction, the vertical deflection $w(x,y)$ satisfies the fourth-order elliptic-type partial differential equation

$$\Delta\Delta w = \frac{f(x,y)}{N} \qquad \Delta\Delta w = \frac{\partial^4 w}{\partial x^4} + 2\frac{\partial^4 w}{\partial x^2\,\partial y^2} + \frac{\partial^4 w}{\partial y^4}$$

where N is a material constant.

(a) The homogeneous equation $\Delta\Delta w = 0$ is called the *biharmonic equation*.   Show that if u is a harmonic function in a region $\mathscr{R}$, then u is a solution of the biharmonic equation in $\mathscr{R}$.

(b) Show that if u, v are harmonic in $\mathscr{R}$, then $w = xu + v$ is biharmonic in $\mathscr{R}$.   Hence the general solution of the biharmonic equation is

$$w = F(x + iy) + xF(x + iy) + G(x - iy) + xG(x - iy)$$

where $i = \sqrt{-1}$ and F, G are arbitrary functions.   *Hint:* Consider the vector identity

$$\Delta(\varphi\psi) = \varphi\,\Delta\psi + \psi\,\Delta\varphi + 2\nabla\varphi \cdot \nabla\psi$$

(c) Let $\bar{\mathscr{R}}$ be the closed rectangle $0 \leq x \leq a, 0 \leq y \leq b$ and let C be its boundary.   Consider the eigenvalue problem

$$\Delta\Delta u + \lambda u = 0 \qquad \text{in } \mathscr{R} \qquad\qquad u = \Delta u = 0 \qquad \text{on } C$$

Show that all the eigenvalues are negative.   *Hint:* The reality of the eigenvalues can be shown by an argument similar to that given for the self-adjoint problem discussed in the text.   To show the eigenvalues are positive, let λ be an eigenvalue and u a corresponding eigenfunction.   Then

$$\int_{\mathscr{R}} u\,\Delta\Delta u\,dx\,dy = -\lambda \int_{\mathscr{R}} u^2\,dx\,dy$$

Now in the Green's formula for the operator Δ replace v by u and $\nabla^2 u$ by $\Delta\Delta u$.   Then

$$\int_{\mathscr{R}} (u\,\Delta\Delta u - |\Delta u|^2)\,dx\,dy = \int_c \left[u\,\frac{\partial}{\partial n}(\Delta u) - \Delta u\,\frac{\partial u}{\partial n} \right] ds = 0$$

since $u = \Delta u = 0$ on C.   Hence

$$\lambda \int_{\mathscr{R}} u^2\,dx\,dy = -\int_{\mathscr{R}} |\Delta u|^2\,dx\,dy \leq 0$$

and $\lambda = 0$ if, and only if, $\Delta u = 0$ in $\mathscr{R}$.   But if $\Delta u = 0$ in $\mathscr{R}$, then $u = 0$ everywhere on $\bar{\mathscr{R}}$.

(d) Since $\lambda < 0$, let $\lambda = -\omega^2$, ω real and positive.   Consider that

$$(\Delta - \omega)(\Delta + \omega) = (\Delta + \omega)(\Delta - \omega) = \Delta^2 - \omega^2 = \Delta\Delta - \omega^2$$

Thus if u satisfies

$$\Delta u + \omega u = 0 \qquad \text{in } \mathscr{R} \qquad\qquad u = \Delta u = 0 \qquad \text{on } C$$

then u is an eigenfunction of the eigenvalue problem in **c** corresponding to eigenvalue $-\omega^2$. But by Example 3-5 the eigenfunctions of this problem are $\varphi_{nm}(x,y) = \sin(n\pi x/a) \times \sin(m\pi y/b)$ with corresponding eigenvalues

$$\omega_{nm} = \frac{n^2\pi^2}{a^2} + \frac{m^2\pi^2}{b^2} \qquad n = 1, 2, \ldots\,; m = 1, 2, \ldots$$

Hence show that the values $\lambda_{nm} = -\omega_{nm}^2$ are eigenvalues of the problem in **c**, with φ_{nm} corresponding eigenfunctions; that is,

$$\Delta\Delta\varphi_{nm} - \omega_{nm}^2\varphi_{nm} = 0 \qquad \varphi_{nm} = \Delta\varphi_{nm} = 0 \qquad \text{on } C$$

The completeness of the orthogonal sequence $\{\varphi_{nm}\}$ implies that these are all the eigenvalues.

(e) If there are no vertical deflections and no bending moments along the edge of the rectangular plate, the deflection w is the solution of the boundary-value problem

$$\Delta\Delta w = \frac{f(x,y)}{N} \qquad \text{in } \mathscr{R} \qquad\qquad w = \Delta w = 0 \qquad \text{on } C$$

(these boundary conditions are often termed the *Navier conditions*).   Assume the solution is

$$w(x,y) = \sum_{n=1}^{\infty} \sum_{m=1}^{\infty} c_{nm}\varphi_{nm}(x,y)$$

Then $w = \Delta w = 0$ on C.   Since w is a solution of the differential equation, it follows that

$$\sum_{n=1}^{\infty} \sum_{m=1}^{\infty} c_{nm} \omega_{nm}{}^2 \varphi_{nm}(x,y) = \frac{f(x,y)}{N}$$

Let p, q be a fixed pair of positive integers.   Follow the procedure of Example 3-5, and use the orthogonality properties of the sequence $\{\varphi_{nm}\}$ to show that

$$c_{pq} = \frac{A_{pq}}{N\omega_{pq}{}^2} = \frac{4}{N\omega_{pq}{}^2 ab} \int_0^a \int_0^b f(x,y) \sin \frac{n\pi x}{a} \sin \frac{m\pi y}{b} \, dx \, dy$$

Hence the solution

$$w(x,y) = \frac{1}{Nab} \sum_{n=1}^{\infty} \sum_{m=1}^{\infty} \frac{A_{nm}}{\omega_{nm}{}^2} \sin \frac{n\pi x}{a} \sin \frac{m\pi y}{b}$$

(f) Show that if the plate is uniformly loaded so that $f(x,y) = f_0$, a constant, the deflection is

$$w(x,y) = \frac{16 f_0}{\pi^6 N} \sum_{n=1}^{\infty} \sum_{m=1}^{\infty} \frac{\sin [(2n-1)\pi x/a] \sin [(2m-1)\pi y/b]}{(2n-1)(2m-1)\beta_{nm}}$$

where

$$\beta_{nm} = \left[\frac{(2n-1)^2}{a^2} + \frac{(2m-1)^2}{b^2} \right]^2$$

If the plate is square, show that the deflection at the center is approximately $4 f_0 a^4/\pi^6 N$.

4 THE WAVE EQUATION

4-1 INTRODUCTION

The prototype of a class of hyperbolic equations and one of the most important differential equations of mathematical physics is the *wave equation*

$$\frac{\partial^2 u}{\partial t^2} = c^2 \, \Delta u \tag{4-1}$$

The symbol Δu denotes the laplacian of u, and c is a real positive constant. Often the variable t has the significance of time. Equation (4-1) is called the *one-, two-,* or *three- dimensional wave equation* according as the laplacian is one-, two-, or three-dimensional. This differential equation occurs in such diverse fields of study as electromagnetic theory, hydrodynamics, acoustics, elasticity, and quantum theory. It is important in these subjects because solutions of the equation furnish a mathematical description of waves and propagation of effects. Physically the solutions may represent waves of electric or magnetic intensity, waves of acoustic pressure, transverse or longitudinal displacement waves in a solid, or other phenomena. A solution of Eq. (4-1) is often termed a *wave function*.

The *inhomogeneous wave equation* is

$$\frac{\partial^2 u}{\partial t^2} - c^2 \, \Delta u = F \tag{4-2}$$

From a physical point of view the given function F represents a known *external driving force*, or a *field source*, which is independent of the *state variable u*. However F may be a function of the space variables as well as time t. A related equation of importance in many applications is

$$\frac{\partial^2 u}{\partial t^2} + 2\gamma \, \frac{\partial u}{\partial t} - c^2 \, \Delta u = F \tag{4-3}$$

where γ denotes a real positive constant. Equation (4-3) is called the *wave equation with damping term*. As a result of the presence of the first derivative there exist solutions of Eq. (4-3) which represent waves whose amplitudes decrease exponentially with increasing time. Such waves appear when there are present within the system dissipative forces (frictional, etc.) which effect an energy loss in the propagated wave. A hyperbolic equation which

160

includes Eqs. (4-2) and (4-3) as special cases is the *telegrapher's equation*

$$\frac{\partial^2 u}{\partial t^2} + 2\gamma \frac{\partial u}{\partial t} + \beta u - c^2 \Delta u = F \tag{4-4}$$

where β and γ denote real constants.

In most problems arising in mathematical physics a function is sought which satisfies prescribed auxiliary conditions as well as the given differential equation. In a type of problem which occurs frequently the state of the system is known completely at an initial instant, i.e., at $t = 0$. Then a solution of the differential equation in some region of space and time is desired such that the solution reduces to the known values at $t = 0$. Usually the solution is determined for $t \geq 0$, so that the state of the system is found for all subsequent time. Such a problem is termed an *initial-value problem*. Now it may happen that the region in space over which the solution is desired is bounded, and the values of the solution are prescribed (known) on the boundary for all $t \geq 0$ as well as at $t = 0$ for all points inside the boundary of the spatial region. A problem of this type is called a *boundary- and initial-value problem*. In this chapter problems of these two types are considered for the wave equation in one, two, and three space dimensions.

4-2 ONE-DIMENSIONAL WAVE EQUATION. INITIAL-VALUE PROBLEM

The homogeneous wave equation in one dimension is

$$\frac{\partial^2 u}{\partial t^2} = c^2 \frac{\partial^2 u}{\partial x^2} \tag{4-5}$$

Observe that Eq. (4-5) has constant coefficients. Thus it follows from the results of Sec. 2-2 that the general solution of Eq. (4-5) is

$$u = \varphi(x - ct) + \psi(x + ct) \tag{4-6}$$

where φ, ψ are arbitrary functions. Assume now that φ is a given function. Then

$$u(x,t) = \varphi(x - ct)$$

represents a wave traveling to the right with speed c whose shape does not change as it moves. To see this let t denote time, and consider the state of affairs at $t = 0$. The function $u(x,0) = \varphi(x)$ is called the *wave profile*. At time $t_1 = 1/c$ we have $u(x,t_1) = \varphi(x - 1)$. Let $x' = x - 1$; then $\varphi(x - 1) = \varphi(x')$. Hence the graph of $u = u(x,t_1)$ is the graph of the wave profile translated one unit to the right along the x axis. At the instant $t_2 = 2/c$ the graph of $u = u(x,t_2)$ is the graph of the wave profile translated two units to the right.   Accordingly as t increases, the wave profile moves to the right along the x axis (see Fig. 4-1).   In particular, at time $t = 1$ we have $u(x,1) = \varphi(x - c)$.   In one unit of time the profile has moved c units to the right. Thus the constant c is the *wave velocity* (or speed of propagation).   In the same manner it is seen that

$$u(x,t) = \varphi(x + ct)$$

represents the wave profile $u(x,0) = \varphi(x)$ traveling to the left along the x axis with speed c.   Hence the general solution (4-6) represents the superposition of two arbitrary wave profiles traveling with a common speed but in opposite directions along the x axis.   Each wave is unchanged in form as it moves.

Let k denote an arbitrary real parameter.   It is clear that

$$u = \varphi[k(x - ct)] + \psi[k(x + ct)]$$

is a solution of the wave equation for any pair of functions φ, ψ.   Let $\omega = kc$; then the solution can be rewritten

$$u = \varphi(kx - \omega t) + \psi(kx + \omega t) \tag{4-7}$$

Conversely, a function of this form is a solution of Eq. (4-5) only if $\omega = kc$. It follows that traveling waves with a speed different from c cannot be described by solutions of the one-dimensional wave equation (4-5).   The function ζ defined by

$$\zeta(x,t) = kx - \omega t \tag{4-8}$$

is called the *phase*.   From Eq. (2-35) the characteristic curves of the wave equation are the straight lines

$$x \pm ct = \text{const}$$

in the xt plane.   Thus curves of constant phase in one dimension are characteristic curves.

Plane Harmonic Waves

Simple but very useful wave functions are obtained from Eq. (4-7) if φ and ψ are chosen to be exponential functions with imaginary argument.   A

Figure 4-1

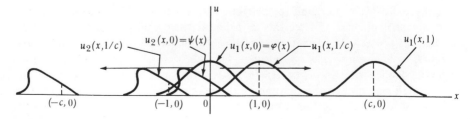

wave function of the form

$$u = Ae^{ikx \pm i\omega t} \qquad \omega = kc; \ i = \sqrt{-1}$$

where A is a constant, is called a *plane harmonic wave* (also monochromatic wave). Such exponential-type solutions were derived previously in Example 2-6. Since the real and imaginary parts of u are also solutions of Eq. (4-5), it follows that the functions

$$u = A \cos (kx \pm \omega t) \qquad u = A \sin (kx \pm \omega t)$$

represent plane waves. The constant A is called the *amplitude,* and ω is termed the *angular frequency.* Often k is called the *wave number.* The variation of each of the above functions is simple harmonic in time at a fixed point x and simple harmonic in space at a fixed time t. Recall that the space frequency $k = \omega/c$. Consider the function

$$u = A \sin (kx - \omega t) \tag{4-9}$$

This represents the sinusoidal profile $u(x,0) = A \sin kx$ traveling to the right along the x axis with speed c (see Fig. 4-2). Note that the dimension of k is reciprocal length. The value $\lambda = 2\pi/k$ is called the *wavelength,* and $1/\lambda$ is called the *wave number.* The wave number is the number of waves per unit of distance along the x axis. Consider an observer stationed at a fixed point x_0. Since

$$u\left(x_0, t + \frac{\lambda}{c}\right) = A \sin \left[kx_0 - \omega\left(t + \frac{\lambda}{c}\right)\right]$$

$$= A \sin \left[kx_0 - \omega\left(t + \frac{2\pi}{kc}\right)\right] = A \sin (kx_0 - \omega t) = u(x_0,t)$$

it is clear that exactly one complete wave passes the observer in time $T = \lambda/c$. This interval of time is called the *period* of the wave. The *frequency* $f = 1/T$ is the number of waves passing a fixed point in unit time. Observe that

Figure 4-2

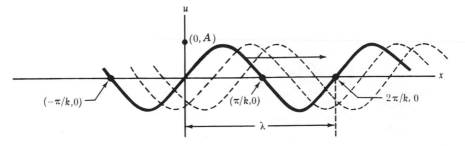

$\omega = 2\pi f$. A fundamental relation between the frequency, wavelength, and speed of propagation is

$$c = f\lambda \tag{4-10}$$

The function $u = A \cos(kx - \omega t)$ represents the wave train just described except that the phase differs by $\pi/2$.

Not every solution of the wave equation represents a traveling wave. Consider the superposition of two traveling sinusoidal waves which have the same amplitude, speed, and frequency but which travel in opposite directions

$$u = A \sin[k(x - ct)] + A \sin[k(x + ct)]$$
$$= (2A \cos kct) \sin kx$$

This is a *standing wave*. The profile $u(x,0) = 2A \sin kx$ is not propagated; instead, the amplitude factor $2A \cos kct = 2A \cos \omega t$ varies sinusoidally with frequency ω (see Fig. 4-3). The points $x_n = n\pi/k$, $n = 0, \pm 1, \pm 2, \ldots$, are called *nodes*. Observe that $u(x_n,t) = 0$ for all t.

Solution of the Initial-value Problem

The initial-value problem

$$u_{tt} - c^2 u_{xx} = F(x,t) \qquad -\infty < x < \infty; t \geq 0$$
$$u(x,0) = f(x) \qquad u_t(x,0) = g(x) \qquad -\infty < x < \infty \tag{4-11}$$

is the Cauchy problem for the one-dimensional wave equation where the curve on which the data f and g are prescribed is the x axis. The given function F is assumed continuous in x and t, and f and g are assumed twice continuously differentiable and continuously differentiable, respectively. An expression for the solution u in terms of the given functions $F, f,$ and g can be derived as follows. Let (x_0,t_0) be a fixed point, with $t_0 > 0$, but otherwise arbitrarily chosen. Let $\mathcal{D}$ be the region in the xt plane bounded by the characteristics

$$x - ct = x_0 - ct_0 \qquad x + ct = x_0 + ct_0$$

Figure 4-3

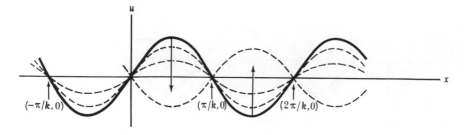

through (x_0, t_0) and the segment $[x_0 - ct_0, x_0 + ct_0]$ of the x axis (see Fig. 4-4). Recall the definition of a self-adjoint second-order differential operator given in Sec. 2-6. It is easy to verify that the operator

$$L = -c^2 D_x{}^2 + D_t{}^2$$

is self-adjoint. Comparison with Eq. (2-80) shows that $n = 2$, and

$$A_{11} = -c^2 \qquad A_{12} = A_{21} = 0 \qquad A_{22} = 1 \qquad C = 0$$

Note that $x_1 = x$, $x_2 = t$. If Green's formula (2-81) is written in terms of L, then

$$\iint_{\mathscr{D}} (\psi L \varphi - \varphi L \psi) \, dx \, dt = \int_C (P_1 \nu_x + P_2 \nu_t) \, ds \qquad (4\text{-}12)$$

holds for every pair of twice continuously differentiable functions ψ, φ on $\mathscr{D}$. In Eq. (4-12) the line integral is taken in the positive sense (counterclockwise) around the triangle C which forms the boundary of $\mathscr{D}$. From Eq. (2-82) the functions

$$P_1 = A_{11}(\psi \varphi_x - \varphi \psi_x) = -c^2(\psi \varphi_x - \varphi \psi_x)$$
$$P_2 = A_{22}(\psi \varphi_t - \varphi \psi_t) = \psi \varphi_t - \varphi \psi_t$$

The direction cosines of the exterior normal along C are denoted by ν_x, ν_t. Choose $\psi = 1$, $\varphi = u$ in Green's formula. Then

$$\iint_{\mathscr{D}} F(x,t) \, dx \, dt = \int_C (-c^2 u_x \nu_x + u_t \nu_t) \, ds \qquad (4\text{-}13)$$

since $Lu = F$. Now the line integral around C equals the sum of the line integrals along C_1, C_2, and C_3, where C_1 is the base of the triangle and C_2

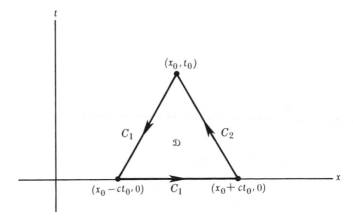

Figure 4-4

and C_3 are the remaining sides. On C_1 the direction cosines $\nu_x = 0$, $\nu_t = -1$, while on C_2 and C_3 the direction cosines are

$$\nu_x = \frac{1}{(1 + c^2)^{\frac{1}{2}}} \qquad \nu_t = \frac{c}{(1 + c^2)^{\frac{1}{2}}}$$

$$\nu_x = \frac{-1}{(1 + c^2)^{\frac{1}{2}}} \qquad \nu_t = \frac{c}{(1 + c^2)^{\frac{1}{2}}}$$

respectively. Also $ds = dx$ on C_1, and

$$ds = -(1 + c^2)^{\frac{1}{2}} \frac{dx}{c} = (1 + c^2)^{\frac{1}{2}} \, dt$$

$$ds = -(1 + c^2)^{\frac{1}{2}} \frac{dx}{c} = -(1 + c^2)^{\frac{1}{2}} \, dt$$

on the oriented contours C_2 and C_3, respectively. From the foregoing it follows that

$$\int_{C_1} (-c^2 u_x \nu_x + u_t \nu_t) \, ds = -\int_{x_0-ct_0}^{x_0+ct_0} u_t \, dx = -\int_{x_0-ct_0}^{x_0+ct_0} g(x) \, dx$$

$$\int_{C_2} (-c^2 u_x \nu_x + u_t \nu_t) \, ds = c \int_{C_2} (u_x \, dx + u_t \, dt) = c \int_{C_2} du$$

$$= c[u(x_0,t_0) - u(x_0 + ct_0, 0)] = c[u(x_0,t_0) - f(x_0 + ct_0)]$$

$$\int_{C_3} (-c^2 u_x \nu_x + u_t \nu_t) \, ds = -c \int_{C_3} du = c[u(x_0,t_0) - f(x_0 - ct_0)]$$

$$\int_{C} (-c^2 u_x \nu_x + u_t \nu_t) \, ds = c[2u(x_0,t_0) - f(x_0 + ct_0) - f(x_0 - ct_0)]$$

$$- \int_{x_0-ct_0}^{x_0+ct_0} g(x) \, dx$$

Thus

$$u(x_0,t_0) = \tfrac{1}{2}[f(x_0 + ct_0) + f(x_0 - ct_0)] + \frac{1}{2c} \int_{x_0-ct_0}^{x_0+ct_0} g(x) \, dx$$

$$+ \frac{1}{2c} \iint_{\mathscr{D}} F(x,t) \, dx \, dt$$

Since (x_0,t_0) was an arbitrary chosen point, it follows that the solution must have the representation

$$u(x,t) = \tfrac{1}{2}[f(x + ct) + f(x - ct)] + \frac{1}{2c} \int_{x-ct}^{x+ct} g(\xi) \, d\xi$$

$$+ \frac{1}{2c} \iint_{\mathscr{D}} F(\xi,\eta) \, d\xi \, d\eta \quad (4\text{-}14)$$

Conversely, if f is twice continuously differentiable, if g is continuously differentiable, and if F has continuous second partial derivatives with respect to x and t, Eq. (4-14) defines a twice continuously differentiable function u which satisfies all the conditions of problem (4-11). Consequently Eq. (4-14) yields the unique solution of the initial-value problem for the one-dimensional wave equation.

Let

$$v(x,t) = \tfrac{1}{2}[f(x+ct) + f(x-ct)] + \frac{1}{2c}\int_{x-ct}^{x+ct} g(\xi)\,d\xi \qquad (4\text{-}15)$$

$$w(x,t) = \frac{1}{2c}\iint_{\mathscr{D}} F(\xi,\eta)\,d\xi\,d\eta \qquad (4\text{-}16)$$

Then the solution of problem (4-11) is the superposition

$$u = v + w$$

where the function v is the solution of the initial-value problem for the homogeneous wave equation

$$v_{tt} - c^2 v_{xx} = 0 \qquad -\infty < x < \infty;\ t \geq 0$$

$$v(x,0) = f(x) \qquad v_t(x,0) = g(x) \qquad -\infty < x < \infty \qquad (4\text{-}17)$$

and w is the solution of the inhomogeneous wave equation with homogeneous initial conditions

$$w_{tt} - c^2 w_{xx} = F(x,t) \qquad -\infty < x < \infty;\ t \geq 0$$

$$w(x,0) = 0 \qquad w_t(x,0) = 0 \qquad -\infty < x < \infty \qquad (4\text{-}18)$$

The function v defined in Eq. (4-15) is called *D'Alembert's solution* of the initial-value problem (4-17). In order to discuss the nature of this solution let

$$u_1(x,t) = \tfrac{1}{2}[f(x+ct) + f(x-ct)] \qquad (4\text{-}19)$$

Then u_1 is the solution of the problem when the prescribed function g is the identically zero function. It is represented here as superposition of two traveling waves which have the same profile and speed but which travel in opposite directions along the x axis. Evidently the initial values $f(x)$ are *propagated* to distant points as time goes on. Indeed if f vanishes outside of a finite interval about the origin and x_0 is a point sufficiently far distant, then $u_1(x_0,t) = 0$ during an interval of time $0 \leq t < t_0$. Here t_0 is time of arrival at x_0 of the leading edge of the outgoing wave. Furthermore there is a time $t_1 > t_0$ such that for $t > t_1$ the value $u_1(x_0,t) = 0$, and so the wave has passed the point x_0. This is illustrated in Fig. 4-1.

Let h be a primitive of the prescribed function g in problem (4-16). Then the second term in D'Alembert's solution can be written

$$u_2(x,t) = \frac{1}{2c} \left[h(x + ct) - h(x - ct) \right] \qquad (4\text{-}20)$$

Observe that u_2 is the unique solution of the problem when the prescribed function f is identically zero. This function can be viewed as the superposition of the traveling waves $h(x + ct)/2c$ and $-h(x - ct)/2c$. The wave $-h(x - ct)/2c$ is the reflection across the x axis of the wave $h(x - ct)/2c$ and has the profile $-h(x)/2c$. In this case it is the integral of the initial values $u_t(x,0) = g(x)$ which is propagated, and this implies an important difference in the nature of the propagation. Even if g vanishes outside of a bounded interval, the primitive h need not, and hence the waves $h(x + ct)/2c$, $-h(x - ct)/2c$ will be of infinite extent. The effect of the initial values of u_t, which are initially confined to some interval about the origin, is such that once the value of u at a distant point x_0 is influenced by the incoming wave, it is influenced for all subsequent time t.

Significance of the Characteristic Curves

The pair of characteristic curves of the wave equation which pass through a given point (x_0,t_0) in the xt plane have the equations

$$x - ct = x_0 - ct_0 \qquad x + ct = x_0 + ct_0 \qquad (4\text{-}21)$$

These are sketched in Fig. 4-5. The characteristics intercept the segment $[x_0 - ct_0, x_0 + ct_0]$ of the x axis. D'Alembert's solution shows that the value $u(x_0,t_0)$ of a solution u of the wave equation is uniquely determined by the initial values $u(x,0)$, $u_t(x,0)$ along the segment $x_0 - ct_0 \leq x \leq x_0 + ct_0$. For this reason the interval $[x_0 - ct_0, x_0 + ct_0]$ is termed the *interval of dependence of the point* (x_0,t_0). Note that the initial values $u(x,0)$, $u_t(x,0)$ can

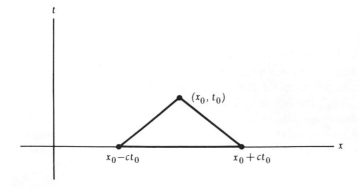

Figure 4-5

be modified arbitrarily outside this interval without changing the value $u(x_0,t_0)$.

From another point of view, let $[a,b]$ be a given interval of the x axis. Draw the characteristics shown in Fig. 4-6a through the points $(a,0)$, $(b,0)$ in the xt plane. The shaded region $\mathscr{D}$ bounded by these lines is called the *domain of determinacy* of $[a,b]$. The significance of this region is as follows. Let u be a solution of the wave equation (4-5). If (x_0,t_0) is a point of $\mathscr{D}$, the interval of dependence of (x_0,t_0) is contained in $[a,b]$. Accordingly the values of u are uniquely determined everywhere in $\mathscr{D}$ by the initial values $u(x,0)$, $u_t(x,0)$ for $a \le x \le b$. In particular, if $u(x,0) = 0$, $u_t(x,0) = 0$ for $a \le x \le b$, then u vanishes identically in $\mathscr{D}$. Equivalently, two solutions of Eq. (4-5) whose initial values coincide for $a \le x \le b$ will coincide in value throughout the region (even though they may differ outside of $\mathscr{D}$). Note also that the initial values $u(x,0)$, $u_t(x,0)$ of a solution can be altered arbitrarily at points of the x axis outside of $[a,b]$, and the values of u in $\mathscr{D}$ remain unchanged.

In Fig. 4-6b the shaded region $\mathscr{R}$ is called the *domain of influence of the interval* $[a,b]$. The boundaries of $\mathscr{R}$ are the half-characteristics shown. If u is a solution of the wave equation and (x_0,t_0) is a point of $\mathscr{R}$, the value $u(x_0,t_0)$ is influenced by the initial values $u(x,0)$, $u_t(x,0)$ in $[a,b]$. This follows from the fact that the interval of dependence of (x_0,t_0) overlaps $[a,b]$. On the other hand, the initial values of u on $[a,b]$ in no way affect the values of u at points outside $\mathscr{R}$. Accordingly the significance of the characteristics is that they form the boundaries of the domain of determinacy and of the domain of influence. The characteristics which form these boundaries can be thought of as the paths in space-time, i.e., in phase space, of signals which leave the end points a, b of the interval at $t = 0$ and travel with speed c to the left and right along the x axis. Finally, if a perturbation in the initial values is made, say for values x in the interval $[x_0 - \epsilon, x_0 + \epsilon]$ about a given point x_0, the perturbation will at a subsequent time $t_1 > 0$ have influenced

Figure 4-6

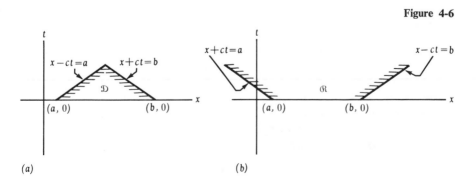

(a)   (b)

all those points in the domain of influence of $[x_0 - \epsilon, x_0 + \epsilon]$ which are bounded above by the horizontal line $t = t_1$.

Initial-value Problem Is Properly Posed

In the class of data described above the existence and uniqueness of solution of the initial-value problem (4-11) is deduced from the representation (4-14). This representation also shows that in an arbitrary fixed time interval $0 \le t \le T$ the solution depends continuously on the data. Suppose u_1 is the solution corresponding to data f_1 and g_1 and u_2 is the solution corresponding to data f_2 and g_2. Then if

$$|f_1(x) - f_2(x)| < \epsilon \qquad |g_1(x) - g_2(x)| < \epsilon \qquad -\infty < x < \infty$$

it follows that, for all x, and for $0 \le t \le T$,

$$|u_1(x,t) - u_2(x,t)| = \left| \tfrac{1}{2}[f_1(x + ct) + f_1(x - ct) - f_2(x + ct) - f_2(x - ct)] \right.$$

$$\left. + \frac{1}{2c} \int_{x-ct}^{x+ct} [g_1(\xi) - g_2(\xi)]\, d\xi \right|$$

$$\le \tfrac{1}{2}[|f_1(x + ct) - f_2(x + ct)| + |f_1(x - ct) - f_2(x - ct)|]$$

$$+ \frac{1}{2c} \int_{x-ct}^{x+ct} |g_1(\xi) - g_2(\xi)|\, d\xi < \tfrac{1}{2}(\epsilon + \epsilon) + \frac{1}{2c} 2cT\epsilon$$

Accordingly the initial-value problem (4-11) is properly posed.

4-3   VIBRATING STRING.   SEPARATION OF VARIABLES

The vibrating string with fastened ends furnishes the classical example of a boundary- and initial-value problem involving the one-dimensional wave equation. A homogeneous thin string, perfectly flexible and under uniform tension, lies in its equilibrium position along the x axis. The ends of the string are fastened at $x = 0$, $x = b > 0$. The string is pulled aside and released (not necessarily with zero speed). It is desired to determine the subsequent motion of the string.

Mathematical Model

The motion of the string is assumed purely transverse, and the tension T is in the direction of the tangent to the string. Choose axes as shown in Fig. 4-7, and let $u(x,t)$ denote the transverse displacement at time t of the point on the string whose abscissa is x. In addition to the tension an external force transverse $F(x,t)$ (force/unit mass) may act on this string, e.g., gravity. Let Δs be an element of the string; then the net transverse force on Δs due to the tension is approximately

$$T_0(u_x|_{x_2} - u_x|_{x_2}) \qquad T_0 = T_1 \cos \alpha_1 = T_2 \cos \alpha_2$$

where x_1, x_2 are the abscissas of the end points of the element. Here the assumption of small slopes along the deflection curve is made. The equation of motion of the element Δs is

$$T_0\left(\frac{\partial u}{\partial x}\bigg|_{x_1} - \frac{\partial u}{\partial x}\bigg|_{x_2}\right) + \rho\,\Delta s\,F(\bar{x},t) = \rho\,\Delta s\,\frac{\overline{\partial^2 u}}{\partial t^2}$$

where $\overline{\partial^2 u/\partial t^2}$ denotes the acceleration of the center of mass, $F(\bar{x},t)$ is the mean intensity of the external force acting, and ρ is the linear density of the string. Let $\Delta x = x_2 - x_1$. Then $\Delta s \approx \Delta x$ to within first-order terms in Δx. If the equation of motion is divided through by Δx and the limit of both sides is taken as $\Delta x \to 0$, the resulting equation is

$$\frac{\partial^2 u}{\partial t^2} - c^2\frac{\partial^2 u}{\partial x^2} = F(x,t) \qquad 0 \le x \le b;\, t \ge 0 \tag{4-22}$$

where

$$c = \sqrt{\frac{T_0}{\rho}}$$

Since the ends of the string are fastened, the boundary conditions are

$$u(0,t) = 0 \qquad u(b,t) = 0 \qquad t \ge 0 \tag{4-23}$$

The string is released from a known position and with known speed. Hence there are the initial conditions

$$u(x,0) = f(x) \qquad u_t(x,0) = g(x) \qquad 0 \le x \le b \tag{4-24}$$

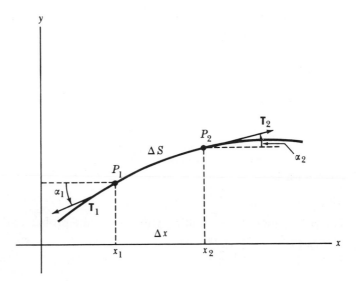

Figure 4-7

where f, g are prescribed functions. The problem is to determine a solution of the partial differential equation (4-22) which satisfies the boundary conditions (4-23) and initial conditions (4-24). It should be noted that physical considerations dictate that u be continuous in x and t together. Thus the boundary conditions (4-23) imply

$$u(0,0) = f(0) = 0 \qquad u(b,0) = f(b) = 0$$

Also the fastened ends imply the relations

$$u_t(0,0) = g(0) = 0 \qquad u_t(b,0) = g(b) = 0$$

Formal Series Solution when External Forces Are Absent

In the event there is no external driving force on the string the function F vanishes identically. In this case one speaks of *free vibrations*. The subsequent motion of the string is described by the solution u of the problem

$$
\begin{aligned}
u_{tt} - c^2 u_{xx} &= 0 \qquad 0 \le x \le b;\, t \ge 0 \\
u(0,t) &= 0 \qquad u(b,t) = 0 \qquad t \ge 0 \\
u(x,0) &= f(x) \qquad u_t(x,0) = g(x) \qquad 0 \le x \le b
\end{aligned}
\tag{4-25}
$$

A formal solution of this problem can be derived by separation of variables. Assume a separable solution

$$u(x,t) = X(x)T(t) \tag{4-26}$$

of the homogeneous wave equation. If the partial derivatives are calculated and substitution is made into the wave equation, the relation

$$X\ddot{T} = c^2 X'' T$$

is obtained. Here dots and primes refer to derivatives taken with respect to t and x, respectively. Division by XT yields

$$\frac{X''}{X} = -\lambda = \frac{\ddot{T}}{c^2 T}$$

where λ is a separation constant. Thus the factors X, T must satisfy

$$X'' + \lambda X = 0 \qquad \ddot{T} + c^2 \lambda T = 0$$

respectively. If the boundary conditions of problem (4-25) are imposed on the assumed form of solution (4-26), it follows that the boundary conditions on X must be

$$X(0) = 0 \qquad X(b) = 0$$

The eigenvalues of the resulting Sturm-Liouville problem are

$$\lambda_n = \frac{n^2 \pi^2}{b^2} \qquad n = 1, 2, \ldots$$

and the corresponding eigenfunctions are

$$X_n = \sin \frac{n\pi x}{b}$$

The general solution of the differential equation for T is

$$T_n(t) = A_n \cos \frac{n\pi ct}{b} + B_n \sin \frac{n\pi ct}{b}$$

where A_n, B_n are arbitrary constants. Thus separable solutions of the homogeneous wave equation which satisfy the homogeneous boundary conditions of problem (4-25) have the form

$$u_n(x,t) = \sin \frac{n\pi x}{b}\left(A_n \cos \frac{n\pi ct}{b} + B_n \sin \frac{n\pi ct}{b} \right) \qquad n = 1, 2, \ldots \qquad (4\text{-}27)$$

The functions u_n are called the *normal modes* of vibration, and the numbers

$$\omega_n = \frac{n\pi c}{b} \qquad n = 1, 2, \ldots \qquad\qquad\qquad (4\text{-}28)$$

are called the *characteristic* (or normal) *frequencies*. Observe that for each fixed positive integer n, the nth normal mode u_n is the solution of problem (4-25) corresponding to the initial conditions

$$f(x) = A_n \sin \frac{n\pi x}{b} \qquad g(x) = \omega_n B_n \sin \frac{n\pi x}{b}$$

that is, sinusoidal initial displacement and initial distribution of speed along the length of the string. It is easily verified that the expression for u_n can be rewritten

$$u_n(x,t) = C_n \sin \frac{n\pi x}{b} \cos (\omega_n t - \epsilon_n) \qquad\qquad\qquad (4\text{-}29)$$

where the amplitude C_n and phase ϵ_n are given by

$$C_n = (A_n{}^2 + B_n{}^2)^{\frac{1}{2}} \qquad \epsilon_n = \tan^{-1}\frac{B_n}{A_n}$$

Clearly then each normal mode is a standing wave. Note that the characteristic frequencies are independent of initial conditions, being determined solely by the homogeneous boundary conditions in problem (4-25).

A superposition

$$u(x,t) = \sum_{n=1}^{\infty} \sin \frac{n\pi x}{b} (A_n \cos \omega_n t + B_n \sin \omega_n t) \qquad\qquad (4\text{-}30)$$

of normal modes will satisfy the boundary conditions as well as the wave equation, provided the series converges in a suitable fashion. Assume that this is the case. Then the function u defined by the series (4-30) will satisfy the initial conditions of problem (4-25) if

$$u(x,0) = \sum_{n=1}^{\infty} A_n \sin \frac{n\pi x}{b} = f(x) \qquad 0 \leq x \leq b \tag{4-31}$$

$$u_t(x,0) = \sum_{n=1}^{\infty} \omega_n B_n \sin \frac{n\pi x}{b} = g(x) \qquad 0 \leq x \leq b \tag{4-32}$$

Equations (4-31) and (4-32) suggest that the series on the left is the Fourier sine series of the function on the right in each equation. Indeed if f, g are piecewise smooth on $[0,b]$, the coefficients A_n, B_n must be given by

$$A_n = \frac{2}{b} \int_0^b f(x) \sin \frac{n\pi x}{b} \, dx \qquad n = 1, 2, \ldots \tag{4-33}$$

$$B_n = \frac{2}{b\omega_n} \int_0^b g(x) \sin \frac{n\pi x}{b} \, dx \qquad n = 1, 2, \ldots \tag{4-34}$$

the Fourier sine coefficients of f and g/ω_n, respectively. The formal solution of problem (4-25) is given by the series (4-30), where the coefficients A_n, B_n are defined in Eqs. (4-33) and (4-34).

Formal Series Solution when External Forces Occur

Consider now the problem

$$
\begin{aligned}
&u_{tt} - c^2 u_{xx} = F(x,t) && 0 \leq x \leq b; t \geq 0 \\
&u(0,t) = 0 \quad u(b,t) = 0 && t \geq 0 \\
&u(x,0) = 0 \quad u_t(x,0) = 0 && 0 \leq x \leq b
\end{aligned}
\tag{4-35}
$$

This problem has homogeneous boundary and initial conditions. Physically it corresponds to the case where an external driving force is applied and the string is released from rest from the equilibrium position. To derive a formal series solution assume that

$$u(x,t) = \sum_{n=1}^{\infty} \varphi_n(t) \sin \frac{n\pi x}{b} \tag{4-36}$$

where the functions φ_n are to be determined subsequently. Evidently if the series on the right in Eq. (4-36) is suitably convergent, the function so defined satisfies the boundary conditions. It will satisfy the initial conditions if

$$\varphi_n(0) = 0 \qquad \dot{\varphi}_n(0) = 0 \qquad n = 1, 2, \ldots \tag{4-37}$$

In order to determine the form of the φ_n calculate the derivatives u_{tt}, u_{xx} and substitute into the inhomogeneous wave equation. The result is

$$\sum_{n=1}^{\infty} [\ddot{\varphi}_n(t) + \omega_n^2 \varphi_n(t)] \sin \frac{n\pi x}{b} = F(x,t)$$

where dots denote differentiation with respect to t. Let k be a fixed, but otherwise arbitrarily chosen, positive integer. Multiply both sides of the preceding equation by $\sin (k\pi x/b)$, integrate with respect to x from $x = 0$ to $x = b$, and interchange the order of summation and integration. From the orthogonality properties of the functions $\sin (n\pi x/b)$, $n = 1, 2, \ldots$, it follows that

$$\ddot{\varphi}_k(t) + \omega_k^2 \varphi_k(t) = F_k(t) \tag{4-38}$$

where $\omega_k = k\pi c/b$ and

$$F_k(t) = \frac{2}{b} \int_0^b F(x,t) \sin \frac{k\pi x}{b} \, dx \tag{4-39}$$

The solution of the differential equation which satisfies zero initial conditions is

$$\varphi_k(t) = \frac{1}{\omega_k} \int_0^t F_k(\xi) \sin [\omega_k(t - \xi)] \, d\xi \tag{4-40}$$

Hence the formal solution of problem (4-35) is

$$u(x,t) = \sum_{n=1}^{\infty} \left\{ \frac{1}{\omega_n} \int_0^t F_n(\xi) \sin [\omega_n(t - \xi)] \, d\xi \right\} \sin \frac{n\pi x}{b} \tag{4-41}$$

where the functions F_n are defined by Eq. (4-39) with n replacing k.

By the linearity of the differential operator appearing in the wave equation and the linearity of the boundary- and initial-value conditions, the solution of the problem consisting of Eqs. (4-22) to (4-24) can be obtained by superposition. Thus if v is a solution of problem (4-25) and w is a solution of problem (4-35), then

$$u = v + w$$

is a solution of the problem consisting of Eqs. (4-22) to (4-24). The formal solution is

$$u(x,t) = \sum_{n=1}^{\infty} (A_n \cos \omega_n t + B_n \sin \omega_n t) \sin \frac{n\pi x}{b}$$

$$+ \sum_{n=1}^{\infty} \left\{ \frac{1}{\omega_n} \int_0^t F_n(\xi) \sin [\omega_n(t - \xi)] \, d\xi \right\} \sin \frac{n\pi x}{b} \tag{4-42}$$

where the coefficients A_n, B_n are given by Eqs. (4-33) and (4-34) and $\omega_n = n\pi c/b$. As mentioned before, the function v represented by the first of the series in Eq. (4-42) constitutes the free vibrations of the string, i.e., the motion in the absence of any external forces. It is due to the prescribed initial conditions and the tensile forces within the string. On the other hand, the function w represented by the second of the series in Eq. (4-42) constitutes the *forced motion* of the string. This is the motion due to the external forces only, since it starts from quiescent initial conditions. The motion of the string in the general case is the superposition of the free vibrations and the forced motion.

EXAMPLE 4-1 The idealized string with fastened ends described in the text has its midpoint pulled transversely aside a distance h and is then released from rest. In addition to the tensile force the force of gravity acts on the string. Determine the subsequent motion.

The axes are chosen as shown in Fig. 4-8. The initial conditions are

$$u(x,0) = \begin{cases} \dfrac{2hx}{b} & 0 \le x \le \dfrac{b}{2} \\[2mm] \dfrac{2h(b-x)}{b} & \dfrac{b}{2} \le x \le b \end{cases}$$

$$u_t(x,0) = 0 \qquad 0 \le x \le b$$

The boundary conditions are given by Eq. (4-23). The external force/unit mass is

$$F(x,t) = -g \qquad g = \text{gravitational constant}$$

From Eqs. (4-33) and (4-34) the coefficients

$$A_n = \frac{2}{b} \left\{ \int_0^{b/2} \frac{2hx}{b} \sin \frac{n\pi x}{b} \, dx + \int_{b/2}^b \frac{2h(b-x)}{b} \sin \frac{n\pi x}{b} \, dx \right.$$

$$= \frac{8h}{n^2\pi^2} \sin \frac{n\pi}{2} \qquad n = 1, 2, \ldots$$

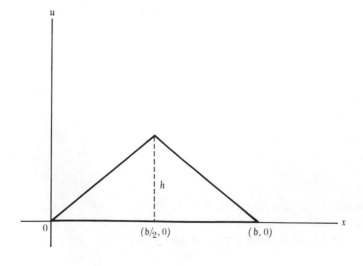

0 $(b/2, 0)$ $(b, 0)$

Figure 4-8

while $B_n = 0$ for all n.   Substitution of $-g$ into Eq. (4-39) yields

$$F_k(t) = \frac{2}{b} \int_0^b (-g) \sin \frac{k\pi x}{b} \, dx = \frac{2g}{k\pi} (\cos k\pi - 1) \qquad k = 1, 2, \ldots$$

Hence

$$\varphi_k(t) = \frac{2g}{k\pi \omega_k} (\cos k\pi - 1) \int_0^t \sin [\omega_k(t - \xi)] \, d\xi$$

$$= \frac{2g}{k\pi \omega_k^2} (\cos k\pi - 1)(1 - \cos \omega_k t)$$

From Eq. (4-42) the formal solution is

$$u(x,t) = \frac{8h}{\pi^2} \sum_{n=1}^{\infty} \frac{(-1)^{n+1}}{(2n - 1)^2} \sin \frac{(2n - 1)\pi x}{b} \cos \frac{(2n - 1)\pi ct}{b}$$

$$- \frac{4b^2 g}{\pi^3 c^2} \sum_{n=1}^{\infty} \frac{1}{(2n - 1)^3} \sin \frac{(2n - 1)\pi x}{b} \left[1 - \cos \frac{(2n - 1)\pi ct}{b} \right]$$

The first series represents the motion of the string if it is released from rest with initial displacement $f(x)$ at $t = 0$ and neglecting gravity.   The second series represents the motion of the string when released from rest in its equilibrium position along the axis and with the external force of gravity present.

The solution derived above is termed a *formal solution* inasmuch as it has not been proved that the series converges and represents a function which actually does satisfy all the conditions of the problem.   The inequalities

$$\left| \frac{\sin [(2n - 1)\pi x/b] \cos [(2n - 1)\pi ct/b]}{(2n - 1)^2} \right| \le \frac{1}{(2n - 1)^2}$$

$$\left| \frac{\sin [(2n - 1)\pi x/b]\{1 - \cos [(2n - 1)\pi ct/b]\}}{(2n - 1)^3} \right| \le \frac{2}{(2n - 1)^3}$$

hold for all x and t.   Accordingly by the Weierstrass M test the series converges for all x and t and converges uniformly on $0 \le x \le b$, $0 \le t \le T$ for each fixed $T > 0$.   This implies that the series defines a continuous function u on $0 \le x \le b$, $t \ge 0$ (indeed u is defined and continuous for all x, t, periodic in x for each fixed t, and periodic in t for each fixed x).   Clearly u satisfies the boundary conditions.   The fact that u satisfies the first initial condition follows from the uniform convergence of

$$u(x,0) = \sum_{n=1}^{\infty} A_n \sin \frac{n\pi x}{b} \qquad 0 \le x \le b$$

where the A_n are the Fourier coefficients of f with respect to the orthogonal sequence $\{\sin (n\pi x/b)\}$, and the uniform convergence of this Fourier series of f to $f(x)$, $0 \le x \le b$. The series obtained by differentiating once with respect to t converges and at $t = 0$ converges to zero.   Hence $u_t(x,0) = 0$, $0 \le x \le b$.   Unfortunately, however, u does not satisfy the wave equation, and so it is not a solution of the problem.   Note that the series obtained by differentiating termwise twice with respect to x fails to converge.

Remarks on the Formal Series Solution

. Example 4-1 illustrates the fact that the formal series derived by the separation-of-variables technique need not be a solution of the boundary- and initial-value problem. Implicit in the preceding statement is the definition of the phrase "solution of the partial differential equation." The classical definition requires of the function continuous derivatives up to the order of the partial differential equation. For the wave equation this means that the solution must have continuous second partial derivatives with respect to the independent variables. Now in regard to the problem in Example 4-1, and, indeed, whenever the separation-of-variables technique is applied to boundary- and initial-value problems, the following important question arises: if the formal series fails to produce a classical solution, can a classical solution be obtained by another method? The answer is no; in fact, a classical solution cannot exist. To see this for the case of the wave equation suppose that u is a twice continuously differentiable solution of problem (4-25). If t is fixed, then u, considered as a function of x, can be expanded in a Fourier sine series on the interval $[0,b]$:

$$u(x,t) = \sum_{n=1}^{\infty} C_n(t) \sin \frac{n\pi x}{b} \qquad 0 \leq x \leq b$$

with uniform and absolute convergence, where the Fourier coefficients

$$C_n(t) = \frac{2}{b} \int_0^b u(x,t) \sin \frac{n\pi x}{b} \, dx$$

Integrate twice by parts and make use of $u(0,t) = u(b,t) = 0$. Then

$$C_n(t) = -\frac{2b}{n^2\pi^2} \int_0^b u_{xx}(x,t) \sin \frac{n\pi x}{b} \, dx$$

Since $u_{xx} = u_{tt}/c^2$,

$$C_n(t) = -\frac{2b}{n^2\pi^2 c^2} \int_0^b u_{tt}(x,t) \sin \frac{n\pi x}{b} \, dx$$

Now let t be variable, $t \geq 0$. Since u is twice continuously differentiable, differentiation of the original expression for C_n twice with respect to t gives

$$\ddot{C}_n(t) = \frac{2}{b} \int_0^b u_{tt}(x,t) \sin \frac{n\pi x}{b} \, dx$$

Comparison shows that $\ddot{C}_n + \omega_n^2 C_n = 0$. Hence

$$C_n(t) = a_n \cos \omega_n t + b_n \sin \omega_n t$$

Again by reference to the original expression for C_n one has

$$C_n(0) = \frac{2}{b} \int_0^b u(x,0) \sin \frac{n\pi x}{b} \, dx$$

$$= \frac{2}{b} \int_0^b f(x) \sin \frac{n\pi x}{b} \, dx = A_n$$

Thus $a_n = A_n$.   In the same way it follows that $b_n = B_n$ for all n.   Accordingly u must be given by Eq. (4-30); that is, u is obtainable by the separation-of-variables technique.

The requirement that the solution be twice continuously differentiable is often an excessive requirement, particularly in problems having a physical origin and where discontinuities are present in the solution or in its derivatives.   In this connection the concept of a *generalized solution* of a partial differential equation enters into the study of partial differential equations and boundary- and initial-value problems.   It is not possible in this introductory text to develop the important concepts of *generalized function* and generalized solution in a systematic way.   Suffice it to say that formal series solutions often define generalized solutions, and the formal manipulations of such functions which are often carried out in practice can be rigorously justified.   For further study of the subject the reader is referred to the bibliography in Appendix 3.

Existence and Uniqueness of Solution

If stronger hypotheses are placed on the given functions f, g, and F in the problem defined by Eqs. (4-22) to (4-24), the formal solution (4-42) can be shown to be the solution in the classical sense.   For example, if f is four times continuously differentiable, if g is three times continuously differentiable, if F is three times continuously differentiable in x and t, and if

$$f(0) = f''(0) = f(b) = f''(b) = g(0) = g''(0) = g(b) = g''(b) = 0$$

$$F(0,t) = F_{xx}(0,t) = F(b,t) = F_{xx}(b,t) = 0 \qquad t \geq 0$$

then Eq. (4-42) defines a twice continuously differentiable function which satisfies all conditions of the boundary- and initial-value problem (see Prob. 43).   It is interesting to note that uniqueness of solution can be proved under much weaker hypotheses.   Moreover, the proof is motivated by physical considerations.   From the expression for the kinetic energy of a mass element dm it follows that the kinetic energy of an element ds of the vibrating string is approximately $\rho u_t^2 \, dx/2$.   The net kinetic energy of motion is then

$$\tfrac{1}{2}\rho \int_0^b u_t^2 \, dx$$

The (elastic) potential energy of the string is

$$\frac{T_0}{2} \int_0^b u_x^2 \, dx$$

(see Prob. 20).   Hence the sum of the kinetic and potential energy is

$$E(t) = \frac{\rho}{2} \int_0^b u_t^2 \, dx + \frac{T_0}{2} \int_0^b u_x^2 \, dx = \frac{\rho}{2} \int_0^b (u_t^2 + c^2 u_x^2) \, dx$$

The time rate of change of the total energy is

$$\frac{dE}{dt} = \rho \int_0^b (u_t u_{tt} + c^2 u_x u_{xt}) \, dx$$

Recall that the solution u is required to have continuous second derivatives. Thus the following identity holds:

$$\frac{\partial}{\partial x} (u_t u_x) = u_t u_{xx} + u_x u_{tx} = u_t u_{xx} + u_x u_{xt}$$

Substitution into the integrand gives

$$\frac{dE}{dt} = \rho \int_0^b u_t (u_{tt} - c^2 u_{xx}) \, dx + \rho c^2 \int_0^b \frac{\partial}{\partial x} (u_t u_x) \, dx$$

$$= \rho \int_0^b u_t F(x,t) \, dx + \rho c^2 [u_t(b,t)u_x(b,t) - u_t(0,t)u_x(0,t)] \qquad (4\text{-}43)$$

Equation (4-43) shows that the rate of change of total energy is the sum of the energy rate of change due to the external force and the energy rate due to the tensile forces acting at the ends $x = 0$, $x = b$ of the string (recall that $c^2 = T_0/\rho$).   In the case of fastened ends no work is done on the ends of the string, and the second term on the right in Eq. (4-43) vanishes.   If in addition there is no external force acting on the string, Eq. (4-43) shows that dE/dt is identically zero, and hence E is constant, $t \geq 0$.   Accordingly in the case where the motion of the string is given by the solution u of problem (4-25), the energy is the constant

$$E = E(0) = \frac{\rho}{2} \int_0^b \{[g(x)]^2 + c^2 [f'(x)]^2\} \, dx$$

that is, the sum of the initial kinetic and potential energies.

In the boundary- and initial-value problem defined by Eqs. (4-22) to (4-24) assume that the given functions f and g are twice continuously differentiable and F is continuous.   To prove there is at most one solution suppose v and w are solutions, and let

$$\psi = v - w$$

By the linearity of the differential equation and auxiliary conditions it follows that ψ is a solution of the homogeneous wave equation and satisfies homogeneous boundary and initial conditions. Thus ψ describes the motion of the string released from rest when in its equilibrium position and with no external forces present. It is expected that no subsequent displacement occurs, and this follows from the preceding expression for the energy. Since $E(0) = 0$, the energy at any time is

$$E = \frac{\rho}{2} \int_0^b (\psi_t{}^2 + c^2\psi_x{}^2)\, dx = 0$$

The integrand is continuous and nonnegative. Hence the vanishing of the integral implies

$$\psi_t = 0 \qquad \psi_x = 0 \qquad 0 \leq x \leq b; \, t \geq 0$$

Thus, $\psi(x,t) = \text{const}$, $0 \leq x \leq b$, $t \geq 0$. But $\psi(x,0) = 0$. Accordingly

$$\psi(x,t) = v(x,t) - w(x,t) = 0 \qquad 0 \leq x \leq b; \, t \geq 0$$

This completes the proof of the uniqueness of solution.

In the xt plane the curve C on which data are given is the polygonal U-shaped curve consisting of the nonnegative half of the t axis, the segment of the x axis with end points $x = 0$, $x = b$, and the part of the line $x = b$ for which $t \geq 0$ (see Fig. 4-9). The region in which the solution is sought is the unbounded region whose boundary is C and in which $t \geq 0$. The initial data (4-24) prescribe the value of u and its normal derivative u_t along the portion of C coinciding with the x axis. The boundary conditions (4-23) prescribe the value of u along the remaining portions of C. It will be seen in the subsequent paragraph why the normal derivative of u, that is, u_x, cannot be prescribed on the vertical parts of C. The data given along the segment of the x axis together with the value of u prescribed on the vertical segments serve to determine u_x on the vertical segments.

Significance of the Characteristic Curves

Recall that the characteristic curves of the one-dimensional wave equation are the lines

$$x \pm ct = \text{const}$$

in the xt plane. Through each point of the xt plane there pass exactly two characteristics. A solution u of the homogeneous wave equation (4-5) has the following useful property along the characteristics. Let

$$p = u_x \qquad q = u_t$$

Then

$$cp + q = \text{const}$$

holds along each characteristic $x + ct =$ const (where it is understood the values of the constants need not be the same), and

$$cp - q = \text{const}$$

holds along each characteristic $x - ct =$ const. To show that the first relation holds, consider that the differential

$$
\begin{aligned}
d(cp + q) &= cu_{xx}\, dx + cu_{xt}\, dt + u_{tx}\, dx + u_{tt}\, dt \\
&= (cu_{xx} + u_{xt})\, dx + (cu_{xt} + u_{tt})\, dt \\
&= (cu_{xx} + u_{xt})(-c\, dt) + (cu_{xt} + u_{tt})\, dt \\
&= (-c^2 u_{xx} + u_{tt})\, dt \\
&= 0
\end{aligned}
$$

along $x + ct =$ const.   The second relation is shown to hold in the same way.   Now by means of these properties and the diagram in Fig. 4-9 it can be seen that the characteristics are the paths in space-time along which the initial values of the solution of problem (4-25) are propagated.   Select a value $t_0 > 0$, and draw the horizontal line $t = t_0$.   Choose a point $P(x_0, t_0)$ on this line such that $0 < x_0 < b$.   Draw the pair of characteristics through $P(x_0, t_0)$.   As shown in Fig. 4-9, let these intersect the vertical lines $x = 0$, $x = b$ at P_1 and P_2, respectively.   Draw the characteristic $x + ct =$ const through P_1 and the characteristic $x - ct =$ const through P_2.   Draw the remaining characteristics, as shown in the figure, until the points P_4, P_5 on the x axis are obtained.   It is convenient to let the subscripts $1, \ldots, 5$ refer to evaluation at the points $P_1, \ldots, P_5$, respectively.   By what was shown in

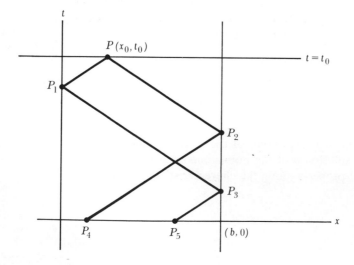

Figure 4-9

the preceding paragraph the relation

$$cp - q = cp_5 - q_5 = \text{const}$$

holds along the line segment $P_5 P_3$.   Now $u(b,t) = 0$ holds on $x = b$.   Hence $q_3 = 0$.   Thus

$$cp_3 = cp_5 - q_5$$

By the same type of reasoning it follows that

$$cp_1 = cp_3$$

On the segment $P_1 P$ the relation

$$cp_1 = cp - q = \text{const}$$

holds.   Hence

$$cu_x(x_0,t_0) - u_t(x_0,t_0) = cp_5 - q_5$$

By a similar procedure the equation

$$cu_x(x_0,t_0) + u_t(x_0,t_0) = cp_4 - q_4$$

is obtained.   These two equations show that the values of u_x, u_t at $P(x_0,t_0)$ are uniquely determined in terms of the values of u_x, u_t at P_4, P_5.   But these latter values are known because of the prescription of u and u_t at $t = 0$, $0 \leq x \leq b$.   Since x_0 was arbitrarily chosen to within the requirement $0 < x_0 < b$, and since u_x, u_t are known at $x = 0$, $x = b$, for $t \geq 0$ in virtue of the boundary conditions, it follows that u_x, u_t are uniquely determined at $t = t_0$ for $0 \leq x \leq b$.   Now $u(0,t_0) = 0$.   Hence

$$u(x,t_0) = \int_0^x u_x(\xi,t_0)\, d\xi \qquad 0 \leq x \leq b$$

and the values of u are known at $t = t_0$.   Since $t_0 > 0$ was arbitrarily chosen, it follows that the initial data given at $t = 0$ for $0 \leq x \leq b$, together with the value of u along $x = 0$, $x = b$ for $t \geq 0$, uniquely determine the solution u for $t > 0$, $0 \leq x \leq b$.

Several techniques for the numerical integration of hyperbolic equations, subject to prescribed initial conditions, are based on properties of the characteristic curves illustrated for the wave equation in the preceding paragraphs. These are termed the *method of characteristics*.   In the case of the inhomogeneous wave equation the relation

$$cp + q = \text{const} \qquad \text{on } x + ct = a = \text{const}$$

is replaced by

$$cp + q = \int F(a - ct, t)\, dt + \text{const}$$

and similarly for the remaining relation.   Of course in the general hyperbolic case the characteristics need not be straight lines in the xt plane, and this may increase the difficulty of numerical integration along a characteristic.

4-4 TWO-DIMENSIONAL WAVE EQUATION. INITIAL-VALUE PROBLEM

The two-dimensional homogeneous wave equation is

$$u_{tt} = c^2 \, \Delta u \tag{4-44}$$

where Δ denotes the two-dimensional laplacian.   The laplacian is written in terms of the coordinate system most convenient to the boundary-value problem at hand.   If rectangular coordinates are used, Eq. (4-44) becomes

$$u_{tt} = c^2(u_{xx} + u_{yy}) \tag{4-45}$$

Solutions of Eq. (4-45) involving arbitrary functions are easily obtained. Indeed Eq. (4-6) defines a solution for each choice of functions φ, ψ.   In the context of this chapter such solutions constitute a special class of two-dimensional waves, namely, waves with direction of propagation parallel to the x axis.   Wave functions of a more general type can be derived as follows. Assume that

$$u = \eta(x,y,t)\varphi[\zeta(x,y,t)] \tag{4-46}$$

satisfies Eq. (4-45) for each choice of φ; that is, η, ζ constitute a functionally invariant pair of Eq. (4-45).   Now Eqs. (2-16) to (2-18) imply that the functions η, ζ must satisfy the system of partial differential equations

$$\begin{aligned} \zeta_t{}^2 - c^2 \, |\nabla \zeta|^2 &= 0 \\ 2\eta_t\zeta_t + \eta\zeta_{tt} &= c^2(2\nabla\eta \cdot \nabla\zeta + \eta \, \Delta\zeta) \\ \eta_{tt} &= c^2 \, \Delta\eta \end{aligned} \tag{4-47}$$

Here ∇ denotes the two-dimensional gradient operator.   Conversely, if η, ζ satisfy the system, then for each choice of φ the function u defined by Eq. (4-46) is a wave function.   Observe that ζ must satisfy

$$Q(\zeta) = \zeta_x{}^2 + \zeta_y{}^2 - \frac{\zeta_t{}^2}{c^2} = 0 \tag{4-48}$$

the characteristic equation of (4-45).   Recall the occurrence of the characteristic equation in Examples 2-3 and 2-14.   Often the function ζ is termed the *phase* of the wave function u.   Accordingly surfaces of constant phase in xyt space are characteristic surfaces of the homogeneous wave equation.

Let

$$\mathbf{n} = n_x\mathbf{i} + n_y\mathbf{j} \tag{4-49}$$

be a unit vector in the xy plane, so that

$$n_x^2 + n_y^2 = 1 \tag{4-50}$$

Let

$$\mathbf{r} = x\mathbf{i} + y\mathbf{j}$$

be the variable position vector in the plane. It is easy to verify that the linear function

$$\zeta = \mathbf{n} \cdot \mathbf{r} \pm ct \tag{4-51}$$

satisfies the characteristic equation (4-48) and in addition is a solution of Eq. (4-45). In order to determine η such that η, ζ constitute a functionally invariant pair of Eq. (4-45), substitute ζ into the second equation of the system (4-47). Since $\nabla \zeta = \mathbf{n}$, this equation reduces to

$$c\mathbf{n} \cdot \nabla \eta = \pm \eta_t$$

Also η must satisfy the wave equation. It can be shown (see Prob. 46) that these conditions imply that η must be of the form

$$\eta = (x \pm n_x ct)F(\mathbf{n} \cdot \mathbf{r} \pm ct) + G(\mathbf{n} \cdot \mathbf{r} \pm ct)$$

It is left to the reader to verify that for each choice of twice differentiable φ and ψ, the function

$$u = (x \pm n_x ct)\varphi(\mathbf{n} \cdot \mathbf{r} \pm ct) + \psi(\mathbf{n} \cdot \mathbf{r} \pm ct) \tag{4-52}$$

is a solution of Eq. (4-45).

Plane Waves

A wave function of the form

$$u = \varphi(\mathbf{n} \cdot \mathbf{r} \pm ct) \tag{4-53}$$

is called a *plane wave*. Such a function represents a wave profile moving across the xy plane in the direction of $\mathbf{n}$ (or $-\mathbf{n}$) with speed c. The shape of the wave is unchanged as it moves. To see this, consider the rotation of axes

$$x' = n_x x + n_y y \qquad y' = -n_y x + n_x y$$

through the angle $\theta = \tan^{-1}(n_y/n_x)$. Equation (4-53) is transformed into

$$u = \varphi(x' \pm ct)$$

and this function represents the two-dimensional wave profile $u = \varphi(x')$ traveling with speed c in the direction of the x' axis. At a fixed time t the line in the xy plane whose equation is

$$\zeta = \mathbf{n} \cdot \mathbf{r} - ct = \text{const}$$

is perpendicular to **n**.   Such a line is a line of constant phase and is termed a *wavefront* of the wave

$$u = \varphi(\mathbf{n} \cdot \mathbf{r} - ct)$$

As t increases, the wavefront $\zeta = $ const moves with speed c in the direction of **n**.   Thus **n** defines the direction of propagation (see Fig. 4-10).   Similar considerations apply to the wave function

$$u = \varphi(\mathbf{n} \cdot \mathbf{r} + ct)$$

whose direction of propagation is $-\mathbf{n}$.
  Let

$$\mathbf{k} = k_x \mathbf{i} + k_y \mathbf{j}$$

be a given vector, and let

$$\omega = |\mathbf{k}| c \tag{4-54}$$

From the preceding paragraph it follows that

$$u = \varphi(\mathbf{k} \cdot \mathbf{r} \pm \omega t) \tag{4-55}$$

is a solution of the wave equation and represents the two-dimensional wave profile

$$u(r,0) = \varphi(\mathbf{k} \cdot \mathbf{r})$$

traveling in the direction of **k** (or $-\mathbf{k}$) with speed c.   Often **k** is called the

Figure 4-10

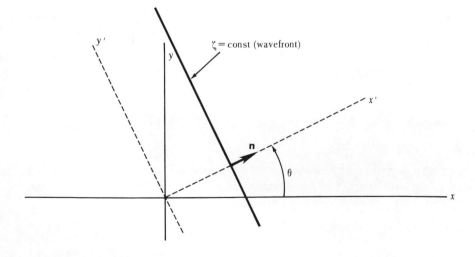

propagation vector of the wave.   In many applications plane harmonic waves

$$u = Ae^{i(\mathbf{k}\cdot\mathbf{r}\pm\omega t)} \qquad i = \sqrt{-1}$$

are useful. Here the amplitude A is constant. The real-valued wave functions

$$u = A \cos (\mathbf{k}\cdot\mathbf{r} \pm \omega t) \qquad u = A \sin (\mathbf{k}\cdot\mathbf{r} \pm \omega t)$$

represent sinusoidal waves traveling in the direction of $\mathbf{k}$.

If φ is a twice differentiable function of ζ such that φ does not vanish identically, and if $\mathbf{k}$ is an arbitrary vector, then a necessary condition for the function u defined by Eq. (4-55) to be a wave function is that Eq. (4-54) hold. Thus, apart from the trivial case, the plane waves (4-55) must travel with constant speed c in order to be wave functions.   However there exist particular solutions of Eq. (4-45) which represent traveling waves having speed less than c. In such waves the amplitude factor η is not constant (see Prob. 47).

Initial-value Problem for the Homogeneous Equation.   Poisson-Parseval Formula

The initial-value problem for the homogeneous two-dimensional wave equation is

$$\begin{aligned}
u_{tt} &= c^2(u_{xx} + u_{yy}) & -\infty < x, y < \infty; t \geq 0 \\
u(x,y,0) &= f(x,y) & -\infty < x, y < \infty \\
u_t(x,y,0) &= g(x,y) & -\infty < x, y < \infty
\end{aligned} \qquad (4\text{-}56)$$

A visualization of the problem is obtained if t is a parameter representing time.   Then, for each fixed t,

$$u = u(x,y,t)$$

is the equation of a surface in xyu space. As t varies, the surface changes shape.   The first initial condition fixes the shape of the surface at $t = 0$. The second initial condition prescribes the rate of change of the height u of the surface at $t = 0$.   The wave equation states that the local departure from the mean ordinate, i.e., the value of Δu at the point, is proportional to the acceleration u_{tt}.

If f has continuous third partial derivatives and g has continuous second partial derivatives, the problem has a unique solution.   A representation of the solution is

$$u(x,y,t) = \frac{1}{2\pi c}\frac{\partial}{\partial t}\int_0^{ct}\int_0^{2\pi}\frac{f(x + r\cos\theta, y + r\sin\theta)}{\sqrt{c^2t^2 - r^2}}\, r\, dr\, d\theta$$

$$+ \frac{1}{2\pi c}\int_0^{ct}\int_0^{2\pi}\frac{g(x + r\cos\theta, y + r\sin\theta)}{\sqrt{c^2t^2 - r^2}}\, r\, dr\, d\theta \qquad (4\text{-}57)$$

Equation (4-57) is called the *Poisson-Parseval formula*. It is derived in Sec. 4-6 as a consequence of a similar formula in three dimensions. The representation is the two-dimensional analog of D'Alembert's solution (4-15). In fact, if f and g are independent of y, Eq. (4-57) reduces to Eq. (4-15) (see Prob. 49).

Characteristic Cone.   Uniqueness of Solution

Uniqueness of solution for problem (4-56) can be proved with the aid of the characteristic-cone concept. Let (x_0, y_0, t_0) be a fixed but otherwise arbitrarily chosen point in xyt space, with $t_0 > 0$. The equation

$$\zeta = n_x x + n_y y - ct = n_x x_0 + n_y y_0 - ct_0$$

represents a one-parameter family of characteristic planes passing through the point (x_0, y_0, t_0) provided n_x, n_y are real parameters such that $n_x{}^2 + n_y{}^2 = 1$. This family envelopes a conical surface K with vertex at (x_0, y_0, t_0), called the *characteristic cone* at (x_0, y_0, t_0). The generators of K make the angle $\gamma = \tan^{-1}(1/c)$ with the t axis. The surface K intercepts a circle of radius ct_0 in the xy plane with center at (x_0, y_0). The circle bounds a region $\mathcal{R}$ which forms the base of a cone $\mathcal{V}$ whose vertex is at (x_0, y_0, t_0) (see Fig. 4-11). The lateral surface of $\mathcal{V}$ is the portion of K lying between the planes $t = 0$ and $t = t_0$.

Suppose now that v, w are solutions of problem (4-56), and let

$$\psi = v - w$$

We wish to prove that ψ vanishes identically in the half space $t \geq 0$. First, note that ψ is a solution of the homogeneous wave equation in the cone $\mathcal{V}$,

Figure 4-11

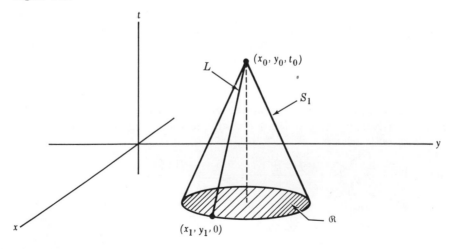

and ψ vanishes on the base $\mathscr{R}$.   Second, consider the identity

$$-2\psi_t\left(\psi_{xx} + \psi_{yy} - \frac{\psi_{tt}}{c^2}\right) = \frac{\partial}{\partial x}(-2\psi_t\psi_x) + \frac{\partial}{\partial y}(-2\psi_t\psi_y)$$

$$+ \frac{\partial}{\partial t}\left(\psi_x^2 + \psi_y^2 + \frac{\psi_t^2}{c^2}\right)$$

The left side of this equation is identically zero in $\mathscr{V}$.   Hence the volume integral over $\mathscr{V}$ of the right-hand member has the value zero.   By means of the divergence theorem [see Eq. (2-76)] the volume integral is transformed into a surface integral over the bounding surface S of $\mathscr{V}$.   The result is the equation

$$\iint_S \left[-2\psi_t\psi_x \cos\alpha - 2\psi_t\psi_y \cos\beta + \left(\psi_x^2 + \psi_y^2 + \frac{\psi_t^2}{c^2}\right)\cos\gamma\right] dS = 0$$

where $\cos\alpha$, $\cos\beta$, $\cos\gamma$ are the direction cosines of the exterior normal on S.   The surface S is composed of the circular region $\mathscr{R}$ in the xy plane and the lateral surface S_1 of the cone.   The function ψ together with its derivatives ψ_t, ψ_x vanish in $\mathscr{R}$.   Thus the previous equation holds with S replaced by S_1.   Multiplication of this new equation by $\cos\gamma$ gives

$$\iint_{S_1} \left[\left(\psi_x^2 + \psi_y^2 + \frac{\psi_t^2}{c^2}\right)\cos^2\gamma \right.$$

$$\left. - 2\psi_t\psi_x \cos\alpha \cos\gamma - 2\psi_t\psi_y \cos\beta \cos\gamma\right] dS = 0$$

Recall that on S_1 the value $\cos\gamma = c/\sqrt{1 + c^2} \neq 0$.   If the term

$$\psi_t^2(\cos^2\alpha + \cos^2\beta)$$

is added and subtracted in the integrand, then the equation

$$\iint_{S_1} \left[(\psi_x \cos\gamma - \psi_t \cos\alpha)^2 + (\psi_y \cos\gamma - \psi_t \cos\beta)^2 \right.$$

$$\left. + \psi_t^2\left(\frac{\cos^2\gamma}{c^2} - \cos^2\alpha - \cos^2\beta\right)\right] dS = 0$$

is obtained.   But

$$\cos^2\alpha + \cos^2\beta = 1 - \cos^2\gamma = \frac{1}{1 + c^2} = \frac{\cos^2\gamma}{c^2}$$

Hence

$$\iint\limits_{S_1} [(\psi_x \cos \gamma - \psi_t \cos \alpha)^2 + (\psi_y \cos \gamma - \psi_t \cos \beta)^2] \, dS = 0$$

The integrand is continuous and nonnegative. Accordingly the fact that the integral is zero implies that

$$\frac{\psi_x}{\cos \alpha} = \frac{\psi_y}{\cos \beta} = \frac{\psi_t}{\cos \gamma}$$

These relations state that at each point on S_1 the gradient vector $\nabla \psi$ is normal to S_1. Hence ψ must have a constant value on S_1. Since ψ has the value zero on the circle which forms the boundary of the base $\mathscr{R}$, and since ψ is continuous, it follows that ψ is zero on S_1. Thus

$$\psi(x_0,y_0,t_0) = 0$$

Recall that (x_0,y_0,t_0) was arbitrarily chosen to within the proviso $t_0 > 0$. Accordingly ψ vanishes identically on the half space $t > 0$. But ψ is zero as well on the plane $t = 0$. Hence ψ is identically zero for $t \geq 0$ and $v = w$ on that half space.

Several consequences of the Poisson-Parseval formula and the uniqueness proof should be noted. First, a solution u of Eq. (4-45) is uniquely determined at a point (x_0,y_0,t_0), $t_0 > 0$, by the values of u and u_t on the base $\mathscr{R}$ of the cone $\mathscr{V}$ with vertex at (x_0,y_0,t_0). Accordingly (and in analogy with the one-dimensional case) the circular region $\mathscr{R}$ is called the *region of dependence* of the point (x_0,y_0,t_0). The values of u and/or u_t at points of the xy plane exterior to $\mathscr{R}$ may be altered arbitrarily, and the value $u(x_0,y_0,t_0)$ remains unchanged. Now let S be a region of the xy plane. Then the set of all points (x,y,t), $t > 0$, whose region of dependence is contained in S is termed the *domain of determinacy* of S. This volume (in xyt space) is the union of all cones $\mathscr{V}$ such that (1) the axis of $\mathscr{V}$ is parallel to the t axis, (2) the base $\mathscr{R}$ of $\mathscr{V}$ is contained in S, and (3) the lateral surface of the cone makes the angle $\gamma = \tan^{-1}(1/c)$ with the t axis. The values of a solution u of Eq. (4-45) are uniquely determined at points of the domain of determinacy by the values of u and u_t on S. In particular, if both u and u_t vanish on S, then u vanishes on the domain of determinacy. Equivalently, if u, v are wave functions whose initial data coincide on S, then $u = v$ everywhere in the domain of determinacy. It should be noted here that these uniqueness properties hold for the more general Cauchy problem, where the initial data are given on a smooth surface in xyt space, where the surface is such that the normal is nowhere parallel to the xy plane. For the method of proof applies to the volume bounded by the characteristic cone and the portion of the surface intercepted by the cone, provided the divergence theorem is applicable to this volume.

Initial-value Problem for the Inhomogeneous Equation

It is shown in Sec. 4-6 that the solution of the initial-value problem

$$u_{tt} - c^2(u_{xx} + u_{yy}) = F(x,y,t) \qquad -\infty < x, y < \infty; t \ge 0$$

$$u(x,y,0) = 0 \qquad -\infty < x, y < \infty \tag{4-58}$$

$$u_t(x,y,0) = 0 \qquad -\infty < x, y < \infty$$

is

$$u(x,y,t) = \frac{1}{2\pi c} \iiint_{\mathscr{G}} \frac{F(\xi,\eta,\tau)}{\sqrt{c^2(t-\tau)^2 - \rho^2}}\, d\xi\, d\eta\, d\tau \tag{4-59}$$

where $\mathscr{G}$ is the region defined by the inequalities

$$0 \le \tau \le t \qquad \rho^2 \le c^2(t-\tau)^2 \qquad \rho^2 = (x-\xi)^2 + (y-\eta)^2$$

The initial-value problem with inhomogeneous initial conditions as well as an inhomogeneous differential equation has as its solution the superposition

$$u = v + w$$

where v denotes the solution of problem (4-56) and w denotes the solution of problem (4-58).

4-5 BOUNDARY- AND INITIAL-VALUE PROBLEM FOR THE TWO-DIMENSIONAL WAVE EQUATION

Let $\mathscr{R}$ be a region in the xy plane bounded by a simple closed piecewise smooth curve C, and let $\bar{\mathscr{R}}$ denote $\mathscr{R}$ together with C (see Fig. 4-12). In this section problems of the type

$$u_{tt} - c^2 \Delta u = F(x,y,t) \qquad (x,y) \text{ in } \bar{\mathscr{R}}; t \ge 0$$

$$B(u) = 0 \qquad \text{on } C; t \ge 0 \tag{4-60}$$

$$u(x,y,0) = f(x,y) \qquad u_t(x,y,0) = g(x,y) \qquad \text{in } \bar{\mathscr{R}}$$

are considered.   Here F, f, and g are given functions and $B(u) = 0$ symbolizes one of the following boundary conditions

$$u = 0 \qquad \qquad \text{on } C \qquad \text{Dirichlet} \tag{4-61}$$

$$\frac{\partial u}{\partial n} = 0 \qquad \qquad \text{on } C \qquad \text{Neumann} \tag{4-62}$$

$$\frac{\partial u}{\partial n} + \sigma u = 0 \qquad \text{on } C \qquad \text{mixed} \tag{4-63}$$

where σ is a given positive constant.

Method of Eigenfunctions

The formal technique of solution outlined here may be termed the *method of eigenfunctions*. For concreteness assume the boundary condition is the Dirichlet condition. In the wave equation there appears the operator Δ. Recall from Sec. 3-5 that in the eigenvalue problem

$$\Delta\varphi + \lambda\varphi = 0 \quad \text{in } \mathscr{R} \qquad \varphi = 0 \quad \text{on } C \tag{4-64}$$

for Δ subject to the Dirichlet condition on C the eigenvalues constitute a real sequence $\{\lambda_n\}$ of positive numbers, and corresponding to each eigenvalue λ_n there is (at least one) real-valued twice continuously differentiable eigenfunction φ_n such that

$$\Delta\varphi_n + \lambda_n\varphi_n = 0 \quad \text{in } \mathscr{R} \qquad \varphi_n = 0 \quad \text{on } C \tag{4-65}$$

The sequence $\{\varphi_n\}$ of eigenfunctions has the orthogonality property

$$\iint_{\mathscr{R}} \varphi_n\varphi_m \, dA = 0 \qquad n \neq m \tag{4-66}$$

and is complete in the class of all functions which are continuous on $\bar{\mathscr{R}}$ and vanish on C. Since each twice continuously differentiable function on $\bar{\mathscr{R}}$ which vanishes on C can be expanded in its Fourier series relative to the orthogonal set $\{\varphi_n\}$, it is expected that, for each fixed t, the solution of

Figure 4-12

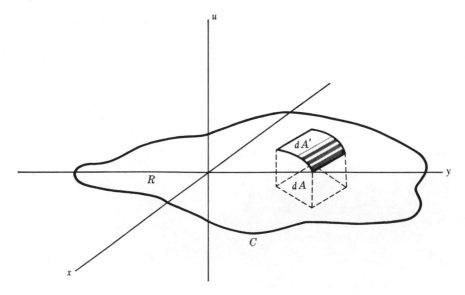

problem (4-60) can be so represented.   Thus one writes

$$u(x,y,t) = \sum_{n=1}^{\infty} C_n(t)\varphi_n(x,y) \tag{4-67}$$

where the $C_n(t)$ are to be determined.   Formal substitution of the series into the wave equation and use of the differential equation satisfied by φ_n yield

$$\sum_{n=1}^{\infty} [\ddot{C}_n(t) + \omega_n{}^2 C_n(t)]\varphi_n(x,y) = F(x,y,t) \tag{4-68}$$

where dots denote derivatives with respect to t and

$$\omega_n = c\sqrt{\lambda_n} \qquad n = 1, 2, \ldots \tag{4-69}$$

Fix $m \geq 1$, multiply both sides of Eq. (4-68) by φ_m, integrate both sides over $\mathscr{R}$, and formally interchange the order of summation and integration.   Then, by virtue of the orthogonality property (4-66), the result is

$$\ddot{C}_m(t) + \omega_m{}^2 C_m(t) = F_m(t) \tag{4-70}$$

where

$$F_m(t) = \frac{1}{\|\varphi_m\|^2} \iint_{\mathscr{R}} F(x,y,t)\varphi_m(x,y)\, dA \tag{4-71}$$

$$\|\varphi_m\|^2 = \iint_{\mathscr{R}} |\varphi_m|^2\, dA \tag{4-72}$$

Now the series (4-67) formally satisfies the initial conditions of problem (4-60) if

$$\sum_{n=1}^{\infty} C_n(0)\varphi_n(x,y) = f(x,y) \qquad \sum_{n=1}^{\infty} \dot{C}_n(0)\varphi_n(x,y) = g(x,y) \tag{4-73}$$

in $\bar{\mathscr{R}}$.   Again by the orthogonality property of the eigenfunctions it follows that

$$C_m(0) = \frac{1}{\|\varphi_m\|^2} \iint_{\mathscr{R}} f\varphi_m\, dA \tag{4-74}$$

$$\dot{C}_m(0) = \frac{1}{\|\varphi_m\|^2} \iint_{\mathscr{R}} g\varphi_m\, dA \tag{4-75}$$

Now the general solution of Eq. (4-70) is

$$C_m(t) = A_m \cos \omega_m t + B_m \sin \omega_m t + \frac{1}{\omega_m} \int_0^t F_m(\xi) \sin [\omega_m(t - \xi)]\, d\xi$$

Impose the initial conditions (4-74) and (4-75) on the general solution; then

$$A_m = \frac{1}{\|\varphi_m\|^2} \iint_{\mathcal{R}} f\varphi_m \, dA \qquad B_m = \frac{1}{\omega_m \|\varphi_m\|^2} \iint_{\mathcal{R}} g\varphi_m \, dA \tag{4-76}$$

This holds for $m = 1, 2, \ldots$. Hence the formal series solution of problem (4-60) is

$$u = v + w$$

where

$$v = \sum_{n=1}^{\infty} (A_n \cos \omega_n t + B_n \sin \omega_n t)\varphi_n(x,y) \tag{4-77}$$

$$w = \sum_{n=1}^{\infty} \left\{ \frac{1}{\omega_n} \int_0^t F_n(\xi) \sin [\omega_n(t - \xi)] \, d\xi \right\} \varphi_n(x,y) \tag{4-78}$$

The coefficients are defined by Eq. (4-76), and the functions F_n are defined in Eq. (4-71).

The function v is the formal solution of the problem

$$v_{tt} - c^2 \, \Delta v = 0 \qquad \text{in } \mathcal{R}; \, t \geq 0$$

$$B(v) = 0 \qquad \text{on } C; \, t \geq 0 \tag{4-79}$$

$$v(x,y,0) = f(x,y) \qquad v_t(x,y,0) = g(x,y) \qquad \text{in } \mathcal{R}$$

involving the homogeneous wave equation. The function w is the formal solution of the problem

$$w_{tt} - c^2 \, \Delta w = F(x,y,t) \qquad \text{in } \mathcal{R}; \, t \geq 0$$

$$B(w) = 0 \qquad \text{on } C; \, t \geq 0 \tag{4-80}$$

$$w(x,y,0) = 0 \qquad w_t(x,y,0) = 0 \qquad \text{in } \mathcal{R}$$

for the inhomogeneous wave equation with homogeneous boundary and initial conditions.

Energy Integral. Uniqueness of Solution

There is at most one solution of problem (4-60). The proof of uniqueness can be made with the aid of an energy integral (as in the one-dimensional problem) provided C is such that the divergence theorem is applicable. Define the energy integral

$$E(t) = \frac{1}{2} \iint_{\mathcal{R}} [\psi_t^2 + c^2(\psi_x^2 + \psi_y^2)] \, dx \, dy \tag{4-81}$$

If a function ψ is twice continuously differentiable, then the energy rate is

$$\dot{E}(t) = \iint\limits_{\mathscr{R}} [\psi_t \psi_{tt} + c^2(\psi_x \psi_{xt} + \psi_y \psi_{yt})]\, dx\, dy \qquad (4\text{-}82)$$

In Green's identity

$$\iint\limits_{\mathscr{R}} v\, \Delta u\, dx\, dy + \iint\limits_{\mathscr{R}} \nabla u \cdot \nabla v\, dx\, dy = \int_C v\, \frac{\partial u}{\partial n}\, ds$$

set $u = \psi$, $v = \psi_t$.   Then

$$\dot{E}(t) = \iint\limits_{\mathscr{R}} \psi_t(\psi_{tt} - c^2\, \Delta\psi)\, dx\, dy + c^2 \int_C \psi_t \frac{\partial\psi}{\partial n}\, ds \qquad (4\text{-}83)$$

Assume now that u_1 and u_2 are solutions of problem (4-60).   By the pervasive linearity of things the difference

$$\psi = u_1 - u_2$$

is a solution of the homogeneous wave equation and satisfies the homogeneous boundary condition as well as homogeneous initial conditions.   Write the energy integral in terms of ψ.   It follows from the results of the preceding paragraph that if the boundary condition is the Dirichlet condition or the Neumann condition, then $\dot{E}(t) = 0$, and hence E has a constant value.   But if the energy integral is evaluated at $t = 0$, it is seen that $E(0) = 0$.   Hence $E(t) = 0, t \geq 0$.   Since the integrand of the energy integral is continuous and nonnegative, the vanishing of the integral for $t \geq 0$ implies that the derivatives ψ_x, ψ_y, and ψ_t are zero at each point of $\bar{\mathscr{R}}$ for $t \geq 0$.   Accordingly ψ is constant in value.   But $\psi(x,y,0) = 0$.   Thus ψ is the identically zero function, and $u_1 = u_2$.   Now suppose that the boundary condition is the mixed condition (4-63).   Then

$$\dot{E}(t) = -c^2\sigma \int_C \psi\psi_t\, ds = -\frac{c^2\sigma}{2} \frac{\partial}{\partial t} \int_C \psi^2\, ds \qquad (4\text{-}84)$$

and so

$$E(t) = -\frac{\sigma c^2}{2} \int_C \psi^2\, ds + \text{const}$$

As before, the initial conditions which are satisfied by ψ imply $E(0) = 0$. Hence

$$E(t) = -\frac{\sigma c^2}{2} \int_C \psi^2\, ds \leq 0 \qquad (4\text{-}85)$$

But from the original definition it is clear that $E(t) \geq 0$.   Thus E must be identically zero, and the remainder of the proof is the same.

Remarks on the Formal Series Solution

The remarks concerning solutions made in Sec. 4-3 (following Example 4-1) apply here.   It is emphasized that the existence of a solution of the boundary- and initial-value problem for the two-dimensional wave equation has not been demonstrated.   Indeed, without placing further hypotheses on the given functions the series (4-77) and (4-78) need not be convergent, let alone define a twice continuously differentiable function.   The question of the existence of a solution for an arbitrary domain cannot be taken up in this text.   It can be shown, with some additional hypotheses on the region $\mathcal{R}$, that if f is four times continuously differentiable, if g is three times continuously differenti- able, on $\bar{\mathcal{R}}$, and if

$$f = \Delta f = 0 \qquad g = \Delta g = 0 \qquad \text{on } C$$

then the series (4-77) defines a twice continuously differentiable function on $\bar{\mathcal{R}}$ which satisfies the boundary and initial conditions and the wave equation of problem (4-79).   Such a proof utilizes an estimate of the order of magni- tude of the eigenvalues λ_n for sufficiently large n (see Ref. 3).   For a rec- tangular region the eigenvalues can be obtained explicitly, and in Prob. 65 the outline of the proof of the existence of a classical solution is given following this method.

The set of eigenfunctions $\{\varphi_n\}$ also have the following property.   If φ is a twice continuously differentiable function on $\bar{\mathcal{R}}$ which satisfies the boundary conditions of the problem, the Fourier series of φ relative to the orthogonal set $\{\varphi_n\}$ converges uniformly to φ on $\bar{\mathcal{R}}$.   If this property is assumed, it is readily proved, and in the same manner as in Sec. 4-3, that if a classical solution of the problem exists, then it must be obtainable by the method of eigenfunctions.   This implies, of course, that if the series fails to define a classical solution, no classical solution exists.

A problem which has inhomogeneous boundary conditions may, in some cases at least, be reduced to a problem having homogeneous boundary conditions.   For example, instead of the homogeneous condition $u = 0$ on C, suppose that the boundary condition of problem (4-60) is

$$u = h(t) \qquad \text{on } C \tag{4-86}$$

where h is a given twice differentiable function.   If v is constructed such that

$$v_{tt} - c^2 \, \Delta v = F - h_{tt} \qquad \text{in } \mathcal{R}; \, t \geq 0$$

$$v = 0 \qquad \text{on } C; \, t \geq 0$$

$$v(x,y,0) = f(x,y) - h(0) \qquad v_t(x,y,0) = g(x,y) - \dot{h}(0) \qquad \text{in } \mathcal{R}$$

then $u = v + h$ is a solution of the original problem.   The solution must be unique, since the difference of two solutions is a solution of the homogeneous wave equation which satisfies the homogeneous boundary condition and the homogeneous initial conditions.   However, as shown above, the only function which satisfies these conditions is the trivial function.

Vibrating Membrane

As an example of a physical problem which leads to a boundary- and initial-value problem for the two-dimensional wave equation, consider a thin elastic membrane under tension which occupies the region $\mathscr{R}$ in its equilibrium configuration.   The edge of the membrane is fastened along C.   The membrane is given an initial transverse displacement and then at time $t = 0$ is released (not necessarily with zero speed).   Subsequently the membrane executes transverse vibrations.   The problem is to describe the motion.

In order to obtain a mathematical model which is amenable to linear analysis several simplifying assumptions are made.   The membrane has negligible thickness and consists of a homogeneous material of density ρ (mass/unit area).   No resistance to bending is offered; however, a change in shape accompanied by changes in elementary areas is resisted by elastic restoring forces.   The tensile force exerted on a surface element by an adjacent surface element is directed perpendicular to the common edge and lies in the tangent plane to the membrane surface at that point.   The magnitude of the tensile force is proportional to the length of the common edge. Thus, if $\Delta A'$ is a surface element of the deformed membrane (see Fig. 4-12), and Δs is the length of the arc element forming the common edge, then the magnitude of the tensile force acting on $\Delta A'$ due to the adjacent element is

$$\Delta F = T \, \Delta s$$

The coefficient of proportionality T is called the *tension*.   It is also assumed that the horizontal component of the tensile force is a constant independent of position, time, and the shape of the membrane surface.   The transverse displacement at time t of the point on the membrane with coordinates (x,y) is denoted by $u(x,y,t)$.   The further assumption is made that the motion is purely transverse with small amplitudes and slopes.   Thus

$$u^2 \ll 1 \qquad u_x{}^2 \ll 1 \qquad u_y{}^2 \ll 1 \tag{4-87}$$

These assumptions make it possible to linearize the equation of motion of the membrane, and hence greatly simplify the analysis.

Let $\Delta A = \Delta x \, \Delta y$ be an area element of the membrane in its equilibrium configuration.   Let $\Delta A'$ be the corresponding surface element into which ΔA is deformed at time $t > 0$.   Then

$$\Delta A' = \Delta A \sec \gamma$$

where γ is the angle which the normal to $\Delta A'$ makes with the positive u axis. Now

$$\sec \gamma = (1 + u_x{}^2 + u_y{}^2)^{1/2} \approx 1 \qquad \Delta A' \approx \Delta A$$

Also, by the type of argument used in the derivation of the equation of motion of the vibrating string, it follows from the assumptions (4-87) that the net transverse component of tensile force acting on $\Delta A'$ is

$$T_0\left(\frac{\partial u}{\partial x}\bigg|_{x_2} - \frac{\partial u}{\partial x}\bigg|_{x_1}\right)\Delta y + T_0\left(\frac{\partial u}{\partial y}\bigg|_{y_2} - \frac{\partial u}{\partial y}\bigg|_{y_1}\right)\Delta x$$

Here T_0 denotes the (constant) horizontal component of the tensile force, and $x_2 - x_1 = \Delta x$, $y_2 - y_1 = \Delta y$. Let $F(x,y,t)$ denote an external transverse force/unit mass which acts at the point (x,y) on the membrane at time t. Then the equation of motion of the element $\Delta A'$ is

$$\rho \, \Delta x \, \Delta y \, \overline{\frac{\partial^2 u}{\partial t^2}} = T_0\left[\left(\frac{\partial u}{\partial x}\bigg|_{x_2} - \frac{\partial u}{\partial x}\bigg|_{x_1}\right)\Delta y + \left(\frac{\partial u}{\partial y}\bigg|_{y_2} - \frac{\partial u}{\partial y}\bigg|_{y_1}\right)\Delta x\right]$$
$$+ \rho F(x,y,t) \, \Delta x \, \Delta y$$

where $\overline{\partial^2 u/\partial t^2}$ denotes the acceleration of the center of mass. If this result is divided through by $\Delta x \, \Delta y$ and the limit is taken as $\Delta x \to 0$, $\Delta y \to 0$, the result is

$$u_{tt} - c^2 u_{xx} = F(x,y,t) \qquad c = \sqrt{\frac{T_0}{\rho}}$$

The fact that the edge is fastened along C is expressed by the Dirichlet boundary condition. Assume that the motion of the membrane is started by an initial displacement and subsequent release, not necessarily with zero speed, at $t = 0$. The initial conditions are those expressed in problem (4-60). The function f prescribes the displacement of the membrane at $t = 0$, and g describes the initial speed. Since the membrane does not tear, the function u is required to be continuous in x, y, and t. Accordingly the boundary and initial conditions taken together imply

$$f(x,y) = 0 \qquad \text{on } C \tag{4-88}$$

If the speed u_t is also required to be continuous, the condition

$$g(x,y) = 0 \qquad \text{on } C \tag{4-89}$$

must also be satisfied.

Characteristic Frequencies. Normal Modes of Vibration

If no external force acts, the subsequent motion of the membrane is given by the solution of the problem

$$u_{tt} - c^2 \Delta u = 0 \quad \text{in } \mathscr{R}; t \geq 0$$

$$u = 0 \quad \text{on } C; t \geq 0$$

$$u(x,y,0) = f(x,y) \qquad u_t(x,y,0) = g(x,y) \quad \text{in } \bar{\mathscr{R}}$$

From Eq. (4-77) the formal series solution is

$$u = \sum_{n=1}^{\infty} (A_n \cos \omega_n t + B_n \sin \omega_n t) \varphi_n(x,y)$$

The values ω_n defined in Eq. (4-69) are called the *characteristic frequencies* (also normal frequencies or natural frequencies) of vibration. Note these are determined by the boundary condition and are independent of initial conditions. The functions

$$u_n = (A_n \cos \omega_n t + B_n \sin \omega_n t) \varphi_n(x,y) \qquad n = 1, 2, \ldots$$

are termed the *normal modes of vibration*. They are the standing-wave solutions of the homogeneous wave equation which satisfy the boundary condition on C. Observe that the motion of the membrane is given as a superposition of normal modes.

Circular Membrane. Green's Function for the Problem

If $\mathscr{R}$ is a rectangular region, the eigenfunctions can be obtained explicitly (see Probs. 52 to 54). The case where $\mathscr{R}$ is a circular region is considered in Example 4-2, and some additional aspects are also discussed.

EXAMPLE 4-2 A circular elastic membrane in its equilibrium configuration under tension occupies the area in the xy plane defined by the inequalities

$$0 \leq r \leq a \qquad 0 \leq \theta \leq 2\pi$$

where (r,θ) are polar coordinates. The edge of the membrane is fastened along the circle $r = a$. The membrane is given an initial transverse displacement and speed

$$u(r,\theta,0) = f(r,\theta) \qquad u_t(r,\theta,0) = g(r,\theta)$$

and then is released. An external transverse force $F(r,\theta,t)$ per unit mass acts on the membrane. Determine the subsequent motion.

The boundary condition is

$$u(a,\theta,t) = 0 \qquad 0 \leq \theta \leq 2\pi; t \geq 0$$

In polar coordinates the inhomogeneous wave equation is

$$\frac{\partial^2 u}{\partial t^2} - c^2 \left(\frac{\partial^2 u}{\partial r^2} + \frac{1}{r} \frac{\partial u}{\partial r} + \frac{1}{r^2} \frac{\partial^2 u}{\partial \theta^2} \right) = F(r,\theta,t)$$

To obtain the normal modes of vibration consider the homogeneous wave equation

$$\frac{\partial^2 u}{\partial t^2} = c^2 \left(\frac{\partial^2 u}{\partial r^2} + \frac{1}{r} \frac{\partial u}{\partial r} + \frac{1}{r^2} \frac{\partial^2 u}{\partial \theta^2} \right)$$

The time dependence is split off by assuming

$$u = \varphi(r,\theta) T(t)$$

Then φ must satisfy the scalar Helmholtz equation in polar coordinates

$$\varphi_{rr} + \frac{\varphi_r}{r} + \frac{\varphi_{\theta\theta}}{r^2} + \lambda\varphi = 0$$

where λ is a separation constant. The boundary condition for u implies the boundary condition

$$\varphi(a,\theta) = 0 \qquad 0 \leq \theta \leq 2\pi$$

The real-valued eigenfunctions of the problem are derived in Example 3-6. They are

$$\varphi_{nk}^{(e)} = J_n\left(\frac{\xi_{nk}r}{a}\right) \cos n\theta \qquad \varphi_{nk}^{(0)} = J_n\left(\frac{\xi_{nk}r}{a}\right) \sin n\theta$$

corresponding to the eigenvalues

$$\lambda_{nk} = \frac{\xi_{nk}^2}{a^2} \qquad n = 0, 1, \ldots ; k = 1, 2, \ldots$$

where the ξ_{nk} are, for each fixed n, the positive zeros of $J_n(\xi)$ arranged in ascending order. The characteristic frequencies of the circular membrane with fastened edge are the numbers

$$\omega_{nk} = \frac{c\xi_{nk}}{a} \qquad n = 0, 1, \ldots ; k = 1, 2, \ldots$$

The normal modes of vibration are the functions

$$u_{nk} = [A_{nk}\varphi_{nk}^{(e)}(r,\theta) + B_{nk}\varphi_{nk}^{(0)}(r,\theta)] \cos \omega_{nk}t + [C_{nk}\varphi_{nk}^{(e)}(r,\theta) + D_{nk}\varphi_{nk}^{(0)}(r,\theta)] \sin \omega_{nk}t$$

These are continuous single-valued separable forms of solution of the homogeneous wave equation which vanish on the circle $r = a$. They satisfy the initial conditions

$$u_{nk}(r,\theta,0) = A_{nk}\varphi_{nk}^{(e)}(r,\theta) + B_{nk}\varphi_{nk}^{(0)}(r,\theta)$$

$$u_{nkt}(r,\theta,0) = \omega_{nk}[C_{nk}\varphi_{nk}^{(e)}(r,\theta) + D_{nk}\varphi_{nk}^{(0)} (r,\theta)]$$

If, for some fixed $p \geq 0$, $q \geq 1$, the membrane is started with the initial displacement $u_{pq}(r,\theta,0)$ and speed $u_{pqt}(r,\theta,0)$, then by the uniqueness of solution it follows that the subsequent motion of the membrane is given by $u_{pq}(r,\theta,t)$.

From Eqs. (4-77) and (4-78) and the properties of the eigenfunctions $\varphi_{nk}^{(e)}$, $\varphi_{nk}^{(0)}$ it follows that the motion of the membrane is given by

$$u = v + w$$

where

$$v = \sum_{n=0}^{\infty} \sum_{k=1}^{\infty} \{[A_{nk}\varphi_{nk}^{(e)}(r,\theta) + B_{nk}\varphi_{nk}^{(0)}(r,\theta)] \cos \omega_{nk}t + [C_{nk}\varphi_{nk}^{(e)}(r,\theta) + D_{nk}\varphi_{nk}^{(0)}(r,\theta)] \sin \omega_{nk}t\}$$

$$w = \sum_{n=0}^{\infty} \sum_{k=1}^{\infty} \frac{J_n(\xi_{nk}r/a)}{\omega_{nk}} \left(\left\{ \int_0^t F_{nk}^{(e)}(\xi) \sin [\omega_{nk}(t - \xi)] \, d\xi \right\} \cos n\theta \right.$$

$$\left. + \left\{ \int_0^t F_{nk}^{(0)}(\xi) \sin [\omega_{nk}(t - \xi)] \, d\xi \right\} \sin n\theta \right)$$

$$A_{0k} = \frac{1}{\pi a^2 J_1^2(\xi_{0k})} \int_0^a \int_0^{2\pi} J_0\left(\frac{\xi_{0k}r}{a}\right) f(r,\theta)r \, dr \, d\theta \qquad k = 1, 2, \ldots$$

$$A_{nk} = \frac{2}{\pi a^2 J_{n+1}^2(\xi_{nk})} \int_0^a \int_0^{2\pi} J_n\left(\frac{\xi_{nk}r}{a}\right) \cos n\theta \, f(r,\theta)r \, dr \, d\theta \qquad n = 1, 2, \ldots; k = 1, 2, \ldots$$

The expression for B_{nk} is identical to the one given for A_{nk} except that $\sin n\theta$ replaces $\cos n\theta$. Also the expressions for C_{nk}, D_{nk} are the same as those for A_{nk}, B_{nk}, respectively, except that ω_{nk} is present in the denominator of the factor which multiplies the integral and the given function g replaces f. The function $F_{nk}^{(e)}$ is defined by

$$F_{0k}^{(e)}(t) = \frac{1}{a^2 J_1^2(\xi_{0k})} \int_0^a \int_0^{2\pi} F(r,\theta,t)J_0\left(\frac{\xi_{0k}r}{a}\right)r \, dr \, d\theta \qquad k = 1, 2, \ldots$$

$$F_{nk}^{(e)}(t) = \frac{1}{\pi a^2 J_{n+1}^2(\xi_{nk})} \int_0^a \int_0^{2\pi} F(r,\theta,t)J_n\left(\frac{\xi_{nk}r}{a}\right) \cos n\theta \, r \, dr \, d\theta$$

$$n = 1, 2, \ldots; k = 1, 2, \ldots$$

Identical formulas define the functions $F_{nk}^{(0)}(t)$, $n = 1, 2, \ldots$, $k = 1, 2, \ldots$, except that $\sin n\theta$ replaces $\cos n\theta$ in the integrand. These expressions can be derived by the method set forth in the text. Assume an expansion of the form

$$u = \sum_{n=0}^{\infty} \sum_{k=1}^{\infty} [C_{nk}^{(e)}(t)\varphi_{nk}^{(e)}(r,\theta) + C_{nk}^{(0)}(t)\varphi_{nk}^{(0)}(r,\theta)]$$

and then obtain the equations which determine the coefficients $C_{nk}^{(e)}$, $C_{nk}^{(0)}$ by means of the orthogonality properties of the functions $\{\varphi\}_{nk}^{(e)}$, $\{\varphi_{nk}^{(0)}\}$.

Circularly symmetric modes of vibrations are the modes

$$u_{0k} = J_0\left(\frac{\xi_{0k}r}{a}\right)(A_{0k} \cos \omega_{0k}t + B_{0k} \sin \omega_{0k}t) \qquad k = 1, 2, \ldots$$

These are the only modes present if the initial displacement, speed, and external force are independent of θ. This follows from the expressions for the coefficients A_{nk}, B_{nk}, C_{nk}, D_{nk} and functions $F_{nk}^{(e)}$, $F_{nk}^{(0)}$, for $n \geq 1$. In this event the subsequent motion is circularly symmetric, given by

$$u(r,t) = \sum_{k=1}^{\infty} J_0\left(\frac{\xi_k r}{a}\right)(A_k \cos \omega_k t + B_k \sin \omega_k t)$$

$$+ \sum_{k=1}^{\infty} \left\{ \frac{1}{\omega_k} \int_0^t F_k(\xi) \sin [\omega_k(t - \xi)] \, d\xi \right\} J_0\left(\frac{\xi_k r}{a}\right)$$

where $\{\xi_k\}$ denotes the sequence of positive zeros of $J_0(\xi)$, $\omega_k = c\xi_k/a$, $k = 1, 2, \ldots$, and

$$A_k = \frac{2}{a^2 J_1^2(\xi_k)} \int_0^a J_0\left(\frac{\xi_k r}{a}\right) f(r) r \, dr$$

$$B_k = \frac{2}{a^2 \omega_k^2 J_1^2(\xi_k)} \int_0^a J_0\left(\frac{\xi_k r}{a}\right) g(r) r \, dr$$

$$F_k(t) = \frac{2}{a^2 J_1^2(\xi_k)} \int_0^a F(r,t) J_0\left(\frac{\xi_k r}{a}\right) r \, dr$$

$k = 1, 2, \ldots$.   As an example suppose the membrane is released from rest with initial shape

$$f(r) = h\left(1 - \frac{r^2}{a^2}\right)$$

where h is a positive constant, and suppose the force due to gravity acts on the membrane. Then

$$F(r,t) = -g$$

g a positive constant.   By the properties of the Bessel functions it follows that

$$A_k = \frac{8}{\xi_k^3 J_1(\xi_k)} \qquad k = 1, 2, \ldots$$

$$F_k(t) = \frac{-2g}{\xi_k J_1(\xi_k)} \qquad k = 1, 2, \ldots$$

Hence the subsequent motion is given by

$$u = 8 \sum_{k=1}^{\infty} \left[\frac{J_0(\xi_k r/a) \cos \omega_k t}{\xi_k^3 J_1(\xi_k)}\right] + \frac{2ga^2}{c^2} \sum_{k=1}^{\infty} \left[J_0\left(\frac{\xi_k r}{a}\right) \frac{\cos \omega_k t - 1}{\xi_k^3 J_1(\xi_k)}\right]$$

Recall the properties of the δ function described in Sec. 3-5.   Here, in conformity with the remarks made there, the properties of the δ function are used in a purely formal manner with the understanding that the operations can be rigorously justified.   A concentrated impulse force (per unit magnitude) of unit magnitude applied at time $t = \tau$ at the point (r_0, θ_0) on the membrane can be represented

$$F(r,\theta,t) = \frac{\delta(r - r_0)\delta(\theta - \theta_0)\delta(t - \tau)}{r}$$

Then

$$F(r,\theta,t) = 0 \qquad (r,\theta,t) \neq (r_0,\theta_0,\tau)$$

and

$$\int_0^t \int_0^a \int_0^{2\pi} F(r,\theta,t) r \, dr \, d\theta \, dt = 1 \qquad t > \tau$$

To determine the "response" of the initially quiescent membrane to the impulsive blow the expressions derived in the previous paragraph are used.   From the properties of the

δ function it follows that

$$F_{nk}^{(e)}(t) = \frac{\epsilon_n \delta(t - \tau)}{\pi a^2 J_{n+1}^2(\xi_{nk})} \left[\int_0^a J_n\left(\frac{\xi_{nk}r}{a}\right) \delta(r - r_0)\, dr \right] \left[\int_0^{2\pi} \cos n\theta\, \delta(\theta - \theta_0)\, d\theta \right]$$

$$= \frac{\epsilon_n J_n(\xi_{nk}r_0/a) \cos n\theta_0\, \delta(t - \tau)}{\pi a^2 J_{n+1}^2(\xi_{nk})}$$

where $\epsilon_0 = 1$ and $\epsilon_n = 2$ if $n \geq 1$. The expressions for the $F_{nk}^{(0)}$ are the same as those above except that $\sin n\theta_0$ replaces $\cos n\theta_0$. Substitution into the previous expression for the solution yields the series

$$G(r,\theta,t;r_0,\theta_0,\tau) = \sum_{n=0}^{\infty} \sum_{k=1}^{\infty} \frac{\epsilon_n J_n(\xi_{nk}r/a) J_n(\xi_{nk}r_0/a)}{\pi a^2 \omega_{nk} J_{n+1}^2(\xi_{nk})} \cos\left[n(\theta - \theta_0)\right] \sin\left[\omega_{nk}(t - \tau)\right]$$

This is the motion due to the impulse. It can be shown that the series converges on $\bar{\mathcal{R}}$ for all t and so defines a continuous function. Clearly G satisfies the boundary condition. Also

$$G(r,\theta,\tau;r_0,\theta_0,\tau) = 0 \qquad \text{all } (r,\theta)$$

Note the symmetry property

$$G(r_0,\theta_0,t;r,\theta,\tau) = G(r,\theta,t;r_0,\theta_0,\tau)$$

This is just the statement of reciprocity of effects: at time t the deflection of (r_0,θ_0) due to a unit impulse force applied at (r,θ) is the same as the deflection at (r,θ) due to a unit impulse force applied at (r_0,θ_0). In order to simulate the motion of the membrane completely it is necessary to define G as being identically zero for $t < \tau$.

The function G is called the *Green's function* of the boundary- and initial-value problem for the circular membrane with fastened edge. Although it is not a solution of the homogeneous wave equation for $t \neq \tau$ and hence is not a solution of the generic problem in the classical sense, the function G has a number of interesting and useful properties. For example, the solution which satisfies the boundary condition and quiescent initial conditions has the representation

$$u(r,\theta,t) = \int_0^t \int_0^a \int_0^{2\pi} G(r,\theta,t;r_0,\theta_0,\tau) F(r_0,\theta_0,\tau) r_0\, dr_0\, d\theta_0\, d\tau$$

To verify that this furnishes the solution the series is substituted into the integrand, and the operations of summation and integration are formally interchanged. The steps are omitted here. The integral representation of the solution may be viewed as a superposition of the effects at the point (r,θ) at time t due to impulsive forces of magnitude $F(r_0,\theta_0,\tau)$ which are continuously distributed over the points (r_0,θ_0) of the membrane from time zero to time t.

Assume now that the membrane is at rest in its equilibrium position. At time $t = 0$ the membrane is struck in such a way that the point (r_0,θ_0) receives a velocity impulse of unit magnitude. Then the initial conditions are

$$u(r,\theta,0) = 0 \qquad u_t(r,\theta,0) = \frac{\delta(r - r_0)\delta(\theta - \theta_0)}{r}$$

If no external forces act, the motion is given by the formal series solution derived previously, where

$$\dot{A}_{nk} = B_{nk} = 0 \qquad n = 0, 1, 2, \ldots; k = 1, 2, \ldots$$

and

$$C_{nk} = \frac{\epsilon_p}{\pi a^2 \omega_{nk} J_{n+1}^2(\xi_{nk})} \left[\int_0^a J_n\left(\frac{\xi_{nk}r}{a}\right) \delta(r - r_0) \, dr \right] \left[\int_0^{2\pi} \cos n\theta \, \delta(\theta - \theta_0) \, d\theta \right]$$

$$= \frac{\epsilon_n J_n(\xi_{nk}r_0/a) \cos n\theta_0}{\pi a^2 \omega_{nk} J_{n+1}^2(\xi_{nk})}$$

$$D_{nk} = \frac{J_n(\xi_{nk}r_0/a) \sin n\theta_0}{\pi a^2 \omega_{nk} J_{n+1}^2(\xi_{nk})}$$

Thus the response of the membrane to the velocity impulse is

$$G(r,\theta,t;r_0,\theta_0,0) = \sum_{n=0}^{\infty} \sum_{m=1}^{\infty} \frac{\epsilon_n J_n(\xi_{nm}r/a) J_n(\xi_{nm}r_0/a)}{\pi a^2 \omega_{nm} J_{n+1}^2(\xi_{nm})} \cos [n(\theta - \theta_0)] \sin \omega_{nm} t$$

It can be verified by the method used previously that the solution of the original problem in the case where no external force acts can be written as

$$u(r,\theta,t) = \int_0^a \int_0^{2\pi} G_t(r,\theta,t;r_0,\theta_0,0) f(r_0,\theta_0) r_0 \, dr_0 \, d\theta_0 + \int_0^a \int_0^{2\pi} G(r,\theta,t;r_0,\theta_0,0) g(r_0,\theta_0) r_0 \, dr_0 \, d\theta_0$$

Now superposition yields the solution of the general problem. Observe from the foregoing that the solution is expressed explicitly in terms of the given data and the Green's function of the problem. Thus the problem has been inverted.

If the density ρ and the horizontal component of tension T_0 are functions of position on the membrane (but are independent of time), a more general hyperbolic equation is obtained for the equation of motion, namely,

$$\frac{\partial^2 u}{\partial t^2} - \frac{1}{\rho}\left[\frac{\partial}{\partial x}\left(T_0 \frac{\partial u}{\partial x}\right) + \frac{\partial}{\partial y}\left(T_0 \frac{\partial u}{\partial y}\right)\right] = F(x,y,t)$$

In this event separation of variables in the corresponding homogeneous equation yields the self-adjoint elliptic equation

$$\frac{\partial}{\partial x}\left(T_0 \frac{\partial \varphi}{\partial x}\right) + \frac{\partial}{\partial y}\left(T_0 \frac{\partial \varphi}{\partial y}\right) + \lambda \rho \varphi = 0$$

since $T_0(x,y) > 0$, $\rho(x,y) > 0$ on $\mathscr{R}$. If T_0 and ρ are twice continuously differentiable on $\mathscr{R}$, the theorems on the existence of eigenvalues and the orthogonality properties of the eigenfunctions of a linear self-adjoint elliptic operator with boundary condition $B(u) = 0$ on C apply. The method of eigenfunctions then leads to the formal series solution of the problem of the vibrating membrane with variable tension and density.

4-6   INITIAL-VALUE PROBLEM FOR THE THREE-DIMENSIONAL WAVE EQUATION

The homogeneous wave equation in three dimensions is

$$u_{tt} = c^2 \, \Delta u \tag{4-90}$$

where Δ is the three-dimensional laplacian.   In this section the solutions of lower-dimensional wave equations are viewed as particular solutions of Eq. (4-90) when the wave functions are independent of one of the space variables.   For example, if f is a twice continuously differentiable function, then

$$u = f(x - ct)$$

is a solution of Eq. (4-90) and represents a three-dimensional plane wave traveling in the x direction in xyz space.   The wavefronts are planes normal to the x axis.   Such solutions are useful in problems where it is known that the variation of u in the y and z directions can be neglected.   In general three-dimensional plane waves are represented by functions

$$u = f(\mathbf{n} \cdot \mathbf{r} - \omega t) \qquad \omega = c\,|\mathbf{n}|$$

where

$$\mathbf{n} = n_x\mathbf{i} + n_y\mathbf{j} + n_z\mathbf{k}$$

is a given constant vector, $|\mathbf{n}|$ its length, and

$$\mathbf{r} = x\mathbf{i} + y\mathbf{j} + z\mathbf{k}$$

is the variable position vector in space.   The wavefronts are planes normal to $\mathbf{n}$, and these are surfaces of constant phase.   The wavefronts travel with speed c in the direction of $\mathbf{n}$.   Similarly

$$u = g(\mathbf{n} \cdot \mathbf{r} + \omega t) \qquad \omega = c\,|\mathbf{n}|$$

represents a plane wave traveling in the direction of $-\mathbf{n}$.

The *initial-value problem*

$$u_{tt} - c^2\,\Delta u = F(x,y,z,t)$$
$$u(x,y,z,0) = f(x,y,z) \qquad u_t(x,y,z,0) = g(x,y,z) \tag{4-91}$$

is of importance in a number of applications.   Here u is a field or state variable whose values $u(x,y,z,t)$ over all space and time $t \geq 0$ describe the phenomena under consideration.   The function F is given.   It represents a known field source, or a driving force, which is independent of the field u. The initial field and the initial time rate of change of the field are assumed to be known.   These are represented by the functions f and g, respectively. The problem is to determine the field throughout space for all time $t \geq 0$.

Uniqueness of Solution

The proof of the uniqueness of solution of problem (4-91) can be constructed using the characteristic cone in $xyzt$ space and in the same manner as in Sec. 4-4 for the two-dimensional problem.   The proof is omitted here.   To demonstrate the existence of a solution use is made of the spherical mean introduced in Sec. 3-2. Recall Eq. (3-9) and the properties of the spherical mean. Let $P(x,y,z)$ be a point and let $S(P,r)$ denote the sphere of radius r about P as center. The spherical mean of the given function g in problem (4-91) is the function

$$\bar{g}(x,y,z;r) = \frac{1}{4\pi r^2} \iint\limits_{S(P,r)} g(x + n_1 r, y + n_2 r, z + n_3 r) \, dS \qquad (4\text{-}92)$$

where n_1, n_2, n_3 are the direction cosines of the exterior normal $\mathbf{n}$ on $S(P,r)$ and dS is the surface element of integration. Of interest are the values of the spherical mean taken over spheres $S(P,ct)$ of radius $r = ct$. Assume that g is twice continuously differentiable. Then so are $\bar{g}$ and the function ψ defined by

$$\psi(x,y,z,t) = t\bar{g}(x,y,z;ct) \qquad (4\text{-}93)$$

Indeed ψ is a solution of the homogeneous wave equation and satisfies the initial conditions

$$\psi(x,y,z,0) = 0 \qquad \psi_t(x,y,z,0) = g(x,y,z)$$

The first initial condition follows from Eq. (4-93) and the fact that

$$\lim_{t \to 0} \bar{g}(x,y,z;ct) = \bar{g}(x,y,z;0) = g(x,y,z)$$

To establish the second initial condition let

$$\xi = x + n_1 ct \qquad \eta = y + n_2 ct \qquad \zeta = z + n_3 ct$$

Introduce spherical coordinates r, θ, φ with origin at P, as in Sec. 3-2. Then the direction cosines of $\mathbf{n}$ are given by Eq. (3-13), and the surface element is

$$dS = c^2 t^2 \sin \theta \, d\theta \, d\varphi$$

Differentiate Eq. (4-93) with respect to t, obtaining

$$\frac{\partial \psi}{\partial t} = \bar{g}(x,y,z;ct) + t \frac{\partial \bar{g}}{\partial t} \qquad (4\text{-}94)$$

Let

$$G(x,y,z,t) = \frac{1}{4\pi} \int_0^{2\pi} \int_0^{\pi} (g_\xi n_1 + g_\eta n_2 + g_\zeta n_3) \sin \theta \, d\theta \, d\varphi$$

$$= \frac{1}{4\pi c^2 t^2} \iint\limits_{S(P,ct)} (g_\xi n_1 + g_\eta n_2 + g_\zeta n_3) \, dS \qquad (4\text{-}95)$$

Observe that G is a continuous function of its arguments and

$$\lim_{t \to 0} G(x,y,z,t)$$

exists.   Since

$$\frac{\partial \bar{g}}{\partial t} = cG(x,y,z,t) \tag{4-96}$$

it follows that

$$\psi_t(x,y,z,0) = \lim_{t \to 0} \psi_t(x,y,z,t) = \lim_{t \to 0} \bar{g}(x,y,z;ct) + \lim_{t \to 0} ctG(x,y,z,t) = g(x,y,z)$$

To show ψ is a solution of the homogeneous wave equation note that

$$\Delta \psi = \frac{t}{4\pi} \int_0^{2\pi} \int_0^{\pi} (g_{\xi\xi} + g_{\eta\eta} + g_{\zeta\zeta}) \sin \theta \, d\theta \, d\varphi \tag{4-97}$$

Now ψ_{tt} is calculated from Eqs. (4-94) and (4-96).

$$\psi_{tt} = 2\frac{\partial \bar{g}}{\partial t} + t\frac{\partial^2 \bar{g}}{\partial t^2} = 2cG(x,y,z,t) + ctG_t(x,y,z,t)$$

The derivative G_t is calculated from Eq. (4-95). First, however, G is re-written, by means of the divergence theorem, as a volume integral over the spherical region $V(P,ct)$ bounded by the sphere $S(P,ct)$.

$$G(x,y,z,t) = \frac{1}{4\pi c^2 t^2} \iiint\limits_{V(P,ct)} (g_{\xi\xi} + g_{\eta\eta} + g_{\zeta\zeta}) \, d\tau$$

$$= \frac{1}{4\pi c^2 t^2} \int_0^{ct} \int_0^{2\pi} \int_0^{\pi} (g_{\xi\xi} + g_{\eta\eta} + g_{\zeta\zeta}) r^2 \sin \theta \, dr \, d\theta \, d\varphi$$

Then

$$\frac{\partial G}{\partial t} = -\frac{2}{4\pi c^2 t^3} \iiint\limits_{V(P,ct)} (g_{\xi\xi} + g_{\eta\eta} + g_{\zeta\zeta}) \, d\tau$$

$$+ \frac{1}{4\pi c^2 t^2} \frac{\partial}{\partial t} \left[\int_0^{ct} \int_0^{2\pi} \int_0^{\pi} (g_{\xi\xi} + g_{\eta\eta} + g_{\zeta\zeta}) r^2 \sin \theta \, dr \, d\theta \, d\varphi \right]$$

$$= -\frac{2}{t} G(x,y,z,t) + \frac{c^3 t^2}{4\pi c^2 t^2} \int_0^{2\pi} \int_0^{\pi} (g_{\xi\xi} + g_{\eta\eta} + g_{\zeta\zeta}) \sin \theta \, d\theta \, d\varphi$$

$$= -\frac{2}{t} G(x,y,z,t) + \frac{c}{t} \Delta \psi$$

Accordingly

$$\psi_{tt} = 2cG(x,y,z,t) - 2cG(x,y,z,t) + c^2 \Delta \psi = c^2 \Delta \psi$$

Poisson's Formula

In direct analogy with the procedure above, let $\bar{f}$ denote the spherical mean over spheres $S(P,ct)$ of the given function f in problem (4-91).   Define the function φ by

$$\varphi(x,y,z,t) = t\bar{f}(x,y,z;ct) \tag{4-98}$$

Assume that f is three times continuously differentiable. Then so is φ, and by the same reasoning as used heretofore it follows that φ is a solution of the homogeneous wave equation, and

$$\varphi(x,y,z,0) = 0 \qquad \varphi_t(x,y,z,0) = f(x,y,z) \tag{4-99}$$

Now consider the function φ_t. This function is twice continuously differentiable, and

$$\frac{\partial^2 \varphi_t}{\partial t^2} = \frac{\partial}{\partial t}(\varphi_{tt}) = \frac{\partial}{\partial t}(c^2 \, \Delta\varphi) = c^2 \, \Delta\varphi_t$$

Thus φ_t satisfies the wave equation. Also

$$\varphi_{tt}(x,y,z,0) = 0$$

since

$$\varphi_{tt}(x,y,z,0) = (c^2 \, \Delta\varphi)\big|_{t=0} = c^2[\varphi_{xx}(x,y,z,0) + \varphi_{yy}(x,y,z,0) + \varphi_{zz}(x,y,z,0)]$$
$$= 0$$

by the first condition in Eq. (4-99).

Let

$$u = \psi + \varphi_t \tag{4-100}$$

where ψ and φ_t are defined in the preceding paragraphs. Then u is twice continuously differentiable and, by the superposition principle, satisfies the wave equation. Moreover

$$u(x,y,z,0) = \psi(x,y,z,0) + \varphi_t(x,y,z,0) = f(x,y,z)$$
$$u_t(x,y,z,0) = \psi_t(x,y,z,0) + \varphi_{tt}(x,y,z,0) = g(x,y,z)$$

Accordingly u is a solution of the problem

$$u_{tt} = c^2(u_{xx} + u_{yy} + u_{zz})$$
$$u(x,y,z,0) = f(x,y,z) \qquad u_t(x,y,z,0) = g(x,y,z) \tag{4-101}$$

The solution (4-100) can be rewritten

$$u(x,y,z,t) = \frac{\partial}{\partial t}[t\bar{f}(x,y,z;ct)] + t\bar{g}(x,y,z;ct)$$

$$= \frac{1}{4\pi}\frac{\partial}{\partial t}\int_0^{2\pi}\int_0^{\pi} tf(x + n_1 ct, \, y + n_2 ct, \, z + n_3 ct)\sin\theta \, d\theta \, d\varphi$$

$$+ \frac{1}{4\pi}\int_0^{2\pi}\int_0^{\pi} tg(x + n_1 ct, \, y + n_2 ct, \, z + n_3 ct)\sin\theta \, d\theta \, d\varphi \tag{4-102}$$

where

$$n_1 = \sin \theta \cos \varphi \qquad n_2 = \sin \theta \sin \varphi \qquad n_3 = \cos \theta$$

Equation (4-102) is called *Poisson's formula*. It can also be derived by direct integration of the wave equation and imposition of the initial conditions.

Propagation of Disturbances

In common with D'Alembert's solution of the one-dimensional initial-value problem Poisson's formula implies the propagation of an initial disturbance throughout space. Suppose g is identically zero. Then the disturbance at (x,y,z) at time t is

$$u(x,y,z,t) = \frac{1}{4\pi c^2} \frac{\partial}{\partial t} \iint\limits_{S(P,ct)} \frac{f(x + n_1 ct, \, y + n_2 ct, \, z + n_3 ct)}{t} \, dS \tag{4-103}$$

Assume the initial disturbance f is localized; that is, f is identically zero outside of a small sphere $\bar{\mathscr{R}}$ about a point Q. Inside $\bar{\mathscr{R}}$ the function f takes on nonzero values. Let $P(x,y,z)$ be a fixed point outside $\bar{\mathscr{R}}$ at which the values of u are to be calculated at subsequent times. The point P is referred to as an observation point. Recall that the integral in Eq. (4-103) is taken over the spherical surface $S(P,ct)$ of radius ct about P. There is a shortest distance d from P to the sphere $\bar{\mathscr{R}}$. If $0 < ct < d$, the surface $S(P,ct)$ does not intersect $\bar{\mathscr{R}}$, and the value of u at P is zero. Accordingly there is a time interval during which no effect is observed at P. As t increases, the radius ct increases until at some point in time the surface $S(P,ct_1)$ intersects $\bar{\mathscr{R}}$. At this time u takes on nonzero values at P (in general), and so the effect is observed at P. The *wavefront* of the disturbance at time t_1 is defined to be the locus of all points at which the effect is now first manifest, i.e., the set of boundary points of the undisturbed region. The observation point P lies on the wavefront at time t_1. As t increases, $S(P,ct)$ eventually includes $\bar{\mathscr{R}}$ in its interior. Then the value of u at P is again zero, and the trailing wavefront has passed P. The property of the three-dimensional wave equation just described is called *Huygens' principle*. An examination of the structure of the Poisson-Parseval formula (4-57) together with a review of the discussion in Sec. 4-4 shows that Huygens' principle does not hold for the two-dimensional wave equation.

Solution of the Problem

Consider the initial-value problem

$$u_{tt} - c^2(u_{xx} + u_{yy} + u_{zz}) = F(x,y,z,t)$$

$$u(x,y,z,0) = 0 \qquad u_t(x,y,z,0) = 0 \tag{4-104}$$

Let F be twice continuously differentiable. Then the solution is

$$u(x,y,z,t) = \frac{1}{4\pi} \int_0^t \tau \, d\tau \int_0^{2\pi} \int_0^{\pi} F(x + n_1 ct, \, y + n_2 ct, \, z + n_3 ct, \, t - \tau)$$

$$\times \sin \theta \, d\theta \, d\varphi \quad (4\text{-}105)$$

To derive Eq. (4-105) let τ be a fixed, but otherwise arbitrarily chosen, non-negative number. Solve the problem

$$\psi_{tt} = c^2(\psi_{xx} + \psi_{yy} + \psi_{zz})$$

$$\psi(x,y,z,\tau) = 0 \qquad \psi_t(x,y,z,\tau) = F(x,y,z,\tau)$$

In this problem the initial conditions are prescribed at $t = \tau$ rather than at $t = 0$. By a linear change of variable in Eq. (4-102) it follows that the solution of the problem is

$$\psi(x,y,z,t;\tau) = (t - \tau)\bar{F}[x,y,z;c(t - \tau)]$$

$$= \frac{t - \tau}{4\pi} \int_0^{2\pi} \int_0^{\pi} F[x + n_1 c(t - \tau), \, y + n_2 c(t - \tau), \, z + n_3 c(t - \tau), \, \tau]$$

$$\times \sin \theta \, d\theta \, d\varphi$$

Now let τ be a real nonnegative parameter, and consider the function u defined by

$$u(x,y,z,t) = \int_0^t \psi(x,y,z,t;\tau) \, d\tau \qquad (4\text{-}106)$$

Clearly $u(x,y,z,0) = 0$. Also

$$\frac{\partial u}{\partial t} = \psi(x,y,z,t;t) + \int_0^t \frac{\partial \psi}{\partial t} \, d\tau = \int_0^t \frac{\partial \psi}{\partial t} \, d\tau$$

so that $u_t(x,y,z,0) = 0$. Further

$$\frac{\partial^2 u}{\partial t^2} = \psi_t(x,y,z,t;t) + \int_0^t \frac{\partial^2 \psi}{\partial t^2} \, d\tau$$

$$= F(x,y,z,t) + c^2 \int_0^t \Delta\psi \, d\tau$$

$$= F(x,y,z,t) + c^2 \, \Delta u$$

Accordingly u is a solution of problem (4-104). The form given in Eq. (4-105) results when a linear change of variable is made in the integral in Eq. (4-106). Equation (4-105) can be written in yet another form. Make the change of variable $r = c(t - \tau)$ in the integral.   Then

$$u(x,y,z,t) = \frac{1}{4\pi c^2} \int_0^{ct} \int_0^{2\pi} \int_0^{\pi} F\left(x + n_1 r, y + n_2 r, z + n_3 r, t - \frac{r}{c}\right) r \sin \theta \, dr \, d\theta \, d\varphi$$

$$= \frac{1}{4\pi c^2} \int_0^{ct} \int_0^{2\pi} \int_0^{\pi} \frac{F(x + n_1 r, y + n_2 r, z + n_3 r, t - r/c)}{r} r^2 \sin \theta \, dr \, d\theta \, d\varphi$$

$$= \frac{1}{4\pi c^2} \iiint_{V(P,ct)} \frac{F(x + n_1 r, y + n_2 r, z + n_3 r, t - r/c)}{r} \, d\tau$$

$$= \frac{1}{4\pi c^2} \iiint_{V(P,ct)} \frac{F(\xi, \eta, \zeta, t - r/c)}{r} \, d\xi \, d\eta \, d\zeta \tag{4-107}$$

where $\xi = x + n_1 r$, $\eta = y + n_2 r$, and $\zeta = z + n_3 r$ and $V(P,ct)$ is the spherical region bounded by the sphere $S(P,ct)$ of radius ct about $P(x,y,z)$.   In this form the solution is called a *retarded potential*.   Note that the time dependence of the integrand involves the time

$$t^* = t - \frac{r}{c}$$

Hence $t = t^* + r/c$, and u is evaluated at a later instant.   The difference $t - t^* = r/c$ is the time required for a concentrated source at $Q(\xi,\eta,\zeta)$ to influence the field at P.   As in the case of Poisson's formula, it follows from Eq. (4-107) that if the source is localized, there must be a time interval during which no effect is observed at a distant observation point P.   As soon as the sphere $V(P,ct)$ intersects the domain containing the sources, the field at u changes, and the wavefront passes P.   However, in distinction with the field described by Poisson's formula, there is no trailing wavefront.   The influence of the source is permanent unless F vanishes at some subsequent time.   A basic difference exists between the nature of the propagation of initial values and the propagation of the effects of sources distributed in space and time.

   The solution of problem (4-91) can now be obtained by superposition. Let v denote the solution of problem (4-101) given in Eq. (4-102), and let w be the solution of problem (4-104) given in Eq. (4-105).   Then the solution of problem (4-91) is given by

$$u = v + w$$

Derivation of the Poisson-Parseval Formula

   Solutions of the wave equation which are independent of z are called *cylindrical wave functions*.   Such solutions are constant in value on cylinders whose generators are parallel to the z axis (at a fixed time t).   In problem (4-101) let the given functions f and g be independent of z.   Then problem (4-101) is equivalent to problem (4-56).   Consider the solution (4-102) with

f and g independent of z.   Rewritten as an integral over the sphere $S(P,ct)$, the solution is

$$u(x,y,t) = \frac{1}{4\pi c^2} \frac{\partial}{\partial t} \iint_{S(P,ct)} \frac{f(x + n_1 ct, y + n_2 ct)}{t} \, dS$$

$$+ \frac{1}{4\pi c^2} \iint_{S(P,ct)} \frac{g(x + n_1 ct, y + n_2 ct)}{t} \, dS$$

This form can be transformed into the Poisson-Parseval formula (4-57) as follows.   Choose $S(P,ct)$ to be the surface of the sphere of radius ct about the point $P(x,y,0)$ in the xy plane.   Let

$$\xi = x + n_1 ct \qquad \eta = y + n_2 ct \qquad \zeta = n_3 ct \tag{4-108}$$

where $n_1 = \sin \theta \cos \varphi$, $n_2 = \sin \theta \sin \varphi$, and $n_3 = \cos \theta$.   Then the equations in (4-108) are the parametric equations of the surface $S(P,ct)$.   Choose a local cartesian coordinate system $\xi' \eta' \zeta'$ as shown in Fig. 4-13.   Let A_{ct} be the projection of $S(P,ct)$ onto the $\xi' \eta'$ plane.   The surface element dS is related to the area element $d\xi' \, d\eta'$ in the $\xi' \eta'$ plane by

$$dS = d\xi' \, d\eta' \sec \gamma$$

where γ is the angle which the normal to dS makes with the positive ζ' axis. Let

$$r^2 = (\xi - x)^2 + (\eta - y)^2 = (\xi')^2 + (\eta')^2$$

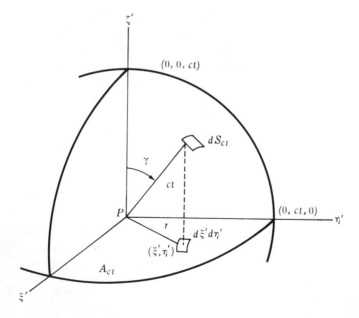

Figure 4-13

Then

$$\cos \gamma = \frac{(c^2t^2 - r^2)^{1/2}}{ct}$$

The first of the surface integrals is transformed into an integral over A_{ct}

$$\iint_{S(P,ct)} \frac{f(x + n_1ct, y + n_2ct)}{t} \, dS = 2c \iint_{A_{ct}} \frac{f(x + \xi', y + \eta')}{\sqrt{c^2t^2 - r^2}} \, d\xi' \, d\eta'$$

The second integral can be similarly transformed.   Thus

$$u(x,y,t) = \frac{1}{2\pi c} \frac{\partial}{\partial t} \iint_{A_{ct}} \frac{f(x + \xi', y + \eta')}{\sqrt{c^2t^2 - r^2}} \, d\xi' \, d\eta'$$

$$+ \frac{1}{2\pi c} \iint_{A_{ct}} \frac{g(x + \xi', y + \eta')}{\sqrt{c^2t^2 - r^2}} \, d\xi' \, d\eta'$$

The domain of integration A_{ct} is the disk in the plane defined by the inequality

$$(\xi')^2 + (\eta')^2 \le c^2t^2$$

Now r, φ are polar coordinates in the $\xi'\eta'$ plane so

$$x + \xi' = x + r \cos \varphi \qquad y + \eta' = y + r \sin \varphi$$

The Poisson-Parseval formula (4-57) follows immediately.   By a similar procedure Eq. (4-59) is deducible from Eq. (4-107) when F is independent of z.   Thus the solution of the two-dimensional problem can be obtained from the solution of the three-dimensional problem.   As shown in Prob. 49, D'Alembert's formula (4-15) for the solution of the one-dimensional problem is a consequence of the Poisson-Parseval formula.   Thus solutions of lower-dimensional problems are derivable from the expressions for the solutions of higher-dimensional problems.   This procedure is called the *method of descent*.

Spherically Symmetric Wave Functions

In *radiation problems* the field is often described by a solution of the wave equation in which the source F is confined to a small neighborhood of a fixed point, say the origin.   Also it may happen that the source is a spherically symmetric function.   In this case the influence of the source is propagated outward from the origin, and uniformly with respect to direction.   The magnitude of the effect at a distant point depends only on the distance r and on time.   *Spherically symmetric wave functions* are useful in describing such fields.   At points in space where no sources are present the spherically symmetric wave functions satisfy

$$\frac{\partial^2 u}{\partial t^2} = c^2 \left(\frac{\partial^2 u}{\partial r^2} + \frac{2}{r} \frac{\partial u}{\partial r} \right) \tag{4-109}$$

Equation (4-109) is obtained by writing the homogeneous wave equation in spherical coordinates r, θ, φ, and neglecting the angular dependence. Let $v = ru$; then the equation in v is

$$\frac{\partial^2 v}{\partial t^2} = c^2 \frac{\partial^2 v}{\partial r^2}$$

The general solution of this equation is

$$v = \varphi(r - ct) + \psi(r + ct)$$

where φ, ψ are arbitrary functions. Hence the general solution of Eq. (4-109) is

$$u = \frac{1}{r}\left[\varphi(r - ct) + \psi(r + ct)\right] \tag{4-110}$$

The function

$$u_1 = \frac{1}{r}\,\varphi(r - ct)$$

represents a spherically symmetric wave propagated radially outward from the origin with speed c. It is termed an *outgoing wave*. The surfaces of constant phase (wavefronts) are spherical surfaces about the origin whose radii increase with speed c. The function

$$u_2 = \frac{1}{r}\,\psi(r + ct)$$

represents a spherically symmetric wave propagated radially inward. The factor $1/r$ in these expressions implies that the amplitudes of the waves tend to zero with increasing distance r from the origin, provided φ and ψ are uniformly bounded functions. An example of a spherically symmetric wave function which is useful in many applications is

$$u = \frac{1}{r}\,e^{i(kr - \omega t)} \qquad k = \omega c$$

At points where sources are present the wave function satisfies the inhomogeneous equation

$$\frac{\partial^2 u}{\partial t^2} - c^2\left(\frac{\partial^2 u}{\partial r^2} + \frac{2}{r}\frac{\partial u}{\partial r}\right) = F(r,t) \tag{4-111}$$

where F represents the source density. Suppose the initial conditions are

$$u(r,0) = f(r) \qquad u_t(r,0) = g(r) \tag{4-112}$$

As before, let $v = ru$. Then the differential equation becomes

$$\frac{\partial^2 v}{\partial t^2} - c^2 \frac{\partial^2 v}{\partial r^2} = rF(r,t)$$

and the initial conditions are

$$v(r,0) = rf(r) \qquad v_t(r,0) = rg(r)$$

The solution of this problem is immediately obtainable from Eqs. (4-15) and (4-16). Hence the solution of Eq. (4-111) which satisfies the initial conditions (4-112) is

$$u(r,t) = \frac{1}{2r} [(r + ct)f(r + ct) + (r - ct)f(r - ct)]$$

$$+ \frac{1}{2cr} \int_{r-ct}^{r+ct} \xi g(\xi)\, d\xi + \frac{1}{2cr} \iint_{\mathscr{D}_{rt}} \xi F(\xi,\eta)\, d\xi\, d\eta \quad (4\text{-}113)$$

where $\mathscr{D}_{rt}$ is the triangular-shaped region in the $\xi\eta$ plane bounded by the interval $[r - ct, r + ct]$ of the ξ axis and the lines

$$\xi - c\eta = r - ct \qquad \xi + c\eta = r + ct$$

through (r,t).

4-7 SPHERICAL WAVES. CYLINDRICAL WAVES

Let S be a simple closed piecewise smooth surface in xyz space. Let $\mathscr{V}$ be the region interior to S, and let $\overline{\mathscr{V}} = \mathscr{V} + S$. The boundary- and initial-value problem for the three-dimensional wave equation is

$$u_{tt} - c^2\, \Delta u = F(x,y,z,t) \qquad \text{in } \overline{\mathscr{V}};\ t \geq 0$$

$$B(u) = 0 \qquad \text{on } S;\ t \geq 0 \tag{4-114}$$

$$u(x,y,z,0) = f(x,y,z) \qquad u_t(x,y,z,0) = g(x,y,z) \qquad \text{on } \overline{\mathscr{V}}$$

Here $B(u) = 0$ symbolizes one of the three types of boundary conditions considered in Sec. 4-5, i.e., Dirichlet, Neumann, or mixed. The method of eigenfunctions extends immediately to this problem. With the exception that the eigenfunctions $\{\varphi_n(x,y,z)\}$ involve one additional independent variable, the resulting equations are identical to those written in Sec. 4-5, and they are not repeated here. The proof that there exists at most one solution can be made in the same way using the energy integral

$$E(t) = \tfrac{1}{2} \iiint (\psi_t^2 + c^2\, |\nabla\psi|^2)\, dx\, dy\, dz \tag{4-115}$$

In order to derive the series solution explicitly in a given problem one must obtain all the eigenfunctions. Unfortunately this is either very difficult or impossible in most cases. In previous sections the technique of separation of variables has been used in a few problems to derive eigenfunctions. Recall, however, that the laplacian takes on different forms in the various coordinate systems. Thus the separation-of-variables technique is limited to those problems where the coordinate system is such that the Helmholtz equation is *separable:* the partial differential equation can be split into ordinary differential equations on assuming a particular form of solution. Only in relatively few coordinate systems is this possible. For a discussion of the conditions of separability of the Helmholtz equation (and Laplace's equation) see Ref. 8. Separation is possible in spherical and cylindrical coordinates, and the remainder of this section is devoted to these two systems.

Separable Solutions in Spherical Coordinates

In spherical coordinates the scalar Helmholtz equation is

$$\frac{1}{r^2}\frac{\partial}{\partial r}\left(r^2\frac{\partial \psi}{\partial r}\right) + \frac{1}{r^2 \sin \theta}\frac{\partial}{\partial \theta}\left(\sin \theta \frac{\partial \psi}{\partial \theta}\right) + \frac{1}{r^2 \sin^2 \theta}\frac{\partial^2 \psi}{\partial \varphi^2} + \lambda \psi = 0 \qquad (4\text{-}116)$$

Here (r,θ,φ) are the spherical coordinates shown in Fig. 3-3. Equation (4-116) is separable. If $\lambda = 0$, it reduces to Laplace's equation. The separable solutions of Laplace's equation in spherical coordinates were derived in Sec. 3-4. Hence the assumption is made that $\lambda \neq 0$. Assume a solution of Eq. (4-116) of the form

$$\psi = R(r)Y(\theta,\varphi) \qquad (4\text{-}117)$$

The resulting equations are

$$(r^2 R')' + (\lambda r^2 - \mu)R = 0 \qquad (4\text{-}118)$$

$$\Delta^* Y + \mu Y = 0 \qquad (4\text{-}119)$$

where μ is a separation constant, primes denote derivatives with respect to r, and Δ^* is the surface laplacian defined by Eq. (3-59). In Eq. (4-118) make the change of dependent variable

$$w = \sqrt{r}\, R$$

Then the equation in w is

$$r^2 w'' + r w' + (\lambda r^2 - \mu - \tfrac{1}{4})w = 0 \qquad (4\text{-}120)$$

Now make the change of independent variable

$$s = \sqrt{\lambda}\, r$$

Then

$$s^2 \frac{d^2w}{ds^2} + s\frac{dw}{ds} + (s^2 - \mu - \tfrac{1}{4})w = 0$$

This last is Bessel's equation with $\alpha^2 = \mu + \tfrac{1}{4}$. Accordingly the general solution of Eq. (4-118) is

$$R = \frac{1}{\sqrt{r}}[AJ_\alpha(\sqrt{\lambda}\,r) + BY_\alpha(\sqrt{\lambda}\,r)] \qquad \alpha = (\mu + \tfrac{1}{4})^{\frac{1}{2}} \tag{4-121}$$

where J_α, Y_α are Bessel functions of the first and second kind, respectively, and A, B are arbitrary constants. In Eq. (4-119) μ is an arbitrary constant. Hence let β be arbitrary, and write

$$\mu = \beta(\beta + 1)$$

Then $\alpha^2 = \mu + \tfrac{1}{4} = (\beta + \tfrac{1}{2})^2$. Equation (4-19) becomes

$$\Delta^* Y + \beta(\beta + 1) Y = 0 \tag{4-122}$$

The reader should distinguish between the symbol Y used in Eq. (4-122) and the symbol Y_α used in Eq. (4-121) to denote the Bessel function of the second kind. The distinction is also made in the sequel. Observe that Eq. (4-122) is identical to Eq. (3-60) except that β replaces the α of that equation. It follows from the discussion of the separation of variables in Eq. (3-60), and the radial dependence shown in Eq. (4-121), that separable solutions of Eq. (4-116) are (for $\lambda \neq 0$)

$$\psi_{\beta\lambda\nu}(r,\theta,\varphi) = \frac{1}{\sqrt{r}}[AJ_{\beta+\frac{1}{2}}(\sqrt{\lambda}\,r) + BY_{\beta+\frac{1}{2}}(\sqrt{\lambda}\,r)]$$

$$\times [CP_\beta{}^\nu(\cos\theta) + DQ_\beta{}^\nu(\cos\theta)](E\cos\nu\varphi + F\sin\nu\varphi) \tag{4-123}$$

provided $\nu \neq 0$. Here $J_{\beta+\frac{1}{2}}$, $Y_{\beta+\frac{1}{2}}$ denote the Bessel functions of the first and second kinds, respectively, of order $\beta + \tfrac{1}{2}$; $P_\beta{}^\nu$, $Q_\beta{}^\nu$ are the associated Legendre functions, and the constants $A, B, C, D, E, F, \lambda(\neq 0)$, β, and $\nu(\neq 0)$, are arbitrary. If $\nu = 0$, it is understood that the φ dependence is

$$c_1 + c_2\varphi$$

where c_1, c_2 are arbitrary constants.

As a particular case let $\mathscr{V}$ be the sphere of radius a about the origin, and let S be its surface. Then the solution ψ of the Helmholtz equation must be continuous and single-valued for

$$0 \leq r \leq a \qquad 0 \leq \theta \leq \pi \qquad 0 \leq \varphi \leq 2\pi$$

Since the Bessel function of the second kind $Y_{\beta+\frac{1}{2}}$ is singular at $r = 0$, it is necessary to set $B = 0$ in the separable form of solution shown in Eq. (4-123). Also, ψ must be periodic in φ, of period 2π, and continuous at the poles $\theta = 0, \theta = \pi$. As in Sec. 3-4, the foregoing requirements restrict the separation constants β and ν to the nonnegative integers

$$\beta = n \qquad n = 0, 1, \ldots \qquad\qquad \nu = q \qquad q = 0, 1, \ldots, n$$

Set $\beta = n$ in Eq. (4-122). Then the linearly independent solutions which satisfy the continuity and periodicity requirements are the surface harmonics

$$Y_{nq}(\theta,\varphi) = P_n{}^q(\cos\theta)(E_{nq}\cos q\varphi + F_{nq}\sin q\varphi) \qquad q = 0, 1, \ldots, n$$

If $n = 0$, this reduces to a constant. Set $\mu = n$ in Eq. (4-120). Then the solution which satisfies the continuity requirement at $r = 0$ is the Bessel function $J_{n+\frac{1}{2}}(\sqrt{\lambda}\,r)$ of order half an odd integer. Of frequent use in boundary-value problems involving spherical regions are the *spherical Bessel functions of the first kind*

$$j_n(r) = \sqrt{\frac{\pi}{2r}}\, J_{n+\frac{1}{2}}(r) \qquad n = 0, 1, \ldots \tag{4-124}$$

Thus the continuous and periodic separable forms of solution of the Helmholtz equation in $\overline{\mathscr{V}}$ are

$$\psi_{n\lambda q}(r,\theta,\varphi) = j_n(\sqrt{\lambda}\,r)\,Y_{nq}(\theta,\varphi) \qquad n = 0, 1, \ldots\,; q = 0, 1, \ldots, n$$

Recall that the eigenvalues are real and nonnegative for the three types of boundary conditions under consideration. Hence let

$$\lambda = k^2 \qquad k \text{ real}$$

For each fixed n, k, and q in the admissible ranges the function

$$\begin{aligned} u_{nkq}(r,\theta,\varphi,t) &= \psi_{nkq}(r,\theta,\varphi)(A_{nkq}\cos kct + B_{nkq}\sin kct) \\ &= j_n(kr)P_n{}^q(\cos\theta)(E_{nq}\cos q\varphi + F_{nq}\sin q\varphi)(A_{nkq}\cos kct + B_{nkq}\sin kct) \end{aligned}$$

is a spherical wave function which is continuous for $r \geq 0$ and periodic in φ, of period 2π. *Axially symmetric* separable wave functions are those which are independent of φ:

$$u_{nk}(r,\theta,t) = j_n(kr)P_n(\cos\theta)(A_{nk}\cos kct + B_{nk}\sin kct) \qquad n = 0, 1, \ldots\,; k \text{ real}$$

Here P_n is the Legendre polynomial of order n defined in Eq. (59) of Appendix 2. *Spherically symmetric* wave functions are

$$u_k(r,t) = j_0(kr)(A_k\cos kct + B_k\sin kct) \qquad k \text{ real}$$

From the formulas listed in Sec. 4 of Appendix 2 one obtains

$$j_0(r) = \frac{\sin r}{r}$$

Accordingly the spherically symmetric separable wave functions for $r \geq 0$ can be written in the form

$$u_k(r,t) = \frac{\sin kr}{kr} (A_k \cos kct + B_k \sin kct)$$

Boundary- and Initial-value Problem for a Sphere. Green's Function for the Problem

The following example illustrates the manner in which the boundary condition prescribed on S determines the eigenvalues. Also, several additional aspects are discussed.

EXAMPLE 4-3 Let $\bar{\mathscr{V}}$ be the sphere of radius a about the origin, and let S denote its surface. Solve the problem

$$u_{tt} - c^2 \Delta u = F(r,\theta,\varphi,t) \qquad \text{in } \bar{\mathscr{V}}; t \geq 0$$

$$u = \sin \omega t \qquad \text{on } S; t \geq 0$$

$$u(r,\theta,\varphi,0) = f(r,\theta,\varphi) \qquad u_t(r,\theta,\varphi,0) = g(r,\theta,\varphi) \qquad \text{in } \bar{\mathscr{V}}$$

where F, f, and g are given functions, and ω is a real positive constant, not a multiple of $\pi c/a$.

From the text it follows that

$$w(r,t) = \frac{a}{r} \frac{\sin (\omega r/c)}{\sin (\omega a/c)} \sin \omega t$$

is a solution of the homogeneous wave equation in $\bar{\mathscr{V}}$. Also, w satisfies the prescribed boundary condition on S. Now suppose v is a function which satisfies the inhomogeneous wave equation in $\bar{\mathscr{V}}$, $v = 0$ on S for $t \geq 0$, and the initial conditions

$$v(r,\theta,\varphi,0) = f(r,\theta,\varphi) \qquad v_t(r,\theta,\varphi,0) = g(r,\theta,\varphi) - w_t(r,0)$$

Then the superposition

$$u = v + w$$

is a solution of the original problem. The function v is constructed as a series in the eigenfunctions of the problem.

Recall from the text that the continuous and periodic separable solutions of the Helmholtz equation in $\bar{\mathscr{V}}$ are

$$\psi_{n\lambda q}(r,\theta,\varphi) = j_n(\sqrt{\lambda}\, r) Y_{nq}(\theta,\varphi)$$

where λ is real and nonnegative. The boundary condition on S is

$$\psi_{n\lambda q}(a,\theta,\varphi) = 0 \qquad 0 \leq \theta \leq \pi; 0 \leq \varphi \leq 2\pi$$

This implies

$$_n(\sqrt{\lambda}\, a) = 0 \qquad n = 0, 1, \ldots$$

For each fixed n let $\{\xi_{nm}\}$, $m = 1, 2, \ldots$ be the real positive zeros of $J_{n+\frac{1}{2}}(\xi)$ ordered according to increasing magnitude. Then the eigenvalues in the present case are

$$\lambda_{nm} = \frac{\xi_{nm}^2}{a^2} \qquad n = 0, 1, \ldots; \; m = 1, 2, \ldots$$

Corresponding to the eigenvalue λ_{nm} there are the $2n + 1$ linearly independent eigenfunctions

$$\psi_{nmq}^{(e)} = j_n\left(\frac{\xi_{nm}r}{a}\right) Y_{nq}^{(e)}(\theta, \varphi) \qquad q = 0, 1, \ldots, n$$

$$\psi_{nmq}^{(0)} = j_n\left(\frac{\xi_{nm}r}{a}\right) Y_{nq}^{(0)}(\theta, \varphi) \qquad q = 1, \ldots, n$$

where the surface harmonics $Y_{nq}^{(e)}$, $Y_{nq}^{(0)}$ are defined by Eq. (3-68). The eigenfunctions have the orthogonality properties

$$\int_0^a \int_0^{2\pi} \int_0^\pi \psi_{nmq}^{(e)} \psi_{n'm'q'}^{(e)} r^2 \sin\theta \, dr \, d\theta \, d\varphi = 0 \qquad (n,m,q) \neq (n',m',q')$$

and the same holds with (e) replaced by (0)

$$\int_0^a \int_0^{2\pi} \int_0^\pi \psi_{nmq}^{(e)} \psi_{n'm'q'}^{(e)} r^2 \sin\theta \, dr \, d\theta \, d\varphi = 0 \qquad \text{all } (n,m,q), (n',m',q')$$

Also

$$\|\psi_{nmq}^{(e)}\|^2 = \int_0^a \int_0^{2\pi} \int_0^\pi (\psi_{nmq}^{(e)})^2 r^2 \sin\theta \, dr \, d\theta \, d\varphi$$

$$= \epsilon_q \frac{\pi^2 a^3}{4\xi_{nm}} \frac{(n+q)!}{(n-q)!} [J_{n+\frac{3}{2}}(\xi_{nm})]^2$$

where $\epsilon_0 = 2$, $\epsilon_q = 1$ for $q \geq 1$. The evaluation of the integral follows from Eqs. (3-71) and (3-72). The same result holds for $\|\psi_{nmq}^{(0)}\|$, $q \geq 1$. The set of eigenfunctions

$$\{\psi_{nmq}^{(e)}, \psi_{nmq}^{(0)}\}$$

is complete in the space of all functions which are continuous on $\overline{\mathscr{V}}$. A proof of this property can be constructed along the lines of the proof given in Example 3-5 provided the completeness properties of the individual sets

$$\left\{J_{n+\frac{1}{2}}\left(\frac{\xi_{nm}r}{a}\right)\right\} \qquad \{Y_{nq}^{(e)}(\theta, \varphi), Y_{nq}^{(0)}(\theta, \varphi)\}$$

on appropriate function spaces are assumed. Once the completeness is established, it follows immediately that the set $\{\lambda_{nm}\}$ constitutes all the eigenvalues of the problem. Moreover any eigenfunction corresponding to an eigenvalue λ_{nm} must be a linear combination of the functions

$$\psi_{nm0}^{(e)}, \psi_{nm1}^{(e)}, \ldots, \psi_{nmn}^{(e)} \qquad \psi_{nm1}^{(0)}, \ldots, \psi_{nmn}^{(0)}$$

The *normal modes* of vibration of the present problem are the ∞^3 functions

$$u_{nmq}^{(e)} = \psi_{nmq}^{(e)}(r, \theta, \varphi)(A_{nmq}^{(e)} \cos \omega_{nm}t + B_{nmq}^{(e)} \sin \omega_{nm}t)$$

$$u_{nmq}^{(0)} = \psi_{nmq}^{(0)}(r, \theta, \varphi)(A_{nmq}^{(0)} \cos \omega_{nm}t + B_{nmq}^{(0)} \sin \omega_{nm}t)$$

$$n = 0, 1, \ldots \qquad m = 1, 2, \ldots \qquad q = 0, 1, \ldots, n$$

where the *characteristic frequencies* are

$$\omega_{nm} = c\sqrt{\lambda_{nm}} = \frac{c\xi_{nm}}{a} \qquad n = 0, 1, \ldots ; m = 1, 2, \ldots$$

The normal modes are the continuous single-valued separable wave functions in $\bar{\mathscr{V}}$ which vanish on S. They represent the standing waves in the sphere which have a node on the surface S. The initial assumption regarding ω made in the statement of the original problem is now clear: the prescribed frequency of the field on S must not coincide with a characteristic frequency.

Consider the problem

$$v_{tt} = c^2\, \Delta v \qquad \text{in } \bar{\mathscr{V}}; t \geq 0$$

$$v = 0 \qquad \text{on } S; t \geq 0$$

$$v(r,\theta,\varphi,0) = f(r,\theta,\varphi) \qquad v_t(r,\theta,\varphi,0) = g(r,\theta,\varphi) - w_t(r,0)$$

To solve this problem assume a superposition of normal modes

$$v = \sum_{n=0}^{\infty} \sum_{m=1}^{\infty} \sum_{q=0}^{n} (u_{nmq}^{(e)} + u_{nmq}^{(0)})$$

The first initial condition is satisfied if

$$\sum_{n=0}^{\infty} \sum_{m=1}^{\infty} \sum_{q=0}^{n} [A_{nmq}^{(e)}\psi_{nmq}^{(e)}(r,\theta,\varphi) + A_{nmq}^{(0)}\psi_{nmq}^{(0)}(r,\theta,\varphi)] = f(r,\theta,\varphi)$$

By the now familiar method which makes use of the orthogonality properties of the eigenfunctions it follows that the coefficients

$$A_{nmq}^{(e)} = \frac{1}{\|\psi_{nmq}^{(e)}\|^2} \int_0^a \int_0^{2\pi} \int_0^{\pi} f(r,\theta,\varphi)\psi_{nmq}^{(e)}(r,\theta,\varphi)r^2 \sin\theta\, dr\, d\theta\, d\varphi$$

$$n = 0, 1, \ldots ; m = 1, 2, \ldots ; q = 0, 1, \ldots, n$$

The same result holds for $A_{nmq}^{(0)}$, $q \geq 1$. In the same manner satisfaction of the second initial condition determines the coefficients $B_{nmq}^{(e)}$, $B_{nmq}^{(0)}$.

Now consider the problem

$$v_{tt} - c^2\, \Delta v = F(r,\theta,\varphi,t) \qquad \text{in } \bar{\mathscr{V}}; t \geq 0$$

$$v = 0 \qquad \text{on } S; t \geq 0$$

$$v(r,\theta,\varphi,0) = 0 \qquad v_t(r,\theta,\varphi,0) = 0 \qquad \text{in } \bar{\mathscr{V}}$$

Assume a solution of the form

$$v = \sum_{n=0}^{\infty} \sum_{m=1}^{\infty} \sum_{q=0}^{\infty} [h_{nmq}^{(e)}(t)\psi_{nmq}^{(e)}(r,\theta,\varphi) + h_{nmq}^{(0)}(t)\psi_{nmq}^{(0)}(r,\theta,\varphi)]$$

where the functions $h_{nmq}^{(e)}$, $h_{nmq}^{(0)}$ are to be determined. As in Example 4-2 the assumed form is substituted into the inhomogeneous wave equation. Then the orthogonality properties of the eigenfunctions together with the requirements

$$h_{nmq}^{(e)}(0) = \dot{h}_{nmq}^{(e)}(0) = 0 \qquad h_{nmq}^{(0)}(0) = \dot{h}_{nmq}^{(0)}(0) = 0$$

lead to the expressions

$$h_{nmq}^{(e)}(t) = \frac{1}{\omega_{nm}} \int_0^t F_{nmq}^{(e)}(\xi) \sin[\omega_{nm}(t - \xi)]\, d\xi$$

where

$$F_{nmq}^{(e)}(t) = \frac{1}{\|\psi_{nmq}^{(e)}\|^2} \int_0^a \int_0^{2\pi} \int_0^\pi F(r,\theta,\varphi,t)\psi_{nmq}^{(e)}(r,\theta,\varphi)r^2 \sin\theta \, dr \, d\theta \, d\varphi$$

Replace (e) by (0) throughout; then these same formulas define the functions $h_{nmq}^{(0)}$. Thus the series solution of the problem formulated at the start of this paragraph is obtained. Superposition of this series and the series derived in the preceding paragraph yields the formal solution of the original problem in v.

To derive the eigenfunction representation of the Green's function of the present problem choose

$$F(r,\theta,\varphi,t) = \frac{\delta(r - r_0)\delta(\theta - \theta_0)\delta(\varphi - \varphi_0)\delta(t - \tau)}{r^2 \sin\theta}$$

a concentrated source of unit strength located at (r_0,θ_0,φ_0) and applied at time $\tau > 0$. As in Example 4-2, the properties of the δ function are here used in a purely formal manner. See also the remarks made in Sec. 3-5. Evaluation of the integrals derived in the previous paragraph yields

$$G(r,\theta,\varphi,t;r_0,\theta_0,\varphi_0,\tau) = \sum_{n=0}^\infty \sum_{m=1}^\infty \sum_{q=0}^\infty$$

$$\times \left\{ \frac{j_n(\xi_{nm}r/a)j_n(\xi_{nm}r_0/a)}{\omega_{nm} \|\psi_{nmq}\|^2} P_n{}^q(\cos\theta)P_n{}^q(\cos\theta_0) \cos[q(\varphi - \varphi_0)] \sin[\omega_{nm}(t - \tau)] \right\}$$

$t > \tau$. Define G to be identically zero, $t \le \tau$. Then G satisfies homogeneous initial conditions. Also G satisfies the homogeneous boundary conditions on S. By means of the Green's function the solution of the more general problem

$$u_{tt} - c^2 \Delta u = F(r,\theta,\varphi,t) \quad \text{in } \mathscr{V}; t \ge 0$$
$$u = H(\theta,\varphi,t) \quad \text{on } S; t \ge 0$$
$$u(r,\theta,\varphi,0) = f(r,\theta,\varphi) \quad u_t(r,\theta,\varphi,0) = g(r,\theta,\varphi) \quad t > 0$$

can be expressed in terms of integrals involving the given functions F, H, f, and g. The representation is

$$u = \int_0^t d\tau \iiint_{\mathscr{V}} G(r,\theta,\varphi,t;r_0,\theta_0,\varphi_0,\tau)F(r_0,\theta_0,\varphi_0,\tau) \, dV_0$$

$$+ \iiint_{\mathscr{V}} G_t(r,\theta,\varphi,t;r_0,\theta_0,\varphi_0,0)f(r_0,\theta_0,\varphi_0) \, dV_0 + \iiint_{\mathscr{V}} G(r,\theta,\varphi,t;r_0,\theta_0,\varphi_0,0)g(r_0,\theta_0,\varphi_0) \, dV_0$$

$$- c^2 \int_0^t d\tau \iint_S H(\theta_0,\varphi_0,\tau) \frac{\partial G}{\partial r_0}\bigg|_{r_0=a} dS_0$$

where $dV_0 = r_0{}^2 \sin\theta_0 \, dr_0 \, d\theta_0 \, d\varphi_0$ is the volume element of integration in $\mathscr{V}$ and $dS_0 = a^2 \sin\theta_0 \, d\theta_0 \, d\varphi_0$ is the surface element of integration on S. The following is a strictly heuristic derivation of the formula. Rewrite the boundary- and initial-value problem in terms of the variables, r_0, θ_0, φ_0, and τ. Then u satisfies

$$u_{\tau\tau} - c^2 \Delta u = F(r_0,\theta_0,\varphi_0,\tau) \quad \text{in } \mathscr{V}; \tau \ge 0$$

where it is understood that Δ is written in terms of the variables r_0, θ_0, φ_0. Let (r,θ,φ) be a fixed but otherwise arbitrary point of $\overline{\mathscr{V}}$, and let $t > 0$ be an arbitrary time.   Now the Green's function satisfies

$$G_{\tau\tau} - c^2 \, \Delta G = \frac{\delta(r_0 - r)\delta(\theta_0 - \theta)\delta(\varphi_0 - \varphi)\delta(\tau - t)}{r_0{}^2 \sin \theta_0}$$

Multiply the first equation by G, the second by u, subtract, and then integrate the result over $\mathscr{V}$ and with respect to τ, from 0 to t, thus obtaining the relation

$$\int_0^t d\tau \iiint_{\mathscr{V}} (Gu_{\tau\tau} - uG_{\tau\tau}) \, dV_0 - c^2 \int_0^t d\tau \iiint_{\mathscr{V}} (G \, \Delta u - u \, \Delta G) \, dV_0$$

$$= \int_0^t d\tau \iiint_{\mathscr{V}} G(r,\theta,\varphi,t;r_0,\theta_0,\varphi_0,\tau)F(r_0,\theta_0,\varphi_0,\tau) \, dV_0 - u(r,\theta,\varphi,t)$$

But

$$\int_0^t d\tau \iiint_{\mathscr{V}} (Gu_{\tau\tau} - uG_{\tau\tau}) \, dV_0 = \iiint_{\mathscr{V}} dV_0 \int_0^t \frac{\partial}{\partial \tau} (Gu_\tau - uG_\tau) \, d\tau$$

$$= \iiint_{\mathscr{V}} (Gu_\tau - uG_\tau) \Big|_0^t \, dV_0 = -\iiint_{\mathscr{V}} G \Big|_{\tau=0} g \, dV_0 + \iiint_{\mathscr{V}} G_\tau \Big|_{\tau=0} f \, dV_0$$

$$\iiint_{\mathscr{V}} (G \, \Delta u - u \, \Delta G) \, dV_0 = \iint_S \left(G \frac{\partial u}{\partial r_0} - u \frac{\partial G}{\partial r_0} \right) dS_0 = -\iint_S H(\theta_0,\varphi_0,\tau) \frac{\partial G}{\partial r_0} \Big|_{r_0=a} dS_0$$

Observe that $G_\tau = -G_t$.   Insertion of these last expressions into the first relation yields the formula for $u(r,\theta,\varphi,t)$.   If the boundary condition is

$$\frac{\partial u}{\partial r} = H(\theta,\varphi,t) \qquad \text{on } S; \, t \geq 0$$

that is, of the Neumann type rather than the Dirichlet type, the integral

$$-c^2 \int_0^t d\tau \iint_S H(\theta_0,\varphi_0,\tau) \frac{\partial G}{\partial r_0} \Big|_{r_0=a} dS_0$$

is replaced by

$$c^2 \int_0^t d\tau \iint_S G \Big|_{r_0=a} H(\theta_0,\varphi_0,\tau) \, dS_0$$

Separable Solutions in Cylindrical Coordinates

The Helmholtz equation in cylindrical coordinates is

$$\frac{\partial^2 \psi}{\partial r^2} + \frac{1}{r} \frac{\partial \psi}{\partial r} + \frac{1}{r^2} \frac{\partial^2 \psi}{\partial \theta^2} + \frac{\partial^2 \psi}{\partial z^2} + \lambda\psi = 0 \qquad\qquad (4\text{-}125)$$

If $\lambda = 0$, Eq. (4-125) reduces to Laplace's equation (3-36). The separable forms of solution of Laplace's equation in cylindrical coordinates are derived in Sec. 3-3. Hence $\lambda \neq 0$ is assumed in the following. Suppose

$$\psi = \varphi(r,\theta)Z(z)$$

is a solution of Eq. (4-125). Then the factors must satisfy

$$\varphi_{rr} + \frac{1}{r}\,\varphi_r + \frac{1}{r^2}\,\varphi_{\theta\theta} + \beta^2\varphi = 0 \tag{4-126}$$

$$Z'' + \alpha^2 Z = 0 \tag{4-127}$$

where α, β are arbitrary real separation constants, and

$$\lambda = \alpha^2 + \beta^2$$

The general solution of Eq. (4-127) is

$$Z = \begin{cases} E \cos \alpha z + F \sin \alpha z & \alpha \neq 0 \\ E + Fz & \alpha = 0 \end{cases}$$

Observe that Eq. (4-126) is identical in form to Eq. (3-37) except that β^2 has replaced the λ in Eq. (3-37). Accordingly the results of that section imply the existence of the separable solutions

$$\psi_{\nu\beta\alpha} = [AJ_\nu(\beta r) + BY_\nu(\beta r)](C \cos \nu\theta + D \sin \nu\theta)(E \cos \alpha z + F \sin \alpha z) \tag{4-128}$$

of Eq. (4-126), where ν, β, and α are arbitrary real nonnegative constants. Here it is understood that if $\nu = 0$, the trigonometric function of θ in Eq. (4-128) is replaced by the linear function $C + D\theta$. A similar convention applies if $\alpha = 0$. An examination of the separation-of-variables procedure in Sec. 3-3 shows that if $\beta = 0$, the separable solutions

$$\psi_{\nu\alpha} = (Ar^\nu + Br^{-\nu})(C \cos \nu\theta + D \sin \nu\theta)(E \cos \alpha z + F \sin \alpha z) \tag{4-129}$$

result, and $\lambda = \alpha^2$.

Assume $\lambda = \beta^2 + \alpha^2 > 0$, and let $k = \sqrt{\lambda}$. If $\beta^2 > 0$, the functions

$$u_{\nu\beta\alpha} = \psi_{\nu\beta\alpha}(r,\theta,z)e^{\pm ikct} \qquad i = \sqrt{-1} \tag{4-130}$$

where the $\psi_{\nu\beta\alpha}$ are defined by Eq. (4-128), are separable wave functions in cylindrical coordinates in some region of space. If $\beta = 0$, the functions

$$u_{\nu\alpha} = \psi_{\nu\alpha}(r,\theta,z)e^{\pm iact} \qquad i = \sqrt{-1} \tag{4-131}$$

where the $\psi_{\nu\alpha}$ are defined in Eq. (4-129), are wave functions. If the region of space of interest allows $0 \leq \theta \leq 2\pi$ and single-valued continuous solutions are desired, then v is quantized to integral values: $v = n, n = 0, 1, \ldots$ Also, if the region includes the z axis, that is, $r = 0$, then $B = 0$ is chosen in the separable forms $\psi_{v\beta\alpha}$, $\psi_{v\alpha}$.

EXAMPLE 4-4 Let $\overline{\mathcal{V}}$ denote the right-angled cylindrical and annular wedge defined by the inequalities

$$0 < a \le r \le b \qquad 0 \le \theta \le \frac{\pi}{2} \qquad 0 \le z \le h$$

where r, θ, z are cylindrical coordinates. Solve the problem

$$u_{tt} - c^2 \Delta u = F(r,\theta,z,t) \qquad \text{in } \overline{\mathcal{V}}; \, t \ge 0$$

$$\frac{\partial u}{\partial n} = 0 \qquad \text{on } S; \, t \ge 0$$

$$u(r,\theta,\tau,0) = f(r,\theta,z) \qquad u_t(r,\theta,z,0) = g(r,\theta,z) \qquad \text{in } \overline{\mathcal{V}}$$

where S denotes the surface bounding $\overline{\mathcal{V}}$ and $\partial u / \partial n$ is the derivative in the direction of the exterior normal to S.

Observe first of all that $\lambda = 0$ is an eigenvalue of the eigenvalue problem, with $\psi_0 = 1$ a corresponding real-valued eigenfunction. Now assume $\lambda > 0$, and consider the separable solutions of Eq. (4-125) derived in the text. On the portions of the planes $\theta = 0$, $\theta = \pi/2$ which form part of S the boundary condition is

$$\frac{\partial \psi}{\partial \theta} = 0 \qquad \theta = 0 \qquad \theta = \frac{\pi}{2}$$

Hence

$$D = 0 \qquad \sin \frac{v\pi}{2} = 0$$

which implies

$$v = 2n \qquad n = 0, 1, \ldots$$

On the base and top there are the boundary conditions

$$\frac{\partial \psi}{\partial z} = 0 \qquad z = 0; \, z = h$$

Thus

$$F = 0 \qquad \sin \alpha h = 0$$

and so

$$\alpha = \frac{q^2 \pi^2}{h^2} \qquad q = 1, 2, \ldots$$

On the remaining sides there are the boundary conditions

$$\frac{\partial \psi}{\partial r} = 0 \qquad r = a; \, r = b$$

Consider first the separable solutions in Eq. (4-129). If the boundary conditions at $r = a$ and $r = b$ are imposed on the radial factor

$$R = Ar^\nu + Br^{-\nu}$$

it follows that $A = B = 0$, and the trivial solution results. However if $\nu = 0$, the radial factor $R = $ constant satisfies the boundary conditions. Hence there are the eigenfunctions

$$\psi_q = \cos\frac{q\pi z}{h} \qquad q = 0, 1, \ldots$$

Now impose the boundary conditions at $r = a$ and $r = b$ on the radial factor in the separable form (4-128). The resulting equations are

$$AJ'_{2n}(\beta a) + BY'_{2n}(\beta a) = 0$$

$$AJ'_{2n}(\beta b) + BY'_{2n}(\beta b) = 0$$

In order to obtain nontrivial solutions it is necessary that the determinant be

$$J'_{2n}(\beta a)\, Y'_{2n}(\beta b) - J'_{2n}(\beta b)\, Y'_{2n}(\beta a) = 0$$

This transcendental equation determines the permissible values of β. For each fixed $n = 0, 1, \ldots$ there exists a real sequence $\{\beta_{nm}\}$ of values such that

$$J'_{2n}(\beta_{nm}a)\, Y'_{2n}(\beta_{nm}b) - J'_{2n}(\beta_{nm}b)\, Y'_{2n}(\beta_{nm}a) = 0 \qquad n = 0, 1, \ldots; m = 1, 2, \ldots;$$

$$0 < \beta_{n1} < \cdots < \beta_{nm} < \beta_{n,m+1} < \cdots; \lim_{m \to \infty} \beta_{nm} = +\infty$$

Accordingly

$$\lambda_{nmq} = \beta_{nm}{}^2 + \frac{q^2\pi^2}{h^2} \qquad n = 0, 1, \ldots; m = 1, 2, \ldots; q = 0, 1, \ldots$$

are eigenvalues with corresponding real-valued eigenfunctions

$$\psi_{nmq} = [J_{2n}(\beta_{nm}r)\, Y'_{2n}(\beta_{nm}a) - Y_{2n}(\beta_{nm}r)J'_{2n}(\beta_{nm}a)]\cos 2n\theta \cos\frac{q\pi z}{h}$$

The set of eigenfunctions

$$\psi_0 = 1 \qquad \psi_q = \cos\frac{q\pi z}{h} \qquad \psi_{nmq}(r,\theta,z) \qquad n = 0, 1, \ldots; m = 1, 2, \ldots; q = 0, 1, \ldots$$

are mutually orthogonal over the volume $\mathscr{V}$ and constitute (to within multiplicative factors) all the real-valued eigenfunctions of the problem. There are the corresponding normal modes of vibration

$$u_0 = c_1 + c_2 t$$

$$u_q = \cos\frac{q\pi z}{h}\left(A_q \cos\frac{q\pi ct}{h} + B_q \sin\frac{q\pi ct}{h}\right) \qquad q = 1, 2, \ldots$$

$$u_{nmq} = \psi_{nmq}(r,\theta,z)(A_{nmq}\cos\omega_{nmq}t + B_{nmq}\sin\omega_{nmq}t) \qquad n = 0, 1, \ldots; m = 1, 2, \ldots;$$

$$q = 0, 1, \ldots$$

where $\omega_{nmq} = c\sqrt{\lambda_{nmq}}$.

By the method of the text and making use of the orthogonality properties of the complete set of eigenfunctions derived above, one obtains the solution of the problem

$$v_{tt} = c^2 \Delta v \qquad \text{in } \bar{\mathcal{V}}; t \ge 0$$

$$\frac{\partial v}{\partial n} = 0 \qquad \text{on } S; t \ge 0$$

$$u(r,\theta,z,0) = f(r,\theta,z) \qquad u_t(r,\theta,z,0) = g(r,\theta,z) \qquad \text{in } \bar{\mathcal{V}}$$

in the form

$$u = A_0 + B_0 t + \sum_{q=1}^{\infty} \cos \frac{q\pi z}{h} \left(A_q \cos \frac{q\pi ct}{h} + B_q \sin \frac{q\pi ct}{h} \right)$$

$$+ \sum_{n=0}^{\infty} \sum_{m=1}^{\infty} \sum_{q=0}^{\infty} \psi_{nmq}(r,\theta,z)(A_{nmq} \cos \omega_{nmq} t + B_{nmq} \sin \omega_{nmq} t)$$

where

$$A_0 = \frac{4}{\pi(b^2 - a^2)h} \int_a^b \int_0^{\pi/2} \int_0^h f(r,\theta,z)r \, dr \, d\theta \, dz$$

$$B_0 = \frac{4}{\pi(b^2 - a^2)h} \int_a^b \int_0^{\pi/2} \int_0^h g(r,\theta,z)r \, dr \, d\theta \, dz$$

$$A_q = \frac{8}{\pi h(b^2 - a^2)} \int_a^b \int_0^{\pi/2} \int_0^h f(r,\theta,z) \cos \frac{q\pi z}{h} r \, dr \, d\theta \, dz \qquad q = 1, 2, \ldots$$

$$A_{nmq} = \frac{1}{\|\psi_{nmq}\|^2} \int_a^b \int_0^{\pi/2} \int_0^h f(r,\theta,z)\psi_{nmq}(r,\theta,z)r \, dr \, d\theta \, dz$$

$$B_{nmq} = \frac{1}{\omega_{nmq}\|\psi_{nmq}\|^2} \int_a^b \int_0^{\pi/2} \int_0^h g(r,\theta,z)\psi_{nmq}(r,\theta,z)r \, dr \, d\theta \, dz$$

$$n = 0, 1, \ldots \qquad m = 1, 2, \ldots \qquad q = 0, 1, \ldots$$

where

$$\|\psi_{nmq}\|^2 = \frac{\epsilon_n \epsilon_q}{2\pi\beta_{nm}^2} \left\{ \left(1 - \frac{n^2}{b^2\beta_{nm}^2}\right) \left[\frac{J_n'(\beta_{nm}a)}{J_n'(\beta_{nm}b)}\right]^2 - \left(1 - \frac{n^2}{a^2\beta_{nm}^2}\right) \right\}$$

$$\epsilon_0 = 2; \epsilon_p = 1 \text{ if } p \ge 1$$

To derive the formal series solution of the problem

$$w_{tt} - c^2 \Delta w = F(r,\theta,z,t) \qquad \text{in } \bar{\mathcal{V}}; t \ge 0$$

$$\frac{\partial w}{\partial n} = 0 \qquad \text{on } S; \qquad t \ge 0$$

$$w(r,\theta,z,0) = 0 \qquad w_t(r,\theta,z,0) = 0 \qquad \text{in } \bar{\mathcal{V}}$$

assume

$$w = C_0(t) + \sum_{q=1}^{\infty} C_q(t) \cos \frac{q\pi z}{h} + \sum_{n=0}^{\infty} \sum_{m=1}^{\infty} \sum_{q=0}^{\infty} C_{nmq}(t)\psi_{nmq}(r,\theta,z)$$

where the functions $C_0(t)$, $C_q(t)$, and $C_{nmq}(t)$ are to be determined by the method set forth in the text.   One obtains

$$C_0(t) = \int_0^t d\xi \int_0^\xi F_0(\eta)\, d\eta \qquad c_q(t) = \frac{1}{\omega_q} \int_0^t F_q(\xi) \sin\left[\omega_q(t - \xi)\right] d\xi$$

$$C_{nmq}(t) = \frac{1}{\omega_{nmq}} \int_0^t F_{nmq}(\xi) \sin\left[\omega_{nmq}(t - \xi)\right] d\xi$$

where

$$F_0(t) = \frac{4}{\pi h(b^2 - a^2)} \int_a^b \int_0^{\pi/2} \int_0^h F(r,\theta,z,t)r\, dr\, d\theta\, dz$$

$$F_q(t) = \frac{8}{\pi h(b^2 - a^2)} \int_a^b \int_0^{\pi/2} \int_0^h F(r,\theta,z,t)\cos\frac{q\pi z}{h} r\, dr\, d\theta\, dz \qquad q = 1, 2, \ldots$$

$$F_{nmq}(t) = \frac{1}{\|\psi_{nmq}\|^2} \int_a^b \int_0^{\pi/2} \int_0^h F(r,\theta,z,t)\psi_{nmq}(r,\theta,z)r\, dr\, d\theta\, dz$$

$$n = 0, 1, \ldots\,;\; m = 1, 2, \ldots\,;\; q = 0, 1, \ldots$$

Now the superposition

$$u = v + w$$

yields the solution of the original problem.

PROBLEMS

Sec. 4-2

1   (a) A solution of the homogeneous wave equation (4-5) of the form

$$u(x,y) = \psi(x)e^{\pm i\omega t} \qquad i = \sqrt{-1}\,;\; \omega \text{ real and positive}$$

is called *harmonic time-dependent*.   Substitute into Eq. (4-5), and show that the *amplitude factor* ψ must satisfy

$$\psi'' + k^2\psi = 0 \qquad k^2 = \frac{\omega^2}{c^2}$$

where primes denote derivatives with respect to x.   Hence show that harmonic time-dependent wave functions are of the form

$$u(x,y) = Ae^{i(kw \pm \omega t)} \qquad k = \pm\frac{\omega}{c}\,;\; A = \text{const}$$

(b) Assume a solution of Eq. (4-5) of the form

$$u(x,t) = \psi(x)e^{\pm \omega t} \qquad \omega \text{ real and positive}$$

and derive wave functions of the form

$$u(x,t) = Ae^{kx \pm \omega t} \qquad k = \pm\frac{\omega}{c}\,;\; A = \text{const}$$

(c) Let f be a twice continuously differentiable function.   Show that

$$u(x,t) = f(x - ct) - f(-x - ct)$$

is a wave function which satisfies the boundary condition

$$u(0,t) = 0$$

If, in addition, f is an even function of its argument, $f(-\zeta) = f(\zeta)$, then u satisfies the initial condition

$$u(x,0) = 0$$

(d) Construct a nontrivial real-valued wave function which satisfies the stated condition.
   (i) u is harmonic time-dependent and $u(0,t) = 0$ all t.
   (ii) u is harmonic time-dependent, $u(0,t) = 0$ all t, and $u(x,0) = 0$ all x.
   (iii) $u(0,t) = 0$, $\lim\limits_{x \to +\infty} u(x,t) = 0$, $\lim\limits_{x \to -\infty} u(x,t) = 0$.
   (iv) $u_x(0,t) = 0$ all t and $u_t(x,0) = 0$ all x.
   (v) $(u_x + \alpha u)|_{x=0} = 0$ all t, α a real constant.

2   The one-dimensional homogeneous wave equation with damping is

$$u_{tt} + 2\gamma u_t - c^2 u_{xx} = 0$$

where γ is a real positive constant [recall Eq. (4-3)]. Solutions of this equation may represent waves whose amplitudes are *damped out* with increasing time or waves which are *attenuated* as they travel.

(a) Assume a solution of the form $u = e^{-\gamma t} v(x,t)$. Show that v must satisfy

$$v_{tt} - \gamma^2 v - c^2 v_{xx} = 0$$

To obtain particular solutions of this equation assume

$$v = \psi(x)e^{\pm i\omega t} \qquad i = \sqrt{-1}; \; \omega \text{ real}$$

and show that ψ must satisfy

$$\psi'' + k^2\psi = 0 \qquad k^2 = \frac{\gamma^2 + \omega^2}{c^2}$$

Hence derive particular solutions of the homogeneous damped wave equation of the form

$$u = Ae^{-\gamma t}e^{\pm i(kx - \omega t)} \qquad k = \pm \frac{(\omega^2 + \gamma^2)^{\frac{1}{2}}}{c}$$

where A is a constant. These represent traveling waves with amplitude $A' = Ae^{-\gamma t}$ and with speed

$$c' = \frac{\omega}{k} = \frac{c\omega}{(\omega^2 + \gamma^2)^{\frac{1}{2}}} < c$$

(b) Assume a solution of the form $u = \psi(x)e^{-i\omega t}$, ω a real and positive constant. Show that ψ must satisfy

$$\psi'' - \alpha^2\psi = 0 \qquad \alpha^2 = -\frac{\omega^2 + 2i\omega\gamma}{c^2}$$

Write $\alpha = a + ib$, a, b real. Calculate α^2, equate real and imaginary parts, and obtain

$$a^2 - b^2 = -\frac{\omega^2}{c^2} \qquad ab = -\frac{\omega\gamma}{c^2}$$

Show that $b = \mp k(\omega)$ and $a = \pm \omega\gamma/c^2 k(\omega)$, where

$$k(\omega) = \frac{\sqrt{2}\,\omega\gamma}{c[\omega(\omega^2 + 4\gamma^2)^{\frac{1}{2}} - \omega^2]^{\frac{1}{2}}} > 0$$

Hence derive the particular solutions

$$u = Ae^{-h(\omega)x}e^{i(kx-\omega t)} \qquad h(\omega) = \frac{[\omega(\omega^2 + 4\gamma^2)^{\frac{1}{2}} - \omega^2]^{\frac{1}{2}}}{\sqrt{2}c}$$

The amplitude $A' = Ae^{-h(\omega)x} \to 0$ as $x \to +\infty$. Thus as the wave progresses to the right, the wave is attenuated. The speed

$$c' = \frac{\omega}{k} = \frac{c^2 h(\omega)}{\gamma}$$

depends on the frequency ω. In a superposition of such waves with different frequencies each component has a different speed.

(c) Derive the solutions

$$u = Ae^{-\gamma t}e^{kx-\omega t} \qquad k = \pm\frac{(\omega^2 - \gamma^2)^{\frac{1}{2}}}{c} ; \omega^2 > \gamma^2$$

$$u = Ae^{ikx}e^{-(\gamma+\omega)t} \qquad k = \pm\frac{(\gamma^2 - \omega^2)^{\frac{1}{2}}}{c} ; \gamma^2 > \omega^2$$

3 Use D'Alembert's solution (4-15) to construct the solution of the initial-value problem (4-17) with the given initial data. Note that f, g may not satisfy the differentiability conditions assumed in the text at every point. Nevertheless verify that the function obtained by applying D'Alembert's formula satisfies the initial conditions. Show also that the wave equation is satisfied except possibly at points along the characteristic curves of Eq. (4-5). Sketch the solution at times $t = 0$, $t = 1/c$, and $t = 2/c$.

(a) $f(x) = e^{-x}, g(x) = 0, -\infty < x < \infty$

(b) $f(x) = 1/(1 + x^2), g(x) = 0, -\infty < x < \infty$

(c) $f(x) = A \sin \omega x, g(x) = B \cos \mu x, -\infty < x < \infty$

(d) $f(x) = 0, g(x) = A \sinh ax, -\infty < x < \infty$

(e) $f(x) = 1, |x| \leq 1, f(x) = 0, |x| > 1; g(x) = 0, -\infty < x < \infty$

(f) $f(x) = 1 - |x|, |x| \leq 1, f(x) = 0, |x| > 1; g(x) = 0, -\infty < x < \infty$

(g) $f(x) = \cos x, |x| < \pi/2, f(x) = 0, |x| \geq \pi/2; g(x) = 0, -\infty < x < \infty$

(h) $f(x) = 0, -\infty < x < \infty; g(x) = 1, |x| < \epsilon, g(x) = 0, |x| \geq \epsilon, \epsilon > 0$ a constant

4 Recall the rule for differentiation of an integral with respect to a parameter. Show that if f is twice continuously differentiable and g is continuously differentiable, D'Alembert's solution (4-15) satisfies all the conditions of the initial-value problem (4-17).

5 Construct the solution of the initial-value problem (4-11) if the data are as described.

(a) $F(x,t) = 1, f(x) = \sin \omega x, g(x) = 0$

(b) $F(x,t) = xt, f(x) = g(x) = 0$

(c) $F(x,t) = 4x + t, f(x) = 0, g(x) = \cosh bx$

(d) $F(x,t) = A \sin \omega x \sin \mu t, f(x) = g(x) = 0$

(e) $F(x,t) = A \sin (kx - \omega t), f(x) = g(x) = 0, k = \omega c$

6 Verify that the function w defined by Eq. (4-16) is a solution of problem (4-18).

7 Consider the boundary- and initial-value problem

$u_{tt} = c^2 u_{xx}$ $x > 0; t > 0$

$u(x,0) = f(x)$ $u_t(x,0) = g(x)$ $x \geq 0$

$u(0,t) = 0$ $t \geq 0$

Note that in order for the data to agree at $(0,0)$ it is necessary that $f(0) = 0$, $g(0) = 0$
(see Fig. P4-1). The given functions f, g are defined only for $x \geq 0$; however, assume for
the moment that f, g are defined in some manner on $(-\infty,\infty)$, and apply D'Alembert's
formula (4-15) to the problem. Impose the condition along $x = 0$, and obtain

$$\tfrac{1}{2}[f(-ct) + f(ct)] + \frac{1}{2c} \int_{-ct}^{ct} g(\xi)\, d\xi = 0 \qquad t \geq 0$$

This equation will hold provided f, g are defined such that

$$f(ct) = -f(-ct) \qquad \int_{-t}^{t} g(\xi)\, d\xi = 0 \qquad t > 0$$

Hence, the appropriate procedure is to extend f and g as *odd functions* on $(-\infty,\infty)$:

$$f(x) = -f(-x) \qquad g(x) = -g(-x) \qquad x < 0$$

Now D'Alembert's formula defines the solution of the problem

$$u(x,t) = \begin{cases} \tfrac{1}{2}[f(x - ct) + f(x + ct)] + \dfrac{1}{2c} \displaystyle\int_{x-ct}^{x+ct} g(\xi)\, d\xi & 0 \leq ct \leq x \\[3ex] \tfrac{1}{2}[-f(ct - x) + f(x + ct)] + \dfrac{1}{2c} \displaystyle\int_{ct-x}^{x+ct} g(\xi)\, d\xi & 0 \leq x \leq ct \end{cases}$$

Verify that u satisfies the boundary condition along $x = 0$ as well as the initial conditions.
Note that u satisfies the wave equation except possibly along the characteristics $x = \pm ct$

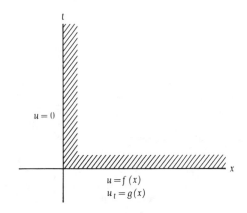

$u = 0$

$u = f(x)$
$u_t = g(x)$

Figure P4-1

in the xt plane.   Use this formula to construct the solution of the boundary- and initial-value problem corresponding to the following prescribed data.   Sketch the solution at $t = 0, t = 1/c, t = 3/2c, t = 2/c$.

(a) $f(x) = 1; g(x) = -\cos x, x \geq 0$

(b) $f(x) = x^2; g(x) = 0, x \geq 0$

(c) $f(x) = 1, 0 \leq x < 1, f(x) = 0, x \geq 1; g(x) = 0, x \geq 0$

(d) $f(x) = 0, 0 \leq x < 1, f(x) = 2(x - 1), 1 \leq x \leq \frac{3}{2}$
$f(x) = 2(2 - x), \frac{3}{2} \leq x \leq 2, f(x) = 0, x > 2; g(x) = 0, x \geq 0$

8   Consider the following boundary- and initial-value problem

$$u_{tt} = c^2 u_{xx} \qquad x > 0; t > 0$$
$$u(x,0) = f(x) \qquad u_t(x,0) = g(x) \qquad x \geq 0$$
$$u(0,t) = h(t) \qquad t \geq 0$$

Assume f, h are twice continuously differentiable and g is continuously differentiable, on $[0,+\infty)$.   Note that if the boundary and initial values agree at the corner $(0,0)$ in the xt plane, the necessary conditions $h(0) = f(0), h'(0) = g(0)$ are implied in the data.   Let v denote the function constructed as the solution in Prob. 7.   Suppose w is a solution of the wave equation (except possibly along the characteristic $x = ct$) such that

$$w(x,0) = 0 \qquad w_t(x,0) = 0 \qquad x \geq 0 \qquad w(0,t) = h(t) \qquad t \geq 0$$

Then $u = v + w$ satisfies the conditions of the boundary- and initial-value problem first proposed, except that the wave equation may not be satisfied along the characteristic. To construct w assume $w = \varphi(x - ct), x \neq ct$.   Impose the boundary condition at $x = 0$; then

$$\varphi(-ct) = h(t) \qquad t \geq 0$$

Let $\xi = -ct$, so that $\varphi(\xi) = h(-\xi/c)$ and

$$\varphi(x - ct) = h\left(t - \frac{x}{c}\right)$$

Define w as

$$w(x,t) = 0 \qquad 0 \leq ct \leq x \qquad w(x,t) = h\left(t - \frac{x}{c}\right) \qquad 0 \leq x \leq ct$$

Verify that w has the desired properties at $t = 0$ and satisfies the wave equation, except possibly along $x = ct, t \geq 0$.   Use the results of these considerations to construct the solution of the boundary- and initial-value problem for each of the following cases.

(a) $f(x) = e^{-x}, g(x) = 1, h(t) = 1$

(b) $f(x) = x, g(x) = 0, h(t) = \sin t$

(c) $f(x) = g(x) = 0, h(t) = 1, 0 \leq t \leq T, h(t) = 0, t > T$, where T is a given constant

(d) $f(x) = g(x) = 0, h(t) = A \sin \omega t$

9   Consider the problem

$$u_{tt} - c^2 u_{xx} = 0 \qquad u = f(x) \qquad u_t = g(x) \qquad \text{on } C$$

where C is the characteristic $t = x/c$ in the xt plane.   Attempt to fit the general solution (4-6) of the wave equation to the prescribed data.   Show that no solution exists unless the given functions f, g satisfy the relation

$$g(x) = \frac{cf'(x)}{2} + a$$

for some constant a. Show that if f, g satisfy such a relation, there are infinitely many distinct solutions of the form

$$u = \varphi(x - ct) + f\left(\frac{x + ct}{2}\right) - \varphi(0)$$

where φ is any twice continuously differentiable function such that $\varphi'(0) = -a/c$.

10 Let k be a fixed real number such that $k > 1/c$. Consider the following problem involving the homogeneous wave equation:

$$u_{tt} = c^2 u_{xx} \qquad x > 0; \; 0 < t < kx$$

$$u(x,0) = f(x) \qquad u_t(x,0) = g(x) \qquad x \geq 0$$

$$u(x,kx) = h(x) \qquad x \geq 0$$

Here the curve in the xt plane on which the data are prescribed consists of the nonnegative x axis and the half line $t = kx$, $x \geq 0$ (see Fig. P4-2). Let v denote the function constructed as the solution in Prob. 7. Then v satisfies the conditions prescribed on $t = 0$. Also

$$v(x,kx) = G(x) = \tfrac{1}{2}\{-f[(kc - 1)x] + f[(kc + 1)x]\} + \frac{1}{2c}\int_{(kc-1)x}^{(kc+1)x} g(\xi)\, d\xi$$

It is desired to construct the solution of the present problem by superposition:

$$u = v + w$$

Evidently w must satisfy the homogeneous wave equation and the initial conditions

$$w(x,0) = w_t(x,0) = 0 \qquad w(x,kx) = h(x) - G(x) \qquad x \geq 0$$

To construct w assume

$$w(x,t) = 0 \qquad 0 \leq ct \leq x \qquad\qquad w(x,t) = \varphi(x - ct) \qquad 0 \leq x \leq ct$$

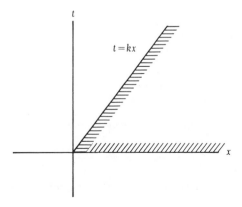

Figure P4-2

where φ is a function to be determined. Then w satisfies the necessary conditions along $t = 0$. To satisfy the remaining condition it is necessary that

$$\varphi(x) = h(x) - G(x) \qquad x \geq 0$$

Hence

$$\varphi(\xi) = h\left(\frac{\xi}{1 - kc}\right) - G\left(\frac{\xi}{1 - kc}\right)$$

and

$$\varphi(x - ct) = h\left(\frac{x - ct}{1 - kc}\right) - G\left(\frac{x - ct}{1 - kc}\right)$$

Complete the remaining steps of the derivation and show that the solution of the problem is

$$u(x,t) = \begin{cases} v(x,t) & 0 \leq ct \leq x \\ h\left(\dfrac{x - ct}{1 - kc}\right) + \tfrac{1}{2}\{f[\beta(x,t)] - f[\alpha(x,t)]\} + \dfrac{1}{2c}\displaystyle\int_{\alpha}^{\beta} g(\xi)\,d\xi & 0 \leq x \leq ct \end{cases}$$

where

$$\alpha(x,t) = \frac{1 + kc}{(1 - kc)(x - ct)} \qquad \beta(x,t) = x + ct$$

Verify that u satisfies the conditions of the problem. Note that u is a solution of the homogeneous wave equation, except possibly along the characteristic $x = ct$. Write down the solution of the Cauchy problem if

$$f(x) = g(x) = 0 \qquad h(x) = A \sin \omega x \qquad x \geq 0$$

where $A > 0$ is a constant. Discuss the behavior of u along the characteristic $x = ct$ for this case.

11 Recall the discussion in Sec. 2-3. The general homogeneous linear second-order hyperbolic equation with constant coefficients in two independent variables x, t is

$$Lu = Au_{xx} + 2Bu_{xt} + Cu_{tt} + Du_x + Eu_t + Fu = 0$$

where $\Delta = B^2 - AC > 0$. Let a be a fixed real positive constant, and suppose that for arbitrary choice of function f

$$u = f(x - at)$$

is a solution of the differential equation. Utilize the arbitrariness of f to show that in this case the relations

(i) $F = 0$ (ii) $D - aE = 0$ (iii) $Ca^2 - 2Ba + A = 0$

must hold. Conversely suppose $F = 0$ and the coefficients are such that

(iv) $AE^2 - 2BDE + CD^2 = 0$

Show that plane-wave solutions of arbitrary shape and with common speed $a = D/E$ exist. On the other hand, if $F \neq 0$, or if $F = 0$, $D \neq 0$, $E \neq 0$, and **iv** does not hold, or if $F = D = 0$, $E \neq 0$ (or $F = E = 0$, $D \neq 0$), the equation does not admit plane-wave solutions of arbitrary profile having a common speed. Show that if $C > 0$ and

$$F = D = E = 0 \qquad a_{1,2} = \frac{B \pm \sqrt{\Delta}}{C}$$

plane-wave solutions of arbitrary profile having common speed a_1 (or a_2) exist.

12 **(a)** The system of first-order linear partial differential equations

$$\frac{\partial v}{\partial x} + L\frac{\partial i}{\partial t} + Ri = 0 \qquad \frac{\partial i}{\partial x} + C\frac{\partial v}{\partial t} + Gv = 0$$

occurs in the theory of electric transmission lines; they are called the *transmission-line equations*. Here $v(x,t)$ denotes the voltage and $i(x,t)$ the current, at distance x along the line at time t. The constants L, C, R, G are real and nonnegative. They denote inductance, capacitance, resistance, and conductance (per unit length). Differentiate the equations with respect to x and t in an appropriate fashion, and deduce that if v, i are solutions of the transmission-line equations, then v must satisfy the *telegrapher's equation*

$$LCv_{tt} + (RC + LG)v_t + RGv - v_{xx} = 0$$

[see Eq. (4-4)]. Show that i satisfies this equation also. If $L > 0$ and $C > 0$, the equation is hyperbolic. If $L = 0$, or if $C = 0$, the equation is parabolic. Assume $L > 0$ and $C > 0$. If $R = 0$ and $G \neq 0$, show that the telegrapher's equation can be rewritten

$$v_{tt} + bv_t - c^2 v_{xx} = 0$$

which is the damped-wave equation, where

$$c = \frac{1}{\sqrt{LC}}$$

If $R = G = 0$, then the wave equation results.

(b) Utilize the results of Prob. 11 and show that if $R \neq 0$, or if $G \neq 0$, the telegrapher's equation does not admit traveling-wave solutions

$$v = f(x - at)$$

of arbitrary profile having common speed a. In turn this implies the damped-wave equation does not admit traveling waves of arbitrary profile.

(c) A transmission line (assumed to be characterized by the telegrapher's equation) is called *distortionless* if the line admits traveling waves

$$v = e^{-\mu t}f(x - at)$$

of arbitrary profile having common speed a, for some real nonnegative constant μ. If $\mu > 0$, the waves are damped out exponentially with increasing time. Show that a necessary and sufficient condition for a distortionless line is

$$RC = LG$$

In this event show that the common speed is $a = 1/\sqrt{LC}$.

13 **(a)** Let L be the linear hyperbolic operator defined by

$$Lu = \frac{\partial^2 u}{\partial x \, \partial y} + a(x,y)\frac{\partial u}{\partial x} + b(x,y)\frac{\partial u}{\partial y} + c(x,y)u$$

Assume the coefficients are twice continuously differentiable in some region $\mathcal{R}$ of the xy plane. Verify from Eq. (2-74) that the adjoint of L is the linear hyperbolic operator L^*

defined by

$$L^*v = \frac{\partial^2 v}{\partial x\, \partial y} - \frac{\partial}{\partial x}(av) - \frac{\partial}{\partial y}(bv) + cv$$

Utilize Lagrange's identity to show that if u, v are twice continuously differentiable, then

$$vLu - uL^*v = \frac{\partial Q}{\partial x} - \frac{\partial P}{\partial y}$$

where

$$P = \tfrac{1}{2}(uv_x - vu_x) - buv \qquad Q = \tfrac{1}{2}(vu_y - uv_y) + auv$$

Let Γ be a simple closed curve in $\mathscr{R}$, and let $\mathscr{R}_1$ denote the region enclosed by Γ. Green's formula (2-77) states that if u, v are as described above, then

$$\iint\limits_{\mathscr{R}_1} (vLu - uL^*v)\, dx\, dy = \int_{\Gamma} P\, dx + Q\, dy$$

where the line integral is taken in the positive sense around Γ.

(b) (*Riemann's method*) Let C be a given smooth curve lying in $\mathscr{R}$ and such that any line parallel to a coordinate axis intersects C in at most one point. Thus C is noncharacteristic. Consider the Cauchy problem

$$Lu = F(x,t) \qquad (x,y) \text{ in } \mathscr{R}$$

$$u = f(x) \qquad \frac{\partial u}{\partial n} = g(x) \qquad \text{on } C$$

where F is a given continuous function, f is twice continuously differentiable, and g is continuously differentiable. The symbol $\partial u / \partial n$ denotes the derivative of u in the direction of the normal $\mathbf{n}$ to C. Let $P_0(x_0, y_0)$ be a point not on C. Then Riemann's method of deriving the expression for the solution of the Cauchy problem at (x_0, y_0) is as follows. Construct the closed curve Γ as shown in Fig. P4-3. The segments BP_0, AP_0 are parallel

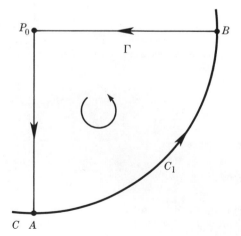

Figure P4-3

to the coordinate axes.   Let $\mathscr{R}_1$ denote the region enclosed by Γ.   If u is the solution of the Cauchy problem, then by Green's formula derived in a above

$$\iint_{\mathscr{R}_1} vF \, dx \, dy = \iint_{\mathscr{R}_1} uL^*v \, dx \, dy + \int_{\Gamma} P \, dx + Q \, dy$$

for any twice continuously differentiable function v.   The appropriate choice for v is now derived.   Rewrite P, Q as

$$P = -\frac{(uv)_x}{2} + u(v_x - bv) \qquad Q = \frac{(uv)_y}{2} - u(v_y - av)$$

The line integral around Γ is calculated as follows.   Along the characteristic BP_0

$$\int_{BP_0} P \, dx + Q \, dy = \int_B^{P_0} P \, dx = \frac{(uv)|_B - (uv)|_{P_0}}{2} + \int_B^{P_0} u(v_x - bv) \, dx$$

Along the characteristic $P_0 A$

$$\int_{P_0 A} P \, dx + Q \, dy = \int_{P_0}^A Q \, dy = \frac{(uv)|_A - (uv)|_{P_0}}{2} - \int_{P_0}^A u(v_y - av) \, dy$$

Thus

$$\int_{\Gamma} P \, dx + Q \, dy = -(uv)|_{P_0} + \tfrac{1}{2}[(uv)|_B + (uv)|_A]$$

$$+ \int_B^{P_0} u(v_x - bv) \, dx - \int_{P_0}^A u(v_y - av) \, dy + \int_{C_1} P \, dx + Q \, dy$$

where C_1 is the portion of C lying between the points A, B.   Assume now that $v(x,y;x_0,y_0)$ is a function satisfying the following conditions:

$$L^*v = 0 \qquad \text{in } \mathscr{R}_1$$

$$v_x = bv \qquad \text{on } y = y_0 \qquad\qquad v_y = av \qquad \text{on } x = x_0$$

$$v(x_0,y_0;x_0,y_0) = 1$$

The function v is called the *Riemann Green's function* of the Cauchy problem.   From the expressions derived above it follows that

$$u|_{P_0} = \tfrac{1}{2}[(uv)|_A + (uv)|_B] + \int_{C_1} P \, dx + Q \, dy - \iint_{\mathscr{R}_1} vF \, dx \, dy$$

Since the right-hand member involves only known functions, the value $u(x_0,y_0)$ is now determined.

14   (a) Let L be the hyperbolic operator defined by

$$Lu = u_{xy}$$

Show that L is self-adjoint: $L^* = L$.   Show that the Riemann Green's function $v(x,y;x_0,y_0)$ associated with L must satisfy $Lv = 0$ and

$$v_x = 0 \qquad \text{on } y = y_0 \qquad\qquad v_y = 0 \qquad \text{on } x = x_0$$

$$v(x_0,y_0;x_0,y_0) = 1$$

Clearly the function $v(x,y;x_0,y_0) = 1$, all x, y, has the requisite properties.

(b) Let C be the line $y = x$ in the xy plane. Consider the Cauchy problem

$$u_{xy} = F(x,y) \qquad u|_C = f(x) \qquad \frac{\partial u}{\partial n}\bigg|_C = g(x)$$

where F, f, g are given functions and $\partial u/\partial n$ denotes the derivative of u in the direction of the normal $\mathbf{n}$ to C. Thus $\partial u/\partial n = (u_y - u_x)/\sqrt{2}$. Apply the results of Prob. 13, and derive the solution of the Cauchy problem in the form

$$u(x_0,y_0) = \tfrac{1}{2}[f(x_0) + f(y_0)] + \frac{1}{\sqrt{2}} \int_{x_0}^{y_0} g(x)\, dx - \iint_{\mathscr{R}_1} F(x,y)\, dx\, dy$$

where $\mathscr{R}_1$ is the triangular region in the xy plane bounded by C and the lines $x = x_0$, $y = y_0$ through (x_0,y_0).

15 (a) In the initial-value problem (4-11) make the change of independent variables

$$\xi = x - ct \qquad \eta = x + ct$$

Show that the transformed problem is

$$u_{\xi\eta} = G(\xi,\eta) \qquad G(\xi,\eta) = \frac{-F[(\eta + \xi)/2,\,(\eta - \xi)/2c]}{4c^2}$$

$$u|_C = f(\xi) \qquad \frac{\partial u}{\partial n}\bigg|_C = \frac{g(\xi)}{\sqrt{2}c}$$

where C is the line $\eta = \xi$ in the $\xi\eta$ plane and $\partial u/\partial n$ denotes the derivative of u in the direction of the normal $\mathbf{n}$ to C. Show that the transformation is such that the upper half plane $t \geq 0$ in the xt plane is mapped onto the half plane $\eta \geq \xi$ in the $\xi\eta$ plane. Is the transformed problem equivalent to problem (4-11)? Why?

(b) Use the results of Prob. 14 together with the result in **a** above to derive the solution of problem (4-11).

16 (a) Let L be the hyperbolic operator defined by

$$Lu = u_{xy} + au$$

where $a > 0$ is a constant. Show that L is self-adjoint: $L^* = L$. Show that the Riemann Green's function $v(x,y;x_0,y_0)$ associated with L must satisfy $Lv = 0$ and

$$v_x = 0 \qquad \text{on } y = y_0 \qquad v_y = 0 \qquad \text{on } x = x_0$$

$$v(x_0,y_0;x_0,y_0) = 1$$

(b) To determine the Riemann Green's function assume

$$v = \varphi[(x - x_0)(y - y_0)]$$

where φ is a function to be determined. Show this assumed form for v satisfies the required conditions on $y = y_0$ and on $x = x_0$. Substitute v into $Lv = 0$ and show that φ must satisfy

$$s\varphi''(s) + \varphi'(s) + a\varphi(s) = 0 \qquad s = (x - x_0)(y - y_0)$$

Make the change of variable $t = 2\sqrt{as}$, and show that the preceding differential equation becomes

$$t\ddot{\varphi} + \dot{\varphi} + t\varphi = 0$$

where dots denote differentiation with respect to t. This is Bessel's equation of order zero. Hence

$$\varphi(s) = J_0(2\sqrt{as})$$

and so

$$v = J_0[2\sqrt{a(x - x_0)(y - y_0)}]$$

is the Riemann Green's function of L.

(c) Let C be the line $y = x$ in the xy plane. Consider the Cauchy problem

$$u_{xy} + au = F(x,y) \qquad u|_C = f(x) \qquad \left.\frac{\partial u}{\partial n}\right|_C = g(x)$$

where F, f, g are given functions and $\partial u/\partial n$ denotes the derivative of u in the direction of the normal $\mathbf{n}$ to C. Thus $\partial u/\partial n = (u_y - u_x)/\sqrt{2}$. Apply the results of Prob. 13 and part b of the present problem, and derive the solution of the Cauchy problem in the form

$$u(x_0,y_0) = \tfrac{1}{2}[f(x_0) + f(y_0)] + \frac{1}{\sqrt{2}}\int_{x_0}^{y_0} J_0(\mu)g(x)\,dx$$

$$+ a(x_0 - y_0)\int_{x_0}^{y_0}\frac{J_0'(\mu)}{\mu}f(x)\,dx - \iint_{\mathscr{R}_1} J_0(2\sqrt{as})F(x,y)\,dx\,dy$$

where $s = (x - x_0)(y - y_0)$, $\mu = 2\sqrt{a(x - x_0)(x - y_0)}$, and $\mathscr{R}_1$ is the triangular region in the xy plane bounded by C and the lines $x = x_0$, $y = y_0$ through (x_0,y_0).

17 (a) The one-dimensional telegrapher's equation [recall Eq. (4-4)] is

$$u_{tt} + 2\gamma u_t + \beta u - c^2 u_{xx} = F(x,t)$$

where the constants γ, β, c are such that

$$\gamma > 0 \qquad \beta \geq 0 \qquad c > 0$$

Make the change of dependent variable

$$v(x,t) = e^{\gamma t}u(x,t)$$

and show that the resulting equation is

$$v_{tt} + (\beta - \gamma^2)v - c^2 v_{xx} = e^{\gamma t}F(x,t)$$

Note that if $\beta = \gamma^2$, the wave equation results. Now make the change of independent variables

$$\xi = x - ct \qquad \eta = x + ct$$

and show that the differential equation is transformed into

$$v_{\xi\eta} + av = H(\xi,\eta)$$

where

$$a = \frac{\gamma^2 - \beta}{4c^2} \qquad H(\xi,\eta) = -e^{\gamma(\eta-\xi)/2c} \frac{F[(\eta + \xi)/2, (\eta - \xi)/2c]}{4c^2}$$

(b) Consider the problem of solving the telegrapher's equation subject to the initial conditions

$$u(x,0) = f(x) \qquad u_t(x,0) = g(x) \qquad -\infty < x < \infty$$

where f, g are given functions. Show that this initial-value problem is equivalent to the Cauchy problem

$$v_{\xi\eta} + av = H(\xi,\eta)$$

$$v = f(\xi) \qquad \frac{\partial v}{\partial n} = \frac{\gamma f(\xi) + g(\xi)}{\sqrt{2}c} \qquad \text{on } C$$

where C is the line $\eta = \xi$ in the $\xi\eta$ plane and $\partial v/\partial n$ is the derivative of v in the direction of the normal $\mathbf{n}$ to C. The constant a and the function H are as described in **a** above.

(c) Assume $\gamma^2 > \beta$, so that $a > 0$. Apply the results of Prob. 16 and part **b** of the present problem, and derive the solution of the initial-value problem for the one-dimensional telegrapher's equation in the form

$$u(x,t) = e^{-\gamma t} v(x,t)$$

where

$$v(x,t) = \tfrac{1}{2}[f(x - ct) + f(x + ct)] + \frac{1}{2c} \int_{x-ct}^{x+ct} J_0(\mu)[\gamma f(s) + g(s)]\, ds - 2act$$

$$\times \int_{x-ct}^{x+ct} \frac{J_0'(\mu)}{\mu} f(s)\, ds + \frac{1}{2c} \iint_{\mathscr{R}_1} J_0\{2\sqrt{a[(s-x)^2 - c^2(\tau - t)^2]}\} e^{\gamma \tau} F(s,\tau)\, ds\, d\tau$$

where

$$a = \frac{\gamma^2 - \beta}{4c^2} \qquad \mu = 2\sqrt{a(x-s)^2 - c^2 t^2}$$

and $\mathscr{R}_1$ is the domain of determinacy of the interval $[x - ct, x + ct]$.

Sec. 4-3

18 **(a)** The idealized string described in the text has fastened ends at $x = 0$, $x = b$. In the case where the string is released from rest and no external force acts, the subsequent displacement is given by Eq. (4-30) with $B_n = 0$, $n = 1, 2, \ldots$. Use the identity

$$\sin(a + b) + \sin(a - b) = 2 \sin a \cos b$$

and show that the displacement is given by

$$u(x,t) = \frac{1}{2} \sum_{n=1}^{\infty} A_n \sin \frac{n\pi(x - ct)}{b} + \frac{1}{2} \sum_{n=1}^{\infty} A_n \sin \frac{n\pi(x + ct)}{b}$$

Now let f_0 be the function which is the odd periodic extension of f to $(-\infty,\infty)$, with period $2b$. That is,

$$f_0(x) = f(x) \qquad 0 \le x \le b \qquad\qquad f_0(x) = -f(-x) \qquad -b \le x < 0$$
$$f_0(x + 2b) = f_0(x) \qquad -\infty < x < \infty$$

Assume f is continuous and piecewise smooth and

$$f(0) = f(b) = 0$$

State why it is true that the displacement of the freely vibrating string is given by

$$u(x,t) = \tfrac{1}{2}[f_0(x - ct) + f_0(x + ct)] \qquad 0 \le x \le b; t \ge 0$$

Interpret this result graphically in terms of a superposition of traveling waves. Relate to D'Alembert's solution (4-15).

(b) Consider the freely vibrating string which is released from its equilibrium position, so that $f(x) = 0, 0 \le x \le b$, and with speed $g(x), 0 \le x \le b$. Assume g is not identically zero, continuous, and piecewise smooth and

$$g(0) = g(b) = 0$$

Let g_0 be the odd periodic extension of g to $(-\infty, \infty)$ with period $2b$. Show that

$$u_t(x,t) = \sum_{n=1}^{\infty} \omega_n B_n \sin \frac{n\pi x}{b} \cos \frac{n\pi ct}{b}$$

$$= \tfrac{1}{2}[g_0(x - ct) + g_0(x + ct)]$$

and hence

$$u(x,t) = \frac{1}{2c} \int_{x-ct}^{x+ct} g_0(\xi) \, d\xi$$

Relate to D'Alembert's solution (4-15).

19 The string with fastened ends at $x = 0$, $x = b$ executes free vibrations after release from the initial displacement $f(x)$ and with initial speed $g(x)$. Determine the series expression for the subsequent displacement $u(x,t)$.
(a) $f(x) = 4hx(b - x)/b^2, g(x) = 0, 0 \le x \le b$ (h = const)
(b) $f(x) = 10 \sin (\pi x/b), g(x) = 0, 0 \le x \le b$
(c) $f(x) = A \sin \omega x, g(x) = 1, 0 \le x \le b$ (ω is a constant and is not an integral multiple of π/b)
(d) $f(x) = 0, 0 \le x \le b; g(x) = v_0, b/2 - \epsilon \le x \le b/2 + \epsilon, g(x) = 0$ elsewhere (ϵ is a small positive constant)

20 **(a)** The kinetic energy of an element ds of the vibrating string is

$$\frac{(dm)u_t^2}{2} \simeq \frac{(\rho \, dx)u_t^2}{2}$$

where ρ is the (constant) linear density. Hence the total kinetic energy of motion is

$$K = \frac{\rho}{2} \int_0^b u_t^2 \, dx$$

To obtain the expression for the potential energy consider an element dx of the string when the string lies in its equilibrium configuration along the x axis. Let ds be the length of the element at a subsequent time $t > 0$. The change in length is

$$ds - dx = (1 + u_x^2)^{1/2} \, dx - dx \simeq u_x^2 \frac{dx}{2}$$

if higher-order terms in u_x^2 are neglected. The extension occurs in the presence of an elastic restoring force of magnitude T_0. Hence the increase in potential energy is

$$T_0 u_x^2 \frac{dx}{2}$$

The net potential energy is

$$V = \frac{T_0}{2} \int_0^b u_x^2 \, dx$$

The total energy is $E = K + V$.

(b) Show that in a traveling wave $u = f(x \pm ct)$, where $c = \sqrt{T_0/\rho}$, the kinetic energy equals the potential energy.

(c) The nth normal mode of vibration of the string with fastened ends is given by Eq. (4-27). Show that u_n can be rewritten

$$u_n(x,t) = C_n \sin \frac{n\pi x}{b} \cos \left(\frac{n\pi ct}{b} - \epsilon_n \right)$$

where the *amplitude* C_n and *phase* ϵ_n are given by

$$C_n = (A_n^2 + B_n^2)^{\frac{1}{2}} \qquad \tan \epsilon_n = \frac{B_n}{A_n}$$

Obtain the expressions

$$K_n(t) = \frac{T_0 \pi^2 n^2 C_n^2 \sin^2 (n\pi ct/b - \epsilon_n)}{4b}$$

$$V_n(t) = \frac{T_0 \pi^2 n^2 C_n^2 \cos^2 (n\pi ct/b - \epsilon_n)}{4b}$$

for the kinetic and potential energy of the nth mode. Hence the total energy is

$$E_n = K_n + V_n = \frac{T_0 \pi^2 n^2 C_n^2}{4b} = \frac{T_0 b \omega_n^2 C_n^2}{4c^2} = \frac{M \omega_n^2 C_n^2}{4}$$

where M is the total mass of the string. Observe that the energy of the nth mode is proportional to the square of the amplitude and proportional to the square of frequency.

(d) Subsequent to release from an initial displacement $f(x)$ with initial velocity $g(x)$ the string with fastened ends executes free vibrations. Use the series (4-30) and show that the energy is

$$E = \sum_{n=1}^{\infty} E_n = \frac{M}{4} \sum_{n=1}^{\infty} \omega_n^2 C_n^2$$

that is, the sum of the individual energies of the normal modes.

21 (a) The string with fastened ends at $x = 0$, $x = b$ executes forced vibrations under the external force per unit mass

$$F(t) = F_0 \sin \omega t$$

where F_0, ω are given positive constants.   The string is released from rest with zero displacement.   Use Eq. (4-41) to obtain an expression for the displacement $u(x,t)$.   Distinguish the cases **(i)** $\omega \neq \omega_k$, all k; **(ii)** $\omega = \omega_k$, some k.   Case ii illustrates *resonance* for the vibrating string with fastened ends.

(b) Obtain the expression for $u(x,t)$ if

$$f(x) = 1 - \cos\frac{2m\pi x}{b} \qquad g(x) = 0; 0 \leq x \leq b$$

where m is a given positive integer, and the external driving force per unit mass is

$$F(x,t) = F_0 e^{-t} \sin\frac{\pi x}{b}$$

where F_0 is a positive constant.

(c) Obtain the expression for $u(x,t)$ if f, g are identically zero and the external force per unit mass is

$$F(x,t) = \begin{cases} F_0 & x_0 - \epsilon < x < x_0 + \epsilon; 0 < t < \delta \\ 0 & x \text{ not in } (x_0 - \epsilon, x_0 + \epsilon) \text{ or } t > \delta \end{cases}$$

Here F_0 is a constant, and ϵ, δ are small positive numbers.   This represents a force of magnitude F_0 confined to the interval $(x_0 - \epsilon, x_0 + \epsilon)$ about the point x_0 and of duration δ.   In the expression for the resulting displacement u let $\epsilon \to 0$, $\delta \to 0$, and obtain the motion due to a concentrated force applied to the point $x = x_0$ at time $t = 0$.

22 (a) Recall the properties of the unit impulse function given in Sec. 3-5.   A concentrated impulsive force per unit mass and of unit magnitude applied at time $t = \tau$ to the point $x = \xi$ on the string can be represented

$$F(x,t) = \delta(x - \xi)\delta(t - \tau)$$

Thus

$$F(x,t) = 0 \qquad (x,t) \neq (\xi,\tau)$$

and

$$\int_0^t dt \int_0^b F(x,t)\, dx = 1 \qquad t > \tau$$

Let the ends of the string be fastened, and assume the initial conditions are zero.   Use Eq. (4-41) and the properties of the δ function to obtain the formal series

$$G(x,t;\xi,\tau) = \frac{2}{\pi c} \sum_{n=1}^{\infty} \frac{1}{n} \sin\frac{n\pi x}{b} \sin\frac{n\pi \xi}{b} \sin\left[\frac{n\pi c}{b}(t - \tau)\right]$$

for the subsequent displacement.   It can be shown that the series converges for all values of x, t, ξ, τ.   Note that G, as a function of x, satisfies the boundary conditions

$$G(0,t;\xi,\tau) = 0 \qquad G(b,t;\xi,\tau) = 0 \qquad t \geq 0$$

Also

$$G(x,\tau;\xi,\tau) = 0 \qquad\qquad 0 \leq x \leq b$$
$$G(\xi,t;x,\tau) = G(x,t;\xi,\tau) \qquad \text{all } x, \xi$$
$$G(x,\tau;\xi,t) = -G(x,t;\xi,\tau) \qquad \text{all } t, \tau$$

The function G is called the *Green's function* of the boundary- and initial-value problem embodied in Eqs. (4-22) to (4-24). The function G is not a solution of the wave equation in the usual classical sense; however, it has many useful properties.

(b) Verify that the solution (4-41) of problem (4-35) can be written

$$u(x,t) = \int_0^t \int_0^b G(x,t;\xi,\tau)F(\xi,\tau) \, d\xi \, d\tau$$

by substitution of the series expression for the Green's function and a formal interchange of the operations of summation and integration so as to obtain the series solution (4-41). Interpret the integral in a physical way.

(c) The string with fastened ends is at rest in its equilibrium configuration along the x axis. At time $t = 0$ the string is struck in such a way that the point $x = \xi$ receives a velocity impulse of unit magnitude. Thus the initial conditions are

$$u(x,0) = 0 \qquad u_t(x,0) = \delta(x - \xi) \qquad 0 \le x \le b$$

Apply Eqs. (4-30), (4-33), and (4-34) in a formal manner, and use the properties of the δ function to show that the subsequent motion is given by

$$G(x,t;\xi,0) = \frac{2}{\pi c} \sum_{n=1}^{\infty} \frac{1}{n} \sin \frac{n\pi x}{b} \sin \frac{n\pi \xi}{b} \sin \frac{n\pi c t}{b}$$

(d) Substitute the series expression for $G(x,t;\xi,0)$, and formally interchange the operations of summation and integration to verify that the solution of problem (4-25) in the case where f is identically zero on $[0,b]$ is given by

$$u(x,t) = \int_0^b G(x,t;\xi,0)g(\xi) \, d\xi$$

where g is the initial velocity. Interpret this result in a physical way.

(e) Let $G(x,t;\xi,\tau)$ be the Green's function obtained in **a** above. Substitute the series expression and formally interchange the operations of summation and integration to verify that the series solution (4-30) of problem (4-25) is given by

$$u(x,t) = \int_0^b G_t(x,t;\xi,0)f(\xi) \, d\xi + \int_0^b G(x,t;\xi,0)g(\xi) \, d\xi$$

where f is the initial displacement and g is the initial velocity.

23   A pair of concentrated impulsive forces, of equal magnitude F_0 but oppositely directed, is applied transversely to the interior points of trisection of the string with fastened ends. Obtain the formal series expression for the subsequent displacement. Show that the midpoint of the string remains at rest.

24 (a) Consider the boundary- and initial-value problem embodied in Eqs. (4-22) to (4-24). Suppose the external force is independent of t.

$$F(x,t) = F_0(x)$$

Choose $v(x)$ such that

$$v''(x) = \frac{-F_0(x)}{c^2} \qquad v(0) = v(b) = 0$$

Let $w(x,t)$ be the solution of problem (4-25) with $f(x)$ replaced by $f(x) - v(x)$. Verify that the superposition

$$u = w + v$$

satisfies Eqs. (4-22) to (4-24).

(b) Solve the problem in Example 4-1 of the text by the method outlined in a.

(c) Use a slight modification of the method outlined in a to derive a formal solution of the boundary- and initial-value problem.

$$u_{tt} - c^2 u_{xx} = F_0 \sin \omega x \qquad 0 \le x \le b; \, t \ge 0$$

$$u(0,t) = h_0 \qquad u(b,t) = h_1 \qquad t \ge 0$$

$$u(x,0) - x(b \quad x) \qquad u_t(x,0) = 0 \qquad 0 \le x < b$$

where F_0, ω, h_0, h_1 are given positive real constants.

25 (a) Consider the boundary- and initial-value problem embodied in Eqs. (4-22), (4-23), and (4-24) when the driving function has the form

$$F(x,t) = F_1(x) \sin \omega t$$

where $F_1(x)$ is a given function and ω is a real positive constant. Assume a particular solution of Eq. (4-22) of the form

$$v(x,t) = X(x) \sin \omega t$$

Show that X must satisfy

$$X'' + \mu^2 X = \frac{-F_1(x)}{c^2} \qquad \mu = \frac{\omega}{c}$$

Conversely, if X is a solution of this ordinary differential equation such that

$$X(0) = X(b) = 0$$

then the function v is a solution of Eq. (4-22) which satisfies the boundary conditions (4-23). Suppose $w(x,t)$ is a solution of problem (4-25) with $g(x)$ replaced by $g(x) - \omega X(x)$; show that the superposition

$$u = v + w$$

satisfies Eqs. (4-22) to (4-24).

(b) It should be noted that the method outlined in a assumes a solution of the two-point boundary-value problem

$$X'' + \mu^2 X = \frac{-F_1(x)}{c^2} \qquad X(0) = 0 \qquad X(b) = 0 \qquad \mu = \frac{\omega}{c}$$

exists. However, a solution of the problem may or may not exist. Prove that (i) if ω is not a characteristic frequency ($\omega \ne \omega_k$, $k = 1, 2, \ldots$), a solution of the two-point problem exists; (ii) if $\omega = \omega_m = m\pi c/b$ for some positive integer m (that is, the case of *resonance*, ω coincides with a characteristic frequency), no solution of the two-point problem exists unless

$$\int_0^b F_1(\xi) \sin \frac{m\pi\xi}{b} \, d\xi = 0$$

In this event there are infinitely many distinct solutions.

(c) Solve Prob. 21a by the method outlined in **a** of the present problem.

(d) Solve the boundary- and initial-value problem

$$u_{tt} - c^2 u_{xx} = F_0 \sin vx \cos \omega t \qquad 0 \le x \le b; t \ge 0$$
$$u(0,t) = h_0 \qquad u(b,t) = h_1 \qquad t \ge 0$$
$$u(x,0) = x(b - x) \qquad u_t(x,0) = 0 \qquad 0 \le x \le b$$

where F_0, v, ω, h_0, h_1 are given real positive constants. Distinguish the cases

(i) $\omega \ne n\pi c/b, n = 1, 2, \ldots$

(ii) $\omega = m\pi c/b, m$ a positive integer.

26 If, instead of being fastened, the end $x = 0$ of the string is elastically constrained, the boundary condition is

$$u_x(0,t) - hu(0,t) = 0 \qquad h = \frac{K}{T_0} > 0$$

where K is the spring constant and T_0 is the horizontal component of the tension. Let the end $x = b$ be fixed. Assume the initial conditions are given by Eq. (4-24).

(a) Separate variables in the wave equation, and show the x-dependent factor X must satisfy

$$X'' + \lambda X = 0 \qquad X'(0) - hX(0) = 0 \qquad X(b) = 0$$

where λ is a separation constant. This is a Sturm-Liouville problem with unmixed boundary conditions. Review Sec. 1 of Appendix 2, and appeal to the appropriate theorems in order to verify that the eigenvalues form a real monotone discrete sequence $\{\lambda_n\}$ and that all the eigenvalues are positive. Now, by direct integration of the differential equation and imposition of the boundary conditions, show that $\lambda = 0$ is not an eigenvalue. Derive the eigenfunctions and eigenvalues

$$X_n = \sin [\mu_n(x - b)] \qquad \lambda_n = \mu_n^2 \qquad n = 1, 2, \ldots$$

where the μ_n are the real positive roots of the transcendental equation

$$h \tan b\mu + \mu = 0$$

arranged in increasing order. Sketch the graphs of the equations

$$y = \tan b\mu \qquad y = -\frac{\mu}{h}$$

on the same set of axes, and illustrate graphically the occurrence of the roots of the transcendental equation. Hence verify the discreteness and monotonicity of the sequence of eigenvalues $\{\lambda_n\}$. Note that

$$\frac{(2n - 1)\pi}{2b} < \mu_n < \frac{(2n + 1)\pi}{2b} \qquad n = 1, 2, \ldots$$

and for large n

$$\mu_n \sim \frac{(2n - 1)\pi}{2b}$$

By direct integration show that the eigenfunctions $\{X_n\}$ form an orthogonal sequence on $[0,b]$.

(b) Derive the normal modes

$$u_n(x,t) = \sin [\mu_n(x - b)](A_n \cos \omega_n t + B_n \sin \omega_n t) \qquad n = 1, 2, \ldots$$

where the characteristic frequencies $\omega_n = c\mu_n$, $n = 1, 2, \ldots$.

(c) Show that if the superposition of normal modes

$$u(x,t) = \sum_{n=1}^{\infty} (A_n \cos \omega_n t + B_n \sin \omega_n t) \sin [\mu_n(x - b)]$$

is suitably convergent and satisfies the initial conditions (4-24), the coefficients must be given by

$$A_n = \frac{1}{\alpha_n} \int_0^b f(x) \sin [\mu_n(x - b)] \, dx$$

$$B_n = \frac{1}{\alpha_n \omega_n} \int_0^b g(x) \sin [\mu_n(x - b)] \, dx$$

where

$$\alpha_n = \frac{hb + \cos^2 \mu_n b}{2h} \qquad n = 1, 2, \ldots$$

27   Derive the formal series solution of the problem

$$u_{tt} = c^2 u_{xx} \qquad 0 \le x \le b; t \ge 0$$

$$u_x(0,t) - h_1 u(0,t) = 0 \qquad u_x(b,t) + h_2 u(b,t) = 0 \qquad t \ge 0$$

$$u(x,0) = f(x) \qquad u_t(x,0) = g(x) \qquad 0 \le x \le b$$

where h_1 and h_2 are given positive constants.

28   Instead of being fastened, the end $x = 0$ of the string is forced (by some external means) to undergo a prescribed displacement.   Then the boundary conditions are

$$u(0,t) = h(t) \qquad u(b,t) = 0 \qquad t \ge 0 \tag{1}$$

where h is a given function.   Let the initial conditions be given by Eq. (4-24). Continuity of the spring implies

$$h(0) = f(0) \qquad \dot{h}(0) = g(0)$$

Choose a simple function v which satisfies the boundary conditions (1).   Let w be a solution of Eq. (4-22) with F replaced by

$$F(x,t) - v_{tt} + c^2 v_{xx}$$

Assume also that w satisfies the homogeneous boundary conditions (4-23) and the initial conditions

$$w(x,0) = f(x) - v(x,0) \qquad w_t(x,0) = g(x) - v_t(x,0)$$

Verify that the superposition $u = v + w$ satisfies Eq. (4-22), the boundary conditions (1) and the initial conditions (4-24).   An example of a suitable function v is

$$v(x,t) = \left(1 - \frac{x}{b}\right) h(t)$$

This function interpolates linearly between the value $h(t)$ at $x = 0$ and the value 0 at $x = b$.

29 (a) Solve the boundary- and initial-value problem

$$u_{tt} - c^2 u_{xx} = -g \qquad 0 \le x \le b; t \ge 0$$
$$u(0,t) = h_0 \sin \omega t \qquad u(b,t) = 0$$
$$u(x,0) = 0 \qquad u_t(x,0) = h_0 \omega$$

where g, h_0, ω are given real positive constants and

$$\omega \ne \frac{n\pi c}{b} \qquad n = 1, 2, \ldots$$

(b) Solve the boundary- and initial-value problem

$$u_{tt} - c^2 u_{xx} = x(b - x) \sin vt$$
$$u(0,t) = h_0 \sin \omega t \qquad u(b,t) = h_1 \cos \omega t$$

$$u(x,0) = 0 \qquad u_t(x,0) = \frac{(b - x)h_0\omega}{b_0}$$

where v, h_0, h_1, ω are given real constants and

$$v \ne \frac{n\pi c}{b} \qquad \omega \ne \frac{n\pi c}{b} \qquad n = 1, 2, \ldots$$

30 If the string vibrates in a medium, e.g., air, a frictional force may be present.   In the case where the friction force is proportional to the speed the equation of motion of the idealized string is

$$u_{tt} + 2\gamma u_t - c^2 u_{tt} = F(x,t) \qquad 0 \le x \le b; t \ge 0 \qquad (1)$$

Equation (1) is called the *damped-wave equation*.   Here γ is a known positive constant and $c = \sqrt{T_0/\rho}$.   It was observed in Prob. 2 that solutions of the homogeneous equation may represent traveling waves whose amplitudes are attenuated as they travel.   Assume the ends of the string are fastened.   Then the boundary conditions are given by Eq. (4-23). Let the initial conditions be given by Eq. (4-24).

(a) The *homogeneous damped-wave equation* is (1) with F identically zero.   Separate variables in the homogeneous equation, and derive the *normal modes*

$$u_n(x,t) = e^{-\gamma t} \sin \frac{n\pi x}{b} (A_n \cos \omega_n t + B_n \sin \omega_n t) \qquad n = 1, 2, \ldots$$

where the *normal* (or characteristic) *frequencies*

$$\omega_n = \left(\frac{n^2\pi^2 c^2}{b^2} - \gamma^2\right)^{\frac{1}{2}} \qquad n = 1, 2, \ldots$$

The normal modes are damped oscillatory provided the damping constant γ is sufficiently small.   Henceforth assume this is so.

(b) Consider the homogeneous damped-wave equation subject to the boundary conditions (4-23) and initial conditions (4-24).   Derive the formal series solution

$$u(x,t) = e^{-\gamma t} \sum_{n=1}^{\infty} \sin \frac{n\pi x}{b} (A_n \cos \omega_n t + B_n \sin \omega_n t)$$

where

$$A_n = \frac{2}{b} \int_0^b f(x) \sin \frac{n\pi x}{b} dx \qquad B_n = \frac{\gamma A_n}{\omega_n} + \frac{2}{b\omega_n} \int_0^b g(x) \sin \frac{n\pi x}{b} dx$$

(c) Consider the damped-wave equation (1) subject to the homogeneous boundary conditions (4-23) and the homogeneous initial conditions

$$u(x,0) = 0 \qquad u_t(x,0) = 0 \qquad 0 \le x \le b$$

Assume a solution of the form in Eq. (4-36).   Show that the functions φ_k must satisfy

$$\ddot{\varphi}_k + 2\gamma\dot{\varphi}_k + \frac{k^2\pi^2 c^2}{b^2} \varphi_k = F_k(t) \qquad \varphi_k(0) = \dot{\varphi}_k(0) = 0$$

$k = 1, 2, \ldots$, where the F_k are defined in Eq. (4-39).   Thus derive the solution

$$u(x,t) = e^{-\gamma t} \sum_{n=1}^{\infty} \left\{ \frac{1}{\omega_n} \int_0^t e^{\gamma\xi} \sin [\omega_n(t-\xi)]F_n(\xi) \, d\xi \right\} \sin \frac{n\pi x}{b}$$

31   Derive the formal series solution.

$$u_{tt} + 2\gamma u_t - c^2 u_{xx} = -g \qquad 0 \le x \le b; t \ge 0$$
$$u(0,t) = 0 \qquad u(b,t) = 0 \qquad t \ge 0$$
$$u(x,0) = x(b-x) \qquad u_t(x,0) = 0 \qquad 0 \le x \le b$$

32   With regard to the string described in Prob. 30, a concentrated impulsive force

$$F(x,t) = \delta(x - \xi)\delta(t - \tau)$$

of unit magnitude is applied to the string at time $t = \tau$ and at the point $x = \xi$.   The motion starts from rest with zero displacement.   Show that the subsequent motion is given by

$$G(x,t;\xi,\tau) = \frac{2}{b} e^{-\gamma t} \sum_{n=1}^{\infty} \frac{1}{\omega_n} \sin \frac{n\pi x}{b} \sin \frac{n\pi\xi}{b} \sin [\omega_n(t-\tau)]$$

where the ω_n are the characteristic frequencies derived in Prob. 30.   Use substitution and formal interchange of the operations of integration and summation to show that the solution of the problem of forced motion derived in Prob. 30c can be written

$$u(x,t) = \int_0^t \int_0^b G(x,t;\xi,\tau)F(\xi,\tau) \, d\xi \, d\tau$$

Use this expression and the *Green's function* together with superposition to derive the solution of Prob. 31.   What relation exists between this Green's function and the Green's function obtained in Prob. 22?

33   Give a complete discussion and derive the formal series solution of the problem

$$u_{tt} + 2\gamma u_t - c^2 u_{xx} = F_1(x) \sin \omega t \qquad 0 \le x \le b; t \ge 0$$
$$u(0,t) = h_1 \qquad u(b,t) = h_2 \qquad t \ge 0$$
$$u(x,0) = f(x) \qquad u_t(x,0) = g(x) \qquad 0 \le x \le b$$

Here γ, h_1, h_2, and ω are given real positive constants, and $F_1(x)$, $f(x)$, $g(x)$ are given real-valued functions.

34 In the determination of the voltage distribution $v(x,t)$ along a transmission line of length b the following boundary- and initial-value problem arises.

$$v_{xx} = LCv_{tt} + RCv_t \qquad 0 \le x \le b; \, t \ge 0$$
$$v(0,t) = 0 \qquad v_x(b,t) = 0 \qquad t \ge 0$$
$$v(x,0) = v_0 \qquad v_t(x,0) = 0 \qquad 0 \le x \le b$$

Here L, C, R, and v_0 are given real positive constants, and $L > R^2Cb^2/\pi^2$. Derive the formal series solution

$$v(x,t) = e^{-Rt/2L} \sum_{n=1}^{\infty} A_n \sin \frac{(2n-1)\pi x}{2b} \sin (\omega_n t + \epsilon_n)$$

where

$$A_n = \frac{4v_0}{\pi(2n-1)\sin \epsilon_n} \qquad \tan \epsilon_n = \frac{2L\omega_n}{R}$$

$$\omega_n = \frac{[L\pi^2(2n-1)^2 - R^2Cb^2]^{\frac{1}{2}}}{2bLC^{\frac{1}{2}}} \qquad n = 1, 2, \ldots$$

35 Derive the formal series solution

$$u_{tt} + 2\gamma u_t - c^2 u_{xx} = F_0 \sin vt \qquad 0 \le x \le b; \, t \ge 0$$
$$u(0,t) = 0 \qquad u_x(b,t) = h_1 \sin \omega t \qquad t \ge 0$$
$$u(x,0) = 0 \qquad u_t(x,0) = 0 \qquad 0 \le x \le b$$

Here F_0, γ, v, h_1, ω are given real positive constants.

36 The vibrating string illustrates *transverse waves*: the direction of motion of the individual particles is perpendicular to the direction of propagation of waves. The occurrence of elastic waves in a solid bar, e.g., metal, illustrates *longitudinal waves*: the direction of motion of individual particles is the same as the direction of propagation of waves. Consider a long, slender cylindrical homogeneous solid bar, of uniform cross section, which is at rest with its axis coincident with the x axis. The ends of the bar are at $x = 0$, $x = b > 0$. The rod is assumed to be perfectly elastic, so that if an elongation takes place as a result of the application of external forces at the ends of the bar, tensile forces, directed parallel to the x axis, are set up within the bar. If now the forces are removed, the bar vibrates longitudinally in accordance with the laws of elasticity. Let $\xi(x,t)$ denote the longitudinal displacement at time t of the point in the bar whose equilibrium position was x for $t < 0$. It can be shown that the function ξ must satisfy the homogeneous wave equation

$$\xi_{tt} - c^2 \xi_{xx} = 0 \qquad 0 \le x \le b; \, t \ge 0$$

where $c = \sqrt{E/\rho}$, E = modulus of elasticity, ρ = density. Accordingly waves of longitudinal displacement occur. Assume the end $x = 0$ of the bar is held fixed, and the end $x = b$ is free. Then the boundary conditions are

$$\xi(0,t) = 0 \qquad \xi_x(b,t) = 0 \qquad t \ge 0$$

The second boundary condition is a consequence of the relation $T = EA\xi_x$, where T is the tensile force within the bar at the point x and A is the cross-sectional area of the bar. The initial conditions are

$$\xi(x,0) = \frac{(b_1 - b)x}{b} \qquad \xi_t(x,0) = 0 \qquad 0 \leq x \leq b$$

where b_1 is the length of bar at maximum extension.   Derive a formal series solution of the boundary- and initial-value problem.

37   Derive a formal series solution of the problem

$$\xi_{tt} - c^2 \xi_{xx} = 0 \qquad 0 \leq x \leq b; \, t \geq 0$$

$$\xi(0,t) = 0 \qquad M\xi_{tt}(b,t) - EA\xi_x(b,t) = 0 \qquad t \geq 0$$

$$\xi(x,0) = \frac{(b_1 - b)x}{b} \qquad \xi_t(x,0) = 0 \qquad 0 \leq x \leq b$$

where M, E, A, b_1 are given real positive constants.

38   Let L be the linear operator defined by

$$Lu = A(x)u_{xx} + C(t)u_{tt} + A_1(x)u_x + C_1(t)u_t + [A_0(x) + C_0(t)]u$$

Assume the coefficients are twice continuously differentiable, and $A(x)C(t) < 0$ in the region of the xt plane under consideration.   Then L is a hyperbolic operator.

(a) Consider first the boundary- and initial-value problem

$$Lu = 0 \qquad 0 \leq x \leq b; \, t \geq 0$$

$$a_1 u(0,t) + a_2 u_x(0,t) = 0 \qquad b_1 u(b,t) + b_2 u_x(b,t) = 0 \qquad t \geq 0$$

$$u(x,0) = f(x) \qquad u_t(x,0) = g(x) \qquad 0 \leq x \leq b$$

where $[0,b]$ is a given fixed interval, a_1, a_2, b_1, b_2 are given real constants such that

$$(a_1^2 + a_2^2)(b_1^2 + b_2^2) \neq 0$$

and f, g are given real-valued twice continuously differentiable functions.   Assume a separable solution

$$u(x,t) = X(x)T(t)$$

of $Lu = 0$, and show that the factors X, T must satisfy the linear second-order ordinary differential equations

$$AX'' + A_1 X' + A_0 X = \lambda X$$

$$C\ddot{T} + C_1 \dot{T} + C_0 T = -\lambda T$$

respectively, where λ is a separation constant.   Here primes denote derivatives with respect to x, and dots denote derivatives with respect to t.   Without loss of generality it can be assumed that $A(x) < 0, \, 0 \leq x \leq b$.   Multiply the differential equation satisfied by X by the function

$$\rho(x) = \frac{-\exp\left[\displaystyle\int (A_1/A)\,dx\right]}{A}$$

Show that the self-adjoint equation

$$(pX')' + qX + \lambda \rho X = 0$$

is obtained, where $p = -A\rho$ and $q = -A_0\rho$. Note that $p(x) > 0$ and $\rho(x) > 0$. The appropriate boundary conditions on X are

$$a_1 X(0) + a_2 X'(0) = 0 \qquad b_1 X(b) + b_2 X'(b) = 0$$

Review Sec. 1 of Appendix 2. The problem in X is a regular self-adjoint Sturm-Liouville problem. Accordingly the eigenvalues constitute a real sequence $\{\lambda_n\}$ such that

$$\lambda_1 < \lambda_2 < \cdots < \lambda_n < \lambda_{n+1} < \cdots \qquad \lim_{n \to \infty} \lambda_n = +\infty$$

There are at most a finite number of negative eigenvalues. If $A_0(x) \geq 0$, all eigenvalues are nonnegative. Corresponding to each eigenvalue λ_n is a real-valued eigenfunction φ_n. The sequence $\{\varphi_n\}$ of eigenfunctions forms an orthogonal sequence on $[0,b]$ with weight function ρ

$$\int_0^b \rho(x)\varphi_n(x)\varphi_m(x)\, dx = 0 \qquad n \neq m$$

(b) It can be assumed that $C(t) > 0$, $t \geq 0$. For each positive integer n there exists a fundamental set $T_n^{(1)}(t)$, $T_n^{(2)}(t)$ of the second-order linear equation

$$C\ddot{T} + C_1\dot{T} + (C_0 + \lambda_n)T = 0 \qquad t \geq 0$$

such that $T_n^{(1)}(0) = 1$, $T_n^{(1)'}(0) = 0$, $T_n^{(2)}(0) = 0$, and $T_n^{(2)'}(0) = 1$. Define the *normal modes* of the boundary- and initial-value problem

$$u_n(x,t) = \varphi_n(x)[A_n T_n^{(1)}(t) + B_n T_n^{(2)}(t)] \qquad n = 1, 2, \ldots$$

Each normal mode satisfies $Lu = 0$ as well as the homogeneous boundary conditions of the problem.

(c) Consider a superposition of normal modes

$$u(x,t) = \sum_{n=1}^{\infty} \varphi_n(x)[A_n T_n^{(1)}(t) + B_n T_n^{(2)}(t)]$$

Show that if u is a solution of the boundary- and initial-value problem, the coefficients in the series must be given by

$$A_n = \frac{1}{\|\varphi_n\|^2} \int_0^b \rho(x)f(x)\varphi_n(x)\, dx \qquad B_n = \frac{1}{\|\varphi_n\|^2} \int_0^b \rho(x)g(x)\varphi_n(x)\, dx$$

where

$$\|\varphi_n\|^2 = \int_0^b \rho(x)\varphi_n^2(x)\, dx \qquad n = 1, 2, \ldots$$

(d) Consider now the problem

$$Lu = F(x,t) \qquad 0 \leq x \leq b; \, t \geq 0$$
$$a_1 u(0,t) + a_2 u_x(0,t) = 0 \qquad b_1 u(b,t) + b_2 u_x(b,t) = 0 \qquad t \geq 0$$
$$u(x,0) \doteq 0 \qquad u_t(x,0) = 0 \qquad 0 \leq x \leq b$$

involving homogeneous boundary and initial conditions and the inhomogeneous partial differential equation. Here F is a given real-valued twice continuously differentiable function. Assume a solution of the form

$$u(x,t) = \sum_{n=1}^{\infty} \varphi_n(x)\psi_n(t)$$

where the φ_n are the eigenfunctions obtained in a above and the functions ψ_n are to be determined. Then u satisfies the boundary conditions. It will satisfy the initial conditions if

$$\psi_n(0) = 0 \qquad \dot{\psi}_n(0) = 0 \qquad n = 1, 2, \ldots$$

Substitute u into the partial differential equation, and show that the ψ_n must satisfy

$$C\ddot{\psi}_n + C_1\dot{\psi}_n + (C_0 + \lambda_n)\psi_n = F_n(t) \qquad n = 1, 2, \ldots$$

where

$$F_n(t) = \frac{1}{\|\varphi_n\|^2} \int_0^b \rho(x)F(x,t)\varphi_n(x)\, dx \qquad n = 1, 2, \ldots$$

The initial conditions and the differential equation uniquely determine the ψ_n. With the ψ_n determined this way, the formal series solution of the problem is obtained. Now, using superposition, the formal series solution of the problem

$$Lu = F(x,t) \qquad 0 \le x \le b; t \ge 0$$
$$a_1u(0,t) + a_2u_x(0,t) = 0 \qquad b_1u(b,t) + b_2u_x(b,t) = 0 \qquad t \ge 0$$
$$u(x,0) = f(x) \qquad u_t(x,0) = g(x) \qquad 0 \le x \le b$$

can be derived. If the boundary conditions are inhomogeneous

$$a_1u(0,t) + a_2u_x(0,t) = h_1(t) \qquad b_1u(b,t) + b_2u_x(b,t) = h_2(t) \qquad t \ge 0$$

where h_1, h_2 are given functions, application of the technique outlined in Prob. 28 together with superposition again leads to the formal solution.

39 Derive the formal series solution.

(a) $u_{tt} - c^2u_{xx} - \omega^2u = F_0 \sin \nu t \qquad 0 \le x \le b; t \ge 0$
$$u(0,t) = 0 \qquad u(b,t) = 0 \qquad t \ge 0$$
$$u(x,0) = f(x) \qquad u_t(x,0) = g(x)$$

where c, ω, F_0, ν are given real positive constants, $\omega \ne \nu$.

(b) $\dfrac{u_{tt}}{c^2} - u_{xx} - 2u_x = 0 \qquad 0 \le x \le b; t \ge 0$

$$u(0,t) = 0 \qquad u(b,t) = 0 \qquad t \ge 0$$
$$u(x,0) = f(x) \qquad u_t(x,0) = g(x) \qquad 0 \le x \le b$$
(c) $u_{tt} - c^2x^2u_{xx} = 0 \qquad 0 < a \le x \le b; t \ge 0$
$$u(a,t) = 0 \qquad u(b,t) = 0 \qquad t \ge 0$$
$$u(x,0) = f(x) \qquad u_t(x,0) = g(x) \qquad a \le x \le b$$

40 Derive the formal solution of the problem

$$u_{tt} - c^2[(b^2 - x^2)u_x]_x = 0 \qquad 0 < x < b; t > 0$$
$$u(0,t) = 0 \qquad \lim_{\substack{x \to b \\ x < b}} u(x,t) \text{ exists} \qquad t \ge 0$$
$$u(x,0) = f(x) \qquad u_t(x,0) = g(x) \qquad 0 \ge x \ge b$$

in the form

$$u(x,t) = \sum_{n=1}^{\infty} P_{2n-1}\left(\frac{x}{b}\right)(A_n \cos \omega_n t + B_n \sin \omega_n t)$$

where the P_n are the Legendre polynomials, and the coefficients

$$A_n = \frac{4n-1}{b} \int_0^b f(x) P_{2n-1}\left(\frac{x}{b}\right) dx \qquad B_n = \frac{4n-1}{b\omega_n} \int_0^b g(x) P_{2n-1}\left(\frac{x}{b}\right) dx$$

The characteristic frequencies are

$$\omega_n = c[2n(2n-1)]^{\frac{1}{2}} \qquad n = 1, 2, \ldots$$

41   Derive the formal series solution of the problem

$$u_{tt} - c^2(xu_x)_x = 0 \qquad 0 < x < b; \, t > 0$$

$$\lim_{\substack{x \to 0 \\ x > 0}} u(x,t) \text{ exists} \qquad u(b,t) = 0 \qquad t \geq 0$$

$$u(x,0) = f(x) \qquad u_t(x,0) = g(x) \qquad 0 \leq x \leq b$$

42   (a) Let L be the linear operator defined in Prob. 38. Assume C, C_1 are constants,
$C > 0$, $C_1 \geq 0$. Also assume $A(x) < 0$, $0 \leq x \leq b$. Then L is a hyperbolic operator.
Suppose u is a solution of the homogeneous equation $Lu = 0$. Define the function

$$v(x,t) = u(x,t) \exp\left(\frac{C_1}{2C} t\right)$$

Show that v must satisfy the equation

$$A(x)v_{xx} + A_1(x)v_x + Cv_{tt} + \left(A_0 + C_0 - \frac{C_1{}^2}{4C}\right) v = 0$$

Conversely, if v satisfies this partial differential equation, then u is a solution of $Lu = 0$.
Define the function

$$p(x) = \exp\left(\int \frac{A_1}{A} dx\right)$$

Show that multiplication by p transforms the differential equation in v into
$(pv_x)_x - rv_{tt} - qv = 0$
where

$$r(x) = \frac{-Cp(x)}{A(x)} \qquad q(x,t) = \frac{[C_1{}^2/4C - A_0(x) - C_0(t)]p(x)}{A(x)}$$

Since $p(x) > 0$, $r(x) > 0$, $0 \leq x \leq b$, it follows that the transformed equation is hyperbolic
also.

(b) Consider the problem

$$Lu = A(x)u_{xx} + Cu_{tt} + A_1(x)u_x + C_1u_t + A_0(x)u = 0$$

$$u(x,0) = f(x) \qquad u_t(x,0) = g(x) \qquad 0 \leq x \leq b$$

with boundary conditions one of the following:

$u(0,t) = u(b,t) = 0$	(1)
$u(0,t) = u_x(b,t) = 0$	(2)
$u_x(0,t) = u(b,t) = 0$	(3)
$u_x(0,t) - h_1u(0,t) = 0 \qquad u_x(b,t) + h_1u(b,t) = 0 \qquad h_1 > 0; \, h_2 > 0$	(4)
$u_x(0,t) = u_x(b,t) = 0$	(5)

Assume C, C_1 are constants, $C > 0$, $C_1 \geq 0$, and $A(x) < 0$, $0 \leq x \leq b$, as in part **a** above. In addition assume

$$A_0(x) \geq \frac{C_1^2}{4C} \qquad 0 \leq x \leq b$$

Then the problem can have at most one solution; i.e., if a solution exists, it is unique. The proof of this fact can be made with the results of **a** above. By linearity, the difference of two solutions of the given problem is a solution of the homogeneous problem, i.e., with f and g identically zero. If u denotes this difference and v is as defined in **a**, then v satisfies the homogeneous problem

$$(pv_x)_x - rv_{tt} - qv = 0 \qquad 0 \leq x \leq b; \ t \geq 0$$

$$v(x,0) = 0 \qquad v_t(x,0) = 0 \qquad 0 \leq x \leq b$$

$$B(v) = 0$$

where $B(v) = 0$ is symbolic of one of the boundary conditions (1) to (5), with u replaced by v. Now v can be shown to be the trivial solution as follows. Define the *energy integral*

$$E(t) = \frac{1}{2} \int_0^b (pv_t^2 + rv_t^2 + qv^2)\, dx$$

Since $p(x) > 0$, $r(x) > 0$, and $q(x) \geq 0$, it follows that $E(t) \geq 0$, $t \geq 0$. Differentiation yields

$$\frac{dE}{dt} = \int_0^b (pv_x v_{xt} + rv_t v_{tt} + qvv_t)\, dx$$

$$= pv_x v_t \Big|_0^b - \int_0^b (pv_x)_x - rv_{tt} - qv)v_t\, dx$$

$$= p(b)v_x(b,t)v_t(b,t) - p(0)v_x(0,t)v_t(0,t)$$

Show that if the boundary conditions are either (1), (2), (3), or (5), then dE/dt is zero for $t \geq 0$, and so E has a constant value. Apply the initial conditions and show $E = 0$. In turn prove that $E = 0$ implies v is identically zero. If the boundary conditions are (4), show that

$$\frac{dE}{dt} = -\frac{1}{2}\frac{d}{dt}[h_1 p(0)v^2(0,t) + h_2 p(b)v^2(b,t)]$$

and so

$$E(t) - E(0) = -\tfrac{1}{2}[h_1 p(0)v^2(0,t) + h_2 p(b)v^2(b,t)]$$

By the initial conditions, $E(0) = 0$. Thus $E(t) \leq 0$, $t \geq 0$. In turn this implies $E(t) = 0$, $t \geq 0$, and so v is identically zero.

43 (a) Prove the following existence theorem for problem (4-25). Let f have a continuous fourth derivative and g have a continuous third derivative on $[0,b]$ and such that

$$f(0) = f''(0) = f(b) = f''(b) = 0 \qquad g(0) = g''(0) = g(b) = g''(b) = 0$$

Then the series (4-30), where the coefficients are given by Eqs. (4-33) and (4-34), defines a function u which is twice continuously differentiable in x and t for $0 \le x \le b$, $t \ge 0$ and satisfies the homogeneous wave equation and the boundary and initial conditions of problem (4-25). *Hint:* Use integration by parts and show that

$$A_n = \frac{2b^3}{n^4\pi^4} \int_0^b f^{(4)}(x) \sin \frac{n\pi x}{b}\, dx \qquad B_n = -\frac{2b^3}{n^4\pi^4 c} \int_0^b g^{(3)}(x) \cos \frac{n\pi x}{b}\, dx$$

Hence there exists a constant $M > 0$ such that

$$|A_n| < \frac{M}{n^4} \qquad |B_n| < \frac{M}{n^4} \qquad n = 1, 2, \ldots$$

Now

$$\left| \sin \frac{n\pi x}{b} \left(A_n \cos \frac{n\pi ct}{b} + B_n \sin \frac{n\pi ct}{b} \right) \right| \le |A_n| + |B_n| < \frac{2M}{n^4}$$

Let $T > 0$ be fixed. Use the Weierstrass test in conjunction with the convergent series of constants

$$\sum_{n=1}^{\infty} \frac{2M}{n^4}$$

to prove the series (4-30) converges uniformly for $0 \le x \le b$, $0 \le t \le T$ and so defines a continuous function u there. Since $T > 0$ is arbitrary, the function u defined by the series is continuous for $0 \le x \le b$, $t \ge 0$. Clearly u satisfies the boundary conditions of problem (4-25). Since the series for $u(x,0)$, $u_t(x,0)$ converge to f and g, respectively, it follows that u satisfies the initial conditions. It remains to show that u is a solution of the homogeneous wave equation. Since each normal mode u_n is a solution of the wave equation, one need only show that the series can be differentiated twice with respect to x or twice with respect to t. Show, for example, that the series

$$\sum_{n=1}^{\infty} \frac{\partial^2 u_n}{\partial x^2} = \sum_{n=1}^{\infty} \frac{n^2\pi^2}{b^2} \sin \frac{n\pi x}{b} (A_n \cos \omega_n t + B_n \sin \omega_n t)$$

converges suitably. Do the same for the series obtained by differentiating termwise twice with respect to t.

(b) Prove the following existence theorem for problem (4-35). Let F have continuous third partial derivatives with respect to x and t such that

$$0 = F(0,t) = F(b,t) = F_{xx}(0,t) = F_{xx}(b,t) \qquad t \ge 0$$

Then the series (4-41), with the functions F_n defined as in Eq. (4-39), converges and defines a function u which is twice continuously differentiable for $0 \le x \le b$, $t \ge 0$ and which satisfies problem (4-35). *Hint:* From Eq. (4-37) and the form of the series (4-41) it follows that if the series is suitably convergent, the function u so defined satisfies the homogeneous boundary and initial conditions. Fix $T > 0$, and let

$$M(T) = \max [F_{xxx}(x,t)] \qquad 0 \le x \le b; 0 \le t \le T$$

Use integration by parts to show that

$$F_n(t) \le \frac{2b^3 M(T)}{n^3\pi^3} \qquad n = 1, 2, \ldots$$

and hence

$$|\varphi_n(t)| \leq \frac{2b^3 TM(T)}{n^4\pi^4 c} \qquad n = 1, 2, \ldots$$

Use these inequalities to prove the series (4-41) converges uniformly on $0 \leq x \leq b$, $0 \leq t \leq T$ and so defines a continuous function there. In addition show that the series obtained by differentiating termwise twice with respect to x converges uniformly and

$$u_{xx}(x,t) = \sum_{n=1}^{\infty} \left(-\frac{n^2\pi^2}{b^2} \right) \varphi_n(t) \sin \frac{n\pi x}{b} \qquad 0 \leq x \leq b; 0 \leq t \leq T$$

Since

$$\ddot{\varphi}_n = -\omega_n{}^2\varphi_n + F_n$$

show that

$$|\ddot{\varphi}_n(t)| \leq \omega_n{}^2 |\varphi_n(t)| + |F_n(t)| \leq \frac{2b^2 M(T)(cT + b)}{n^2\pi^2}$$

Thus the series obtained by differentiating termwise twice with respect to t converges uniformly, and

$$u_{tt} = \sum_{n=1}^{\infty} \ddot{\varphi}_n(t) \sin \frac{n\pi x}{b} = \sum_{n=1}^{\infty} [-\omega_n{}^2\varphi_n(t) + F_n(t)] \sin \frac{n\pi x}{b}$$

Hence

$$u_{tt} - c^2 u_{xx} = \sum_{n=1}^{\infty} F_n(t) \sin \frac{n\pi x}{b} = F(x,t) \qquad 0 \leq x \leq b; 0 \leq t \leq T$$

44 Recall Prob. 12 and the telegrapher's equation, which results from the transmission-line equations. In Prob. 17 the initial-value problem for this equation was solved with the aid of the Riemann Green's function. If the transmission line is of finite length, boundary conditions on the voltage and current occur at the ends of the line, in addition to the prescribed initial conditions. In addition an external impressed voltage may be present. The resulting boundary- and initial-value problem in terms of the telegrapher's equation has the form

$$Lu = u_{tt} + 2\gamma u_t + \omega^2 u - c^2 u_{xx} = F(x,t) \qquad a \leq x \leq b; t \geq 0$$
$$a_1 u(a,t) + a_2 u_x(a,t) = h_1(t) \qquad b_1 u(b,t) + b_2 u_x(b,t) = h_2(t) \qquad t \geq 0$$
$$u(x,0) = f(x) \qquad u_t(x,0) = g(x) \qquad a \leq x \leq b$$

Here γ, ω are real nonnegative constants, and c is a real positive constant. If $\omega = 0$, $\gamma = 0$, the wave equation results. If $\omega = 0$, $\gamma > 0$, the damped-wave equation is obtained (see Probs. 30 to 35). Note that the operator L is a special case of the operator discussed in Prob. 38 and also is a special case of the operator considered in Prob. 42. Accordingly the results obtained in those problems apply to the present problem. In particular, if the corresponding homogeneous boundary conditions take one of the forms (1) to (5) stated in Prob. 42, there is at most one solution.

(a) As a special case consider the problem

$$Lu = F(x,t) \qquad a \leq x \leq b; t \geq 0$$
$$u(a,t) = h_1(t) \qquad u_x(b,t) = h_2(t) \qquad t \geq 0$$
$$u(x,0) = f(x) \qquad u_t(x,0) = g(x) \qquad a \leq x \leq b$$

where $\gamma > 0$, $\omega > 0$, and F, h_1, h_2, f, and g are given functions.   First it is desired to remove the inhomogeneity in the boundary conditions.   Consider the function

$$w(x,t) = \frac{x(x - 2b)h_1(t)}{a(a - 2b)} + \frac{ax(a - x)h_2(t)}{a(a - 2b)}$$

This function satisfies the boundary conditions.   If v satisfies

$$Lv = F(x,t) - Lw \qquad a \leq x \leq b; t \geq 0$$

$$v(a,t) = 0 \qquad v_x(b,t) = 0 \qquad t \geq 0$$

$$v(x,0) = f(x) \qquad v_t(x,0) = g(x) \qquad a \leq x \leq b$$

then $u = v + w$ is a formal solution of the problem.

(b) Assume now that the constants ω, γ are such that

$$\gamma^2 < \omega^2 + \left[\frac{\pi c}{2(b - a)}\right]^2$$

Consider the problem

$$Lu = 0 \qquad a \leq x \leq b; t \geq 0$$

$$u(a,t) = 0 \qquad u_x(b,t) = 0 \qquad t \geq 0$$

$$u(x,0) = f(x) \qquad u_t(x,0) = g(x) \qquad a \leq x \leq b$$

Separate variables, and derive the formal series solution

$$u(x,t) = e^{-\gamma t} \sum_{n=1}^{\infty} (A_n \cos \nu_n t + B_n \sin \nu_n t) \sin [\mu_n(x - a)]$$

where

$$A_n = \frac{2}{b - a} \int_a^b f(x) \sin [\mu_n(x - a)] \, dx$$

$$B_n = \frac{\gamma A_n}{\nu_n} + \frac{2}{\nu_n(b - a)} \int_a^b g(x) \sin [\mu_n(x - a)] \, dx$$

$$\mu_n = \frac{(2n - 1)\pi}{2(b - a)} \qquad \nu_n = (\omega_n{}^2 + \omega^2 - \gamma^2)^{\frac{1}{2}}$$

$$\omega_n = c\mu_n \qquad n = 1, 2, \ldots$$

(c) Consider the problem

$$Lu = F(x,t) \qquad a \leq x \leq b; t \geq 0$$

$$u(a,t) = 0 \qquad u_x(b,t) = 0 \qquad t \geq 0$$

$$u(x,0) = 0 \qquad u_t(x,0) = 0 \qquad a \leq x \leq b$$

where F is a given function.   Derive the formal series solution

$$u(x,t) = \sum_{n=1}^{\infty} \left\{ \frac{1}{\nu_n} \int_0^t e^{-\gamma(t-\tau)} \sin [\nu_n(t - \tau)] F_n(\tau) \, d\tau \right\} \sin [\mu_n(x - a)]$$

where

$$F_n(t) = \frac{2}{b-a} \int_a^b F(x,t) \sin [\mu_n(x-a)] \, dx$$

(d) Choose $F(x,t) = \delta(x-\xi)\delta(t-\tau)$, and derive the formal series expression for the Green's function

$$G(x,t;\xi,\tau) = \frac{2}{b-a} e^{-\gamma(t-\tau)} \sum_{n=1}^{\infty} \frac{1}{\nu_n} \sin [\mu_n(x-a)] \sin [\mu_n(\xi-a)] \sin [\nu_n(t-\tau)]$$

Thus, if $H(x,t) = Lw$, where w is the function described in **a**, the formal solution of the problem posed in **a** is

$$u(x,t) = e^{-\gamma t} \sum_{n=1}^{\infty} (A_n \cos \nu_n t + B_n \sin \nu_n t) \sin [\mu_n(x-a)]$$

$$+ \int_0^t \int_a^b G(x,t;\xi,\tau)[F(\xi,\tau) - H(\xi,\tau)] \, d\xi \, d\tau$$

Sec. 4-4

45 (a) A function of the form

$$u = \psi(x,y)e^{\pm i\omega t} \qquad i = \sqrt{-1}; \; \omega \text{ real and positive}$$

is called *harmonic time-dependent*. Show that in order for u to be a solution of the homogeneous wave equation (4-45) it is necessary that the *amplitude factor* satisfy

$$\psi_{xx} + \psi_{yy} + k^2\psi = 0 \qquad k^2 = \frac{\omega^2}{c^2}$$

the *scalar Helmholtz equation* in two dimensions.

(b) Assume a solution of the Helmholtz equation of the form

$$\psi = e^{i(\alpha x + \beta y)} \qquad \alpha, \beta \text{ real constants}$$

and so derive the plane harmonic wave functions

$$u = e^{i(\alpha x \pm \beta y \pm \omega t)} \qquad \alpha^2 + \beta^2 = k^2 = \frac{\omega^2}{c^2}$$

These have sinusoidal variation in the direction of each space axis as well as time.

(c) Assume a solution of the Helmholtz equation of the form

$$\psi = e^{i\alpha x - \beta y} \qquad \alpha, \beta \text{ real constants}$$

and so derive the plane harmonic wave functions

$$u = e^{i(\alpha x \pm \omega t) - \beta y} \qquad \beta^2 = \alpha^2 - k^2 = \alpha^2 - \frac{\omega^2}{c^2}$$

If $\beta > 0$, such a wave function has its amplitude decaying exponentially in the positive y direction.

(d) Derive the wave functions

$$u = e^{\alpha x + \beta y \pm \omega t} \qquad \alpha, \beta \text{ real}; \; \alpha^2 + \beta^2 = \frac{\omega^2}{c^2}$$

(e) Construct a nontrivial real-valued wave function such that

$$\lim_{x \to \infty} u(x,y,t) = 0 \qquad u(x,0,t) = 0 \qquad u(x,y,0) = 0$$

(f) Construct a nontrivial real-valued wave function such that

$$u(x,y,0) = 0 \quad \text{all } x, y \qquad u(0,y,t) = 0 \quad \text{all } y, t \qquad u(x,0,t) = 0 \quad \text{all } x, t$$

46 (a) Assume that for each choice of (suitably differentiable) function φ, the function u defined by Eq. (4-46) is a solution of the wave equation (4-45). Show that the functions η, ζ must satisfy the system (4-47).

(b) Suppose ζ is defined by Eq. (4-51). Show that the function η must satisfy the wave equation and

$$\mathbf{n} \cdot \nabla \eta \mp \frac{\eta_t}{c} = 0$$

where $\nabla \eta = \eta_x \mathbf{i} + \eta_y \mathbf{j}$ is the gradient of η. This is a linear first-order partial differential equation with constant coefficients, the independent variables being x, y, t. Apply the results of Prob. 18, Chap. 1, and show that η must be of the form

$$\eta = f(x \pm cn_x t, y \pm cn_y t)$$

Let $r = x \pm n_x ct$ and $s = y \pm cn_y t$. Substitute the function η into the wave equation, and show that

$$n_y^2 f_{rr} \pm 2 n_x n_y f_{rs} + n_x^2 f_{ss} = 0$$

This is a factorable linear second-order equation. Apply the results of Sec. 2-2, and show that

$$f(r,s) = rF(n_x r \pm n_y s) + G(n_x r \pm n_y s)$$

is the form of f. Thus deduce the general form of solution given in Eq. (4-52).

47 (a) Assume a solution of the wave equation (4-45) of the form

$$u = \eta(x,y) e^{i\zeta(x,y,t)} \qquad i = \sqrt{-1}$$

where η, ζ are real-valued twice continuously differentiable functions. Show that η, ζ must satisfy the system

$$\Delta \eta + \eta \left(\frac{\zeta_t^2}{c^2} - |\nabla \zeta|^2 \right) = 0 \qquad 2 \nabla \eta \cdot \nabla \zeta + \eta \left(\Delta \zeta - \frac{\zeta_{tt}}{c^2} \right) = 0$$

where ∇ is gradient operator, Δ the laplacian, in two dimensions x, y.

(b) With reference to part **a** assume

$$\zeta = \mathbf{k} \cdot \mathbf{r} - \omega t$$

where $\mathbf{k} = k_x \mathbf{i} + k_y \mathbf{j}$ is an arbitrary real vector and ω is an arbitrary real positive constant. Show η must satisfy the scalar Helmholtz equation

$$\Delta \eta + \left(\frac{\omega^2}{c^2} - k^2 \right) \eta = 0 \qquad k = |\mathbf{k}|$$

and also

$$\mathbf{k} \cdot \nabla \eta = 0$$

which is a linear first-order equation with constant coefficients. Call η the *amplitude* and ζ the *phase*. At a fixed time t the lines $\zeta = $ const in the xy plane define a family of curves of constant phase. The *propagation vector* is $\mathbf{k} = \nabla\zeta$, and $\mathbf{k}$ is normal to each line of constant phase. The curves $\eta = $ const define the family of curves of constant amplitude of the wave. Thus the curves of constant phase and the curves of constant amplitude are orthogonal families.

(c) From b and Sec. 1-4 show η must be of the form

$$\eta = f(k_y x - k_x y)$$

Let $s = k_y x - k_x y$. Substitute η into the Helmholtz equation and obtain

$$f''(s) - \mu^2 f(s) = 0 \qquad \mu^2 = 1 - \frac{\omega^2}{k^2 c^2}$$

Thus

$$\eta = e^{\pm\mu(k_y x - k_x y)}$$

(d) Choose a set of values k_x, k_y and a value $\omega > 0$ such that $\omega < kc$. Then (taking the positive root) $0 < \mu < 1$. Now verify that the function

$$u = Ae^{\pm\mu(k_y x - k_x y)}e^{\pm i(\mathbf{k}\cdot\mathbf{r} - \omega t)} \qquad A = \text{const}$$

is a solution of Eq. (4-45). Such a wave function represents a plane harmonic wave with amplitude

$$\eta = Ae^{\pm\mu(k_y x - k_x y)}$$

a function of x and y. The speed of the wave is

$$c' = \frac{\omega}{k} < c$$

For example, if $0 < \omega < c$, the function

$$u = e^{-\mu y}\cos(x - \omega t) \qquad \mu = \frac{(c^2 - \omega^2)^{1/2}}{c}$$

is a wave function and represents a two-dimensional sinusoidal wave profile which travels in a direction parallel to the x axis with speed $c' = \omega < c$. The amplitude $\eta = e^{-\mu y}$ decreases exponentially with increasing y. These examples illustrate *dispersion*: the speed of the wave is a function of the frequency ω.

48 Apply the Poisson-Parseval formula (4-57), and construct a solution of the initial-value problem (4-56) given the functions f and g.

(a) $f(x,y) = 1 \qquad g(x,y) = 0$ (b) $f(x,y) = 0 \qquad g(x,y) = 1$

49 In the initial-value problem (4-56) let the given functions f and g be independent of y. Then it is easy to verify directly that D'Alembert's formula (4-15) furnishes the unique solution. Show that the Poisson-Parseval formula (4-57) reduces to (4-15) in this case. *Hint:* Let $\xi = x + r\cos\theta$ and $\eta = y + r\sin\theta$ in the double integrals in (4-57). Then the domain of integration becomes the disk bounded by the circle

$$(\xi - x)^2 + (\eta - y)^2 = c^2 t^2$$

Suppose f is independent of y. Then

$$\int_0^{ct} \int_0^{2\pi} \frac{f(x + r\cos\theta)}{\sqrt{c^2t^2 - r^2}}\, r\, dr\, d\theta = \int_{x-ct}^{x+ct} f(\xi)\, d\xi \int_{-\mu+y}^{\mu+y} \frac{d\eta}{\sqrt{c^2t^2 - (\xi - x)^2 - (\eta - y)^2}}$$

where $\mu = \sqrt{c^2t^2 - (\xi - x)^2}$. Now carry out the integration with respect to η.

50 **(a)** In plane polar coordinates (r,θ) the homogeneous wave equation (4-44) takes the form

$$\frac{\partial^2 u}{\partial t^2} = c^2 \left(\frac{\partial^2 u}{\partial r^2} + \frac{1}{r}\frac{\partial u}{\partial r} + \frac{1}{r^2}\frac{\partial^2 u}{\partial \theta^2} \right)$$

A solution independent of θ is called a *circularly symmetric wave function*. Consider a function

$$u(r,t) = f(r - ct) \qquad r = (x^2 + y^2)^{\frac{1}{2}}$$

Such a function represents a circularly symmetric traveling wave which is propagated radially outward from the origin with speed c. Show that no solution of the wave equation of this form exists (apart from the trivial case $f = $ const).

(b) Show that there does not exist a nontrivial amplitude function $\eta(r,t)$ such that

$$u = \eta(r,t)f(r - ct)$$

is a solution of the homogeneous wave epuation for arbitrary choice of f.

(c) A solution of the homogeneous wave equation in polar coordinates of the form

$$u = R(r)e^{-i\omega t} \qquad \omega \text{ real and positive; } i = \sqrt{-1}$$

is called *harmonic time-dependent*. On each circle $r = $ const the variation of u is sinusoidal with time t. Substitute into the wave equation, and show the radial dependent factor R must satisfy

$$rR'' + R' + \frac{\omega^2 r R}{c^2} = 0$$

Bessel's equation of order zero. Thus derive the circularly symmetric wave functions

$$u = \left[AJ_0\left(\frac{\omega r}{c}\right) + BY_0\left(\frac{\omega r}{c}\right) \right] e^{-i\omega t}$$

where A, B are arbitrary real constants, and J_0, Y_0 denote the Bessel functions of the first and second kind, respectively, of order zero. Recall that Y_0 has a logarithmic singularity at $r = 0$. Hence if the region in the xy plane in which solutions are desired includes the origin, the choice $B = 0$ is necessary.

51 The homogeneous *two-dimensional damped-wave equation* is

$$u_{tt} + 2\gamma u_t - c^2 \Delta u = 0$$

where γ is a real positive constant and Δu denotes the two-dimensional laplacian of u. Assume a solution of the form

$$u = e^{-\gamma t}v$$

and show that v must satisfy

$$v_{tt} - \gamma^2 v - c^2 \Delta v = 0$$

Assume a solution of this equation of the form

$$v = \psi(x,y)e^{-i\omega t} \qquad \omega \text{ real and positive}; \, i = \sqrt{-1}$$

Show that ψ must satisfy the scalar Helmholtz equation

$$\Delta\psi + \mu^2\psi = 0 \qquad \mu^2 = \frac{\omega^2 + \gamma^2}{c^2}$$

Let

$$\mathbf{k} = k_x\mathbf{i} + k_y\mathbf{j}$$

be a vector in the xy plane such that

$$|\mathbf{k}| = \mu = \frac{(\omega^2 + \gamma^2)^{\frac12}}{c}$$

Show that

$$\psi = e^{i\mathbf{k}\cdot\mathbf{r}} \qquad \mathbf{r} = x\mathbf{i} + y\mathbf{j}$$

satisfies the scalar Helmholtz equation.   Thus derive the solutions

$$u = e^{-\gamma t}e^{-i(\mathbf{k}\cdot\mathbf{r}-\omega t)}$$

of the damped-wave equation.   These represent two-dimensional damped traveling waves which move with speed $c' = \omega/\mu < c$ in the direction of the vector $\mathbf{k}$.

Sec. 4-5

52 (a) Consider problem (4-79) for the freely vibrating membrane with fastened edges when the boundary C is a rectangle.   Assume the membrane at rest occupies the domain defined by

$$0 \le x \le a \qquad 0 \le y \le b$$

The boundary conditions can be written as

$$\begin{aligned} u(x,0,t) &= 0 & u(x,b,t) &= 0 & 0 \le x \le a; \, t \ge 0 \\ u(0,y,t) &= 0 & u(a,y,t) &= 0 & 0 \le y \le b; \, t \ge 0 \end{aligned}$$

If a separable solution of the form $u = \varphi(x,y)T(t)$ is assumed, the boundary conditions on u imply that the space-dependent factor φ must satisfy the boundary conditions

$$\begin{aligned} \varphi(x,0) &= 0 & \varphi(x,b) &= 0 & 0 \le x \le a \\ \varphi(0,y) &= 0 & \varphi(a,y) &= 0 & 0 \le y \le b \end{aligned}$$

These conditions together with the Helmholtz equation (4-64), written in rectangular coordinates, constitute an eigenvalue problem for the operator Δ.   This eigenvalue problem was solved in Example 3-5 in connection with finding an eigenfunction expansion for the solution of Poisson's equation on the rectangle with the same boundary conditions.   Review Example 3-5, and verify that the eigenfunctions of the problem are

$$\varphi_{nm}(x,y) = \sin\frac{n\pi x}{a}\sin\frac{m\pi y}{b} \qquad n = 1, 2, \dots; \, m = 1, 2, \dots$$

corresponding to the eigenvalues

$$\lambda_{nm} = \frac{n^2\pi^2}{a^2} + \frac{m^2\pi^2}{b^2} \qquad n = 1, 2, \dots; \, m = 1, 2, \dots$$

The eigenfunctions are orthogonal over the domain

$$\int_0^a \int_0^b \varphi_{nm}(x,y)\varphi_{pq}(x,y)\, dx\, dy = 0 \qquad (n,m) \neq (p,q)$$

These eigenfunctions are not normalized, since

$$\int_0^a \int_0^b \varphi_{nm}^2(x,y)\, dx\, dy = \frac{ab}{4} \qquad n = 1, 2, \ldots; \; m = 1, 2, \ldots$$

The time-dependent factor corresponding to φ_{nm} is

$$T_{nm}(t) = A_{nm} \cos \omega_{nm}t + B_{nm} \sin \omega_{nm}t$$

where the ω_{nm} are the characteristic frequencies:

$$\omega_{nm} = c\sqrt{\lambda_{nm}} = c\pi \left[\left(\frac{n}{a}\right)^2 + \left(\frac{m}{b}\right)^2 \right]^{\frac{1}{2}} \qquad n = 1, 2, \ldots; \; m = 1, 2, \ldots$$

Observe that in contrast with the vibrating string the vibrating membrane possesses a doubly infinite sequence of characteristic frequencies. The normal modes of vibration are

$$u_{nm}(x,y,t) = (A_{nm} \cos \omega_{nm}t + B_{nm} \sin \omega_{nm}t) \sin \frac{n\pi x}{a} \sin \frac{m\pi y}{b}$$

(b) In order to satisfy the initial conditions consider a superposition of normal modes

$$u(x,y,t) = \sum_{n=1}^{\infty} \sum_{m=1}^{\infty} (A_{nm} \cos \omega_{nm}t + B_{nm} \sin \omega_{nm}t) \sin \frac{n\pi x}{a} \sin \frac{m\pi y}{b}$$

If the series converges suitably, it is clear that the function u so defined satisfies the boundary conditions. The initial conditions are satisfied if

$$u(x,y,0) = \sum_{n=1}^{\infty} \sum_{m=1}^{\infty} A_{nm} \sin \frac{n\pi x}{a} \sin \frac{m\pi y}{b} = f(x,y)$$

$$u_t(x,y,0) = \sum_{n=1}^{\infty} \sum_{m=1}^{\infty} \omega_{nm}B_{nm} \sin \frac{n\pi x}{a} \sin \frac{m\pi y}{b} = g(x,y)$$

$0 \leq x \leq a, 0 \leq y \leq b$. The series on the left must be the double sine series for f and g respectively. It follows that the coefficients A_{nm}, B_{nm} are given by

$$A_{nm} = \frac{4}{ab} \int_0^a \int_0^b f(x,y) \sin \frac{n\pi x}{a} \sin \frac{m\pi y}{b}\, dx\, dy$$

$$B_{nm} = \frac{4}{ab\omega_{nm}} \int_0^a \int_0^b g(x,y) \sin \frac{n\pi x}{a} \sin \frac{m\pi y}{b}\, dx\, dy$$

(c) Assume the membrane is released from rest with the initial displacement

$$u(x,y,0) = Axy(a - x)(b - y) \qquad A = \text{const}$$

Determine the formal series expression for the resulting motion.

(d) The membrane is released from rest with the initial displacement

$$u(x,y,0) = A \sin \frac{\pi x}{a} \sin \frac{\pi y}{b} \qquad A = \text{const}$$

Determine the subsequent motion. What is the speed of the midpoint at time t?

53 **(a)** The rectangular membrane with fastened edges executes forced vibrations under the driving force

$$F(x,y,t) = F_0(x,y) \sin \omega t$$

Assume that $\omega \neq \omega_{nm}$, all n, m, where the ω_{nm} are the characteristic frequencies of the rectangular membrane derived in Prob. 52. The motion starts from rest with zero displacement. Utilize the eigenfunctions obtained in Prob. 52 and Eq. (4-78) to derive the expression for the subsequent motion in the form

$$u(x,y,t) = \sum_{n=1}^{\infty} \sum_{m=1}^{\infty} B_{nm} \frac{\omega \sin \omega_{nm} t - \omega_{nm} \sin \omega t}{\omega_{nm}(\omega^2 - \omega_{nm}{}^2)} \sin \frac{n\pi x}{a} \sin \frac{m\pi y}{b}$$

where

$$B_{nm} = \frac{4}{ab} \int_0^a \int_0^b F_0(x,y) \sin \frac{n\pi x}{a} \sin \frac{m\pi y}{b} \, dx \, dy$$

(b) In part **a** it was assumed that the frequency of the driving force does not coincide with a characteristic frequency. Derive the series solution if $\omega = \omega_{pq}$, p, q a given fixed pair of positive integers. This is the case of *resonance*.

(c) Derive a formal series expression for the motion of the rectangular membrane if the initial conditions are

$$u(x,y,0) = \left(1 - \cos \frac{2\pi x}{a}\right)\left(1 - \cos \frac{2\pi y}{b}\right) \qquad u_t(x,y,0) = 0$$

and the external force of gravity acts on the membrane.

54 **(a)** Recall the properties of the unit impulse function given in Sec. 3-5. A concentrated impulsive force per unit mass and of unit magnitude applied at time $t = \tau$ at the point (ξ,η) on the rectangular membrane can be represented

$$F(x,y,t) = \delta(x - \xi)\delta(y - \eta)\delta(t - \tau)$$

Thus

$$F(x,y,t) = 0 \qquad (x,y,t) \neq (\xi,\eta,\tau)$$

and

$$\int_0^t \int_0^a \int_0^b F(x,y,t) \, dx \, dy \, dt = 1 \qquad t > \tau$$

Let the edge be fastened, and assume zero initial conditions. Use Eq. (4-78), and derive the formal series

$$G(x,y,t;\xi,\eta,\tau) = \frac{4}{ab} \sum_{n=1}^{\infty} \sum_{m=1}^{\infty} \frac{1}{\omega_{nm}} \sin \frac{n\pi x}{a} \sin \frac{n\pi \xi}{a} \sin \frac{m\pi y}{b} \sin \frac{m\pi \eta}{b} \sin [\omega_{nm}(t - \tau)]$$

for the subsequent displacement. Here the ω_{nm} are the characteristic frequencies of the freely vibrating rectangular membrane. It can be shown that the series converges uniformly on the rectangle and defines a continuous function. The function G is called the *Green's function* of the problem. In order to simulate the complete motion the function G is defined to be zero for $t \leq \tau$. Note that G satisfies the boundary condition and has the symmetry

$$G(\xi,\eta,t;x,y,\tau) = G(x,y,t;\xi,\eta,\tau)$$

(b) For the rectangle $0 \le x \le a$, $0 \le y \le b$ the solution (4-78) of problem (4-80) is

$$u(x,y,t) = \sum_{n=1}^{\infty} \sum_{m=1}^{\infty} \left\{ \frac{1}{\omega_{nm}} \int_0^t F_{nm}(\tau) \sin \left[\omega_{nm}(t - \tau) \right] d\tau \right\} \sin \frac{n\pi x}{a} \sin \frac{m\pi y}{b}$$

where

$$F_{nm}(\tau) = \frac{4}{ab} \int_0^a \int_0^b F(x,y,\tau) \sin \frac{n\pi x}{a} \sin \frac{m\pi y}{b} dx\, dy$$

Use formal interchange of the operations of summation and integration, and show the solution can be rewritten

$$u(x,y,t) = \int_0^t \int_0^a \int_0^b G(x,y,t;\xi,\eta,\tau) F(\xi,\eta,\tau)\, d\xi\, d\eta\, d\tau$$

(c) The rectangular membrane with fastened edges is at rest in its equilibrium configuration. At time $t = 0$ the membrane is struck in such a manner that the point (ξ,η) receives a velocity impulse of unit magnitude. Thus the initial conditions are

$$u(x,y,0) = 0 \qquad u_t(x,y,0) = \delta(x - \xi)\delta(y - \eta)$$

Apply the formulas for the coefficients A_{nm}, B_{nm} derived in Prob 52b, together with the properties of the δ function, and show that the subsequent motion is given by

$$u(x,y,t) = G(x,y,t;\xi,\eta,0)$$

(d) Verify that

$$u(x,y,t) = \int_0^a \int_0^b G_t(x,y,t;\xi,\eta,0) f(\xi,\eta)\, d\xi\, d\eta + \int_0^a \int_0^b G(x,y,t;\xi,\eta,0) g(\xi,\eta)\, d\xi\, d\eta$$

furnishes the solution of problem (4-79) for the case of the fastened rectangle by substitution of the series for G and G_t evaluated at $\tau = 0$, formal interchange of the operations of summation and integration, and comparison with the series solution derived in Prob. 52b.

55   If the edge of the membrane is elastically constrained, the boundary condition is

$$\frac{\partial u}{\partial n} + \sigma u = 0 \qquad \text{on } C$$

where $\partial/\partial n$ denotes the derivative in the direction of the exterior normal to C and σ is a positive constant. Let the initial conditions be $u(x,y,0) = f(x,y)$ and $u_t(x,y,0) = g(x,y)$. Assume that a known external force $F(x,y,t)$ per unit mass acts on the membrane. Derive the expression for the motion. Also derive the series expression for the Green's function of the problem.

56   **(a)** If the membrane vibrates in a medium, a frictional force occurs. In the case where the retarding force is proportional to the speed, the equation of motion is

$$u_{tt} + 2\gamma u_t - c^2\, \Delta u = F$$

the *two-dimensional damped-wave equation*. Here γ is a positive constant, $c = \sqrt{T_0/\rho}$, and F is an external force per unit mass. It was observed in Prob. 51 that solutions of the corresponding homogeneous equation may represent two-dimensional traveling waves whose amplitudes decrease with increasing time t. Let the membrane be rectangular:

$0 \le x \le a$, $0 \le y \le b$, with fastened edges along the boundary C. Derive the normal modes

$$u_{nm} = e^{-\gamma t}(A'_{nm} \cos \nu_{nm}t + B'_{nm} \sin \nu_{nm}t) \sin \frac{n\pi x}{a} \sin \frac{m\pi y}{b}$$

and the characteristic frequencies

$$\nu_{nm} = (\omega_{nm}{}^2 - \gamma^2)^{\frac12} \qquad n = 1, 2, \ldots ; \; m = 1, 2, \ldots$$

where the ω_{nm} are the characteristic frequencies of the undamped rectangular membrane with fastened edges (see Prob. 52a) and the A'_{nm}, B'_{nm} are arbitrary constants. Observe that the normal modes are time harmonic if the damping constant $\gamma < \omega_{11}$. Assume henceforth that this is the case.

(b) Let the initial conditions be $u(x,y,0) = f(x,y)$ and $u_t(x,y,0) = g(x,y)$. Derive the series solution

$$u = e^{-\gamma t} \sum_{n=1}^{\infty} \sum_{m=1}^{\infty} (A'_{nm} \cos \nu_{nm}t + B'_{nm} \sin \nu_{nm}t) \sin \frac{n\pi x}{a} \sin \frac{m\pi y}{b}$$

where $A'_{nm} = A_{nm}$ and $B'_{nm} = \gamma A_{nm}/\nu_{nm} + B_{nm}$, the coefficients A_{nm}, B_{nm} being those defined in Prob. 52b, except that ν_{nm} replaces ω_{nm} in the expression for B_{nm}.

(c) Consider the problem

$$u_{tt} + 2\gamma u_t - c^2 \, \Delta u = F(x,y,t)$$

$$u = 0 \qquad \text{on } C$$

$$u(x,y,0) = 0 \qquad u_t(x,y,0) = 0 \qquad (x,y) \text{ in } \overline{\mathscr{R}}$$

Assume a solution

$$u = \sum_{n=1}^{\infty} \sum_{m=1}^{\infty} \psi_{nm}(t) \sin \frac{n\pi x}{a} \sin \frac{m\pi y}{b}$$

Follow the method of the text, and derive the solution

$$u(x,y,t) = e^{-\gamma t}v(x,y,t)$$

$$v(x,y,t) = \sum_{n=1}^{\infty} \sum_{m=1}^{\infty} \left\{ \frac{1}{\nu_{nm}} \int_0^t e^{\gamma t}F_{nm}(\tau) \sin [\nu_{nm}(t - \tau) \, d\tau] \right\} \sin \frac{n\pi x}{a} \sin \frac{m\pi y}{b}$$

where $F_{nm}(t)$ is defined in Prob. 54b.

(d) Derive the *Green's function* of the problem as the series

$$G(x,y,t;\xi,\eta,\tau) = \frac{4}{ab} e^{-\gamma(t-\tau)} \sum_{n=1}^{\infty} \sum_{m=1}^{\infty} \frac{1}{\nu_{nm}} \sin \frac{n\pi x}{a} \sin \frac{n\pi \xi}{a} \sin \frac{m\pi y}{b} \sin \frac{m\pi \eta}{b} \sin [\nu_{nm}(t - \tau)]$$

57 Derive the series expression for the vibrations of a circular membrane of radius a with fastened edge if the external force per unit mass

$$F(r,t) = \varphi(r) \sin \omega t$$

acts on the membrane. Here $\varphi(r)$ is a given continuous function. The membrane is released from rest with zero displacement. Distinguish two cases: **(i)** $\omega \ne \omega_{nm}$, all n, m; **(ii)** $\omega = \omega_{pq}$ for a given pair of integers p, q, $p \ge 0$, $q \ge 1$ (case of resonance). Here the ω_{nm} are the characteristic frequencies derived in Example 4-2. Write the solution for the particular case $\varphi(r) = F_0 = $ const.

58   The circular membrane described in Example 4-2 is released from rest with the initial displacement

$$u(r,\theta,0) = A(a^2 - r^2) \qquad A = \text{const}$$

Instead of being fastened, the edge is elastically constrained.   Thus the boundary condition

$$\frac{\partial u}{\partial r} + \sigma u = 0 \qquad \text{on } r = a$$

replaces the fixed-edge condition.   Here σ is a real positive constant.   Assume the external force of gravity acts on the membrane.   Derive the series expression for the vibrations of the membrane.

59   Assume no external force acts on the surface of the membrane described in Example 4-2.   However, by external means the edge of the membrane is forced to vibrate at a fixed frequency ω, so that the boundary condition

$$u(a,\theta,t) = \varphi(\theta) \sin \omega t \qquad 0 \le \theta \le 2\pi; \, t \ge 0$$

replaces the fixed-edge condition.   Here $\varphi(\theta)$ is a given piecewise smooth function which is periodic, of period 2π.   The motion starts with zero displacement.   Derive the expression for the subsequent vibrations.

60   A thin elastic circular membrane of radius a has its edge fastened.   An external force of magnitude $F_0 \sin \omega t$ (F_0, ω positive constants), confined to a concentric circle of radius $b < a$, is applied to the membrane for $t \ge 0$.   The membrane starts from rest and with zero displacement.   Determine the series expression for the subsequent motion.

61   For the circular membrane described in Example 4-2 assume a frictional force proportional to the speed.   Then the equation of motion is

$$u_{tt} + 2\gamma u_t - c^2 \left(u_{rr} + \frac{u_r}{r} + \frac{u_{\theta\theta}}{r^2} \right) = F(r,\theta,t)$$

where γ is a given positive constant and F is a known external force per unit mass.   Assume the edge of the membrane is fastened.   Give a formal derivation of the solution of the boundary- and initial-value problem.   Derive the series representation of the Green's function of the problem.   Find the motion for the particular case where the membrane is released from rest with zero displacement and

$$F(r,\theta,t) = F_0 \sin \omega t \qquad F_0, \, \omega \text{ real positive constants}$$

62   (a) Let β be a fixed angle, $0 < \beta < 2\pi$, and let $\mathscr{R}$ be the sector whose boundary consists of the ray segments

$$\theta = 0 \qquad 0 \le r \le a \qquad \qquad \theta = \beta \qquad 0 \le r \le a$$

together with the circular arc

$$r = a \qquad 0 \le \theta \le \beta$$

Here r, θ are polar coordinates in the plane.   Assume an elastic membrane occupies $\mathscr{R}$ in its equilibrium configuration and the edge of the membrane is fastened along the boundary C of $\mathscr{R}$.   The membrane is given an initial displacement and speed

$$u(r,\theta,0) = f(r,\theta) \qquad u_t(r,\theta,0) = g(r,\theta)$$

A known exterior force $F(r,\theta,t)$ per unit mass acts on the membrane.   Derive a formal series expression for the subsequent motion.   Also derive the series representation of the Green's function of the problem.

(b) The circular membrane described in Example 4-2 is fastened along the ray segment

$$\theta = 0 \qquad 0 \leq r \leq a$$

as well as on the circle $r = a$.   Use the results of a to obtain a series expression for the subsequent motion of the membrane.

63   Let $\mathscr{R}$ be the region bounded by the concentric circles $r = a$, $r = b$, where $0 < a < b$. An elastic membrane occupies $\mathscr{R}$ in its equilibrium configuration.   The edge of the membrane is fastened along the circles $r = a$, $r = b$.   Give a formal discussion of the problem. Derive the series representation of the Green's function of the problem.   Find the expression for the subsequent motion if the membrane is released from rest with zero displacement and the external force per unit mass $F(r,\theta,t) = F_0 \sin \omega t$ acts on the membrane, where F_0, ω are given positive constants.

64   (a) The kinetic energy of an element dA of the vibrating membrane is $(\rho u_t{}^2 \, dA)/2$, where ρ is the density (mass/unit area).   Hence the total kinetic energy of motion is

$$K = \frac{\rho}{2} \iint\limits_{\mathscr{R}} u_t{}^2 \, dA$$

To obtain an expression for the potential energy consider an element dA when the membrane lies in its equilibrium configuration.   Let dA' be the area of the corresponding element at a subsequent time $t > 0$.   The change in area is

$$dA' - dA = (1 + |\nabla u|^2)^{1/2} \, dA - dA \approx |\nabla u|^2 \frac{dA}{2}$$

This deformation occurs in the presence of an elastic restoring force of magnitude T_0. Thus the potential energy due to tension is

$$T_0 |\nabla u|^2 \frac{dA}{2}$$

The total potential energy is

$$V = \frac{T_0}{2} \iint\limits_{\mathscr{R}} |\nabla u|^2 \, dA$$

The total energy is

$$E = K + V = \frac{1}{2} \iint\limits_{\mathscr{R}} (\rho u_t{}^2 + T_0 |\nabla u|^2) \, dA$$

(b) Show that in a traveling wave

$$u = f(\mathbf{k} \cdot \mathbf{r} - \omega t)$$

where

$$\mathbf{k} = k_1 \mathbf{i} + k_2 \mathbf{j} \qquad \mathbf{r} = x \mathbf{i} + y \mathbf{j}$$

and $\omega = |\mathbf{k}| \, c$, the kinetic energy equals the potential energy.

(c) The nth normal mode of vibration for the membrane with fastened edge is

$$u_n = \varphi_n(x,y)T_n(t) = \varphi_n(x,y)(A_n \cos \omega_n t + B_n \sin \omega_n t)$$
$$= C_n \cos (\omega_n t - \epsilon_n)$$

where

$$C_n = (A_n{}^2 + B_n{}^2)^{\frac{1}{2}} \qquad \tan \epsilon_n = B_n A_n$$

The space factor φ_n satisfies

$$\Delta \varphi_n + \lambda_n \varphi_n = 0 \quad \text{in } \overline{\mathscr{R}} \qquad \varphi_n = 0 \quad \text{on } C$$

The nth characteristic frequency is $\omega_n = c\sqrt{\lambda_n}$. Show that the energy in the nth normal mode is

$$E_n = \frac{\rho \omega_n{}^2 C_n{}^2 \|\varphi_n\|^2}{2}$$

a constant proportional to the square of the characteristic frequency and proportional to the square of the amplitude C_n.

(d) If the edge of membrane is fastened, the initial conditions are

$$u(x,y,0) = f(x,y) \qquad u_t(x,y,0) = g(x,y)$$

and no external forces act on the membrane, then the total energy of motion is the constant

$$E = \frac{\rho}{2} \iint\limits_{\mathscr{R}} \{[g(x,y)]^2 + c^2 |\nabla f|^2\} \, dA$$

Use this result and the result in **c** above to show that the total energy of motion is the sum of the energies of the individual modes

$$E = \sum_{n=1}^{\infty} E_n = \frac{\rho}{2} \sum_{n=1}^{\infty} \omega_n{}^2 C_n{}^2 \|\varphi_n\|^2$$

Hint: For **c**,

$$E_n = \frac{\rho}{2} \iint\limits_{\mathscr{R}} (\dot{T}_n{}^2 \varphi_n{}^2 + c^2 T_n{}^2 |\nabla \varphi_n|^2) \, dA$$

Apply Green's formula derived in Prob. 12a, Chap. 3, and show

$$\iint\limits_{\mathscr{R}} |\nabla \varphi_n|^2 \, dA = \lambda_n \iint \varphi_n{}^2 \, dA$$

so that $E_n = \rho(\dot{T}_n{}^2 + \omega_n{}^2 T_n{}^2) \|\varphi_n\|^2/2$.

65 Prove the following existence theorem for problem (4-79) when $\overline{\mathscr{R}}$ is the rectangle $a \leq x \leq b$, $c \leq y \leq d$. Let the given functions f and g be four times continuously differentiable and such that

$$f = \Delta f = g = \Delta g = 0 \qquad \text{on the boundary } C$$

Then the formal series solution derived in Prob. 52b converges uniformly on $\overline{\mathscr{R}}$ and defines a twice continuously differentiable function u which satisfies the homogeneous wave equation and the boundary and initial conditions. *Hint:* Recall

$$\iint_{\mathscr{R}} \psi \, \Delta\varphi \, dx \, dy = \iint_{\mathscr{R}} \varphi \, \Delta\psi \, dx \, dy$$

holds for every pair of twice continuously differentiable functions on $\overline{\mathscr{R}}$ which vanish on the boundary C. Let φ_{nm} be the eigenfunction derived in Prob. 52a. From Prob. 52b, the fact that $\Delta\varphi_{nm} = -\lambda_{nm}\varphi_{nm}$, and the above, it follows that

$$A_{nm} = \frac{4}{ab} \iint_{\mathscr{R}} f\varphi_{nm} \, dx \, dy = -\frac{4}{ab\lambda_{nm}} \iint_{\mathscr{R}} f \, \Delta\varphi_{nm} \, dx \, dy$$

$$= -\frac{4}{ab\lambda_{nm}} \iint_{\mathscr{R}} \varphi_{nm} \, \Delta f \, dx \, dy = \frac{4}{ab\lambda_{nm}^2} \iint_{\mathscr{R}} \Delta\varphi_{nm} \, \Delta f \, dx \, dy$$

$$= \frac{4}{ab\lambda_{nm}^2} \iint_{\mathscr{R}} \varphi_{nm} \, \Delta\Delta f \, dx \, dy$$

Hence

$$|A_{nm}| \le \frac{4}{ab\lambda_{nm}^2} \iint_{\mathscr{R}} |\varphi_{nm}| \, |\Delta\Delta f| \, dx \, dy \le \frac{M}{\lambda_{nm}^2}$$

where M is a positive constant independent of n and m. Similarly $|B_{nm}| \le M/\omega_{nm}\lambda_{nm}^2 \le M/\lambda_{nm}^2$. Thus

$$|(A_{nm} \cos \omega_{nm}t + B_{nm} \sin \omega_{nm}t)\varphi_{nm}(x,y)| \le |A_{nm}| + |B_{nm}| \le \frac{2M}{\lambda_{nm}^2}$$

for all x, y, t. A dominant convergent series is

$$\sum_{n=1}^{\infty} \sum_{m=1}^{\infty} \frac{2M}{\lambda_{nm}^2} \qquad \lambda_{nm} = \left(\frac{n\pi}{a}\right)^2 + \left(\frac{m\pi}{b}\right)^2$$

Accordingly the series of Prob. 52b converges uniformly on $\overline{\mathscr{R}}$. Now show that the series obtained by differentiating termwise twice with respect to t converges uniformly and the series obtained by differentiating termwise twice with respect to x, and also with respect to y, converges uniformly on $\overline{\mathscr{R}}$. Hence show u satisfies the homogeneous wave equation. Since the series for u converges on $\overline{\mathscr{R}}$ for $t \ge 0$, it is clear that u satisfies the boundary conditions. To show u satisfies the initial conditions one can apply theorems on double Fourier series analogous to those stated in Sec. 2 of Appendix 2. Alternatively the completeness of the set $\{\varphi_{nm}(x,y)\}$ in the space of all functions continuous on $\overline{\mathscr{R}}$ and vanishing on the boundary implies the satisfaction of the initial conditions. Since the series converges uniformly in x and y at $t = 0$,

$$u(x,y,0) = \sum_{n=1}^{\infty} \sum_{m=1}^{\infty} A_{nm}\varphi_{nm}(x,y)$$

Let p, q be a fixed but otherwise arbitrarily chosen pair of positive integers. Multiply both sides of the above equation by φ_{pq}, integrate over $\overline{\mathscr{R}}$, interchange the order of summation and integration, and apply the orthogonality properties of the sequence $\{\varphi_{nm}\}$. The result is

$$A_{pq} = \frac{4}{ab} \iint_{\mathscr{R}} u(x,y,0)\varphi_{pq}(x,y)\, dx\, dy$$

But the coefficients A_{nm} are the Fourier coefficients of f with respect to the φ_{nm}. Hence

$$\iint_{\mathscr{R}} [u(x,y,0) - f(x,y)]\varphi_{nm}(x,y)\, dx\, dy = 0$$

for all n, m. Since f vanishes on the boundary of $\mathscr{R}$, so does the difference $u(x,y,0) - f(x,y)$. Hence the completeness property implies

$$u(x,y,0) = f(x,y) \qquad a \le x \le b;\ c \le y \le d$$

The remaining initial condition is shown in the same way.

66 Prove the following existence theorem for problem (4-80) when $\bar{\mathscr{R}}$ is the rectangle $a \le x \le b, c \le y \le d$. Let the given function F be four times continuously differentiable with respect to the independent variables and such that $F = \Delta F = 0$ on the boundary for $t \ge 0$. Then the formal series solution written in Prob. 54b converges uniformly on $\bar{\mathscr{R}}$ and defines a twice continuously differentiable function u which satisfies the inhomogeneous wave equation and the homogeneous boundary and initial conditions.

Sec. 4-6

67 Let the source function in problem (4-104) be a *point source*, of *strength* $F_0(t)$, located at the fixed point (x_0, y_0, z_0):

$$F(x,y,z,t) = \delta(x - x_0)\delta(y - y_0)\delta(z - z_0)F_0(t)$$

Then

$$F(x,y,z,t) = 0 \qquad (x,y,z) \neq (x_0,y_0,z_0)$$

and

$$\iiint_V F(x,y,z,t)\, dx\, dy\, dz = F_0(t)$$

whenever V includes the point (x_0,y_0,z_0). Use Eq. (4-107) to show that the resulting field at a point $P(x,y,z)$ distinct from (x_0,y_0,z_0) is

$$u(x,y,z,t) = \begin{cases} 0 & 0 \le ct < r \\[2mm] \dfrac{F_0(t - r/c)}{4\pi c^2 r} & ct \ge r \end{cases}$$

where $r^2 = (x - x_0)^2 + (y - y_0)^2 + (z - z_0)^2$. Observe that u defines a spherically symmetric wave. Accordingly, the field due to a time-harmonic point source

$$F(x,y,z,t) = F_0 \delta(x - x_0)\delta(y - y_0)\delta(z - z_0)e^{-i\omega t}$$

which vanishes, together with its first derivative with respect to t, at $t = 0$ is given by

$$u(x,y,z,t) = \begin{cases} 0 & 0 \le ct < r \\[2mm] \dfrac{F_0 e^{-i\omega(t-r/c)}}{4\pi c^2 r} & ct \ge r \end{cases}$$

Show that the field due to an instantaneous point source located at (x_0,y_0,z_0) at time t_0, of strength F_0, is given by

$$u(x,y,z,t) = \begin{cases} 0 & 0 \le ct < r \\ \dfrac{F_0\delta[(t-t_0)-r/c]}{4\pi c^2 r} & ct \ge r \end{cases}$$

68 Let the source function in the two-dimensional problem (4-58) be a point source of strength $F_0(t)$ located at (x_0,y_0):

$$F(x,y,t) = \delta(x-x_0)\delta(y-y_0)F_0(t)$$

Apply Eq. (4-59) and show that the resulting field at a point $P(x,y)$ distinct from (x_0,y_0) is given by

$$u(x,y,t) = \begin{cases} 0 & 0 \le ct < r \\ \dfrac{1}{2\pi c}\displaystyle\int_0^{t-r/c} \dfrac{F_0(\tau)}{\sqrt{c^2(t-\tau)^2 - r^2}}\,d\tau & ct \ge r \end{cases}$$

where $r^2 = (x-x_0)^2 + (y-y_0)^2$. Show that the field due to an instantaneous point source located at (x_0,y_0) at time t_0, of strength $F_0 = $ const, is given by

$$u(x,y,t) = \begin{cases} 0 & 0 \le c(t-t_0) < r \\ \dfrac{F_0}{2\pi c\sqrt{c^2(t-t_0)^2 - r^2}} & c(t-t_0) \ge r \end{cases}$$

69 (a) Let $r^2 = x^2 + y^2 + z^2$. Find a solution of the homogeneous wave equation for $r > 0$ such that

$$u|_{t=0} = \frac{A}{r} \qquad u_t|_{t=0} = B \qquad A, B \text{ real positive constants}$$

(b) Find a solution of the homogeneous wave equation for $r > 0$ such that

$$u|_{t=0} = u_0 \qquad u_t|_{t=0} = \frac{u_1}{r} \qquad u_0, u_1 \text{ real positive constants}$$

70 In spherical coordinates the homogeneous wave equation is

$$\frac{\partial^2 u}{\partial t^2} = \frac{c^2}{r^2}\left[\frac{\partial}{\partial r}\left(r^2\frac{\partial u}{\partial r}\right) + \frac{1}{\sin\theta}\frac{\partial}{\partial\theta}\left(\sin\theta\frac{\partial u}{\partial\theta}\right) + \frac{1}{\sin^2\theta}\frac{\partial^2 u}{\partial\varphi^2}\right]$$

As discussed in the text, a function of the form

$$u = \psi(r - ct)$$

represents a spherical wave propagated radially outward from the origin with speed c. Show that no nontrivial function of this form can be a solution of the homogeneous wave equation. On the other hand, a function

$$u = \eta(r,t)f(r - ct)$$

represents a spherical wave traveling outward from the origin but with amplitude $\eta(r,t)$ a function of position and time. Assume a solution of this form with f arbitrary (twice differentiable).   Show that necessarily

$$\eta(r,t) = \frac{\psi(r - ct)}{r}$$

for some function ψ.   Thus for each choice of ψ the pair of functions

$$\eta = \frac{\psi(r - ct)}{r} \qquad \zeta = r - ct$$

constitutes a functionally invariant pair of the wave equation in spherical coordinates. *Hint:* Show that η must satisfy

$$\frac{\partial^2 \eta}{\partial t^2} = \frac{c^2}{r^2} \frac{\partial}{\partial r}\left(r^2 \frac{\partial \eta}{\partial r}\right) \qquad c \frac{\partial}{\partial r}(r\eta) + \frac{\partial \eta}{\partial t} = 0$$

Let $v = r$.

71   Derive the solution (4-59) of the two-dimensional initial-value problem (4-58) from Eq. (4-107).

Sec. 4-7

72   **(a)** Consider problem (4-114) when $\mathscr{V}$ is the rectangular parallelepiped defined by the inequalities

$$0 \leq x \leq a \qquad 0 \leq y \leq b \qquad 0 \leq z \leq h$$

Let the boundary condition be the Dirichlet condition

$$u = 0 \qquad \text{on } S; t \geq 0$$

where S is the surface which bounds $\mathscr{V}$.   Use separation of variables to derive the ∞^3 eigenfunctions

$$\psi_{nmq}(x,y,z) = \sin \frac{n\pi x}{a} \sin \frac{m\pi y}{b} \sin \frac{q\pi z}{h} \qquad n = 1, 2, \ldots\,; m = 1, 2, \ldots\,; q = 1, 2, \ldots$$

corresponding to the eigenvalues

$$\lambda_{nmq} = \frac{n^2 \pi^2}{a^2} + \frac{m^2 \pi^2}{b^2} + \frac{q^2 \pi^2}{h^2}$$

Show directly that the eigenfunctions have the properties

$$\int_0^a \int_0^b \int_0^h \psi_{nmq}\psi_{n'm'q'}\, dx\, dy\, dz = 0 \qquad (n,m,q) \neq (n',m',q')$$

$$\|\psi_{nmq}\|^2 = \frac{abh}{8}$$

Outline the proof of the fact that (apart from constant factors) the set $\{\psi_{nmq}\}$ constitutes all the eigenfunctions of problem (4-117) in the present case.
(b) Obtain the normal modes of vibration

$$u_{nmq}(x,y,z,t) = \psi_{nmq}(x,y,z)(A_{nmq} \cos \omega_{nmq} t + B_{nmq} \sin \omega_{nmq} t)$$

where the characteristic frequencies

$$\omega_{nmq} = c\pi \left[\left(\frac{n}{a}\right)^2 + \left(\frac{m}{b}\right)^2 + \left(\frac{q}{h}\right)^2 \right]^{1/2}$$

(c) Derive the solution of problem (4-114) when F is identically zero in the form

$$u = \sum_{n=1}^{\infty} \sum_{m=1}^{\infty} \sum_{q=1}^{\infty} (A_{nmq} \cos \omega_{nmq} t + B_{nmq} \sin \omega_{nmq} t) \psi_{nmq}(x,y,z)$$

$$A_{nmq} = \frac{8}{abh} \int_0^a \int_0^b \int_0^h f(x,y,z)\psi_{nmq}(x,y,z) \, dx \, dy \, dz$$

$$B_{nmq} = \frac{8}{abh\omega_{nmq}} \int_0^a \int_0^b \int_0^h g(x,y,z)\psi_{nmq}(x,y,z) \, dx \, dy \, dz$$

(d) Derive the solution of problem (4-114) when f and g are identically zero in the form

$$u = \sum_{n=1}^{\infty} \sum_{m=1}^{\infty} \sum_{q=1}^{\infty} \left\{ \frac{1}{\omega_{nmq}} \int_0^t F_{nmq}(\tau) \sin [\omega_{nmq}(t-\tau)] \, d\tau \right\} \psi_{nmq}(x,y,z)$$

where

$$F_{nmq}(t) = \frac{8}{abh} \int_0^a \int_0^b \int_0^h F(x,y,z,t)\psi_{nmq}(x,y,z) \, dx \, dy \, dz$$

(e) Derive the *Green's function* of the problem

$$G(x,y,z,t;\xi,\eta,\zeta,\tau) = \frac{8}{abh} \sum_{n=1}^{\infty} \sum_{m=1}^{\infty} \sum_{q=1}^{\infty} \frac{\psi_{nmq}(x,y,z)\psi_{nmq}(\xi,\eta,\zeta)}{\omega_{nmq}} \sin [\omega_{nmq}(t-\tau)]$$

for $t > \tau$.   Define G to be identically zero for $t \le \tau$.

(f) Derive the solution of the problem

$$u_{tt} - c^2 \Delta u = A \sin \omega t \qquad \text{in } \bar{\mathscr{V}}; \, t \ge 0$$

$$u = B \sin vt \qquad \text{on } S; \, t \ge 0$$

$$u(x,y,z,0) = 0 \qquad u_t(x,y,z,0) = Bv \qquad \text{in } \bar{\mathscr{V}}$$

where A, B, ω, and v are given real positive constants and $\omega \ne \omega_{nmq}$, $v \ne \omega_{nmq}$, all n, m, q.

73   Let $\bar{\mathscr{V}}$ be the rectangular parallelepiped of Prob. 72.   Derive the formal series solution of the problem

$$u_{tt} + 2\gamma u_t + \beta u - c^2 \Delta u = F_0 \sin \omega t$$

$$\frac{\partial u}{\partial n} = 0 \qquad \text{on } S; \, t \ge 0$$

$$u(x,y,z,0) = \cos \frac{\pi x}{a} \cos \frac{\pi y}{b} \cos \frac{\pi z}{h} \qquad u_t(x,y,z,0) = 0$$

where F_0, ω, and γ are given real positive constants and β is a real nonnegative constant. Assume also

$$\gamma^2 < \omega^2 + \pi^2 c^2 \left(\frac{1}{a^2} + \frac{1}{b^2} + \frac{1}{h^2} \right)$$

74 In the study of small amplitude vibrations of an ideal gas confined to the interior of a rigid spherical surface of radius a about the origin it is shown that the *velocity potential u* (a function such that the velocity vector $\mathbf{v} = -\nabla u$) satisfies

$$u_{tt} = c^2 \, \Delta u \qquad \text{in } \overline{\mathscr{V}}; \; t \geq 0$$

$$\frac{\partial u}{\partial r} = 0 \qquad r = a \qquad t \geq 0$$

$$u(r,0) = f(r) \qquad u_t(r,0) = g(r) \qquad \text{in } \overline{\mathscr{V}}$$

where (r,θ,φ) are spherical coordinates and $\overline{\mathscr{V}}$ is the sphere of radius a about the origin. Derive the formal series solution

$$u(r,t) = A_0 + B_0 t + \frac{1}{r} \sum_{n=1}^{\infty} (A_n \cos \omega_n t + B_n \sin \omega_n t) \sin \frac{\mu_n r}{a}$$

where $\{\mu_n\}$ is the sequence of real positive roots of the transcendental equation

$$\tan \mu = \mu$$

and

$$\omega_n = \frac{c\mu_n}{a} \qquad n = 1, 2, \ldots$$

are the characteristic frequencies of vibration. The coefficients in the series are determined by

$$A_0 = \frac{3}{a^3} \int_0^a r^2 f(r) \, dr \qquad B_0 = \frac{3}{a^3} \int_0^a r^2 g(r) \, dr$$

$$A_n = \frac{2(1 + 1/\mu_n{}^2)}{a} \int_0^a r f(r) \sin \frac{\mu_n r}{a} \, dr \qquad n \geq 1$$

$$B_n = \frac{2(1 + 1/\mu_n{}^2)}{a\mu_n} \int_0^a r g(r) \sin \frac{\mu_n r}{a} \, dr \qquad n \geq 1$$

75 Let r, θ, φ be spherical coordinates and let $\overline{\mathscr{V}}$ denote the hemisphere defined by the inequalities

$$0 \leq r \leq a \qquad 0 \leq \theta \leq \frac{\pi}{2} \qquad 0 \leq \varphi \leq 2\pi$$

Derive the *Green's function* of the problem

$$u_{tt} - c^2 \, \Delta u = F(r,\theta,t) \qquad \text{in } \overline{\mathscr{V}}; \; t \geq 0$$

$$\frac{\partial u}{\partial n} = 0 \qquad \text{on } S; \; t \geq 0$$

$$u(r,\theta,0) = f(r,\theta) \qquad u_t(r,\theta,0) = g(r,\theta) \qquad \text{in } \mathscr{V}$$

where S denotes the surface which bounds $\overline{\mathscr{V}}$ and $\partial u / \partial n$ is the derivative of u in the direction of the exterior normal on S. Obtain the solution of the boundary- and initial-value problem in the particular case

$$F(r,\theta,t) = F_0 \sin \omega t \qquad f(r,\theta) = h(r) \cos \theta \qquad g = 0$$

F_0, ω positive constants.

76 With reference to Prob. 74, assume the gas is confined to the region bounded by concentric spheres of inner radius a and outer radius b. Then the boundary- and initial-value problem for the velocity potential u is

$$u_{tt} = c^2 \Delta u \qquad \text{in } \mathscr{V}; t \geq 0$$

$$\frac{\partial u}{\partial r} = 0 \qquad r = a; r = b; t \geq 0$$

$$u(r,0) = f(r) \qquad u_t(r,0) = g(r) \qquad \text{in } \mathscr{V}$$

where $\mathscr{V}$ is the volume defined by the inequalities

$$0 < a \leq r \leq b \qquad 0 \leq \theta \leq \pi \qquad 0 \leq \varphi \leq 2\pi$$

Show that the eigenvalues for the present case are

$$\lambda_0 = 0 \qquad \lambda_n = \xi_n{}^2 \qquad n = 1, 2, \ldots$$

where $\{\xi_n\}$ is the sequence of positive roots of the transcendental equation

$$\tan [\xi(b - a)] = \frac{\xi(b - a)}{1 + ab\xi^2}$$

Obtain the corresponding eigenfunctions

$$\psi_0 = 1 \qquad \psi_n = \frac{\sin \xi_n r + \alpha_n \cos \xi_n r}{r} \qquad n = 1, 2, \ldots$$

where

$$\alpha_n = \frac{a\xi_n \cos \xi_n a - \sin \xi_n a}{a\xi_n \sin \xi_n a + \cos \xi_n a}$$

Verify that the eigenfunctions are orthogonal on $[a,b]$ with weight function r^2:

$$\int_0^b r^2 \psi_n(r)\psi_m(r) \, dr = 0 \qquad n \neq m$$

The characteristic frequencies of vibration are $\omega_n = c\xi_n$, $n = 1, 2, \ldots$. Derive the formal series solution of the boundary- and initial-value problem in the form

$$u = A_0 + B_0 t + \sum_{n=1}^{\infty} (A_n \cos \omega_n t + B_n \sin \omega_n t)\psi_n(r)$$

where the coefficients are defined by

$$A_0 = \frac{3}{b^3 - a^3} \int_a^b r^2 f(r) \, dr \qquad B_0 = \frac{3}{b^3 - a^3} \int_a^b r^2 g(r) \, dr$$

$$A_n = \frac{1}{\|\psi_n\|^2} \int_a^b r^2 f(r)\psi_n(r) \, dr \qquad B_n = \frac{1}{\omega_n \|\psi_n\|^2} \int_a^b r^2 g(r)\psi_n(r) \, dr \qquad n \geq 1$$

77 Let $\mathscr{V}$ be the cone defined by the inequalities

$$0 \leq r \leq a \qquad 0 \leq \theta \leq \beta \qquad 0 \leq \varphi \leq 2\pi$$

where r, θ, φ are spherical coordinates and β is a fixed angle such that $0 < \beta < \pi/2$. Consider the boundary- and initial-value problem

$$u_{tt} - c^2 \Delta u = F(r,t) \qquad \text{in } \mathscr{V}; \, t \geq 0$$

$$\frac{\partial u}{\partial n} = 0 \qquad \text{on } S; \, t \geq 0$$

$$u(r,\theta,\varphi,0) = f(r) \qquad u_t(r,\theta,\varphi,0) = g(r) \qquad \text{in } \mathscr{V}$$

Derive the form of the eigenfunctions and eigenvalues for this case.   Derive the formal series solution of the boundary- and initial-value problem.

78   Let $\mathscr{V}$ be the cylinder of radius a and altitude h defined by the inequalities

$$0 \leq r \leq a \qquad 0 \leq \theta \leq 2\pi \qquad 0 \leq z \leq h$$

where r, θ, z are cylindrical coordinates.   Consider the boundary- and initial-value problem

$$u_{tt} - c^2 \Delta u = F(r,\theta,z,t) \qquad \text{in } \mathscr{V}; \, t \geq 0$$

$$u = 0 \qquad \text{on } S; \, t \geq 0$$

$$u(r,\theta,z,0) = f(r,\theta,z) \qquad u_t(r,\theta,z,0) = g(r,\theta,z) \qquad \text{in } \mathscr{V}$$

(a)   Derive the eigenfunctions

$$\psi^{(e)}_{nmq} = J_n\left(\frac{\xi_{nm}r}{a}\right) \cos n\theta \sin \left(\frac{q\pi z}{h}\right) \qquad \psi^{(0)}_{nmq} = J_n\left(\frac{\xi_{nm}r}{a}\right) \sin n\theta \sin \frac{q\pi z}{h}$$

corresponding to the ∞^3 eigenvalues

$$\lambda_{nmq} = \frac{\xi_{nm}^2}{a^2} + \frac{q^2\pi^2}{h^2} \qquad n = 0, 1, \ldots; m = 1, 2, \ldots; q = 1, 2, \ldots$$

where ξ_{nm} denotes the mth positive zero of the Bessel function $J_n(\xi)$.   Outline the proof of the fact that these are all the eigenvalues of the problem.   Show that the eigenfunctions have the orthogonality properties

$$\iiint_{\mathscr{V}} \psi^{(e)}_{nmq}\psi^{(e)}_{n'm'q'} \, dV = 0 \qquad \iiint_{\mathscr{V}} \psi^{(0)}_{nmq}\psi^{(0)}_{nmq} \, dV = 0 \qquad (n,m,q) \neq (n',m',q')$$

$$\iiint_{\mathscr{V}} \psi^{(e)}_{nmq}\psi^{(0)}_{n'm'q'} \, dV = 0 \qquad \text{all } (n,m,q), (n',m',q')$$

where $dV = r \, dr \, d\theta \, dz$.   Show also that

$$\|\psi^{(e)}_{nmq}\|^2 = \|\psi^{(0)}_{nmq}\|^2 = \frac{a^2 h [J_n'(\xi_{nm})]^2}{4} \qquad n \geq 1; m \geq 1; q \geq 1$$

$$\|\psi^{(e)}_{0mq}\|^2 = \frac{a^2 h [J_0'(\xi_{0m})]^2}{2} \qquad m \geq 1; q \geq 1$$

(b)   Obtain the ∞^3 normal modes of vibration

$$u^{(e)}_{nmq}(r,\theta,z,t) = \psi^{(e)}_{nmq}(r,\theta,z)(A^{(e)}_{nmq} \cos \omega_{nmq}t + B^{(e)}_{nmq} \sin \omega_{nmq}t)$$

$$u^{(0)}_{nmq}(r,\theta,z,t) = \psi^{(0)}_{nmq}(r,\theta,z)(A^{(0)}_{nmq} \cos \omega_{nmq}t + B^{(0)}_{nmq} \sin \omega_{nmq}t)$$

where the characteristic frequencies are

$$\omega_{nmq} = c\sqrt{\lambda_{nmq}} \qquad n = 0, 1, \ldots ; m = 1, 2, \ldots ; q = 1, 2, \ldots$$

(c) Derive the *Green's function*

$$G(r,\theta,z,t;r_0,\theta_0,z_0,\tau) = \sum_{m=0}^{\infty} \sum_{n=1}^{\infty} \sum_{q=1}^{\infty}$$

$$\frac{J_n(\xi_{nm}r/a)J_n(\xi_{nm}r_0/a)}{\omega_{nmq} \|\psi_{nmq}\|^2} \cos\left[n(\theta - \theta_0)\right] \sin\frac{q\pi z}{h} \sin\frac{q\pi z_0}{h} \sin\left[\omega_{nmq}(t - \tau)\right]$$

(d) Derive the series solution of the boundary- and initial-value problem if $g = 0$ and

$$F(r,\theta,z,t) = F_0 \sin \omega t \qquad f(r,\theta,z) = (a - r)\sin\frac{\pi z}{h}$$

where F_0, ω are given positive constants.

79 Let $\mathcal{V}$ be the cylinder in Prob. 78. Solve the problem

$$u_{tt} = c^2 \Delta u \qquad \text{in } \mathcal{V}; t \geq 0$$

$$\frac{\partial u}{\partial r} = 0 \qquad r = a \qquad\qquad \frac{\partial u}{\partial z} = 0 \qquad z = 0$$

$$\frac{\partial u}{\partial z} = \varphi(r)\cos p\theta \cos \omega t \qquad z = h$$

$$\rho u(r,\theta,z,0) = 0 \qquad u_t(r,\theta,z,0) = 0$$

where φ is a given function, p a given positive integer, and ω a given positive constant.

80 Let S be a simple closed piecewise smooth surface which bounds a region $\mathcal{V}$ of xyz space. Let $\overline{\mathcal{V}}$ denote $\mathcal{V}$ together with S. Assume the divergence theorem is applicable to $\overline{\mathcal{V}}$. Let p_1, p_2, p_3 be given functions, defined and positive-valued on $\overline{\mathcal{V}}$. Assume also these are continuously differentiable on $\overline{\mathcal{V}}$. Define the elliptic operator L by

$$L\varphi = (p_1\varphi_x)_x + (p_2\varphi_y)_y + (p_3\varphi_z)_z$$

for every twice continuously differentiable function φ on $\overline{\mathcal{V}}$. In the particular case where $p_i = 1$, $i = 1, 2, 3$, the operator reduces to the laplacian. Consider the boundary- and initial-value problem.

$$\rho u_{tt} + 2\gamma u_t + \beta u - Lu = F(x,y,z,t) \qquad \text{in } \overline{\mathcal{V}}; t \geq 0$$

$$B(u) = 0 \qquad \text{on } S; t \geq 0$$

$$u(x,y,z,0) = f(x,y,z) \qquad u_t(x,y,z,0) = g(x,y,z) \qquad \text{in } \overline{\mathcal{V}}$$

where ρ is a given positive valued continuously differentiable function on $\overline{\mathcal{V}}$, γ and β are given real nonnegative constants, and F, f, g are given functions. Here $B(u) = 0$ symbolizes one of the three types of boundary conditions

$$u = 0 \qquad \frac{\partial u}{\partial n} = 0 \qquad \frac{\partial u}{\partial n} + \sigma u = 0 \qquad \sigma > 0$$

(a) If $v(x,y,z,t)$ is a twice continuously differentiable function on $\overline{\mathscr{V}}$ and $t \geq 0$, define the vector

$$\mathbf{P} = p_1 v_x \mathbf{i} + p_2 v_y \mathbf{j} + p_3 v_z \mathbf{k}$$

and the *energy integral*

$$E(t) = \frac{1}{2} \iiint_{\mathscr{V}} (\rho v_t{}^2 + \beta v^2 + |\mathbf{P}|^2) \, d\tau$$

where $d\tau$ is the volume element in $\mathscr{V}$. Show that

$$\dot{E}(t) = \iiint_{\mathscr{V}} (\rho v_{tt} + \beta v) v_t \, d\tau + \iiint_{\mathscr{V}} \mathbf{P} \cdot \nabla v_t \, d\tau$$

Now make use of the vector identity

$$\nabla \cdot (w\mathbf{A}) = \nabla w \cdot \mathbf{A} + w \nabla \cdot \mathbf{A}$$

and show that

$$\dot{E}(t) = \iiint_{\mathscr{V}} (\rho v_{tt} + \beta v - Lv) v_t \, d\tau + \iint_{S} v_t \mathbf{P} \cdot \mathbf{n} \, dS$$

In particular if u is a solution of the partial differential equation in the boundary- and initial-value problem, then the energy rate of change is

$$\dot{E}(t) = \iiint_{\mathscr{V}} u_t F(x,y,z,t) \, d\tau - 2\gamma \iiint_{\mathscr{V}} u_t{}^2 \, d\tau + \iint_{S} u_t \mathbf{P} \cdot \mathbf{n} \, dS$$

(b) Prove the following uniqueness theorem for the boundary- and initial-value problem. If the boundary condition is the Dirichlet condition $u = 0$ on S, there can be at most one twice continuously differentiable solution. If $p_1 = p_2 = p_3$ on $\overline{\mathscr{V}}$ and $\beta > 0$, the same is true for the Neumann boundary condition $\partial u/\partial n = 0$. If $p_1 = p_2 = p_3$ on $\overline{\mathscr{V}}$ and $\beta = 0$, a solution is unique to within an additive constant. If $p_1 = p_2 = p_3$ on $\overline{\mathscr{V}}$ and the boundary condition is the mixed condition $\partial u/\partial n + \sigma u = 0$ on S, there is at most one twice continuously differentiable solution. *Hint:* Suppose u_1, u_2 are twice continuously differentiable solutions of the problem. Let $u = u_1 - u_2$. By the linearity of things u satisfies the boundary- and initial-value problem with F, f, and g replaced by zero. Hence the energy rate is

$$\dot{E}(t) = -2\gamma \iiint_{\mathscr{V}} u_t{}^2 \, d\tau + \iint_{S} u_t \mathbf{P} \cdot \mathbf{n} \, dS$$

If the boundary condition is $u = 0$ on S, $t \geq 0$, then $u_t = 0$ on S, $t \geq 0$. Hence

$$\dot{E}(t) \leq 0 \qquad t \geq 0$$

This implies $E(t) \leq E(0)$, $t \geq 0$. But $E(0) = 0$. Hence $E(t) \leq 0$, $t \geq 0$. But an examination of the energy integral shows $E(t) \geq 0$, $t \geq 0$. Hence $E(t) = 0$, $t \geq 0$. Show this implies $u(x,y,z,t) = 0$, (x,y,z) in $\mathscr{V}$, $t \geq 0$. Suppose now $p_1 = p_2 = p_3$ in $\overline{\mathscr{V}}$. Then

$$u_t \mathbf{P} \cdot \mathbf{n} = u_t p \frac{\partial u}{\partial n} \qquad p = p_1 = p_2 = p_3$$

If the boundary condition is $\partial u / \partial n = 0$ on S, then again $E(t) = 0$, $t \geq 0$. Show that if $\beta > 0$, this implies $u = 0$ in $\overline{\mathcal{V}}$, $t \geq 0$. On the other hand, if $\beta = 0$, then $u = \text{const}$ is possible. If $p_1 = p_2 = p_3 = p$ and the boundary condition is the mixed condition, then

$$\dot{E}(t) = -2\gamma \iiint\limits_{\mathcal{V}} u_t{}^2 \, d\tau - \sigma \iint\limits_{S} uu_t \, dS$$

so that

$$E(t) - E(0) = -2\gamma \int_0^t dt \iiint u_t{}^2 \, d\tau - \frac{\sigma}{2} \iint\limits_{S} u^2 \, dS \leq 0$$

Again $E(t) = 0$, $t \geq 0$. Show this implies that $u = 0$ on for $t \geq 0$.

5

THE HEAT EQUATION

5-1 INTRODUCTION

An important equation of parabolic type which describes various diffusion processes, including the flow of heat in a thermal conductor, is

$$\frac{\partial u}{\partial t} - \kappa \, \Delta u = F(x,y,z,t) \tag{5-1}$$

called the *heat* (or diffusion) *equation.* Here κ is a positive constant and F is a given function which represents a source term. Equation (5-1) is called *one-*, *two-*, or *three-dimensional* according as the laplacian is written in one, two, or three dimensions. Observe that Eq. (5-1) differs from the wave equation only in that the second derivative u_{tt} is replaced by the first derivative u_t. However this alteration results in very appreciable changes in the properties of solutions, and hence in their physical interpretation. Solutions of the heat equation do not describe wave phenomena, at least in the usual sense. Disturbances in a field described by a solution of Eq. (5-1) are observed immediately at every point in the domain of definition, even though the disturbance may be initially restricted to a small neighborhood of a given point.

In order to illustrate the manner in which Eq. (5-1) occurs in heat-conduction problems assume S is a simple closed piecewise smooth surface which forms the boundary of a thermally conducting solid. Let $\mathscr{V}$ denote the region interior to S, and let $\overline{\mathscr{V}}$ denote $\mathscr{V}$ together with S. The temperature at a point (x,y,z) at time t (measured on some scale from a fixed reference level) is written $u(x,y,z,t)$

An important quantity in the mathematical theory of heat flow is *flux of heat q.* Let $\Delta S'$ be a surface element with given normal $\mathbf{n}$. Then the flux of heat across $\Delta S'$ is defined as the amount of heat per unit area per unit time flowing across the surface element. A basic law of heat conduction states that the flux across $\Delta S'$ is proportional to the component of the temperature gradient in the direction of the normal:

$$q = -K \, \Delta u \cdot \mathbf{n} = -K \frac{\partial u}{\partial n}$$

The coefficient K is called the *thermal conductivity*; it is a material property, and in general is a function of position and temperature in the conductor. However here K is assumed independent of u. Observe that the equation states that the flux of heat is positive when the temperature decreases in the direction of **n**; that is, heat flows in the direction of decreasing temperature. Assume the temperature u is a twice continuously differentiable function of position and time, for points in $\mathscr{V}$ and for $t > 0$. Let ΔV be an arbitrary sphere in $\mathscr{V}$, with surface ΔS. Then the net flux of heat through ΔS is

$$-\iint_{\Delta S} K \nabla u \cdot \mathbf{n}\, dS = -\iiint_{\Delta V} \nabla \cdot (K \nabla u)\, dV$$

where dV denotes the volume element of integration. The preceding relation follows by the divergence theorem.

The *specific heat* of a substance is the amount of heat required to raise the temperature of a unit mass one unit of temperature. Let c denote the specific heat of the conductor which occupies $\bar{\mathscr{V}}$. If dV is a volume element in $\bar{\mathscr{V}}$ and ρ is the material density, the amount of heat dQ required to increase the temperature of the mass element $dm = \rho\, dV$ to a value u (above the reference level) is

$$dQ = cu\rho\, dV$$

Thus the total heat content of the material in the sphere ΔV when at temperature u is

$$Q = \iiint_{\Delta V} c\rho u\, dV$$

The rate of change of this heat content is

$$\frac{dQ}{dt} = \iiint_{\Delta V} c\rho\, \frac{\partial u}{dt}\, dV$$

provided the boundary of the sphere does not change with time.

It is possible that there are points of ΔV at which heat is being generated or absorbed. Such points are often referred to as source points or sink points. For example, a chemical reaction may take place which liberates heat. Let $F(x,y,z,t)$ denote the amount of heat per unit volume per unit time generated at the point (x,y,z) at time t. The function F is referred to as the *source density*. The time rate of change of the heat content due to the sources is

$$\iiint_{\Delta V} F(x,y,z,t)\, dV$$

The law of conservation of heat is applied to the sphere ΔV. This states that the rate at which the heat content of ΔV changes is equal to the rate at which heat is generated inside ΔV diminished by the flux of heat out through the boundary surface ΔS:

$$\iiint\limits_{\Delta V} c\rho \, \frac{\partial u}{\partial t} \, dV = \iiint\limits_{\Delta V} F(x,y,z,t) \, dV + \iiint\limits_{\Delta V} \nabla \cdot (K \nabla u) \, dV$$

Thus

$$\iiint\limits_{\Delta V} \left[c\rho \, \frac{\partial u}{\partial t} - \nabla \cdot (K \nabla u) - F(x,y,z,t) \right] dV = 0$$

This relation must be true for every sphere ΔV in $\mathscr{V}$.   Hence the temperature must satisfy the parabolic equation

$$c\rho \, \frac{\partial u}{\partial t} - \nabla \cdot (K \nabla u) = F(x,y,z,t) \tag{5-2}$$

in $\mathscr{V}$ for $t > 0$.   This is the generalized heat equation.   If the material is such that c, ρ, and K are constants, let

$$\kappa = \frac{K}{c\rho}$$

Then Eq. (5-2) reduces to Eq. (5-1), provided F is replaced by $F/c\rho$.

A *stationary* (or steady-state) *temperature* distribution is a temperature field which is independent of time.   Such a state may be realized if the sources are time-independent and a sufficiently long time has elapsed from the start of heat flow.   Then $\partial u/\partial t = 0$, and Eq. (5-1) reduces to Poisson's equation (3-8).   If there are no sources within the conductor and the medium is homogeneous, the steady-state temperature satisfies Laplace's equation in $\mathscr{V}$.

A generalization of Eq. (5-2) is

$$c\rho \, \frac{\partial u}{\partial t} - \nabla \cdot (K \nabla u) = F(x,y,z,t,u) \tag{5-3}$$

where c, ρ, and K may be functions of x, y, and z.   Equation (5-3) describes temperature fields in which the source density is temperature-dependent. In general Eq. (5-3) is nonlinear.   However if F is a linear function of the temperature, the partial differential equation is linear.   Suppose

$$F(x,y,z,t,u) = -bu + F_1(x,y,z,t)$$

where b is, in general, a function of x, y, and z.   Then the partial differential equation is

$$c\rho \, \frac{\partial u}{\partial t} + bu - \nabla \cdot (K \nabla u) = F_1(x,y,z,t) \tag{5-4}$$

The information available in a heat-conduction problem usually includes the temperature distribution in the conductor at some instant, say $t = 0$. Then there is given an initial condition

$$u(x,y,z,0) = f(x,y,z) \quad \text{in } \overline{\mathcal{V}} \tag{5-5}$$

which must be satisfied. Note that no attempt is made to prescribe u_t at $t = 0$ in addition to u at $t = 0$. Indeed, if f is twice continuously differentiable, if u is a twice continuously differentiable solution of the heat equation for $t \geq 0$, and if u satisfies (5-5), then

$$u_t(x,y,z,0) = \kappa \, \Delta f + F(x,y,z,0)$$

Thus u_t at $t = 0$ is determined and cannot be prescribed in an arbitrary manner.

Boundary conditions in heat-conduction problems may be of the Dirichlet, Neumann, or mixed type. For example, suppose the temperature on the surface S is a prescribed function of position and (possibly) time. Then the boundary condition has the form

$$u = H(x,y,z,t) \quad \text{on } S; \, t \geq 0 \tag{5-6}$$

which is of the Dirichlet type. A special case occurs when the surface is held at a constant temperature:

$$u = u_0 \quad \text{on } S; \, t \geq 0 \tag{5-7}$$

where u_0 is a constant. A Neumann type of boundary condition is obtained if the flux across S is known:

$$\frac{\partial u}{\partial n} = H(x,y,z,t) \quad \text{on } S; \, t \geq 0 \tag{5-8}$$

Here $\partial u/\partial n$ denotes the derivative of u in the direction of the exterior normal on S. If the surface is insulated, no flux across S is possible. The boundary condition is

$$\frac{\partial u}{\partial n} = 0 \quad \text{on } S; \, t \geq 0 \tag{5-9}$$

A mixed type of boundary condition occurs in the problem of cooling. Assume that the unbounded region exterior to S constitutes a heat sink of infinite capacity, so that the flux of heat out through S does not change the external temperature. Now Newton's law of cooling states that the flux of heat across S is proportional to the difference between the surface temperature and the external temperature. Thus the boundary condition is

$$\frac{\partial u}{\partial n} + h(u - u_0) = 0 \quad \text{on } S; \, t \geq 0 \tag{5-10}$$

where h is a known positive constant and u_0 denotes the temperature of the exterior region. Finally it should be noted that several different boundary conditions may be present on different portions of the surface S.

5-2   INITIAL-VALUE PROBLEMS

In some mathematical models of heat conduction an unbounded region of space is assumed, and there are no boundary conditions. An example is furnished by the model which describes the flow of heat in a long slender homogeneous bar of uniform cross section. The lateral surface of the bar is insulated, and the length is sufficiently great so that the end effect can be neglected. If the bar is initially heated, say over a relatively short interval near the middle, the subsequent heat flow is along the length of the bar, i.e., a one-dimensional heat flow. Choose the x axis as the axis of the bar. Then the temperature $u(x,t)$ satisfies

$$u_t - \kappa u_{xx} = F(x,t) \qquad -\infty < x < \infty; t > 0$$
$$u(x,0) = f(x) \qquad -\infty < x < \infty$$

(5-11)

Here $F(x,t)$ describes the distribution of heat sources contained within the bar, and $f(x)$ describes the initial temperature. Problem (5-11) is an initial-value problem for the one-dimensional heat equation. The data are prescribed along the line $t = 0$ in the xt plane, and the solution is to be determined in the upper half plane $t \geq 0$. It is understood that the initial data are taken on continuously; i.e., if u is a solution, then

$$\lim_{\substack{t \to 0 \\ t > 0}} u(x,t) = f(x) \tag{5-12}$$

Fundamental Solution of the One-dimensional Heat Equation

A function which is very useful in many problems of one-dimensional heat flow is

$$v(x,t;\xi,\tau) = \begin{cases} \dfrac{e^{-(x-\xi)^2/4\kappa(t-\tau)}}{\sqrt{4\pi\kappa(t-\tau)}} & -\infty < x < \infty; t > \tau \\ 0 & -\infty < x < \infty \end{cases} \tag{5-13}$$

A motivation for this form of function is given in Prob. 1. Here ξ and τ are parameters independent of x and t. The function v is called a *fundamental solution* of the homogeneous heat equation

$$u_t = \kappa u_{xx} \tag{5-14}$$

for reasons which are set forth below. First observe that v has the following properties.   Let ξ and τ be fixed.   Then:

1. v is defined for $t \geq \tau$ and is continuous on this half plane except at the point (ξ,τ), where it has an infinite discontinuity:

$$\lim_{\substack{t \to \tau \\ t > \tau}} v(\xi,t;\xi,\tau) = \lim_{\substack{t \to \tau \\ t > \tau}} \frac{1}{\sqrt{4\pi\kappa(t-\tau)}} = +\infty$$

2. v is a solution of the heat equation (5-14) for $t > \tau$

3. $\displaystyle\int_{-\infty}^{\infty} v(x,t;\xi,\tau)\, dx = 1 \qquad t > \tau$

4. $v(\xi,t,x,\tau) = v(x,t;\xi,\tau) \qquad t > \tau$

It is clear that v is continuous and, in fact, continuously differentiable any number of times in the open half plane $t > \tau$.   Now let (x_0,τ) be a point on the boundary of this half plane.   Thus to establish property (1) it is sufficient to show

$$\lim_{\substack{(x,t) \to (x_0,\tau) \\ t > \tau}} v(x,t;\xi,\tau) = 0 \qquad x_0 \neq \xi \tag{5-15}$$

Let δ be chosen such that $0 < \delta < |x_0 - \xi|/2$.   If (x,t) lies in the upper half of the interior of the circle of radius δ about (x_0,τ), then

$$|x - \xi| = |x - x_0 + x_0 - \xi| \geq |x_0 - \xi| - |x - x_0| > 2\delta - \delta = \delta$$

Recall that if h is a sufficiently large positive number, then

$$e^h > \frac{h^2}{2} \qquad \text{and so} \qquad e^{-h} < \frac{2}{h^2}$$

Hence

$$v(x,t;\xi,\tau) < \frac{e^{-\delta^2/4\kappa(t-\tau)}}{\sqrt{4\pi\kappa(t-\tau)}} < \frac{16\kappa^{3/4}(t-\tau)^{3/4}}{\sqrt{\pi}\delta^4}$$

Since

$$\lim_{\substack{t \to \tau \\ t > \tau}} (t - \tau)^{3/4} = 0$$

the continuity (from the interior) of v at a point $(x_0,\tau) \neq (\xi,\tau)$ on the boundary is established.   To show v is a solution of the heat equation for $t > \tau$,

$$v_t = \frac{1}{\sqrt{4\pi\kappa}}\left[-\frac{1}{2(t-\tau)^{3/2}} + \frac{(x-\xi)^2}{4\kappa(t-\tau)^{5/2}}\right] e^{-(x-\xi)^2/4\kappa(t-\tau)}$$

$$v_x = \frac{1}{\sqrt{4\pi\kappa(t-\tau)}}\left[-\frac{x-\xi}{2\kappa(t-\tau)}\right] e^{-(x-\xi)^2/4\kappa(t-\tau)}$$

$$v_{xx} = \frac{1}{\sqrt{4\pi\kappa(t-\tau)}}\left[-\frac{1}{2\kappa(t-\tau)} + \frac{(x-\xi)^2}{4\kappa^2(t-\tau)^2}\right] e^{-(x-\xi)^2/4\kappa(t-\tau)}$$

Thus v satisfies Eq. (5-14).   Property 3 follows immediately if the change of variable

$$\eta = \frac{\xi - x}{\sqrt{4\pi\kappa(t - \tau)}} \tag{5-16}$$

is made in the integrand.   One obtains

$$\int_{-\infty}^{\infty} v(x,t;\xi,\tau)\, dx = \frac{1}{\sqrt{\pi}} \int_{-\infty}^{\infty} e^{-\eta^2}\, d\eta = 1 \tag{5-17}$$

If there are no heat sources within the infinitely long bar, the temperature u satisfies

$$\begin{aligned} u_t &= \kappa u_{xx} & -\infty < x < \infty;\ t > 0 \\ u(x,0) &= f(x) & -\infty < x < \infty \end{aligned} \tag{5-18}$$

If ξ is fixed, the fundamental solution

$$v(x,t;\xi,0) = \frac{e^{-(x-\xi)^2/4\kappa t}}{\sqrt{4\pi\kappa t}} \qquad v(x,0;\xi,0) = 0 \tag{5-19}$$

can be interpreted as the temperature distribution in the bar resulting from an initial temperature f having the properties

$$f(x) = 0 \qquad x \neq \xi \qquad \lim_{x \to \xi} f(x) = +\infty \qquad \int_{-\infty}^{\infty} f(x)\, dx = 1$$

that is, a δ-function type of initial temperature concentrated at the point $x = \xi$.   This interpretation follows from the properties of v discussed in the previous paragraph.   It is emphasized here that in this section, and in the sequel, the properties of the δ function are used in a purely formal manner.   However, the operations can be rigorously justified when the definition of the δ function is based on the concept of a continuous linear functional (see Sec. 3-5).

The fundamental solution (5-13) can be interpreted as the temperature in the bar due to a δ-function type of initial temperature concentrated at $x = \xi$ and applied at time $t = \tau$.   Property 4 of v is then a statement of the reciprocity of effects: the temperature at a point ξ due to a δ-function initial temperature applied at x is the same as the temperature at x due to a δ-function initial temperature applied at ξ.   Note that the total heat content in the bar as a result of this type of initial temperature is

$$Q = c\rho \int_{-\infty}^{\infty} v(x,t;\xi,\tau)\, dx = c\rho = \text{const}$$

at any time $t > \tau$.

Formal Derivation of the Solution of the Initial-value Problem

The fundamental solution is a *Green's function* for the initial-value problem (5-11). To construct an expression for the solution with the aid of v, fix x and t, with $t > 0$. Let ξ and τ be independent variables, $-\infty < \xi < \infty$, $\tau \le t$. Then v as a function of ξ and τ satisfies

$$v_\tau + \kappa v_{\xi\xi} = 0 \qquad \tau < t \qquad \lim_{\substack{\tau \to t \\ \tau < t}} v(x,t;\xi,\tau) = 0 \qquad \xi \ne x$$

The partial differential equation satisfied by v is the adjoint of the heat equation. Rewrite problem (5-11) using ξ and τ instead of x and t, and let $u(\xi,\tau)$ be a solution. Then the relation

$$\int_a^b (uv_{\xi\xi} - vu_{\xi\xi}) \, d\xi = (uv_\xi - vu_\xi)\big|_a^b \qquad \tau < t$$

holds. This is the one-dimensional form of Green's formula. From the partial differential equations satisfied by u and v it follows that

$$\int_a^b (uv_\tau + vu_\tau) \, d\xi = \int_a^b v(x,t;\xi,\tau)F(\xi,\tau) \, d\xi - \kappa(uv_\xi - vu_\xi)\big|_a^b$$

for $\tau < t$. Let $a \to -\infty$, $b \to +\infty$. Assume u and u_ξ are bounded. Then $\lim\limits_{\substack{b \to +\infty \\ a \to -\infty}} (uv_\xi - vu_\xi)\big|_a^b = 0$

Hence

$$\frac{d}{d\tau} \int_{-\infty}^\infty uv \, d\xi = \int_{-\infty}^\infty v(x,t;\xi,\tau)F(\xi,\tau) \, d\xi \qquad \tau < t$$

Integrate both sides of this equation with respect to τ from $\tau = 0$ to $\tau = t - \epsilon$, where ϵ is a small positive number. The result is

$$\int_{-\infty}^\infty uv\big|^{t-\epsilon} d\xi = \int_{-\infty}^\infty v(x,t;\xi,0)f(\xi) \, d\xi + \int_0^{t-\epsilon} d\tau \int_{-\infty}^\infty v(x,t;\xi,\tau)F(\xi,\tau) \, d\xi$$

Now

$$\lim_{\epsilon \to 0} \int_{-\infty}^\infty uv\big|^{t-\epsilon} d\xi = u(x,t)$$

The preceding relation is suggested by the following procedure. Make the change of variable given by Eq. (5-16), and let

$$U(\eta,\tau) = u[x + \eta\sqrt{4\kappa(t-\tau)}, \tau]$$

Then

$$\lim_{\epsilon \to 0} \int_{-\infty}^{\infty} uv|^{t-\epsilon} d\xi = \lim_{\epsilon \to 0} \left[\frac{1}{\sqrt{\pi}} \int_{-\infty}^{\infty} e^{-\eta^2} U(\eta, t - \epsilon) \, d\eta \right]$$

$$= u(x,t) \left(\frac{1}{\sqrt{\pi}} \int_{-\infty}^{\infty} e^{-\eta^2} \, d\eta \right) = u(x,t)$$

Accordingly

$$u(x,t) = \int_{-\infty}^{\infty} v(x,t;\xi,0)f(\xi) \, d\xi + \int_{0}^{t} d\tau \int_{-\infty}^{\infty} v(x,t;\xi,\tau)F(\xi,\tau) \, d\xi \qquad t > 0 \quad (5\text{-}20)$$

Define

$$u(x,0) = f(x) \qquad -\infty < x < \infty$$

Verification of the Solution

The preceding derivation is only heuristic. However, under appropriate hypotheses on the given functions, it is now shown that the improper integrals converge and define a function u which satisfies all the conditions. Moreover

$$u_1(x,t) = \int_{-\infty}^{\infty} v(x,t;\xi,0)f(\xi) \, d\xi \tag{5-21}$$

defines a solution of problem (5-18), and

$$u_2(x,t) = \int_{0}^{t} d\tau \int_{-\infty}^{\infty} v(x,t;\xi,\tau)F(\xi,\tau) \, d\xi \tag{5-22}$$

defines a solution of

$$u_t - \kappa u_{xx} = F(x,t) \qquad -\infty < x < \infty; t > 0$$

$$u(x,0) = 0 \qquad -\infty < x < \infty$$

Assume the function f in problem (5-18) is continuous and uniformly bounded on $(-\infty, \infty)$: there exists a constant $M > 0$ such that

$$|f(x)| < M \qquad -\infty < x < \infty$$

Choose positive numbers x_0, t_0, T such that $t_0 < T$, and let $\mathscr{D}_0$ denote the closed and bounded domain

$$-x_0 \le x \le x_0 \qquad t_0 \le t \le T$$

Then the improper integral in Eq. (5-21) converges uniformly in x and t on $\mathscr{D}_0$. To see this note that

$$|v(x,t;\xi,0)f(\xi)| < \frac{Me^{-(x-\xi)^2/4\kappa T}}{\sqrt{4\pi\kappa t_0}} \le \frac{Me^{-(\xi^2 + 2x_0\xi)/4\kappa T}}{\sqrt{4\pi\kappa t_0}}$$

Since the improper integral

$$\int_{-\infty}^{\infty} e^{-(\xi^2+2x_0\xi)4/\kappa T} \, d\xi$$

exists, the uniform convergence follows by the Weierstrass (dominated-convergence) theorem. In turn the uniform convergence implies that the function u_1 is continuous on $\mathcal{D}_0$. Indeed u_1 is continuously differentiable any number of times in $\mathcal{D}_0$, since the improper integrals obtained by differentiating the integrand with respect to x and/or t also converge uniformly. Recall that the values x_0, t_0, and T were arbitrary to within the stated restrictions. It follows that the function u_1 defined by Eq. (5-21) is continuously differentiable any number of times in the upper half plane $t > 0$. Since differentiation under the integral sign is valid, and since $v(x,t;\xi,0)$ as a function of x and t is a solution of Eq. (5-14) for $t > 0$, it follows that u_1 is a solution of the homogeneous heat equation for $t > 0$. To show that

$$\lim_{\substack{t \to 0 \\ t > 0}} u_1(x,t) = f(x) \qquad -\infty < x < \infty$$

make the change of variable given in Eq. (5-16). Then

$$u_1(x,t) = \frac{1}{\sqrt{\pi}} \int_{-\infty}^{\infty} f(x + \eta\sqrt{4\kappa t}) \, e^{-\eta^2} \, d\eta$$

This is called the *Laplace solution* of problem (5-18). Let x_0 be a fixed value. Then

$$\lim_{t \to 0} u_1(x_0,t) = \frac{1}{\sqrt{\pi}} \int_{-\infty}^{\infty} [\lim_{t \to 0} f(x_0 + \eta\sqrt{4\kappa t})] \, e^{-\eta^2} \, d\eta$$

$$= \frac{f(x_0)}{\sqrt{\pi}} \int_{-\infty}^{\infty} e^{-\eta^2} \, d\eta = f(x_0)$$

The interchange of operations is permissible in view of the uniform convergence of the integral with respect to t. Now define $u_1(x,0) = f(x)$, $-\infty < x < \infty$. Then u_1 is continuous on the half plane $t \geq 0$ and is a solution of problem (5-18). Note that u_1 is uniformly bounded on the closed half plane, since

$$|u_1(x,t)| \leq \frac{1}{\sqrt{\pi}} \int_{-\infty}^{\infty} |f(x + \eta\sqrt{4\kappa t})|e^{-\eta^2} \, d\eta \leq \frac{M}{\sqrt{\pi}} \int_{-\infty}^{\infty} e^{-\eta^2} \, d\eta = M$$

It can be shown that u_1 is the unique solution of problem (5-18) in the class of functions which are bounded on the half plane $t \geq 0$.

Now in regard to Eq. (5-22) define the function w by

$$w(x,t;\tau) = \int_{-\infty}^{\infty} v(x,t;\xi,\tau)F(\xi,\tau) \, d\xi \qquad 0 \leq \tau < t$$

Let F be twice continuously differentiable and bounded on the half plane $t \geq 0$. Then the preceding integral converges and defines a twice continuously differentiable function for $\tau < t$. Moreover

$$\lim_{\substack{\tau \to t \\ \tau > t}} w(x,t;\tau) = F(x,t) \qquad -\infty < x < \infty; t > 0$$

Define

$$w(x,t;t) = F(x,t) \qquad -\infty < x < \infty; t > 0$$

Then

$$u_2(x,t) = \int_0^t w(x,t;\tau)\, d\tau$$

Hence

$$\frac{\partial u_2}{\partial t} = w(x,t;t) + \int_0^t \frac{\partial w}{\partial t}\, d\tau \qquad \frac{\partial^2 u_2}{\partial x^2} = \int_0^t \frac{\partial^2 w}{\partial x^2}\, d\tau$$

and so

$$\frac{\partial u_2}{\partial t} - \kappa \frac{\partial^2 u_2}{\partial x^2} = F(x,t) + \int_0^t (w_t - \kappa w_{xx})\, d\tau = F(x,t)$$

Clearly

$$\lim_{\substack{t \to 0 \\ t > 0}} u_2(x,t) = 0 = u_2(x,0) \qquad -\infty < x < \infty$$

Accordingly u_2 is a solution of the inhomogeneous heat equation for $t > 0$ and satisfies the zero initial condition. It should be noted that u_2 need not be bounded for $t \geq 0$. For example, if $F(x,t) = F_0$, a positive constant, Eq. (5-22) yields

$$u_2(x,t) = F_0 t$$

Thus, although the source density F is uniformly bounded on the half plane $t \geq 0$, the solution u is not.

Observe that Eq. (5-21) implies the following property on the heat equation. Let the initial temperature in the bar be different from zero in a small interval about a given point x_0 and zero outside the interval. Then the subsequent temperature at arbitrarily small time $t > 0$ at distant points is immediately perturbed (in general) away from the zero value. There is no delay of effects. In descriptive terms, the "heat waves" in the bar are propagated with "infinite speed."

EXAMPLE 5-1   The initial temperature in the bar is

$$f(x) = \begin{cases} \dfrac{x+\delta}{\delta^2} & -\delta \leq x < 0 \\[2ex] \dfrac{-x+\delta}{\delta^2} & 0 \leq x \leq \delta \\[2ex] 0 & x < -\delta \text{ or } x > \delta \end{cases}$$

Thus the initial temperature is zero outside the interval $[-\delta,\delta]$. In the interval $[-\delta,\delta]$ the graph of f is a triangular-shaped pulse of altitude $1/\delta$. For small values of δ the initial temperature approximates a δ function. The heat stored in the bar due to the initial temperature is

$$Q_1 = cp \int_{-\infty}^{\infty} f(x)\, dx = c$$

Let the heat sources within the bar be given by

$$F(x,t) = F_0(t)\delta(x - x_0)$$

where $F_0(t)$ is a given function and x_0 is a fixed point. This is a concentrated heat source at $x = x_0$, of strength

$$\int_{-\infty}^{\infty} F(x,t)\, dx = F_0(t)$$

Determine the temperature in the bar for $t > 0$.

From Eq. (5-20) it follows that the subsequent temperature is

$$u(x,t) = u_1(x,t) + u_2(x,t) = \int_{-\delta}^{\delta} v(x,t;\xi,0)f(\xi)\, d\xi + \int_0^t v(x,t;x_0,\tau)F_0(\tau)\, d\tau$$

In the first integral make the change of variable

$$\eta = \frac{\xi - x}{\sqrt{4\kappa t}}$$

Then

$$u_1(x,t) = \frac{1}{\delta^2\sqrt{4\kappa t}}\left[\int_{-\delta}^{0}(\xi + \delta)e^{-(x-\xi)^2/4\kappa t}\, d\xi + \int_0^{\delta}(-\xi + \delta)e^{-(x-\xi)^2/4\kappa t}\, d\xi\right]$$

$$= \frac{1}{\sqrt{\pi}\,\delta^2}\left[\sqrt{4\kappa t}\left(\int_{\eta_1}^{\eta_2} \eta e^{-\eta^2}\, d\eta - \int_{\eta_2}^{\eta_3} \eta e^{-\eta^2}\, d\eta\right)\right.$$

$$\left. + (x + \delta)\int_{\eta_1}^{\eta_2} e^{-\eta^2}\, d\eta + (\delta - x)\int_{\eta_2}^{\eta_3} e^{-\eta^2}\, d\eta\right]$$

where

$$\eta_1 = -\frac{x+\delta}{\sqrt{4\kappa t}} \qquad \eta_2 = \frac{-x}{\sqrt{4\kappa t}} \qquad \eta_3 = \frac{\delta - x}{\sqrt{4\kappa t}}$$

Recall the definition of the *error function*:

$$\text{erf } x = \frac{2}{\sqrt{\pi}} \int_0^x e^{-\eta^2}\, d\eta$$

Thus

$$u_1(x,t) = \frac{1}{\delta^2} \sqrt{\frac{4\kappa t}{\pi}} \, e^{-x^2/4\kappa t} \left(e^{-\delta^2/4\kappa t} \cosh \frac{x\delta}{2\kappa t} - 1 \right)$$

$$+ \frac{1}{2\delta^2} \left[(x+\delta)\left(\mathrm{erf}\,\frac{x+\delta}{\sqrt{4\kappa t}} - \mathrm{erf}\,\frac{x}{\sqrt{4\kappa t}} \right) + (x-\delta)\left(\mathrm{erf}\,\frac{x-\delta}{\sqrt{4\kappa t}} - \mathrm{erf}\,\frac{x}{\sqrt{4\kappa t}} \right) \right]$$

The temperature in the bar due to the point source at $x = x_0$ is

$$u_2(x,t) = \frac{1}{\sqrt{4\pi\kappa}} \int_0^t F_0(\tau) \frac{e^{-(x-x_0)^2/4\kappa(t-\tau)}}{\sqrt{t-\tau}} \, d\tau$$

As a particular case suppose

$$F_0(t) = \delta(t - t_0)$$

where t_0 is a given instant, $t_0 > 0$.   Then

$$u_2(x,t) = \frac{e^{-(x-x_0)^2/4\kappa(t-t_0)}}{\sqrt{4\pi(t-t_0)}} = v(x,t;x_0,t_0)$$

Again there follows the interpretation of the fundamental solution as the temperature in the bar due to a point source of heat applied at $x = x_0$ at time $t = t_0$.
   As another example suppose $F_0(t) = F_0$, a constant.   Let

$$\eta = \frac{|x - x_0|}{\sqrt{4\kappa(t-\tau)}}$$

in the integral above.   Then

$$u_2(x,t) = \frac{F_0 |x - x_0|}{4\kappa\sqrt{\pi}} \int_{\eta_1}^{\infty} \frac{e^{-\eta^2}}{\eta^2} \, d\eta$$

where $\eta_1 = |x - x_0|/\sqrt{4\kappa t}$.   Now

$$\frac{d}{d\eta}\left(-\frac{e^{-\eta^2}}{\eta} \right) = \frac{e^{-\eta^2}}{\eta^2} + 2e^{-\eta^2}$$

Hence

$$u_2(x,t) = F_0 \left[\sqrt{\frac{t}{\pi\kappa}}\, e^{-(x-x_0)^2/4\kappa t} - \frac{|x - x_0|}{2\kappa}\left(1 - \mathrm{erf}\,\frac{|x - x_0|}{\sqrt{4\kappa t}} \right) \right]$$

The initial-value problems for the two- and three-dimensional heat equations are motivated and their solutions established in a similar manner. For example a thin homogeneous plate of uniform thinness and of large extent lies in the xy plane.   The top and bottom sides are insulated.   If edge effects are neglected, one obtains the two-dimensional initial-value problem. The fundamental solution of the two-dimensional homogeneous heat equation is

$$v(x,y,t;\xi,\eta,\tau) = \frac{e^{-[(x-\xi)^2+(y-\eta)^2]/4\kappa(t-\tau)}}{4\pi\kappa(t-\tau)} \tag{5-23}$$

The expressions for the solutions of the two- and three-dimensional initial-value problems are direct generalizations of Eq. (5-20). The solution of the three-dimensional problem

$$u_t - \kappa \Delta u = F(x,y,z,t) \qquad -\infty < x, y, z < \infty; t > 0$$
$$u(x,y,z,0) = f(x,y,z) \qquad -\infty < x, y, z < \infty \qquad (5\text{-}24)$$

is

$$u(x,y,z,t) = \int_{-\infty}^{\infty} \int_{-\infty}^{\infty} \int_{-\infty}^{\infty} v\big|_{\tau=0} f(\xi,\eta,\zeta)\, d\xi\, d\eta\, d\zeta$$
$$+ \int_0^t d\tau \int_{-\infty}^{\infty} \int_{-\infty}^{\infty} \int_{-\infty}^{\infty} vF(\xi,\eta,\zeta,\tau)\, d\xi\, d\eta\, d\zeta \quad (5\text{-}25)$$

where the fundamental solution is

$$v(x,y,z,t;\xi,\eta,\zeta,\tau) = \frac{e^{-[(x-\xi)^2+(y-\eta)^2+(z-\zeta)^2]/4\kappa(t-\tau)}}{[4\pi\kappa(t-\tau)]^{3/2}} \qquad (5\text{-}26)$$

Note that higher-dimensional fundamental solutions are products of one-dimensional fundamental solutions.

5-3 BOUNDARY- AND INITIAL-VALUE PROBLEM FOR THE HEAT EQUATION. METHOD OF EIGENFUNCTIONS

As in Sec. 5-1, let S be a simple closed piecewise smooth surface which forms the boundary of a thermal conductor. Let $\mathscr{V}$ denote the region interior to S and let $\bar{\mathscr{V}}$ denote $\mathscr{V}$ together with S. Assume the initial temperature in the conductor is known, and one of the boundary conditions in Eq. (5-6) to (5-10) applies. Then the boundary- and initial-value problem for the determination of the temperature in the conductor for $t > 0$ is

$$u_t - \kappa \Delta u = F(x,y,z,t) \qquad \text{in } \mathscr{V}; t > 0$$
$$B(u) = H(x,y,z,t) \qquad \text{on } S; t \geq 0 \qquad (5\text{-}27)$$
$$u(x,y,z,0) = f(x,y,z) \qquad \text{in } \bar{\mathscr{V}}$$

Here $B(u) = H$ symbolizes the boundary condition, and f describes the initial temperature. It should be noted that if the boundary condition is given by Eq. (5-7) or (5-10), then without loss of generality it can be assumed that $u_0 = 0$. For if w is a solution of the problem in which u_0 is replaced by zero in the boundary condition, and f is replaced by $f - u_0$ in the initial condition, the superposition

$$u = w + u_0$$

is a solution of the original problem.

The method of eigenfunctions can be applied to problem (5-27), and a formal series solution derived following the procedure set forth in previous chapters.   First separable solutions

$$u = \psi(x,y,z)T(t)$$

of the homogeneous heat equation

$$u_t = \kappa \, \Delta u \tag{5-28}$$

which satisfy the homogeneous boundary condition

$$B(u) = 0 \qquad \text{on } S; \, t \geq 0 \tag{5-29}$$

are derived.   If the assumed form is substituted into Eq. (5-28), the equations

$$\psi \dot{T} = \kappa \, \Delta \psi \, T \qquad \frac{\Delta \psi}{\psi} = -\lambda = \frac{\dot{T}}{T}$$

result, where dots denote derivatives with respect to t.   Hence the factors must satisfy

$$\dot{T} + \kappa \lambda T = 0 \qquad \Delta \psi + \lambda \psi = 0 \tag{5-30}$$

respectively.   In addition ψ must satisfy

$$B(\psi) = 0 \qquad \text{on } S \tag{5-31}$$

Equations (5-30) and (5-31) embody the self-adjoint eigenvalue problem for the elliptic operator Δ, whose various forms, corresponding to various boundary conditions and coordinate systems, are discussed in Chaps. 3 and 4.   For each of the three principal types of boundary conditions (Dirichlet, Neumann, and mixed) the eigenvalues are real, nonnegative, and constitute a monotonic sequence tending to $+\infty$.   Let $\{\lambda_n\}$ be the eigenvalues and $\{\psi_n\}$ a corresponding sequence of real-valued eigenfunctions.   Each ψ_n is a twice continuously differentiable solution of Eq. (5-30), with $\lambda = \lambda_n$, which satisfies the homogeneous boundary condition (5-31) on S.   The $\{\psi_n\}$ form an orthogonal set over $\mathcal{V}$, and moreover the set is complete in the space of all functions which are continuous on $\overline{\mathcal{V}}$.   Corresponding to $\lambda = \lambda_n$ there is the time-dependent factor

$$T_n(t) = e^{-\kappa \lambda_n t}$$

The functions

$$u_n(x,y,z,t) = \psi_n(x,y,z)e^{-\kappa \lambda_n t} \qquad n = 1, 2, \ldots \tag{5-32}$$

are twice continuously differentiable solutions of the homogeneous heat equation in $\overline{\mathcal{V}}$ for all t, and satisfy the homogeneous boundary condition on S.

Formal Series Solution

Assume the boundary condition in problem (5-27) is homogeneous.   In order to derive a formal series solution let

$$u = \sum_{n=1}^{\infty} C_n(t)\psi_n(x,y,z) \tag{5-33}$$

where the C_n are to be determined.   The ψ_n are the eigenfunctions described in the previous paragraph.   Thus if the series converges on $\mathcal{V}$ for $t \geq 0$, the function u satisfies the homogeneous boundary condition on S.   Substitute u into the inhomogeneous heat equation, and interchange the operations of summation and differentiation.   The result is

$$\sum_{n=1}^{\infty} (\dot{C}_n\psi_n - \kappa C_n \Delta\psi_n) = \sum_{n=1}^{\infty} (\dot{C}_n + \kappa\lambda_n C_n)\psi_n = F$$

By the orthogonality of the eigenfunctions it follows that

$$\dot{C}_n + \kappa\lambda_n C_n = F_n(t) \qquad n = 1, 2, \dots \tag{5-34}$$

where

$$F_n(t) = \frac{1}{\|\psi_n\|^2} \iiint_{\mathcal{V}} F(x,y,z)\, dx\, dy\, dz \tag{5-35}$$

The general solution of Eq. (5-34) is

$$C_n(t) = \int_0^t e^{-\kappa\lambda_n(t-\tau)} F_n(\tau)\, d\tau + A e^{-\kappa\lambda_n t}$$

where $A_n = C_n(0)$.   It remains to satisfy the initial condition.   This is satisfied if

$$\sum_{n=1}^{\infty} C_n(0)\psi_n(x,y,z) = \sum_{n=1}^{\infty} A_n\psi_n(x,y,z) = f(x,y,z) \qquad \text{in } \mathcal{V}$$

Again the orthogonality of the eigenfunctions implies

$$A_n = \frac{1}{\|\psi_n\|^2} \iiint_{\mathcal{V}} f(x,y,z)\psi_n(x,y,z)\, dx\, dy\, dz \tag{5-36}$$

Accordingly the formal series solution of problem (5-27) in the case where the boundary condition is homogeneous is

$$u = u_1 + u_2 \tag{5-37}$$

where

$$u_1(x,y,z,t) = \sum_{n=1}^{\infty} A_n\psi_n(x,y,z)e^{-\kappa\lambda_n t} \tag{5-38}$$

$$u_2(x,y,z,t) = \sum_{n=1}^{\infty} \left[\int_0^t e^{-\kappa\lambda_n(t-\tau)} F_n(\tau)\, d\tau \right] \psi_n(x,y,z) \tag{5-39}$$

The coefficients A_n are given by Eq. (5-36), and the functions F_n are defined in Eq. (5-35).

Equation (5-38) gives the formal series solution of the problem

$$u_t = \kappa \,\Delta u \qquad \text{in } \mathscr{V}; t > 0$$

$$B(u) = 0 \qquad \text{on } S; t \geq 0 \qquad u(x,y,z,0) = f(x,y,z) \qquad \text{in } \overline{\mathscr{V}} \tag{5-40}$$

Under suitable hypotheses on f the formal solution constitutes a classical solution and hence establishes an existence theorem. For example, let f be continuous and piecewise continuously differentiable in $\overline{\mathscr{V}}$, and let f satisfy the homogeneous boundary condition on S. Then the series (5-38) converges on $\overline{\mathscr{V}}$ for $t \geq 0$ and defines a function u_1 which is continuous in the four variables x, y, z, and t for $t \geq 0$, twice continuously differentiable in $\mathscr{V}$ for $t > 0$, satisfies the homogeneous heat equation for $t > 0$, and satisfies the homogeneous boundary condition as well as the initial condition in the continuous sense

$$\lim_{\substack{t \to 0 \\ t > 0}} u(x,y,z,t) = f(x,y,z) = u(x,y,z,0) \qquad \text{in } \overline{\mathscr{V}}$$

Equation (5-39) gives the formal series solution of the problem

$$u_t - \kappa \,\Delta u = F(x,y,z,t) \qquad \text{in } \mathscr{V}; t > 0$$

$$B(u) = 0 \qquad \text{on } S; t \geq 0 \qquad u(x,y,z,0) = 0 \qquad \text{in } \overline{\mathscr{V}} \tag{5-41}$$

Convergence of the series (5-38) and (5-39) in the one-dimensional case for Dirichlet boundary conditions is discussed in Sec. 5-4; see also Example 5-2.

Green's Function for the Problem

There exists a *Green's function* for problem (5-27). To obtain the series representation of the Green's function in terms of eigenfunctions let (x_0, y_0, z_0) be a fixed but otherwise arbitrarily chosen point of $\mathscr{V}$, and let $\tau > 0$ be an arbitrary time. In problem (5-41) let the source density be

$$F(x,y,z,t) = \delta(x - x_0)\delta(y - y_0)\delta(z - z_0)\delta(t - \tau)$$

This is a concentrated unit heat source at (x_0, y_0, z_0) which occurs at time $t = \tau$. From Eqs. (5-35) and (5-39) it follows that the temperature distribution in $\overline{\mathscr{V}}$ for $t > \tau$ is

$$G(x,y,z,t;x_0,y_0,z_0,\tau) = \sum_{n=1}^{\infty} \frac{\psi_n(x,y,z)\psi_n(x_0,y_0,z_0)}{\|\psi_n\|^2} e^{-\kappa \lambda_n(t-\tau)} \qquad t > \tau$$

$$= 0 \qquad t \leq \tau \tag{5-42}$$

By means of the Green's function one can express the solution of problem (5-27) in terms of the given functions f, F, and H, that is, *invert the problem*.

The solution is

$$u(x,y,z,t) = \iiint_{\mathscr{V}} G|_{\tau=0} f(x_0,y_0,z_0)\, dV_0$$

$$+ \int_0^t d\tau \iiint GF(x_0,y_0,z_0,\tau)\, dV_0 - \kappa \int_0^t d\tau \iint \left(u \frac{\partial G}{\partial n} - G \frac{\partial u}{\partial n} \right) dS_0 \quad (5\text{-}43)$$

where $dV_0 = dx_0\, dy_0\, dz_0$ is the volume element of integration in $\mathscr{V}$, dS_0 is the surface element of integration, and $\mathbf{n}$ is the exterior normal on S. If the boundary condition is $u = H$ on S, then $G = 0$ on S, and the last term in Eq. (5-43) simplifies to

$$-\kappa \int_0^t d\tau \iint_S H \frac{\partial G}{\partial n}\, dS_0 \qquad\qquad (5\text{-}44)$$

Similarly if the boundary condition is $\partial u/\partial n = H$ on S, the last term reduces to

$$+ \kappa \int_0^t d\tau \iint_S HG\, dS_0 \qquad\qquad (5\text{-}45)$$

If the boundary condition is

$$\frac{\partial u}{\partial n} + hu = H \qquad \text{on } S;\, t \geq 0 \qquad\qquad (5\text{-}46)$$

again the last term is given by (5-45). In the one-dimensional form of problem (5-27) the domain $\mathscr{V}$ is a closed interval $[a,b]$, and the solution has the form

$$u(x,t) = \int_a^b G(x,t;\xi,0)f(\xi)\, d\xi + \int_0^t d\tau \int_a^b G(x,t;\xi,\tau)F(\xi,\tau)\, d\xi$$

$$- \kappa \int_0^t \left(u \frac{\partial G}{\partial \xi} - G \frac{\partial u}{\partial \xi} \right)\bigg|_a^b d\tau \quad (5\text{-}47)$$

A formal derivation of Eq. (5-43) is as follows. Let (x,y,z) be a fixed point in $\mathscr{V}$, $t > 0$ a fixed time. Let (x_0,y_0,z_0) be a variable point in $\mathscr{V}$, and let τ be a nonnegative variable. Then G, as a function of x_0, y_0, z_0 and τ, satisfies the adjoint equation

$$G_\tau + \kappa \Delta G = -\delta(x_0 - x)\delta(y_0 - y)\delta(z_0 - z)\delta(\tau - t)$$

where the laplacian is taken with respect to x_0, y_0, z_0. Now rewrite problem (5-27) using the variables x_0, y_0, z_0, τ instead of x, y, z, t. Let $u(x_0,y_0,z_0,\tau)$ be a solution. Then Green's formula states

$$\iiint_{\mathscr{V}} (u \Delta G - G \Delta u)\, dV_0 = \iint_S \left(u \frac{\partial G}{\partial n} - G \frac{\partial u}{\partial n} \right) dS_0$$

Substitution for Δu and ΔG from the respective partial differential equations and formal use of the property of the δ function yields

$$u(x,y,z,\tau)\delta(\tau - t) = \iiint_{\mathscr{V}} GF\, dV_0 - \frac{\partial}{\partial \tau}\iiint_{\mathscr{V}} uG\, dV_0 - \kappa\iint_{S}\left(u\frac{\partial G}{\partial n} - G\frac{\partial u}{\partial n}\right)dS_0$$

Now integrate both sides of the equation with respect to τ from $\tau = 0$ to $\tau = t$. Since G is zero at $\tau = t$, Eq. (5-43) follows.

The following examples illustrate the method of eigenfunctions for some particular heat-conduction problems.

EXAMPLE 5-2   A slender homogeneous bar of uniform cross section lies along the x axis with ends at $x = 0$, $x = L$. The end $x = 0$ is held at a constant temperature u_0, while the end $x = L$ is insulated. The heat-source density within the bar is

$$F(x,t) = F_0\delta(x - x_0)\cos \omega t$$

where F_0, ω are positive constants and x_0 is a fixed point, $0 < x_0 < L$. Determine the temperature distribution in the bar for $t > 0$, given $u(x,0) = f(x)$.

Assume a separable solution $u = X(x)T(t)$ of the homogeneous heat equation

$$u_t = \kappa u_{xx}$$

Separation of variables shows that the factors must satisfy

$$X'' + \lambda X = 0 \qquad \dot{T} + \kappa\lambda T = 0$$

respectively. The homogeneous form of the boundary conditions in the present problem is

$$u(0,t) = 0 \qquad u_x(L,t) = 0 \qquad t \geq 0$$

Hence X must satisfy

$$X(0) = 0 \qquad X'(L) = 0$$

The eigenvalues of the problem are

$$\lambda_n = \left[\frac{(2n - 1)\pi}{2L}\right]^2 \qquad n = 1, 2, \ldots$$

Corresponding real-valued eigenfunctions are

$$X_n = \sin \frac{(2n - 1)\pi x}{2L} \qquad n = 1, 2, \ldots$$

The time-dependent factor corresponding to $\lambda = \lambda_n$ is

$$T_n = e^{-\kappa(2n-1)^2\pi^2 t/4L^2}$$

The functions

$$u_n(x,t) = \sin \frac{(2n - 1)\pi x}{2L} e^{-\kappa(2n-1)^2\pi^2 t/4L^2} \qquad n = 1, 2, \ldots$$

are the separable solutions of the homogeneous heat equation which satisfy the homogeneous boundary conditions at $x = 0$, $x = L$.

The solution of the original problem is obtained by superimposing solutions of the problems

$$u_t = \kappa u_{xx}$$

$$u(0,t) = u_0 \qquad u_x(0,t) = 0 \qquad u(x,0) = f(x)$$

$$u_t - \kappa u_{xx} = F_0 \delta(x - x_0) \cos \omega t$$

$$u(0,t) = 0 \qquad u_x(0,t) = 0 \qquad u(x,0) = 0$$

From the remarks made at the beginning of this section a solution of the first problem can be obtained in the form

$$v = w + u_0$$

where w satisfies

$$w_t = \kappa w_{xx}$$

$$w(0,t) = 0 \qquad w_x(0,t) = 0 \qquad w(x,0) = f(x) - u_0$$

By Eq. (5-38)

$$w(x,t) = \sum_{n=1}^{\infty} A_n \sin \frac{(2n-1)\pi x}{2L} e^{-\kappa(2n-1)^2 \pi^2 t / 4L^2}$$

Since $\| X_n \|^2 = L/2$, it follows that

$$A_n = \frac{2}{L} \int_0^L [f(x) - u_0] \sin \frac{(2n-1)\pi x}{2L} dx$$

$$= B_n - \frac{4u_0}{(2n-1)\pi}$$

where

$$B_n = \frac{2}{L} \int_0^L f(x) \sin \frac{(2n-1)\pi x}{2L} dx \qquad n = 1, 2, \ldots$$

It is interesting to observe that the series for w [more generally, the series solution (5-38) of problem (5-40)] converges for $t > 0$ under relatively weak hypotheses on f. For example, suppose f is merely bounded and integrable on $[0,L]$. Then the series at $t = 0$ is

$$\sum_{n=1}^{\infty} A_n \sin \frac{(2n-1)\pi x}{2L}$$

and this may very well be divergent. Nevertheless the series for w converges and defines a twice continuously differentiable solution of the homogeneous heat equation on the open half plane $t > 0$. To show this suppose

$$|f(x)| < M \qquad 0 \le x \le L$$

Then

$$|A_n| \le \frac{2}{L} \int_0^L |f(x) - u_0| \left| \sin \frac{(2n-1)\pi x}{2L} \right| dx < 2[M + |u_0|]$$

Choose values t_0, T such that $0 < t_0 < T$, and let $\mathscr{D}_0$ denote the closed domain of the xt plane defined by the inequalities

$$0 \le x \le L \qquad t_0 \le t \le T$$

Then

$$\left| A_n \sin \frac{(2n-1)\pi x}{2L} e^{-\kappa \lambda_n t} \right| < 2(M + |u_0|)e^{-\kappa \lambda_n t_0}$$

By the ratio test the series of positive constants

$$\sum_{n=1}^{\infty} e^{-\kappa \lambda_n t_0}$$

converges. The Weierstrass test then implies that the series for w converges uniformly on $\mathscr{D}_0$ and so defines a continuous function there. Similar reasoning leads to the conclusion that the series resulting from termwise differentiation any number of times with respect to x and/or t converges uniformly on $\mathscr{D}_0$. Thus w is continuously differentiable any number of times on $\mathscr{D}_0$. Moreover

$$\frac{\partial w}{\partial t} - \frac{\partial^2 w}{\partial x^2} = \sum_{n=1}^{\infty} A_n \frac{\partial u_n}{\partial t} - \kappa \sum_{n=1}^{\infty} A_n \frac{\partial^2 u_n}{\partial t^2} = \sum_{n=1}^{\infty} A_n \left(\frac{\partial u_n}{\partial t} - \kappa \frac{\partial^2 u_n}{\partial x^2} \right) = 0$$

Hence w is a solution of the heat equation on $\mathscr{D}_0$. Recall that t_0, T were arbitrarily chosen to within the requirement $0 < t_0 < T$. It follows that w is continuously differentiable any number of times and satisfies the heat equation on the open half plane $t > 0$.

In physical terms the meaning of the results of the previous paragraph is that the mathematical model describes a situation in which any discontinuities in the initial temperature of the bar, or any discontinuities in the initial gradient, are immediately smoothed out.

If f is continuous, piecewise smooth, and $f(0) = u_0, f'(L) = 0$, then by the convergence properties of Fourier sine series it follows that

$$\sum_{n=1}^{\infty} A_n \sin \frac{(2n-1)\pi x}{2L} = f(x) - u_0$$

absolutely and uniformly on $[0,L]$. It is shown in Sec. 5-4 that in this case the series for w converges uniformly on the closed half plane $t \ge 0$ and represents the classical solution of the problem.

In order to obtain a solution of the inhomogeneous heat equation which satisfies the homogeneous boundary and initial conditions substitute the given source density function F into Eq. (5-35). Then by the property of the δ function it follows that

$$C_n(t) = \frac{2F_0 \sin [(2n-1)\pi x_0/2L]e^{-\kappa \lambda_n t}}{L} \int_0^t e^{\kappa \lambda_n \tau} \cos \omega \tau \, d\tau$$

$$= \frac{2F_0 \sin [(2n-1)\pi x_0/2L]\varphi_n(t)}{L(\kappa^2 \lambda_n^2 + \omega^2)}$$

where

$$\varphi_n(t) = \kappa \lambda_n \cos \omega t + \omega \sin \omega t - \kappa \lambda_n e^{-\kappa \lambda_n t}$$

Thus the series corresponding to the series in Eq. (5-39) is

$$u_2(x,t) = \frac{2F_0}{L} \sum_{n=1}^{\infty} \frac{\sin [(2n-1)\pi x_0/2L] \sin [(2n-1)\pi x/2L]\varphi_n(t)}{\kappa^2 \lambda_n^2 + \omega^2}$$

This series converges uniformly for $t \geq 0$, since

$$\left| \frac{\sin \left[(2n-1)\pi x_0/2L\right] \sin \left[(2n-1)\pi x/2L\right]\varphi_n(t)}{\kappa^2\lambda_n{}^2 + \omega^2} \right|$$

$$\leq \frac{\kappa\lambda_n + \omega + \kappa\lambda_n e^{-\kappa\lambda_n t}}{\kappa^2\lambda_n{}^2 + \omega^2} < \frac{\kappa\lambda_n + \omega + \kappa\lambda_n}{\kappa^2\lambda_n{}^2} < \frac{\omega + 2}{\kappa\lambda_n}$$

for sufficiently large n and for $0 \leq x \leq L$, $t \geq 0$.   The uniform convergence follows since $\Sigma[(\omega + 2)/\lambda_n]$ converges.   Hence the series defines a continuous function u_2 on $0 \leq x \leq L$, $t \geq 0$.   Now by the continuity at $t = 0$ and the form of the $\varphi_n(t)$ one obtains

$$\lim_{\substack{t \to 0 \\ t > 0}} u_2(x,t) = u_2(x,0) = 0 \qquad 0 \leq x \leq L$$

Also $u_2(0,t) = 0$, $u_{2x}(L,t) = 0$, $t \geq 0$.   Thus u_2 satisfies the homogeneous boundary and initial conditions.   If $t > 0$, the series obtained by differentiating termwise once with respect to t converges, and the series obtained by differentiating termwise twice with respect to x converges.   However u_2 is not a solution of the inhomogeneous heat equation corresponding to the source density $F(x,t)$.

By Eq. (5-42) the Green's function of the present problem is

$$G(x,t;\xi,\tau) = \frac{2}{L} \sum_{n=1}^{\infty} \sin \frac{(2n-1)\pi\xi}{2L} \sin \frac{(2n-1)\pi x}{2L} e^{-\kappa(2n-1)^2\pi^2(t-\tau)/4L^2}$$

In terms of G the complete solution of the original problem can be written

$$u(x,t) = \int_0^L f(\xi)G|_{\tau=0}\, d\xi + F_0 \int_0^t G(x,t;x_0,\tau)\cos\omega\tau\, d\tau - \kappa u_0 \int_0^t \frac{\partial G}{\partial \xi}\bigg|_{\xi=0} d\xi$$

If the series expression for G is substituted into the integrands and the operations of summation and integration are interchanged, it is easily verified that the resulting series gives

$$u = w + u_2$$

where w and u_2 are the functions defined in the previous paragraphs.

EXAMPLE 5-3   A solid conducting sphere of radius a is placed in a unidirectional field of constant flux q_0.   The initial temperature in the sphere is $f(r)$, where r denotes the distance from the center.   At the surface heat is conducted according to Newton's law of cooling.   The temperature of the exterior region remains zero.   There are no heat sources within.   Determine the temperature distribution in the sphere for $t > 0$.

Choose spherical coordinates with origin at the center of the sphere such that $\theta = 0$ coincides with the direction of the flux field.   Then the flux incident on a surface element exposed to the field and whose normal makes an angle θ with the direction of the field is $q_0 \cos\theta\, dA$.   The boundary condition satisfied by the temperature u is

$$\frac{\partial u}{\partial r} + hu = \begin{cases} h_0 q_0 \cos\theta & r = a;\, 0 \leq \theta \leq \dfrac{\pi}{2} \\[2ex] 0 & r = a;\, \dfrac{\pi}{2} \leq \theta \leq \pi \end{cases}$$

The initial condition is

$$u(r,\theta,\varphi,0) = f(r) \qquad 0 \leq r \leq a$$

The temperature satisfies the homogeneous heat equation

$$\frac{\partial u}{\partial t} = \kappa \left[\frac{1}{r^2} \frac{\partial}{\partial r} \left(r^2 \frac{\partial u}{\partial r} \right) + \frac{1}{r^2 \sin \theta} \frac{\partial}{\partial \theta} \left(\sin \theta \frac{\partial u}{\partial \theta} \right) + \frac{1}{r^2 \sin^2 \theta} \frac{\partial^2 u}{\partial \varphi^2} \right]$$

for $0 \leq r < a, 0 \leq \theta \leq \pi, 0 \leq \varphi \leq 2\pi$.

The expression for the temperature for $t > 0$ can be deduced with the aid of the Green's function of the problem and Eq. (5-43). First the eigenfunctions of the problem are obtained. These are the continuous single-valued solutions of the scalar Helmholtz equation (5-30) in the sphere which satisfy the homogeneous boundary condition

$$\frac{\partial \psi}{\partial r} + h\psi = 0 \qquad r = a$$

The eigenfunctions are derived by separation of variables in the Helmholtz equation written in spherical coordinates. Recall Sec. 4-7. Separable solutions which are single-valued and continuous everywhere in the sphere are

$$\psi_{n\lambda q}^{(e)} = j_n(\sqrt{\lambda}\, r) Y_{nq}^{(e)}(\theta,\varphi) \qquad \psi_{n\lambda q}^{(0)} = j_n(\sqrt{\lambda}\, r) Y_{nq}^{(0)}(\theta,\varphi) \qquad n = 0, 1, \ldots ; q = 0, 1, \ldots, n$$

The spherical Bessel functions j_n are defined in Eq. (4-124), and the spherical harmonics $Y_{nq}^{(e)}$, $Y_{nq}^{(0)}$ are defined in Eq. (3-68). The eigenvalues λ are defined by satisfaction of the homogeneous boundary condition on the surface $r = a$. Thus λ is an eigenvalue if

$$\sqrt{\lambda}\, j_n'(\sqrt{\lambda}\, a) + h j_n(\sqrt{\lambda}\, a) = 0$$

For each fixed $n = 0, 1, \ldots$ let $\{\xi_{nm}\}$ be the positive zeros of the function

$$\xi j_n'(\xi) + ha j_n(\xi)$$

arranged in increasing order. Then the real numbers

$$\lambda_{nm} = \frac{\xi_{nm}^2}{a^2} \qquad n = 0, 1, \ldots ; m = 1, 2, \ldots$$

are eigenvalues. Corresponding to the eigenvalue λ_{nm} there are the $2n + 1$ linearly independent real-valued eigenfunctions

$$\psi_{nmq}^{(e)} = j_n \left(\frac{\xi_{nm} r}{a} \right) Y_{nq}^{(e)}(\theta,\varphi) \qquad \psi_{nmq}^{(0)} = j_n \left(\frac{\xi_{nm} r}{a} \right) Y_{nq}^{(0)}(\theta,\varphi) \qquad q = 0, 1, \ldots$$

This set of eigenfunctions forms an orthogonal set on the sphere. The orthogonality property is expressed by equations which are formally the same as those written in Example 4-3. However the normalization factor here is

$$\| \psi_{nmq}^{(0)} \|^2 = \frac{\pi^2 a^3 \epsilon_q}{\xi_{nm}^3 (2n + 1)} \frac{(n + q)!}{(n - q)!} [\xi_{nm}^2 - n(n + 1) - 4ah + 4a^2 h^2] J_{n+1/2}^2(\xi_{nm})$$

where $\epsilon_0 = 2, \epsilon_q = 1, q \geq 1$. The same result holds for $\| \psi_{nmq}^{(0)} \|^2$ and is computed by means of the definition of the functions j_n, $Y_{nq}^{(e)}$, $Y_{nq}^{(0)}$. The set $\{\psi_{nmq}^{(e)}, \psi_{nmq}^{(0)}\}$ is complete in the space of all functions which are continuous and single-valued in the sphere. In turn this implies that the set $\{\lambda_{nm}\}$ constitutes all the eigenvalues of the present problem. The functions

$$u_{nmq}^{(e)}(r,\theta,\varphi,t) = j_n \left(\frac{\xi_{nm} r}{a} \right) Y_{nq}^{(e)}(\theta,\varphi) e^{-\kappa \lambda_{nm} t}$$

$$u_{nmq}^{(0)}(r,\theta,\varphi,t) = j_n \left(\frac{\xi_{nm} r}{a} \right) Y_{nq}^{(0)}(\theta,\varphi) e^{-\kappa \lambda_{nm} t}$$

are real-valued separable solutions of the homogeneous heat equation which are continuous and single-valued in the sphere and satisfy the homogeneous boundary condition

$$\frac{\partial u}{\partial r} + hu = 0 \qquad r = a$$

By Eq. (5-42) the Green's function is

$$G(r,\theta,\varphi,t;r_0,\theta_0,\varphi_0,\tau) = \sum_{n=0}^{\infty} \sum_{m=1}^{\infty} \sum_{q=0}^{n}$$

$$\times \left\{ \frac{j_n(\xi_{nm}r/a)j_n(\xi_{nm}r_0/a)P_n{}^q(\cos \theta)P_n{}^q(\cos \theta_0)}{\|\psi_{nm}\|^2} \cos[q(\varphi - \varphi_0)]e^{-\kappa\lambda_{nm}(t-\tau)} \right\}$$

for $t > \tau$

and

$$G(r,\theta,\varphi,t;r_0,\theta_0,\varphi_0,\tau) = 0 \qquad \text{for } t \le \tau$$

It follows from Eq. (5-43) that the solution of the problem is

$$u(r,\theta,\varphi,t) = \int_0^a \int_0^{2\pi} \int_0^{\pi} G|_{\tau=0} f(r_0)r_0{}^2 \sin \theta_0 \, dr_0 \, d\theta_0 \, d\varphi_0$$

$$+ \kappa \int_0^t d\tau \int_0^{2\pi} \int_0^{\pi} H(\theta_0)G(r,\theta,\varphi,t;a,\theta_0,\varphi_0,\tau)a^2 \sin \theta_0 \, d\theta_0 \, d\varphi_0$$

where $H(\theta_0) = h_0q_0 \cos \theta_0$, $0 \le \theta_0 \le \pi/2$, $H(\theta_0) = 0$, $\pi/2 \le \theta_0 \le \pi$. The series for $G|_{\tau=0}$ and $G|_{r_0=a}$ are substituted into the integrands, the operations of summation and integration are interchanged, and the properties of the trigonometric functions $\cos q\varphi_0$, $\sin q\varphi_0$, and the Legendre polynomials $P_n(\cos \theta_0)$ are utilized. The result is

$$u(r,\theta,t) = 4\pi \sum_{m=1}^{\infty} A_m j_0\left(\frac{\xi_{0m}r}{a}\right)e^{-\kappa\lambda_{0m}t} + a^2 h_0 q_0 \sum_{n=0}^{\infty} \sum_{m=1}^{\infty} (-1)^{n+1} \frac{1 \cdot 3 \cdots (2n-3)}{2 \cdot 4 \cdots (2n+2)}$$

$$\times \frac{j_{2n}(\xi_{2n,2m})j_{2n}(\xi_{2n,2m}r/a)}{\lambda_{2n,2m} \|\psi_{2n,2m}\|^2} P_{2n}(\cos \theta)(1 - e^{-\kappa\lambda_{2n,2m}t})$$

5-4 MAXIMUM-MINIMUM PRINCIPLE FOR THE HEAT EQUATION. AN EXISTENCE AND UNIQUENESS THEOREM

Let S be a simple closed piecewise smooth surface which bounds a region $\mathscr{V}$ of xyz space, and let $\overline{\mathscr{V}}$ denote $\mathscr{V}$ together with S. An important property of solutions of the homogeneous heat equation

$$u_t = \kappa \, \Delta u \qquad\qquad (5\text{-}48)$$

is expressed in the following theorem.

THEOREM 5-1 (Maximum-minimum principle.) Let u, as a function of the four independent variables x, y, z, t, be continuous on $\overline{\mathscr{V}}$ for $t \ge 0$. Let u be a solution of Eq. (5-48) in $\mathscr{V}$ for $t > 0$. If $T > 0$ is a fixed time, the greatest and least values of u on $\overline{\mathscr{V}}$ for $0 \leq t \leq T$ are attained either in $\mathscr{V}$ at time $t = 0$ or else on the boundary S at some time in the interval $0 \leq t \leq T$.

Proof   If the conclusion stated does not hold, then there exists a $T > 0$ such that at least one of the values

$$M = \max u(x,y,z,t) \qquad (x,y,z) \text{ in } \overline{\mathscr{V}}; \, 0 \leq t \leq T$$

$$m = \min u(x,y,z,y) \qquad (x,y,z) \text{ in } \overline{\mathscr{V}}; \, 0 \leq t \leq T$$

is taken on in $\mathscr{V}$ at some time $\hat{t}$ such that $0 < \hat{t} \leq T$.   It is shown now that the supposition that M is taken on as indicated leads to a contradiction. Define the function v by

$$v(x,y,z,t) = u(x,y,z,t) - \epsilon(t - T) \qquad (x,y,z) \text{ in } \overline{\mathscr{V}}; \, t \geq 0$$

where $\epsilon > 0$ is a constant to be chosen.   Consider that

$$u(x,y,z,t) \leq v(x,y,z,t) \leq u(x,y,z,t) + \epsilon T$$

for (x,y,z) in $\overline{\mathscr{V}}$ and $0 \leq t \leq T$.   Let M_1 be the maximum of u on $\overline{\mathscr{V}}$ at $t = 0$, M_2 the maximum of u on S during the interval $0 \leq t \leq T$, and let M_3 be the larger of the numbers M_1, M_2.   Then the assumption made initially is that $M_3 < M$.   Observe that $v(x,y,z,0) \leq M_3 + \epsilon T$ for all points (x,y,z) in $\overline{\mathscr{V}}$, and $v(x,y,z,t) \leq M_3 + \epsilon T$ for all points on S during the interval $0 \leq t \leq T$.   On the other hand, if (x_1,y_1,z_1) is a point in $\mathscr{V}$ at which u takes on the value M at $t = \hat{t}$, then

$$v(x_1,y_1,z_1,\hat{t}) = M + \epsilon(T - \hat{t}) \geq M$$

Thus if $\epsilon > 0$, but sufficiently small so that

$$M_3 + \epsilon T < M$$

then it is clear that

$$\max v(x,y,z,t) \qquad (x,y,z) \text{ in } \overline{\mathscr{V}}; \, 0 \leq t \leq T$$

is taken on in $\mathscr{V}$ at some point (x_0,y_0,z_0) and at some time t_0 such that $0 < t_0 \leq T$.   Of necessity the inequalities

$$v_{xx}(x_0,y_0,z_0,t_0) \leq 0 \qquad v_{yy}(x_0,y_0,z_0,t_0) \leq 0 \qquad v_{zz}(x_0,y_0,z_0,t_0) \leq 0$$

must hold.   The fact that u satisfies Eq. (5-48) implies

$$v_t - \kappa \, \Delta v = u_t - \epsilon - \kappa \, \Delta u = -\epsilon$$

and so

$$v_t(x_0,y_0,z_0,t_0) = \kappa \, \Delta v(x_0,y_0,z_0,t_0) - \epsilon \leq -\epsilon$$

By continuity a constant $h > 0$ can be chosen sufficiently small so that

$$v_t(x_0,y_0,z_0,t) \leq -\frac{\epsilon}{2} \qquad t_0 - h \leq t \leq t_0$$

Thus if t_1 is a value such that $t_0 - h < t_1 < t_0$, then

$$v(x_0,y_0,z_0,t_0) - v(x_0,y_0,z_0,t_1) = \int_{t_1}^{t_0} v_t(x_0,y_0,z_0,t)\, dt \leq \frac{\epsilon}{2} \int_{t_1}^{t_0} dt < 0$$

so that

$$v(x_0,y_0,z_0,t_0) < v(x_0,y_0,z_0,t_1)$$

But this contradicts the assumption that v takes on its maximum at (x_0,y_0,z_0) at time t_0.   Accordingly the maximum must be taken on as stated in the theorem.   Now observe that $-u$ is continuous on $\overline{\mathscr{V}}$ for $t \geq 0$ and satisfies Eq. (5-48) in $\mathscr{V}$ for $t > 0$, and

$$\max(-u) = \min u$$

Hence if u takes on its minimum in $\mathscr{V}$ at some time in the half-open interval $(0,T]$, then $-u$ takes on its maximum in $\mathscr{V}$ in $(0,T]$.   But as just shown, this leads to a contradiction.   Accordingly the minimum must be taken on as stated in the theorem.

Let the function u having the properties assumed in the theorem represent a temperature field in a solid conductor occupying $\overline{\mathscr{V}}$.   Then the physical interpretation of the maximum-minimum principle is as follows.   During a given time interval $[0,T]$ the temperature inside the conductor cannot exceed the largest value of the initial temperature, nor can it exceed the maximum temperature on the surface for $0 \leq t \leq T$.   A similar statement holds with largest replaced by smallest.

In the case of the one-dimensional homogeneous heat equation

$$u_t = \kappa u_{xx} \qquad\qquad (5\text{-}49)$$

the domain $\overline{\mathscr{V}}$ assumed in Theorem 5-1 reduces to a closed interval $[a,b]$ of the x axis.   The boundary S consists of the end points $x = a$, $x = b$.   Let u be a solution of Eq. (5-49) for

$$a < x < b \qquad t > 0$$

Fix $T > 0$, and let $\overline{\mathscr{D}}$ denote the closed rectangle in the xt plane defined by the inequalities

$$a \leq x \leq b \qquad 0 \leq t \leq T$$

See Fig. 5-1.   Let

$M_0 = \max u(x,0)$ $a \le x \le b$

$M_1 = \max u(a,t)$ $M_2 = \max u(b,t)$ $0 \le t \le T$

$M = \max \{M_0, M_1, M_2\}$

In an analogous fashion define the minimum m of u on the part of the boundary of $\bar{\mathscr{D}}$ consisting of the base and the two vertical sides.   Then Theorem 5-1 states that

$$m \le u(x,t) \le M \qquad \text{in } \bar{\mathscr{D}} \tag{5-50}$$

Moreover the values m, M must be attained either on the base or the vertical sides.

The maximum-minimum principle can be used to prove the uniqueness of solution of problem (5-27).   For simplicity the proof given here is for the one-dimensional problem

$$u_t - \kappa u_{xx} = F(x,t) \qquad a < x < b; t > 0$$
$$u(a,t) = H_1(t) \qquad u(b,t) = H_2(t) \qquad t \ge 0 \tag{5-51}$$
$$u(x,0) = f(x) \qquad a \le x \le b$$

Let the given functions F, H_1, H_2, and f be continuous on their respective domains, and assume

$$H_1(0) = f(a) \qquad H_2(0) = f(b)$$

Then there is at most one solution of problem (5-51) which is continuous in x and t for $a \le x \le b$, $t \ge 0$.   To prove this let u_1, u_2 be functions with the

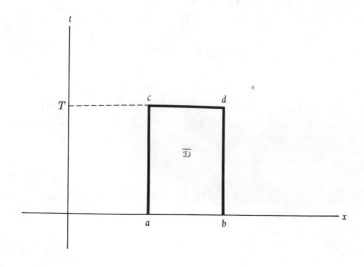

Figure 5-1

properties just described, and let

$$v = u_1 - u_2$$

Then v is a solution of the homogeneous heat equation (5-49) for $a < x < b$ and is continuous in x and t on closed infinite strip $a \le x \le b$, $t \ge 0$, and

$$v(x,0) = 0 \qquad a \le x \le b \qquad\qquad v(a,t) = v(b,t) = 0 \qquad t \ge 0$$

Fix $T > 0$ and let $\bar{\mathscr{D}}$ be the rectangle in Fig. 5-1. Then the inequality (5-50) implies

$$v(x,t) = 0 \qquad \text{in } \bar{\mathscr{D}}$$

Since $T > 0$ was arbitrarily chosen, it follows that

$$u_1(x,t) = u_2(x,t) \qquad a \le x \le b; t \ge 0$$

If, instead, the boundary condition in problem (5-51) is

$$u_x(a,t) - hu(a,t) = H_1(t) \qquad u_x(b,t) + hu(b,t) = H_2(t)$$

where $h \ge 0$ is a constant, then the uniqueness of solution can be proved using a modified form of the maximum-minimum principle.

If the boundary conditions are homogeneous, problem (5-51) becomes

$$u_t - \kappa u_{xx} = F(x,t) \qquad a < x < b; t > 0$$
$$u(a,t) = 0 \qquad u(b,t) = 0 \qquad t \ge 0 \tag{5-52}$$
$$u(x,0) = f(x) \qquad a \le x \le b$$

Separation of variables leads to the formal series solution

$$u = u_1 + u_2 \tag{5-53}$$

where

$$u_1(x,t) = \sum_{n=1}^{\infty} A_n \sin \frac{n\pi(x-a)}{b-a} e^{-\kappa \lambda_n t} \tag{5-54}$$

$$A_n = \frac{2}{b-a} \int_a^b f(x) \sin \frac{n\pi(x-a)}{b-a} \, dx \qquad n = 1, 2, \ldots \tag{5-55}$$

and

$$u_2(x,t) = \sum_{n=1}^{\infty} B_n(t) \sin \frac{n\pi(x-a)}{b-a} \tag{5-56}$$

where

$$B_n(t) = \int_0^t F_n(\tau) e^{-\kappa \lambda_n (t-\tau)} \, d\tau \qquad n = 1, 2, \ldots \tag{5-57}$$

$$F_n(t) = \frac{2}{b-a} \int_a^b F(x,t) \sin \frac{n\pi(x-a)}{a-b} \, dx \qquad n = 1, 2, \ldots \tag{5-58}$$

The eigenvalues of the problem are

$$\lambda_n = \frac{n^2\pi^2}{(b-a)^2} \qquad n = 1, 2, \ldots \tag{5-59}$$

Equation (5-54) is the formal series solution of the problem

$$u_t = \kappa u_{xx} \qquad a < x < b; t > 0$$
$$u(a,t) = 0 \qquad u(b,t) = 0 \qquad t \geq 0 \tag{5-60}$$
$$u(x,0) = f(x) \qquad a \leq x \leq b$$

Assume f is continuous and piecewise smooth on $[a,b]$ and $f(a) = f(b) = 0$. Then the formal series (5-54) converges and defines the classical solution of problem (5-60). To prove this fact recall that

$$\sum_{n=1}^{\infty} A_n \sin \frac{n\pi(x-a)}{b-a} \tag{5-61}$$

where the A_n are defined in Eq. (5-55), is the Fourier series of f relative to the orthogonal set

$$\left\{ \sin \left[\frac{n\pi(x-a)}{b-a} \right] \right\}$$

and the series converges uniformly to f on $[a,b]$ under the present hypotheses. The Cauchy criterion for uniform convergence now implies that given $\epsilon > 0$, there exists an integer $N_\epsilon > 0$, independent of x, such that if $n \geq m \geq N_\epsilon$,

$$\left| \sum_{k=m}^{n} A_k \sin \frac{k\pi(x-a)}{b-a} \right| < \epsilon \qquad a \leq x \leq b$$

Choose $T > 0$, and let $\bar{\mathscr{D}}$ be the rectangle shown in Fig. 5-1. The maximum-minimum principle is now applied to show the series (5-54) converges uniformly in x and t together on $\bar{\mathscr{D}}$. With the integers n, m fixed such that $n \geq m \geq N$, define the function S_{nm} by

$$S_{nm}(x,t) = \sum_{k=m}^{n} A_k \sin \frac{k\pi(x-a)}{b-a} e^{-\kappa\lambda_k t} \qquad a \leq x \leq b; t \geq 0$$

Observe S_{nm} is a twice continuously differentiable solution of Eq. (5-49) on $\bar{\mathscr{D}}$, and

$$S_{nm}(a,t) = 0 \qquad S_{nm}(b,t) = 0 \qquad 0 \leq t \leq T$$

Moreover

$$-\epsilon < S_{nm}(x,0) < \epsilon \qquad a \leq x \leq b$$

By the maximum-minimum principle

$$|S_{nm}(x,t)| < \epsilon \qquad (x,t) \text{ in } \overline{\mathcal{D}}$$

What has been demonstrated is that given $\epsilon > 0$, there exists an integer $N_\epsilon > 0$, independent of x and t, such that if $n \geq m \geq N$, then

$$\left| \sum_{k=m}^{n} A_k \sin \frac{k\pi(x-a)}{b-a} e^{-\kappa \lambda_k t} \right| < \epsilon$$

for $a \leq x \leq b$, $0 \leq t \leq T$. However this is just the Cauchy criterion for the uniform convergence of the series on $\overline{\mathcal{D}}$. Thus the series defines a continuous function u on $\overline{\mathcal{D}}$. Since $T > 0$ was chosen arbitrarily, the series (5-54) defines a continuous function for $a \leq x \leq b$, $t \geq 0$. The series converges at $t = 0$, and $u(x,0) = f(x)$, $a \leq x \leq b$. Since u is continuous,

$$\lim_{\substack{t \to 0 \\ t > 0}} u(x,t) = u(x,0) = f(x) \qquad a \leq x \leq b$$

that is, the initial values are taken on continuously. Clearly u satisfies the homogeneous boundary conditions. It remains to show u satisfies Eq. (5-49). This follows by the same type of argument as that given in the corresponding proof in Example 5-2. The Fourier coefficients of f are bounded:

$$|A_n| \leq 2M \qquad n = 1, 2, \ldots ; \ M = \max f(x); \ a \leq x \leq b$$

Fix $t_0 > 0$. Then the series

$$\sum_{n=1}^{\infty} A_n(-\kappa\lambda_n) \sin \frac{n\pi(x-a)}{b-a} e^{-\kappa\lambda_n t}$$

obtained by termwise differentiation of the series (5-54) with respect to t, is dominated term by term for $a \leq x \leq b$, $t \geq t_0$ by the convergent series of positive constants

$$\sum_{n=1}^{\infty} 2M\kappa\lambda_n e^{-\kappa\lambda_n t_0}$$

Accordingly u has a continuous first derivative with respect to t for $t \geq t_0$. In the same manner it is seen that u has a continuous second derivative with respect to x for $a \leq x \leq b$, $t \geq t_0$. Now since each term is a solution, and since termwise differentiation, once with respect to t and twice with respect to x, is possible, it follows that u is a solution of the homogeneous heat equation for $t \geq t_0$. But $t_0 > 0$ was arbitrarily chosen. Hence u is a solution for $t > 0$.

Under suitable hypotheses the formal series (5-56) converges and defines the solution of the problem

$$u_t - \kappa u_{xx} = F(x,t) \qquad a < x < b; t > 0$$
$$u(a,t) = 0 \qquad u(b,t) = 0 \qquad t \geq 0 \qquad\qquad (5\text{-}62)$$
$$u(x,0) = 0 \qquad a \leq x \leq b$$

Assume F, F_x, F_{xx} are continuous functions of x and t for $a \leq x \leq b$, $t \geq 0$, and $F(a,t) = F(b,t)$, $t \geq 0$. Fix $T > 0$, and let $\overline{\mathscr{D}}$ be the closed rectangle in Fig. 5-1. Then from Eq. (5-58)

$$|F_n(t)| \leq \frac{2}{b-a} \int_a^b |F(x,t)| \, dx \leq 2M_T$$

where

$$M_T = \max F(x,t) \qquad a \leq x \leq b; 0 \leq t \leq T$$

Hence

$$|B_n(t)| \leq 2M_T e^{-\kappa \lambda_n t} \int_0^t e^{\kappa \lambda_n \tau} \, d\tau = \frac{2M_T(1 - e^{-\kappa \lambda_n t})}{\kappa \lambda_n}$$
$$< \frac{2M_T}{\kappa \lambda_n} \qquad 0 \leq t \leq T$$

Thus the series (5-56) is dominated term by term on $\overline{\mathscr{D}}$ by the convergent series

$$\sum_{n=1}^{\infty} \frac{2M_T}{\kappa \lambda_n}$$

This implies the series converges uniformly in x and t on $\overline{\mathscr{D}}$ and defines a continuous function u there. The value $T > 0$ was arbitrarily chosen; hence the series defines a continuous function u for $a \leq x \leq b$, $t \geq 0$. Clearly u satisfies the homogeneous boundary conditions. Also

$$\lim_{\substack{t \to 0 \\ t > 0}} u(x,t) = u(x,0) = 0 \qquad a \leq x \leq b$$

since u is continuous for $t \geq 0$ and $B_n(0) = 0$, all n. Thus u takes on the zero initial values in a continuous manner. It remains to show u satisfies the inhomogeneous heat equation. Integration by parts twice in Eq. (5-58) yields the inequality

$$|F_n(t)| < \frac{K}{\lambda_n} \qquad K = \text{const}$$

Now let t_0 be fixed, $0 < t_0 < T$. Then for $t_0 \leq t \leq T$

$$\dot{B}_n(t) = F_n(t) - \kappa \lambda_n B_n(t)$$

and

$$|B_n(t)| \leq e^{-\kappa\lambda_n t} \left| \int_0^t F_n(\tau) e^{\kappa\lambda_n \tau} \, d\tau \right| < \frac{K e^{-\kappa\lambda_n t}}{\lambda_n} \int_0^t e^{\kappa\lambda_n \tau} \, d\tau \leq \frac{K}{\kappa\lambda_n{}^2}$$

so

$$|\dot{B}_n(t)| \leq |F_n(t)| + \kappa\lambda_n |B_n(t)| < \frac{2K}{\lambda_n}$$

The series of constants

$$\sum_{n=1}^{\infty} \frac{2K}{\lambda_n}$$

converges. Moreover the series resulting from termwise differentiation of the series (5-56) with respect to t is dominated term by term for $t \geq t_0$ by this series of positive constants. Hence the differentiated series converges and converges uniformly on every closed rectangle of the form

$$a \leq x \leq b \qquad t_0 \leq t \leq T$$

In the same way it is seen that termwise differentiation of the series (5-56) twice with respect to x yields a uniformly convergent series on the preceding rectangle. Since the termwise differentiations are permissible, it follows that

$$u_t - \kappa u_{xx} = \sum_{n=1}^{\infty} [\dot{B}_n(t) + \kappa\lambda_n B_n(t)] \sin \frac{n\pi(x-a)}{b-a}$$

$$= \sum_{n=1}^{\infty} F_n(t) \sin \frac{n\pi(x-a)}{b-a} = F(x,t) \qquad a \leq x \leq b; t_0 \leq t \leq T$$

The values $T > t_0 > 0$ were arbitrarily chosen. Hence the function u satisfies the inhomogeneous heat equation for $t > 0$.

THEOREM 5-2 In problem (5-52) let f be continuous and piecewise smooth on $[a,b]$ and such that $f(a) = f(b) = 0$. Assume F is such that F, F_x, F_{xx} are continuous functions of x and t for $a \leq x \leq b$, $t \geq 0$, and

$$F(a,t) = F(b,t) = 0 \qquad t \geq 0$$

Then problem (5-52) has a unique solution. The solution is given by $u = u_1 + u_2$, where u_1 and u_2 are defined by the convergent series (5-54) and (5-56), respectively.

Recall the definition of a properly posed problem given in Sec. 3-5. Then it is clear that problem (5-52) is properly posed in the class of all functions which are continuous on $a \leq x \leq b$, $t \geq 0$, and twice continuously differentiable on $a < x < b$, $t > 0$. The existence and uniqueness of solution was established above. The fact that the solution depends continuously on the data follows from the maximum-minimum principle (and in the same manner in which Theorem 3-5 was proven), or, alternatively, from the representation (5-47) involving the *Green's function*. These remarks about problem (5-52) hold as well for the general problem (5-27).

PROBLEMS

Sec. 5-2

1 Let ξ, τ, ω be fixed values, $\omega > 0$. Verify that the function of x and t defined by

$$u(x,t;\xi,\omega) = \cos\left[\omega(x - \xi)\right]e^{-\kappa\omega^2(t-\tau)}$$

satisfies the one-dimensional homogeneous heat equation. Interpret $t = \tau$ as an "initial time." Then u represents the subsequent temperature in the infinitely long bar due to the initial temperature

$$u(x,\tau;\xi,\tau,\omega) = \cos\left[\omega(x - \xi)\right]$$

This distribution is sinusoidal along the length of the bar with frequency ω. The summation over all frequencies is

$$v(x,t;\xi,\tau) = \frac{1}{\pi}\int_0^\infty \cos\left[\omega(x - \xi)\right]e^{-\kappa\omega^2(t-\tau)}\,d\omega$$

Show that the improper integral converges for $t > \tau$ in such a manner that in calculating v_t and v_{xx} differentiation within the integral with respect to x and t is permissible, and hence show that v is a solution of the heat equation on the open half plane. Now make use of the integration formula

$$\int_0^\infty e^{-x^2}\cos 2bx\,dx = \tfrac{1}{2}\sqrt{\pi}\,e^{-b^2}$$

and obtain the expression (5-13) for the fundamental solution.

2 The initial temperature in the infinitely long bar is

$$f(x) = u_0 \qquad |x| \leq \delta \qquad\qquad f(x) = 0 \qquad |x| > \delta$$

where u_0, δ are given positive constants. There are no heat sources within the bar. Show that the subsequent temperature distribution for $t > 0$ is

$$u(x,t) = \frac{u_0}{2}\left(\operatorname{erf}\frac{x + \delta}{\sqrt{4\pi\kappa t}} - \operatorname{erf}\frac{x - \delta}{\sqrt{4\pi\kappa t}}\right)$$

where the error function is defined by

$$\operatorname{erf} x = \frac{2}{\sqrt{\pi}}\int_0^x e^{-\eta^2}\,d\eta \qquad -\infty < x < \infty$$

3 Solve Prob. 2 if the initial temperature is given by

$$f(x) = \begin{cases} u_1 & -\delta_1 \leq x \leq \delta_1 \\ u_2 & -\delta_2 \leq x < -\delta_1 \text{ or } \delta_1 < x \leq \delta_2 \\ 0 & \text{elsewhere} \end{cases}$$

where the constants u_1, u_2, δ_1, δ_2 are such that

$$0 < u_2 < u_1 \qquad 0 < \delta_1 < \delta_2$$

4   A slender homogeneous conducting bar of uniform cross section extends along the x axis for $x \geq 0$.   The lateral surface is insulated.   The end $x = 0$ is held at temperature $h(t)$ for $t \geq 0$.   The initial temperature is described by the given function $f(x)$, $x \geq 0$. The boundary- and initial-value problem is then

$$u_t - \kappa u_{xx} = F(x,t) \qquad x > 0; t > 0$$
$$u(0,t) = h(t) \qquad t \geq 0 \qquad\qquad u(x,0) = f(x) \qquad x \geq 0$$

To solve the problem let v be the function defined by Eq. (5-13) and define the function w by

$$w(x,t;\xi,\tau) = v(x,t;\xi,\tau) - v(x,t;-\xi,\tau) \qquad -\infty < x < \infty; t > \tau$$
$$w(x,\tau;\xi,\tau) = 0 \qquad -\infty < x < \infty$$

Then w is the *fundamental solution* of the problem of the heat flow in the semi-infinite bar. Observe w is a twice continuously differentiable solution of the homogeneous heat equation for $t > \tau$ and is continuous in x and t everywhere on the half plane $t \geq \tau$ except at the points (ξ,τ), $(-\xi,\tau)$, where it suffers infinite discontinuities.   Show that

$$\lim_{\substack{(x,t)\to(x_0,\tau)\\t>\tau}} w = 0 \qquad x_0 \neq \xi; x_0 \neq -\xi$$

$$\lim_{\substack{t\to\tau\\t>\tau}} w(\xi,t;\xi,\tau) = +\infty \qquad \lim_{\substack{t\to\tau\\t>\tau}} w(-\xi,t;\xi,\tau) = -\infty$$

Note that w is an odd function of x (also odd in ξ), and is symmetric in x and ξ.   Interpret w as the temperature in the infinitely long bar $(-\infty < x < \infty)$ for $t > \tau$ due to an initial temperature

$$f(x) = \delta(x - \xi) - \delta(x + \xi)$$

This is a point source at $x = \xi$ and a point sink at $x = -\xi$.   Observe that

$$w(0,t;\xi,\tau) = 0 \qquad t > \tau$$

Following the method used in the text prior to Eq. (5-20) give a heuristic derivation of the solution of the problem in the form $u = u_1 + u_2 + u_3$, where

$$u_1(x,t) = \int_0^\infty w(x,t;\xi,0)f(\xi)\,d\xi \qquad u_2(x,t) = \kappa \int_0^t h(\tau)\frac{\partial w}{\partial \xi}\Big|_{\xi=0} d\tau$$

$$u_3(x,t) = \int_0^t d\tau \int_0^\infty w(x,t;\xi,\tau)F(\xi,\tau)\,d\xi$$

The function u_1 is the solution of the problem

$$u_t = \kappa u_{xx} \qquad x > 0; t > 0$$
$$u(0,t) = 0 \qquad t > 0 \qquad\qquad u(x,0) = f(x) \qquad x > 0$$

The function u_2 is the solution of the problem

$$u_t = \kappa u_{xx} \qquad x > 0; t > 0$$
$$u(0,t) = h(t) \qquad t > 0 \qquad\qquad u(x,0) = 0 \qquad x > 0$$

Show that u_2 can be rewritten

$$u_2(x,t) = \frac{2}{\sqrt{\pi}} \int_{x/\sqrt{4\kappa t}}^\infty h\left(t - \frac{x^2}{4\kappa\eta^2}\right) e^{-\eta^2}\,d\eta$$

The function u_3 is the solution of the problem

$$u_t - \kappa u_{xx} = F(x,t) \qquad x > 0; t > 0$$

$$u(0,t) = 0 \qquad t > 0 \qquad\qquad u(x,0) = 0 \qquad x > 0$$

5   The problem of heat flow in a semi-infinite slab is mathematically identical to the problem of heat flow in the semi-infinite bar.   Assume that a homogeneous conductor occupies the half space $x \geq 0$ in xyz space.   The temperature on the face $x = 0$ is $h(t)$ at time $t > 0$. The initial temperature in the slab is $f(x)$, $x > 0$.   Thus the subsequent heat flow is one-dimensional.   Assume there are no heat sources within the slab.   Determine the subsequent temperature for each of the following cases.

(a) $h(t) = 0$; $f(x) = u_0$ (u_0 a positive constant)

(b) $h(t) = u_0$; $f(x) = 0$

(c) $h(t) = u_1$; $f(x) = u_2$ (u_1, u_2 nonzero constants)

(d) $h(t) = u_1$, $0 < t < t_1$, $h(t) = u_2$, $t_1 < t < t_2$, $h(t) = 0$ $t > t_2$; $f(x) = 0$

6   The face $x = 0$ of the semi-infinite slab described in Prob. 5 radiates heat into the exterior region $x < 0$ in such a way that the flux across the face is a constant q_0.   The initial temperature is u_0, a constant.   There are no heat sources within the slab.   Show that the subsequent temperature in the slab is

$$u(x,t) = u_0 + \frac{q_0}{K}\left[x\left(\text{erf}\, \frac{x}{\sqrt{4\kappa t}} - 1 \right) + \frac{\sqrt{4\kappa t}}{\pi}\left(e^{-x^2/4\kappa t} - 1 \right) \right]$$

Hint: Recall the flux $q = -Ku_x$.   Show that if u satisfies the homogeneous heat equation, then so does q.   Use the result of Prob. 5b to obtain the appropriate flux q.   Now solve the partial differential equation $u_x = -q/K$ and apply the initial condition.

7   The model which describes the heat flow in the infinitely long bar in which the insulated surface is replaced by a condition of radiation into the exterior region (held at temperature zero) is

$$u_t + bu - \kappa u_{xx} = F(x,t) \qquad -\infty < x < \infty; t > 0$$

$$u(x,0) = f(x) \qquad\qquad -\infty < x < \infty$$

where b is a positive constant.   Show that if w is a solution of problem (5-11), where $F(x,t)$ is replaced by $e^{-bt}F(x,t)$, then

$$u = e^{-bt}w$$

is a solution of the original problem.   Thus the *fundamental solution* of the original problem is

$$\psi(x,t;\xi,\tau) = v(x,t;\xi,\tau)e^{-b(t-\tau)}$$

where v is defined by Eq. (5-13).   Find the temperature in the bar for $t > 0$ if there are no heat sources and the initial temperature is u_0, a positive constant.

8   A thin homogeneous conducting plate of uniform thinness lies in the xy plane.   The plate is of sufficiently large extent so that edge effects can be neglected.   The initial temperature in the plate is

$$u(x,y,0) = \begin{cases} u_0 & |x| < a \text{ and } |y| < a \\ 0 & \text{elsewhere} \end{cases}$$

The heat source density in the plate is

$$F(x,y,t) = F_0 \delta(x - x_0) \delta(y - y_0)$$

where (x_0, y_0) is a given point and F_0 is a positive constant. Find the subsequent temperature in the plate.

9 Let (r, θ, φ) be spherical coordinates. A temperature distribution $u(r,t)$ which is purely radially dependent satisfies

$$u_t - \kappa \frac{(r^2 u_r)_r}{r^2} = F(r,t) \qquad r > 0; t > 0$$

and the initial condition $u(r,0) = f(r)$ for $r \geq 0$. Here F and f are given functions. Let $w = ru$ and obtain

$$w_t - \kappa w_{rr} = rF(r,t) \qquad r > 0; t > 0$$
$$w(0,t) = 0 \qquad t \geq 0 \qquad\qquad w(r,0) = rf(r) \qquad r \geq 0$$

Use the results of Prob. 4 to derive the solution of the original problem in the form

$$u(r,t) = \frac{1}{r\sqrt{4\pi\kappa t}} \int_0^\infty (e^{-(r-r_0)^2/4\kappa(t-\tau)} - e^{-(r+r_0)^2/4\kappa(t-\tau)}) r_0 f(r_0)\, dr_0$$

$$+ \frac{1}{r} \int_0^t \frac{d\tau}{4\pi\kappa(t-\tau)} \int_0^\infty (e^{-(r-r_0)^2/4\kappa(t-\tau)} - e^{-(r+r_0)^2/4\kappa(t-\tau)}) r_0 F(r_0,\tau)\, dr_0$$

$$= \int_0^\infty \int_0^{2\pi} \int_0^\pi \psi(r,t;r_0,0) f(r_0)\, dV_0 + \int_0^t d\tau \int_0^\infty \int_0^{2\pi} \int_0^\pi \psi(r,t;r_0,\tau) F(r_0,\tau)\, dV_0$$

where $dV_0 = r_0^2 \sin\theta_0\, d\theta_0\, d\varphi_0\, dr_0$ is the volume element and ψ is the *fundamental solution*

$$\psi(r,t;r_0,\tau) = \frac{v(r,t;r_0,\tau) - v(r,t;-r_0,\tau)}{8\pi^{3/2} r r_0}$$

where v is defined in Eq. (5-13). Find the solution of the original problem if the source density is zero and the initial temperature in the space is u_0, a positive constant.

10 Show that the solution of the problem

$$u_t = \kappa \left(u_{rr} + \frac{2u_r}{r} \right) \qquad r > a > 0; a = \text{const}$$

$$u(r,0) = u_0 \qquad r \geq a \qquad\qquad u(a,t) = U_0$$

is

$$u(r,t) = \frac{a}{r}(U_0 - u_0)\left(1 - \text{erf}\, \frac{r-a}{\sqrt{4\kappa t}} \right)$$

where erf x is the error function defined in Prob. 2.

Sec. 5-3

11   A slender homogeneous conducting bar of uniform cross section lies along the x axis with ends at $x = 0$, $x = L$.   The lateral surface is insulated.   There are no heat sources within the bar.   The bar has the indicated end conditions and initial temperature.   Derive the series solution for the temperature distribution in the bar for $t > 0$.

(a) Ends are at zero temperature; $f(x) = x$, $0 \leq x \leq L/2$, $f(x) = L - x$, $L/2 \leq x \leq L$

(b) Ends are at zero temperature; $f(x) = 3 \sin 2\pi x/L$, $0 \leq x \leq L$

(c) $u(0,t) = u_1$, $u(L,t) = u_2$, u_1, u_2 nonzero constants; $u(x,0) = f(x)$, $0 \leq x \leq L$

(d) Ends are insulated; $f(x) = x(L - x)$, $0 \leq x \leq L$

(e) End $x = 0$ held at temperature zero, at the end $x = L$ there is a constant flux q_0; $u(x,0) = f(x)$, $0 \leq x \leq L$

12   Instead of being insulated, the lateral surface of the bar described in Prob. 11 radiates heat into the surroundings at temperature zero.   The ends $x = 0$, $x = L$ are held at temperatures u_1 and u_2, respectively.   The initial temperature is $u(x,0) = f(x)$, $0 \leq x \leq L$. Show that the subsequent temperature in the bar is

$$u(x,t) = \varphi(x) + e^{-bt} \sum_{n=1}^{\infty} A_n \sin \frac{n\pi x}{L} e^{-\kappa \lambda_n t}$$

where

$$\varphi(x) = \frac{u_1 \sinh [\sqrt{b}(L - x)/\sqrt{\kappa}] + u_2 \sinh (\sqrt{b}\, x/\sqrt{\kappa})}{\sinh (\sqrt{b}\, L/\sqrt{\kappa})}$$

$$\lambda_n = \frac{n^2 \pi^2}{L^2} \qquad A_n = \frac{2}{L} \int_0^L [f(x) - \varphi(x)] \sin \frac{n\pi x}{L}\, dx \qquad n = 1, 2, \ldots$$

Hint: Find a solution of $u_t + bu = \kappa u_{xx}$ which satisfies the boundary conditions.   Then superimpose with a solution which satisfies homogeneous boundary conditions and an appropriate initial condition.

13   The lateral surface of a slender homogeneous conducting bar of length L is insulated. The temperature at the end $x = 0$ is maintained at $A \sin \omega_1 t$ for $t \geq 0$, where A, ω_1 are positive constants.   The end $x = L$ radiates heat into the exterior region, which is at temperature zero.   The initial temperature in the bar is u_0, a constant.   The heat-source density within the bar is

$$F(x,t) = F_0 \delta(x - x_0) \cos \omega t$$

where x_0 is a given point and F_0, ω are positive constants.   Determine the temperature in the bar for $t > 0$.

14   An infinite slab is bounded by the parallel planes $x = 0$, $x = L$ in xyz space.   The face $x = 0$ is insulated.   Across the face $x = L$ there is a constant inward flux of magnitude q_0.   The initial temperature in the slab is u_0 (a constant).   Show that the temperature in the slab for $t > 0$ is

$$u(x,t) = u_0 + q_0 \left[\frac{\kappa t}{L} + \frac{3x^2 - L^2}{6L} + \frac{2L}{\pi^2} \sum_{n=1}^{\infty} \frac{(-1)^{n+1}}{n^2} \cos \frac{n\pi x}{L} e^{-\kappa n^2 \pi^2 t/L^2} \right]$$

15 The initial temperature of a solid homogeneous conducting cylinder of radius a is

$$u_0\left(1 - \frac{r^2}{a^2}\right) \qquad 0 \leq r \leq a$$

where r is the distance from the axis. The temperature on the lateral surface is zero. Assume the cylinder is infinitely long. There are no heat sources within the cylinder. Determine the temperature distribution in the cylinder for $t > 0$.

16 An infinitely long tube has inner radius a and outer radius b. The material of the tube is homogeneous and contains no heat sources. The inner surface is insulated, and the outer surface is maintained at temperature u_0. The initial temperature is zero. Show that the temperature distribution for $t > 0$ and $a < r < b$ is

$$u(r,t) = u_0 + \pi u_0 \sum_{n=1}^{\infty} J_1^2(\xi_n a) \left[\frac{Y_0(\xi_n h)J_0(\xi_n r) - J_0(\xi_n b) Y_0(\xi_n r)}{J_0^2(\xi_n b) - J_1^2(\xi_n a)}\right] e^{-\kappa \xi_n^2 t}$$

where r is the distance from the axis and $\{\xi_n\}$ is the sequence of positive zeros of the function

$$\varphi(\xi) = J_0(b\xi)\,Y_1(a\xi) - J_1(a\xi)\,Y_0(b\xi)$$

17 A solid homogeneous circular cylinder of radius a and altitude b has its axis coincident with the z axis. The initial temperature is

$$u(r,\theta,z,0) = f(r,\theta,z)$$

where r, θ, z are cylindrical coordinates. The temperature of the base is held constant, at temperature u_0. The top and lateral surfaces are insulated. Determine the temperature in the cylinder for $t > 0$.

18 With reference to the cylinder described in Prob. 17 the top, bottom, and lateral surfaces all radiate heat into the exterior region, which is at temperature zero. The initial temperature

$$u(r,\theta,z,0) = f(r,z)$$

Show that the temperature in the cylinder for $t > 0$ is

$$u = \sum_{n=1}^{\infty} \sum_{m=1}^{\infty} A_{nm} J_0\left(\frac{\xi_n r}{a}\right)[\mu_m \cos \mu_m z + h \sin \mu_m z]e^{-\kappa \lambda_{mn} t}$$

where $\{\xi_n\}$ is the sequence of positive zeros of the function

$$\xi J_0'(\xi) + ha J_0(\xi)$$

and $\{\mu_m\}$ is the sequence of positive roots of the equation

$$\tan b\mu = \frac{2h\mu}{\mu^2 - h^2}$$

and

$$A_{nm} = \frac{1}{\|\psi_{nm}\|^2} \int_0^b \int_0^a rf(r,z)J_0\left(\frac{\xi_n r}{a}\right)[\mu_m \cos \mu_m z + h \sin \mu_m z]\, dr\, dz$$

$$\|\psi_{nm}\|^2 = \frac{a^2}{2\xi_n^2}(a^2 h^2 + \xi_n^2)J_0^2(\xi_n)\left[h + \frac{b}{2}(\mu_m^2 + h^2)\right]$$

$$\lambda_{nm} = \frac{\xi_n^2}{a^2} + \mu_m^2 \qquad n = 1, 2, \ldots \, ; \, m = 1, 2, \ldots$$

19 Let $\overline{\mathscr{V}}$ denote the domain defined by the inequalities

$$0 \le r \le a \qquad 0 \le \theta \le \theta_0 \qquad 0 \le z \le b$$

where r, θ, z are cylindrical coordinates and θ_0 is a given fixed angle, $0 < \theta_0 < \pi$. Let $\mathscr{V}$ denote the interior of $\overline{\mathscr{V}}$. Consider the problem

$$u_t - \kappa \, \Delta u = F(r,\theta,z,t) \qquad \text{in } \overline{\mathscr{V}}; \; t > 0$$

$$\left.\frac{\partial u}{\partial \theta}\right|_{\theta=0} = 0 \qquad \left.\frac{\partial u}{\partial \theta}\right|_{\theta=\theta_0} = 0 \qquad u|_{r=a} = H(\theta,z,t) \qquad t \ge 0$$

$$u(r,\theta,z,0) = f(r,\theta,z) \qquad \text{in } \overline{\mathscr{V}}$$

Given a physical interpretation of the problem, derive the eigenfunctions and then the series representation of the Green's function of the problem. Find the series solution of the problem in the particular case where

$$F = 0 \qquad f = 0 \qquad H(\theta,z) = H_0 = \text{const}$$

20 A homogeneous solid sphere of radius a has the initial temperature distribution

$$a^2 - r^2 \qquad 0 \le r \le a$$

where r denotes distance measured from the center. The surface temperature is maintained at zero. Show that the temperature in the sphere for $t > 0$ is given by

$$u(r,t) = \frac{12a^3}{\pi^3 r} \sum_{n=1}^{\infty} (-1)^{n-1} \frac{\sin (n\pi r/a)}{n^3} e^{-\kappa n^2 \pi^2 t/a^2}$$

21 Solve Prob. 20 if instead of the surface temperature's being zero the surface radiates heat into the exterior region. The temperature of the exterior region is zero.

22 Let r, θ, φ be the usual spherical coordinates (see Fig. 3-3). Let $\overline{\mathscr{V}}$ denote the sphere of radius a about the origin, and let $\mathscr{V}$ be its interior. Consider the problem

$$u_t - \kappa \, \Delta u = F(r,\theta,\varphi,t) \qquad \text{in } \mathscr{V}; \; t > 0$$

$$u(a,\theta,\varphi,t) = u_0 \qquad t \ge 0$$

$$u(r,\theta,\varphi,0) = f(r,\theta,\varphi) \qquad \text{in } \overline{\mathscr{V}}$$

where F and f are given functions and u_0 is a nonzero constant. Derive the formal series solution of the problem. Obtain the solution in the particular case where

$$F(r,\theta,\varphi,t) = \frac{F_0 \delta(r - r_0)\delta(\theta - \theta_0)\delta(\varphi - \varphi_0) \sin \omega t}{r^2 \sin \theta}$$

where F_0, ω are given positive numbers, (r_0,θ_0,φ_0) is a given point, and

$$f(r,\theta,\varphi) = (a^2 - r^2) \sin \theta$$

23 A spherical conducting shell has inner radius a and outer radius b. Through the inner wall from the interior there is a constant flux of heat q_0. The outer wall radiates into the exterior region which is at temperature zero. The initial temperature of the shell is a constant u_0. There are no heat sources within the shell material. Derive the formal series solution of the problem.

CAUCHY-KOWALEWSKI THEOREM (SPECIAL CASE)

The Cauchy problem for a linear second-order partial differential equation in two independent variables is

$$Lz = Az_{xx} + 2Bz_{xy} + Cz_{yy} + Dz_x + Ez_y + Fz = G$$

$$z(x,0) = h(x) \qquad z_y(x,0) = \sigma(x) \tag{1}$$

THEOREM Let the coefficients and given function G be real-valued and analytic on a region $\mathscr{R}$ of the xy plane which includes the origin. Assume $C(x,y) \neq 0$ in $\mathscr{R}$. Given arbitrary real-valued functions h, σ which are analytic functions of x on the segment of the x axis lying in $\mathscr{R}$, there exists a neighborhood $\mathscr{N}$ of $(0,0)$ and one and only one analytic function φ which satisfies the differential equation in $\mathscr{N}$ and satisfies the initial conditions

$$\varphi(x,0) = h(x) \qquad \varphi_y(x,0) = \sigma(x)$$

along the segment of the x axis included in $\mathscr{N}$.

Proof Let

$$a = \frac{A}{C} \qquad b = \frac{2B}{C} \qquad d = \frac{D}{C} \qquad e = \frac{E}{C} \qquad f = \frac{F}{C} \qquad g = \frac{G}{C}$$

Then these are real-valued analytic functions of x, y on $\mathscr{R}$, and the Cauchy problem can be rewritten

$$z_{yy} = -az_{xx} - bz_{xy} - dz_x - ez_y - fz + g \tag{2}$$

$$z(x,0) = h(x) \qquad z_y(x,0) = \sigma(x)$$

Now consider the following initial-value problem for a system of three first-order linear partial differential equations:

$$u_{1y} = u_3$$

$$u_{2y} = u_{3x}$$

$$u_{3y} = -au_{2x} - bu_{3x} - du_2 - eu_3 - fu_1 + g \tag{3}$$

$$u_1(x,0) = h(x) \qquad u_2(x,0) = h'(x) \qquad u_3(x,0) = \sigma(x)$$

The assertion is that problem (3) is equivalent to the Cauchy problem (2) in the sense that there exists a unique solution of (3) if, and only if, there exists a unique solution of (2). To see this suppose first that $z = \varphi(x,y)$ is a solution of the Cauchy problem. Let

$$u_1(x,y) = \varphi(x,y) \qquad u_2(x,y) = \varphi_x(x,y) \qquad u_3(x,y) = \varphi_y(x,y)$$

Then it is easy to verify that u_1, u_2, u_3 are solutions of (3). Conversely, let u_1, u_2, u_3 be solutions of (3), and define φ by

$$\varphi(x,y) = u_1(x,y)$$

Then $\varphi_y = u_3$, and $\varphi_x = u_2$. To show the latter relation observe that

$$\frac{\partial}{\partial y}(\varphi_x - u_2) = \varphi_{xy} - u_{2y} = u_{1xy} - u_{3x} = u_{1yx} - u_{3x} = u_{3x} - u_{3x} = 0$$

Hence

$$\varphi_x(x,y) - u_2(x,y) = q(x)$$

for some function q. But

$$\varphi_x(x,0) - u_2(x,0) = u_{1x}(x,0) - h'(x) = h'(x) - h'(x) = 0$$

Thus $q(x) = 0$, and $\varphi_x(x,y) = u_2(x,y)$. Now the third differential equation of the system implies

$$\varphi_{yy} = -a\varphi_{xx} - b\varphi_{xy} - d\varphi_x - e\varphi_y - f\varphi + g$$

Also

$$\varphi(x,0) = u_1(x,0) = h(x) \qquad \varphi_y(x,0) = u_3(x,0) = \sigma(x)$$

Thus φ is a solution of the Cauchy problem.

Accordingly the existence and uniqueness of the solution of the original Cauchy problem can be shown by demonstrating the existence and uniqueness of solution of the initial-value problem (3). However, problem (3) is a special case of the initial-value problem

$$u_{iy} = \sum_{j=1}^{3} a_{ij}u_{jx} + \sum_{j=1}^{3} b_{ij}u_j + g_i \qquad i = 1, 2, 3$$
$$u_i(x,0) = h_i(x) \qquad\qquad i = 1, 2, 3 \tag{4}$$

Under the hypothesis that the coefficients a_{ij} and b_{ij} and the given functions g_i are analytic on $\mathscr{R}$ it is now shown that there exists a unique analytic solution of problem (4) in a neighborhood of $(0,0)$. In the system of differential equations there appear the linear partial differential operators

$$L_{ij} = a_{ij}\frac{\partial}{\partial x} + b_{ij} \qquad i = 1, 2, 3; j = 1, 2, 3 \tag{5}$$

Problem (4) can be written

$$u_{iy} = \sum_{j=1}^{3} L_{ij}u_j + g_i \qquad i = 1, 2, 3$$
$$u_i(x,0) = h_i(x) \qquad i = 1, 2, 3 \tag{6}$$

Define the analytic functions f_1, f_2, f_3 by

$$f_i = \sum_{j=1}^{3} L_{ij}h_j + g_i \qquad i = 1, 2, 3$$

Then the initial-value problem

$$v_{iy} = \sum_{j=1}^{3} L_{ij}v_j + f_i \qquad i = 1, 2, 3$$
$$v_i(x,0) = 0 \qquad\qquad i = 1, 2, 3 \tag{7}$$

with homogeneous initial conditions is equivalent to (6). Suppose v_1, v_2, v_3 are solutions of (7).   Define the functions u_1, u_2, u_3 by

$$u_i(x,y) = v_i(x,y) + h_i(x) \qquad i = 1, 2, 3$$

Clearly the initial conditions of (6) are satisfied by the u_i.   Also

$$\sum_{j=1}^{3} L_{ij} u_j + g_i = \sum_{j=1}^{3} L_{ij}(v_j + h_j) + g_i$$

$$= \sum_{j=1}^{3} L_{ij} v_j + f_i = v_{iy} = u_{iy}$$

for $i = 1, 2, 3$.   Hence the u_i satisfy (6).   In the same way, if u_1, u_2, u_3 are solutions of (6), and if

$$v_i(x,y) = u_i(x,y) - h_i(x) \qquad i = 1, 2, 3$$

then the v_i satisfy (7).   Henceforth only problem (7) is considered.

The uniqueness of an analytic solution of problem (7) follows from the fact that the coefficients a_{ij}, b_{ij}, the given functions f_i, and the initial data h_i uniquely determine all the partial derivatives at the origin of a solution set v_1, v_2, v_3.   In turn this implies that the Taylor's series expansion about $(0,0)$ of each v_i is uniquely determined and hence also the functions v_i themselves in a neighborhood of the origin.   To show that all the partial derivatives at $(0,0)$ are determined observe first that, for each $i = 1, 2, 3$, the partial derivatives

$$\left. \frac{\partial^k v_i}{\partial x^k} \right|_{(0,0)} = 0 \qquad k = 0, 1, \ldots$$

Here $\partial^0 v_i / \partial x^0$ means simply v_i.   Now this implies that the values of the derivatives

$$\frac{\partial^k}{\partial x^k} \left(\sum_{j=1}^{3} L_{ij} v_j + f_i \right) \qquad k = 0, 1, 2, \ldots$$

at $(0,0)$ are determined, for each $i = 1, 2, 3$, by the coefficients a_{ij} and b_{ij} and the f_i.   Hence, from the differential equations in (7), the partial derivatives

$$\left. \frac{\partial^{k+1} v_i}{\partial x^k \, \partial y} \right|_{(0,0)} \qquad k = 0, 1, \ldots$$

are uniquely determined, for $i = 1, 2, 3$.   The balance of the proof is by induction.   Let l be a fixed positive integer, and assume that the derivatives

$$\left. \frac{\partial^{k+q} v_i}{\partial x^k \, \partial y^q} \right|_{(0,0)} \qquad k = 0, 1, 2, \ldots; q = 0, 1, \ldots, l$$

are uniquely determined, for $i = 1, 2, 3$.   Then

$$\frac{\partial^{k+l+1} v_i}{\partial x^k \, \partial y^{l+1}} = \frac{\partial^{k+l}}{\partial x^k \, \partial y^l} v_{iy} = \frac{\partial^{k+l}}{\partial x^k y^l} \left(\sum_{j=1}^{3} L_{ij} v_j \right) + \frac{\partial^{k+l} f_i}{\partial x^k \, \partial y^l}$$

The value at $(0,0)$ of the first term on the right is determined by the derivatives of the coefficients a_{ij}, b_{ij} together with the previously determined derivatives of v_1, v_2, and v_3 at $(0,0)$.   The second term on the right is determined at $(0,0)$ since the functions f_i are given.   Hence the derivatives

$$\left. \frac{\partial^{k+q} v_i}{\partial x^k \, \partial y^q} \right|_{(0,0)} \qquad k = 0, 1, 2, \ldots; q = 0, 1, \ldots, l+1$$

are uniquely determined, for $i = 1, 2, 3$.   This completes the induction step.

It is clear from the preceding paragraph that the coefficients and given functions and the data determine three formal power series

$$\sum_{k=0}^{\infty} \sum_{l=0}^{\infty} c_{kl}^{(i)} x^k y^l \qquad i = 1, 2, 3 \tag{8}$$

The proof of the existence of an analytic solution of problem (7) is made by showing that these power series converge in a neighborhood of (0,0) and define analytic functions v_1, v_2, v_3 which satisfy (7). The convergence of the power series (8) is shown by constructing a convergent power series

$$\sum_{k=0}^{\infty} \sum_{l=0}^{\infty} C_{kl} x^k y^l \tag{9}$$

having the property that

$$|c_{kl}^{(i)}| \leq C_{kl} \qquad k = 0, 1, \ldots ; l = 0, 1, \ldots$$

for each $i = 1, 2, 3$. The series (9) is said to *majorize* the original series (8).

Given a function u which is analytic at a point, one can construct a function U (not unique) whose power series about the point majorizes the power series for u about the point. Briefly U is said to *majorize* u on the region of convergence of the series. Suppose, for example,

$$u(x,y) = \sum_{k=0}^{\infty} \sum_{l=0}^{\infty} \alpha_{kl} x^k y^l$$

in a neighborhood of (0,0). Then there is a square N_δ defined by

$$|x| \leq \delta \qquad |y| \leq \delta$$

such that the series converges uniformly to u on N_δ. This implies there exists a constant $M > 0$ (dependent on δ) such that

$$|\alpha_{kl} \delta^{k+l}| < M \qquad k = 0, 1, \ldots ; l = 0, 1, \ldots$$

Define the function U by

$$U(x,y) = \frac{M\delta}{\delta - (x + y)} \tag{10}$$

Then U is analytic on the open square

$$|x| < \frac{\delta}{2} \qquad |y| < \frac{\delta}{2}$$

and majorizes u on this open square. To see this note that

$$\left. \frac{\partial^{k+l} U}{\partial x^k \, \partial y^l} \right|_{(0,0)} = \frac{(k + l)!}{\delta^{k+l}} M \qquad k = 0, 1, \ldots ; l = 0, 1, \ldots$$

Hence if $|x| < \delta/2, |y| < \delta/2$, then

$$U(x,y) = \sum_{k=0}^{\infty} \sum_{l=0}^{\infty} \frac{M(k + l)!}{k! \, l! \, \delta^{k+l}} x^k y^l$$

and

$$|\alpha_{kl}| < \frac{M}{\delta^{k+l}} \leq \frac{(k + l)!}{k! \, l!} \frac{M}{\delta^{k+l}} \qquad k = 0, 1, \ldots ; l = 0, 1, \ldots$$

It is clear from the method of construction that, given a set of functions $u_1, \ldots, u_q$, where each u_p is analytic at $(0,0)$, there exists an open square and an analytic function U which majorizes the entire set on the open square.

Let the function U in Eq. (10) and the open square $|x| < \delta/2$, $|y| < \delta/2$ be chosen so as to majorize all the coefficients a_{ij} and b_{ij} and the given functions f_i on the square. Consider the initial-value problem

$$\psi_{iy} = U(x,y)\left[\sum_{j=1}^{3}(\psi_{jx} + \psi_j)\right] + U(x,y) \qquad i = 1, 2, 3$$

$$\psi_i(x,0) = p_i(x) \qquad\qquad\qquad i = 1, 2, 3$$

(11)

where the p_i are as yet unspecified except that each is assumed to majorize zero. Now, by the procedure previously outlined, the function U and data p_i determine power series

$$\sum_{k=0}^{\infty}\sum_{l=0}^{\infty} C_{kl}^{(i)} x^k y^l \qquad i = 1, 2, 3$$

(12)

about $(0,0)$. The assertion is that, for each i, the power series (12) majorizes the corresponding series (8) determined by the initial-value problem (7). This is equivalent to the assertion

$$\left.\frac{\partial^{k+l}\psi_i}{\partial x^k\,\partial y^l}\right|_{(0,0)} \geq \left.\frac{\partial^{k+l}v_i}{\partial x^k\,\partial y^l}\right|_{(0,0)} \qquad k = 0, 1, \ldots\,;\ l = 0, 1, \ldots$$

Again the proof is by induction. Since the p_i majorize zero, one has

$$\left.\frac{\partial^k\psi_i}{\partial x^k}\right|_{(0,0)} = \left.\frac{\partial^k p_i}{\partial x^k}\right|_{(0,0)} \geq 0 = \left.\frac{\partial^k v_i}{\partial x^k}\right|_{(0,0)} \qquad k = 0, 1, \ldots$$

for $i = 1, 2, 3$. From these inequalities and the properties of U and the differential equations in (11) and (7) it follows that

$$\left.\frac{\partial^{k+1}\psi_i}{\partial x^k\,\partial y}\right|_{(0,0)} = \left\{\frac{\partial^k}{\partial x^k}\left[U\sum_{j=1}^{3}(\psi_{jx} + \psi_j)\right] + \left.\frac{\partial^k U}{\partial x^k}\right\}\right|_{(0,0)}$$

$$= \left\{\binom{k}{0}\frac{\partial^k U}{\partial x^k}\sum_{j=1}^{3}(\psi_{jx} + \psi_j) + \binom{k}{1}\frac{\partial^{k-1}U}{\partial x^{k-1}}\frac{\partial}{\partial x}\sum_{j=1}^{3}(\psi_{jx} + \psi_j)\right.$$

$$\left. + \cdots + \binom{k}{k-1}U\frac{\partial^{k-1}}{\partial x^{k-1}}\left[\sum_{j=1}^{3}(\psi_{jx} + \psi_j)\right] + \left.\frac{\partial^k U}{\partial x^k}\right\}\right|_{(0,0)}$$

$$\geq \left[\binom{k}{0}\sum_{j=1}^{3}\left(\left|\frac{\partial^k a_{ij}}{\partial x^k}\right|v_{jx} + \left|\frac{\partial^k b_{ij}}{\partial x^k}\right|v_j\right)\right.$$

$$\left. + \cdots + \binom{k}{k-1}\sum_{j=1}^{3}\left(\left|a_{ij}\frac{\partial^k v_{jx}}{\partial x^k}\right| + \left|b_{ij}\frac{\partial^k v_j}{\partial x^k}\right|\right) + \left.\left|\frac{\partial^k f_i}{\partial x^k}\right|\right]\right|_{(0,0)}$$

$$\geq \left|\binom{k}{0}\sum_{j=1}^{3}\left(\frac{\partial^k a_{ij}}{\partial x^k}v_{jx} + \frac{\partial^k b_{ij}}{\partial x^k}v_j\right)\right.$$

$$\left. + \cdots + \binom{k}{k-1}\sum_{j=1}^{3}\left(a_{ij}\frac{\partial^k v_{jx}}{\partial x^k} + b_{ij}\frac{\partial^k v_j}{\partial x^k}\right) + \left.\frac{\partial^k f_i}{\partial x^k}\right|_{(0,0)}\right.$$

$$= \left.\left|\frac{\partial^k v_{iy}}{\partial x^k}\right|\right|_{(0,0)} = \left.\left|\frac{\partial^{k+1}v_i}{\partial x^k\,\partial y}\right|\right|_{(0,0)} \qquad k = 0, 1, \ldots$$

for $i = 1, 2, 3$.   Let l be a positive integer, and assume that, for each $i = 1, 2, 3$,

$$\left.\frac{\partial^{k+q}\psi_i}{\partial x^k\,\partial y^q}\right|_{(0,0)} \geq \left.\frac{\partial^{k+q}v_i}{\partial x^k\,\partial y^q}\right|_{(0,0)} \qquad k = 0, 1, \ldots\,; q = 1, \ldots, l$$

Now

$$\left.\frac{\partial^{k+l+1}\psi_i}{\partial x^k\,\partial y^{l+1}}\right|_{(0,0)} = \left.\left\{\frac{\partial^{k+l}}{\partial x^k\,\partial y^l}\left[U\sum_{j=1}^{3}(\psi_{jx} + \psi_j) + U\right]\right\}\right|_{(0,0)}$$

Application of Leibniz' rule for calculating the higher-order derivatives of products together with the inequalities

$$\left.\frac{\partial^{k+q}U}{\partial x^k\,\partial y^q}\right|_{(0,0)} \geq \left.\frac{\partial^{k+q}a_{ij}}{\partial x^k\,\partial y^q}\right|_{(0,0)} \qquad \begin{array}{l} k = 0, 1, \ldots\,; q = 0, 1, \ldots, l \\ i = 1, 2, 3; j = 1, 2, 3 \end{array}$$

and a similar set of inequalities involving the b_{ij} and f_i, and the induction hypothesis shows that

$$\left.\frac{\partial^{k+q}\psi_i}{\partial x^k\,\partial y^q}\right|_{(0,0)} \geq \left.\frac{\partial^{k+q}v_i}{\partial x^k\,\partial y^q}\right|_{(0,0)} \qquad k = 0, 1, \ldots\,; q = 0, 1, \ldots, l$$

for $i = 1, 2, 3$.   This completes the induction proof, and so the series (12) majorizes the series (8) for each $i = 1, 2, 3$.

In order to complete the proof of the existence of a solution of problem (7) it is necessary to solve (11) in such a way that the initial values of the solutions majorize zero.   To this end let α be a constant as yet unspecified, except $0 < \alpha < 1$.   Then one can verify that

$$U = \frac{M\delta}{\delta - (x + y/\alpha)}$$

majorizes the function defined in (10) and so majorizes the coefficients and the f_i as assumed in the previous paragraphs.   Let

$$z = x + \frac{y}{\alpha}$$

then the differential equations in problem (11) are

$$\frac{\partial\psi_i}{\partial y} = U(x,y)\left(\frac{\partial\psi_1}{\partial x} + \frac{\partial\psi_2}{\partial x} + \frac{\partial\psi_3}{\partial x} + \psi_1 + \psi_2 + \psi_3\right) \qquad i = 1, 2, 3$$

which are all of the same form.   Assume a solution

$$\psi_i = \psi(z) = \psi\left(x + \frac{y}{\alpha}\right) \qquad i = 1, 2, 3$$

Then ψ must satisfy the ordinary differential equation

$$\frac{1}{\alpha}\frac{d\psi}{dz} = \frac{3M\delta}{\delta - z}\left(\frac{d\psi}{dz} + \psi\right)$$

This can be rewritten

$$\frac{d\psi}{\psi} = \frac{dz}{(\delta - z)/3M\delta\alpha - 1}$$

Let $\beta = 3M\delta\alpha$. Then a particular solution is

$$\psi = \left[\frac{1}{(\delta - z)/\beta - 1}\right]^{\beta}$$

Choose α sufficiently small so that on a neighborhood of $(0,0)$ the inequality

$$\frac{\delta - z}{\beta} - 1 > 0$$

holds. Then ψ is analytic on the neighborhood. It remains only to show that $\psi(x,0)$ majorizes zero. Note that

$$\psi(x,0) = \beta^{-\beta}(\delta - x)^{-\beta}\left(1 - \frac{\beta}{\delta - x}\right)^{-\beta}$$

The binomial expansion of the second factor is

$$1 + \beta(\delta - x)^{-1} + \frac{\beta(\beta + 1)}{2!}(\delta - x)^{-2} + \cdots$$

so that

$$\psi(x,0) = \beta^{-\beta}\left[(\delta - x)^{-\beta} + \beta(\delta - x)^{-(\beta+1)} + \frac{\beta(\beta + 1)}{2!}(\delta - x)^{-(\beta+2)} + \cdots\right.$$

In this expansion the coefficients of the powers $(\delta - x)^{-(\beta+q)}$ are positive. But the coefficients in the expansion of $(\delta - x)^{-(\beta+q)}$ about $x = 0$ are also positive. Hence the coefficients in the power-series expansion of $\psi(x,0)$ about $x = 0$ are all positive, and so $\psi(x,0)$ majorizes zero.

Solutions

$$\psi_i(x,y) = \psi(x,y) \qquad i = 1, 2, 3$$

of the "majorizing" problem (11) have thus been constructed [with $p_i(x) = \psi_i(x,0)$, $i = 1, 2, 3$]. These solutions have power-series expansions about $(0,0)$ which majorize the formal series (8) determined by the coefficients, the given functions f_i, and the initial data. Hence the series (8) converge and define analytic functions v_i in a neighborhood of $(0,0)$. By the manner in which the coefficients $c_k^{(i)}$ were determined it follows that the set v_1, v_2, v_3 is a solution set of problem (7). The equivalence of problems shown in the initial paragraphs then implies the existence of an analytic solution of the original Cauchy problem. The uniqueness of this solution on a neighborhood of $(0,0)$ is a consequence of the uniqueness of solution of problem (7).

2
STURM-LIOUVILLE PROBLEMS. FOURIER SERIES. FOURIER-LEGENDRE AND FOURIER-BESSEL SERIES

1 STURM-LIOUVILLE PROBLEMS

Let x be a real variable, and let the fixed interval $a \leq x \leq b$ of the x axis be denoted by $[a,b]$. Derivatives with respect to x are denoted by primes:

$$y' = \frac{dy}{dx} \qquad y'' = \frac{d^2y}{dx^2}$$

A second-order linear ordinary differential equation in *self-adjoint form* can be written

$$(py')' + qy = F \tag{1}$$

The coefficients p and q are assumed to be real-valued, p is continuously differentiable, and q is continuous, on $[a,b]$. In Eq. (1) there appears the linear *self-adjoint operator*

$$L = \frac{d}{dx}\left(p\frac{d}{dx}\right) + q \tag{2}$$

Accordingly (1) can be written briefly as

$$Ly = F \tag{3}$$

Associated with Eq. (3) is the *homogeneous equation*

$$Ly = 0 \tag{4}$$

Of importance in the study of linear boundary-value problems are linear homogeneous differential equations of the form

$$Ly + \lambda \rho y = 0 \tag{5}$$

where L is a self-adjoint operator, λ is a parameter independent of x, and ρ is a given nontrivial continuous function, $\rho(x) > 0$, on $[a,b]$. Accompanying (5) there are linear two-point boundary conditions

$$B_1(y) = \alpha_1 y(a) + \alpha_2 y'(a) = 0$$
$$B_2(y) = \beta_1 y(b) + \beta_2 y'(b) = 0 \tag{6}$$

called *unmixed boundary* conditions, or

$$B_1(y) = k_1 y(a) + k_2 y(b) = 0$$
$$B_2(y) = m_1 y'(a) + m_2 y'(b) = 0 \tag{7}$$

called *periodic boundary* conditions. Observe these are linear homogeneous equations in the end-point values of y and y'. In (6) and (7) the coefficients are real numbers such that

$$(\alpha_1{}^2 + \alpha_2{}^2)(\beta_1{}^2 + \beta_2{}^2) \neq 0$$

and

$$k_1 k_2 \neq 0 \qquad m_1 m_2 \neq 0 \qquad k_1 m_1 p(b) = k_2 m_2 p(b) \tag{8}$$

Particular cases of unmixed conditions are

$$B_1(y) = y(a) = 0 \qquad B_2(y) = y(b) = 0 \tag{9}$$

called *fixed-end-point conditions*,

$$B_1(y) = y'(a) = 0 \qquad B_2(y) = y'(b) = 0 \tag{10}$$

called *free-end-point conditions*, and

$$B_1(y) = y(a) = 0 \qquad B_2(y) = y'(b) = 0 \tag{11}$$

or

$$B_1(y) = y'(a) = 0 \qquad B_2(y) = y(b) = 0 \tag{12}$$

called *mixed conditions*. One seeks a solution of the differential equation (5) which satisfies one of the sets of boundary conditions (6) to (12).

Equation (5) together with one of the sets of boundary conditions above is called a *Sturm-Liouville system*. It is a homogeneous system, and, in general, the only solution is the *trivial solution*: the function which is identically zero on $[a,b]$. However for certain values of λ there may exist nontrivial solutions. A value λ such that there exists a nontrivial solution φ_λ of the system is called an *eigenvalue*. The corresponding solution φ_λ is called an eigenfunction. The terms *characteristic value* and *characteristic function* are also used in this connection. A *Sturm-Liouville problem* is the problem of determining all the eigenvalues and eigenfunctions of a Sturm-Liouville system. If $p(x) \neq 0, a \leq x \leq b$, the problem is called *regular*. If p vanishes at some point of $[a,b]$, the problem is called *singular*. It is also customary to call a problem singular when the interval is infinite. For example, a problem on $[a, +\infty]$ where the boundary conditions are

$$B_1(y) = y(a) = 0 \qquad B_2(y): \lim_{b \to \infty} y(b) \text{ exists}$$

is singular.

Assume the Sturm-Liouville problem is regular. Then there exists a linearly independent pair of solutions u_λ, v_λ of Eq. (5). Let λ be an eigenvalue and φ_λ a corresponding eigenfunction. Then there exist constants c_1, c_2, not both zero, such that

$$\varphi_\lambda(x) = c_1 u_\lambda(x) + c_2 v_\lambda(x) \qquad a \leq x \leq b \tag{13}$$

Since φ_λ satisfies the boundary conditions and

$$B_i(\varphi_\lambda) = B_i(c_1 u_\lambda + c_2 v_\lambda) = c_1 B_i(u_\lambda) + c_2 B_i(v_\lambda) \qquad i = 1, 2$$

the homogeneous linear equations

$$c_1 B_1(u_\lambda) + c_2 B_1(v_\lambda) = 0 \qquad c_2 B_2(u_\lambda) + c_2 B_2(v_\lambda) = 0 \tag{14}$$

in c_1 and c_2 are satisfied nontrivially. Hence the determinant is

$$\Delta(\lambda) = \begin{vmatrix} B_1(u_\lambda) & B_1(v_\lambda) \\ B_2(u_\lambda) & B_2(v_\lambda) \end{vmatrix} = 0 \tag{15}$$

Conversely, if λ is a value such that $\Delta(\lambda) = 0$, there exists a nontrivial pair of constants c_1, c_2 which satisfy (14), and with c_1, c_2 constructed in this manner the function φ_λ defined by Eq. (13) is an eigenfunction. Equation (15) is called the *characteristic equation* of the problem. Thus a necessary and sufficient condition that λ be an eigenvalue is that λ be a root of the characteristic equation. Often the set of eigenvalues is called the *spectrum* of the problem. The *multiplicity* (or index) of an eigenvalue λ is the number of linearly independent eigenfunctions corresponding to λ. An eigenvalue whose multiplicity is greater than one is called *degenerate*. Otherwise an eigenvalue is called *simple*. Since Eq. (5) is a second-order equation, the multiplicity is at most two.

If the boundary conditions are the unmixed conditions (6), each eigenvalue is simple. To show this let λ be an eigenvalue and φ_λ, ψ_λ corresponding eigenfunctions. Then

$$\alpha_1\varphi_\lambda(a) + \alpha_2\varphi'_\lambda(a) = 0 \qquad \alpha_1\psi_\lambda(a) + \alpha_2\psi'_\lambda(a) = 0$$

are linear homogeneous equations satisfied by the nontrivial values α_1, α_2. Thus the determinant

$$W(a) = \varphi_\lambda(a)\psi'_\lambda(a) - \varphi'_\lambda(a)\psi_\lambda(a) = 0$$

Recall that the Wronskian of φ_λ, ψ_λ is

$$W(x) = \varphi_\lambda(x)\psi'_\lambda(x) - \varphi'_\lambda(x)\psi_\lambda(x)$$

It is known that if the Wronskian of a pair of solutions of the linear homogeneous differential equation (5) vanishes at some point of $[a,b]$, then $W(x) = 0$, $a \le x \le b$. This implies that φ_λ, ψ_λ are linearly dependent on $[a,b]$; that is, λ is a simple eigenvalue.

Let φ, ψ be integrable functions on $[a,b]$. Then φ, ψ are *orthogonal on* $[a,b]$ if

$$\int_a^b \varphi(x)\overline{\psi(x)}\, dx = 0 \tag{16}$$

This embodies the notion of perpendicularity for the class of all functions integrable on $[a,b]$. In Eq. (16) $\overline{\psi(x)}$ denotes the complex conjugate of $\psi(x)$. More generally let

$$\{\varphi_n\} = \{\varphi_1, \varphi_2, \ldots\}$$

be a sequence of integrable functions on $[a,b]$. Then $\{\varphi_n\}$ is called an *orthogonal sequence* on $[a,b]$ if

$$\int_a^b \varphi_n(x)\overline{\varphi_m(x)}\, dx = 0 \qquad n \ne m \tag{17}$$

Thus the members of the sequence are mutually orthogonal on $[a,b]$. If φ is an integrable function, then the *norm* of φ is the real number

$$\|\varphi\| = \left[\int_a^b |\varphi(x)|^2\, dx\right]^{1/2} \tag{18}$$

Observe that $\|\varphi\| \ge 0$ for any φ. Let $\{\varphi_n\}$ be an orthogonal sequence on $[a,b]$ such that

$$\|\varphi_n\| = 1 \qquad n = 1, 2, \ldots \tag{19}$$

Then $\{\varphi_n\}$ is called an *orthonormal sequence* on $[a,b]$, abbreviated ONS.

As an example let $\{\varphi_n\}$ be the sequence defined by

$$\varphi_n(x) = e^{inx} \qquad n = 0, 1, \ldots; i = \sqrt{-1}$$

Then on the interval $[0,2\pi]$

$$\int_0^{2\pi} \varphi_n(x)\overline{\varphi_m(x)}\, dx = \int_0^{2\pi} e^{inx}e^{-imx}\, dx = 0 \qquad n \ne m$$

Thus $\{\varphi_n\}$ is an orthogonal sequence on $[0,2\pi]$. It is not an ONS, however, since

$$\int_0^{2\pi} |\varphi_n(x)|^2\, dx = 2\pi$$

The sequence $\{\psi_n\}$ defined by

$$\psi_n(x) = \frac{1}{\sqrt{2\pi}} e^{inx} \qquad n = 0, 1, 2, \ldots$$

is an ONS on $[0,2\pi]$. In general if $\{\varphi_n\}$ is an orthogonal sequence on $[a,b]$, and $\|\varphi_n\| \neq 0$ all n, an ONS can be constructed from $\{\varphi_n\}$ by defining

$$\psi_n(x) = \frac{\varphi_n(x)}{\|\varphi_n\|} \qquad n = 1, 2, \ldots$$

Let $\{\varphi_n\}$ be an ONS on $[a,b]$. If f is an integrable function on $[a,b]$, the numbers

$$c_n = \int_a^b f(x)\varphi_n(x)\, dx \qquad n = 1, 2, \ldots \tag{20}$$

are called the *Fourier coefficients of f relative to* $\{\varphi_n\}$. The formal series

$$\sum_{n=1}^{\infty} c_n \varphi_n(x) \tag{21}$$

is called the *Fourier series of f relative to* $\{\varphi_n\}$. The series need not be convergent; however, to each integrable function f there corresponds its formal Fourier series (21). Given an integrable function f, one may wish to approximate f by a linear combination

$$a_1\varphi_1 + \cdots + a_n\varphi_n$$

where $a_1, \ldots, a_n$ are constants. Here to "approximate f" means that the value of the square of the norm

$$\|a_1\varphi_1 + \cdots + a_n\varphi_n - f\|^2 = \int_a^b |a_1\varphi_1(x) + \cdots + a_n\varphi_n(x) - f(x)|^2\, dx \tag{22}$$

is to be made small. If n is fixed, the value of (22) is determined by the choice of the coefficients a_i. The following argument shows that the best approximation (for fixed n) is attained by choosing the a_i to be the Fourier coefficients of f.

$$\left\| f - \sum_{k=1}^{n} a_k\varphi_n \right\|^2 = \int_a^b \left(f - \sum_{j=1}^{n} a_j\varphi_j \right)\overline{\left(f - \sum_{k=1}^{n} a_k\varphi_k \right)}\, dx$$

$$= \|f\|^2 - \int_a^b f(x) \sum_{k=1}^{n} \overline{a_k\varphi_k(x)}\, dx - \int_a^b \overline{f(x)} \sum_{j=1}^{n} a_j\varphi_j(x)\, dx$$

$$+ \int_a^b \left[\sum_{j=1}^{n} a_j\varphi_j(x) \right]\left[\sum_{i=1}^{n} \overline{a_k\varphi_k(x)} \right] dx$$

$$= \|f\|^2 - \sum_{k=1}^{n} c_k\bar{a}_k - \sum_{j=1}^{n} a_j\bar{c}_j + \sum_{k=1}^{n} |a_k|^2$$

$$= \|f\|^2 - \sum_{k=1}^{n} |c_k|^2 + \sum_{k=1}^{n} |a_k - c_k|^2 \tag{23}$$

Hence

$$\left\| f - \sum_{k=1}^{n} a_k\varphi_k \right\|^2 \geq \|f\|^2 - \sum_{k=1}^{n} |c_k|^2 \tag{24}$$

for arbitrary choice of the a_i.   Thus a lower bound for the values of (22) for all choices of the a_i is the number on the right in (24), and this lower bound is attained by choosing $a_k = c_k,\ k = 1, \ldots, n$.   Also, from (23), it follows that

$$0 \le \left\| f - \sum_{k=1}^{n} c_k \varphi_k \right\|^2 = \|f\|^2 - \sum_{k=1}^{n} |c_k|^2$$

Accordingly

$$\sum_{k=1}^{n} |c_k|^2 \le \|f\|^2 \tag{25}$$

However, $\|f\|^2$ is a number independent of n, and so the partial sums on the left in (25) are uniformly bounded.   This implies that the series

$$\sum_{k=1}^{\infty} |c_k|^2$$

converges, and, in fact,

$$\sum_{k=1}^{\infty} |c_k|^2 \le \|f\|^2 \tag{26}$$

Relation (26) is called *Bessel's inequality*.   From the equation

$$\left\| f - \sum_{k=1}^{n} c_k \varphi_k \right\|^2 = \|f\|^2 - \sum_{k=1}^{n} |c_k|^2$$

it follows that

$$\lim_{n \to \infty} \left\| f - \sum_{k=1}^{n} c_k \varphi_k \right\|^2 = \int_a^b \left| f(x) - \sum_{k=1}^{n} c_k \varphi_k(x) \right|^2 dx = 0 \tag{27}$$

if, and only if,

$$\sum_{k=1}^{\infty} |c_k|^2 = \|f\|^2 \tag{28}$$

Equation (28) is called *Parseval's formula*.   The Fourier series of f is said to *converge to f in the mean of order two on* $[a,b]$ if Eq. (27) holds.   Hence the holding of Parseval's formula (28) is a necessary and sufficient condition for the convergence of the Fourier series in the mean of order two.

Let $C[a,b]$ denote the class of all functions which are continuous on $[a,b]$.   An ONS $\{\varphi_n\}$ in $C[a,b]$ is said to *span* (or generate) the class $C[a,b]$ if relation (27) holds for each function f in $C[a,b]$.   Thus $\{\varphi_n\}$ has the property that given a continuous function f on $[a,b]$, the Fourier series of f relative to $\{\varphi_n\}$ converges to f in the mean of order two on $[a,b]$.   An ONS $\{\varphi_n\}$ is called *complete* in $C[a,b]$ if

$$\int_a^b f(x) \overline{\varphi_n(x)}\, dx = 0 \qquad n = 1, 2, \ldots \qquad \text{implies} \qquad f(x) = 0 \qquad a \le x \le b \tag{29}$$

It is noted at this point in the literature the terms closed or complete are sometimes used instead of the term span, as used above.   However in this text completeness of an ONS in $C[a,b]$ means the property embodied in (29); i.e., the only function f in the class $C[a,b]$ which is orthogonal to each φ_n in the ONS is the function which is identically zero on $[a,b]$ (the trivial function).

If $\{\varphi_n\}$ is an ONS which spans $C[a,b]$, then $\{\varphi_n\}$ is complete in $C[a,b]$. This fact follows immediately from Parseval's formula. For if f is a continuous function such that

$$c_k = \int_a^b f(x)\overline{\varphi_k(x)}\, dx = 0 \qquad k = 1, 2, \ldots$$

then

$$\int_a^b |f(x)|^2 \, dx = 0$$

The continuity of f then implies $f(x) = 0$, $a \le x \le b$.

One reason for the interest in Sturm-Liouville problems lies in the fact that the eigenfunctions may constitute a complete ONS and so furnish an alternate means of representing arbitrary functions as Fourier series. Before discussing the orthogonality of eigenfunctions it is shown that the eigenvalues of the Sturm-Liouville problem described in the initial paragraphs are real numbers. First note that if u, v are real-valued continuously differentiable functions on $[a,b]$ which satisfy the boundary conditions, then

$$p(vu' - uv')|_a^b = 0 \tag{30}$$

Relation (30) is evident for the boundary conditions (9) to (12), and is easily established in the general case of (6) or (7). Now Green's formula (2-81), expressed in terms of the operator (2), states that

$$\int_a^b (vLu - uLv) \, dx = p(vu' - uv')|_a^b \tag{31}$$

holds for every pair of real-valued twice continuously differentiable functions on $[a,b]$. Equation (31) can be established directly by integration by parts. Also it follows that if u, v are real-valued twice continuously differentiable functions which satisfy the boundary conditions, then

$$\int_a^b vLu \, dx = \int_a^b uLv \, dx \tag{32}$$

This relation expresses the *symmetric* character of the operator L on the space of all real-valued twice continuously differentiable functions on $[a,b]$ which satisfy the boundary conditions of the problem. In turn this implies that

$$\int_a^b v\overline{Lu} \, dx = \int_a^b \bar{u}Lv \, dx \tag{33}$$

holds for every pair of (possibly complex-valued) twice continuously differentiable functions u, v which satisfy the boundary conditions. Suppose now that λ is an eigenvalue and φ_λ is a corresponding eigenfunction of the Sturm-Liouville problem. Set $u = v = \varphi_\lambda$ in Eq. (33). One obtains

$$(\lambda - \bar{\lambda}) \int_a^b \rho(x) \, |\varphi_\lambda(x)|^2 \, dx = 0$$

But

$$\int_a^b \rho(x) \, |\varphi_\lambda(x)|^2 \, dx > 0$$

Hence $\lambda = \bar{\lambda}$; that is, λ is a real number.

Let λ, μ be distinct eigenvalues, and let φ_λ, φ_μ be corresponding eigenfunctions.   Then $\sqrt{\rho}\,\varphi_\lambda, \sqrt{\rho}\,\varphi_\mu$ are orthogonal on $[a,b]$.   Observe

$$L\varphi_\lambda = -\lambda\varphi_\lambda \qquad L\varphi_\mu = -\mu\varphi_\mu$$

and φ_λ, φ_μ satisfy the boundary conditions.   Set $u = \varphi_\lambda, v = \varphi_\mu$ in Eq. (33).   The result is

$$(\lambda - \mu)\int_a^b \rho(x)\varphi_\mu(x)\overline{\varphi_\lambda(x)}\,dx = 0$$

Since $\lambda \neq \mu$, there follows the orthogonality property

$$\int_a^b \sqrt{\rho(x)}\,\varphi_\mu(x)\sqrt{\overline{\rho(x)}}\,\overline{\varphi_\lambda(x)}\,dx = 0 \tag{34}$$

In the particular case where $\rho(x) = 1$, $a \leq x \leq b$, the eigenfunctions are orthogonal on $[a,b]$.

Let the boundary conditions of the Sturm-Liouville problem be (6), where the coefficients are such that

$$\alpha_1\alpha_2 \leq 0 \qquad \beta_1\beta_2 \geq 0$$

or let the boundary conditions be given by (7).   Assume p is real-valued and continuously differentiable and $p(x) > 0$, $a \leq x \leq b$.   Let q be continuous on $[a,b]$, and let ρ be twice continuously differentiable, and $\rho(x) > 0$, on $[a,b]$.   Then the eigenvalues constitute a real monotone sequence $\{\lambda_n\}$ such that

$$\lambda_1 \leq \lambda_2 \leq \cdots \leq \lambda_n \leq \lambda_{n+1} \leq \cdots \qquad \lim_{n\to\infty}\lambda_n = +\infty$$

There are at most a finite number of negative eigenvalues.   If $q(x) \leq 0$, $a \leq x \leq b$, all the eigenvalues are nonnegative.   If the boundary conditions are the unmixed conditions (6), each eigenvalue is simple and appears just once in the sequence

$$\lambda_1 < \lambda_2 < \cdots < \lambda_n < \lambda_{n+1} < \cdots$$

The eigenfunctions can be assumed to be real-valued and such that $\{\sqrt{\rho}\,\varphi_n\}$ constitute an ONS.   The ONS of eigenfunctions is complete in $C[a,b]$ and, indeed, spans $C[a,b]$.   The foregoing properties are stated here without proof.   In this regard see Ref. 3, for example.

Often it is desired to represent a function f as a series in the eigenfunctions $\{\varphi_n\}$.   Suppose

$$f(x) = \sum_{n=1}^\infty c_n\varphi_n(x) \tag{35}$$

In order to determine the necessary form of the coefficients, let m be a fixed positive integer and multiply both sides of Eq. (35) by $\rho(x)\varphi_m(x)$.   Integrate the resulting equation from a to b.

$$\int_a^b \rho(x)f(x)\varphi_m(x)\,dx = \int_a^b \rho(x)\varphi_m(x)\left[\sum_{n=1}^\infty c_n\varphi_n(x)\right]dx$$

Formally interchange the operations of summation and integration, and utilize the orthogonality properties (34).   One obtains

$$c_m = \frac{1}{\|\varphi_m\|^2}\int_a^b \rho(x)f(x)\varphi_m(x)\,dx \qquad m = 1, 2, \ldots \tag{36}$$

where

$$\|\varphi_m\|^2 = \int_a^b \rho(x)\,|\varphi_m(x)|^2\,dx \tag{37}$$

The coefficients c_m defined in Eq. (36) are called the *Fourier coefficients of f relative to the eigenfunctions* $\{\varphi_n\}$, and the series on the right in Eq. (35) is often referred to as the *eigenfunction expansion* of f. Conditions on f sufficient to ensure the convergence of the series are discussed in Ref. 3. For example, if f is continuous with a piecewise continuous derivative on $[a,b]$ and f satisfies the boundary conditions of the problem, Eq. (35) holds for $a \le x \le b$.

2 FOURIER SERIES

Consider the regular self-adjoint Sturm-Liouville problem

$$y'' + \lambda y = 0 \qquad -\pi \le x \le \pi$$
$$B_1(y) = y(-\pi) - y(\pi) = 0 \qquad B_2(y) = y'(-\pi) - y'(\pi) = 0$$

Here the boundary conditions are periodic. A linearly independent pair of solutions of the differential equation is

$$u_\lambda = \cos(\sqrt{\lambda}\,x) \qquad v_\lambda = \sin(\sqrt{\lambda}\,x)$$

The characteristic equation is

$$\Delta(\lambda) = 4\sqrt{\lambda}\,\sin^2(\sqrt{\lambda}\,\pi) = 0$$

The eigenfunction corresponding to the eigenvalue $\lambda_0 = 0$ is $\varphi_0(x) = 1$. If $\lambda \ne 0$, the eigenvalues are the roots of

$$\sin(\sqrt{\lambda}\,\pi) = 0$$

that is,

$$\lambda_n = n^2 \qquad n = 1, 2, \ldots$$

To each eigenvalue λ_n, $n \ge 1$, there corresponds two linearly independent eigenfunctions

$$\cos nx \qquad \sin nx$$

The trigonometric system

$$\begin{aligned} &1, \cos x, \cos 2x, \ldots \\ &\quad \sin x, \sin 2x, \ldots \end{aligned} \tag{38}$$

is one of the first orthogonal systems to be studied. There are the orthogonality properties

$$\int_{-\pi}^{\pi} \cos nx \cos mx\, dx = 0 \qquad \int_{-\pi}^{\pi} \sin nx \sin mx\, dx = 0 \qquad n \ne m, n = 0, 1, \ldots$$

$$m = 0, 1, \ldots$$

and

$$\int_{-\pi}^{\pi} \cos nx \sin mx\, dx = 0 \qquad \text{all } n, m$$

The system (38) is complete in the space of all functions which are continuous on the interval $[-\pi,\pi]$. Thus, if f is continuous on $[-\pi,\pi]$ and

$$\int_{-\pi}^{\pi} f(x) \cos nx \, dx = 0 \qquad n = 0, 1, \ldots$$

$$\int_{-\pi}^{\pi} f(x) \sin nx \, dx = 0 \qquad n = 1, 2, \ldots$$

then $f(x) = 0$, $-\pi \le x \le \pi$.

Let f be an integrable function on $[-\pi,\pi]$. It is customary to call the numbers

$$a_0 = \frac{1}{\pi} \int_{-\pi}^{\pi} f(x) \, dx \qquad a_n = \frac{1}{\pi} \int_{-\pi}^{\pi} f(x) \cos nx \, dx$$

$$b_n = \frac{1}{\pi} \int_{-\pi}^{\pi} f(x) \sin nx \, dx \qquad n = 1, 2, \ldots \tag{39}$$

the *Fourier coefficients of f* relative to the trigonometric system (38). The formal series

$$\frac{a_0}{2} + \sum_{k=1}^{\infty} (a_k \cos kx + b_k \sin kx) \tag{40}$$

is called the *Fourier series of f* on $[-\pi,\pi]$. Note that the first term is just the average value of f on $[-\pi,\pi]$.

The formal series (40) of an integrable function always exists; however, it need not converge. In fact there are continuous functions whose Fourier series diverge at points of $[-\pi,\pi]$. Observe that each term in the series (40) is periodic, of period 2π. Accordingly if the series converges on $[-\pi,\pi]$, it converges for all x. Outside the interval $[-\pi,\pi]$ the series represents the periodic extension of the function to which it converges on $[-\pi,\pi]$.

A function f which is continuously differentiable on an interval is called *smooth* on that interval. A function f defined on the interval $[a,b]$ is called *piecewise* (or sectionally) *smooth on $[a,b]$* if there is a partition

$$a = x_0 < x_1 < \cdots < x_q = b$$

such that f is smooth on each open subinterval (x_k,x_{k+1}) and at the end points the right- and left-hand limits

$$f(x_k+) \qquad f(x_{k+1}-) \qquad f'(x_k+) \qquad f'(x_{k+1}-)$$

of f and f' exist, for $k = 0, 1, \ldots, q - 1$. Observe that a piecewise smooth function need not be continuous. Suppose f is piecewise smooth on $[-\pi,\pi]$. Then the Fourier series (40) of f converges to f on each subinterval of $(-\pi,\pi)$ on which f is continuous. If ξ is a point of $(-\pi,\pi)$ at which f is discontinuous, the series converges to the value

$$\frac{f(\xi+) + f(\xi-)}{2}$$

that is, the average of the left- and right-hand limits at ξ. At the end points $x = \pm\pi$ the series converges to the value

$$\frac{f(\pi-) + f(-\pi+)}{2}$$

In the particular case where f is continuous and piecewise smooth on $[-\pi,\pi]$, and $f(-\pi) = f(\pi)$, the Fourier series of f converges uniformly to f on $[-\pi,\pi]$. The proof of the foregoing statements is given in Ref. 1.

If f is piecewise continuous on $[-\pi,\pi]$, the Fourier series of f can be integrated term by term, and the resulting series converges, even though the original series diverges. In the case of termwise differentiation, however, the situation is different. A Fourier series can be convergent, but the series which results from termwise differentiation may diverge.

Let $[-c,c]$ be a given interval. The real-valued orthogonal system of eigenfunctions of the regular Sturm-Liouville problem

$$y'' + \lambda y = 0 \qquad -c \leq x \leq c$$
$$y(-c) - y(c) = 0 \qquad y'(-c) - y'(c) = 0$$

is

$$1, \cos\frac{\pi x}{c}, \cos\frac{2\pi x}{c}, \ldots, \cos\frac{n\pi x}{c}, \ldots$$

$$\sin\frac{\pi x}{c}, \sin\frac{2\pi x}{c}, \ldots, \sin\frac{n\pi x}{c}, \ldots \tag{41}$$

This is a generalization of the system (38) and reduces to that system if $c = \pi$. If f is an integrable function on $[-c,c]$, the real numbers

$$a_0 = \frac{1}{c}\int_{-c}^{c} f(x)\,dx \qquad a_n = \frac{1}{c}\int_{-c}^{c} f(x)\cos\frac{n\pi x}{c}\,dx$$

$$b_n = \frac{1}{c}\int_{-c}^{c} f(x)\sin\frac{n\pi x}{c}\,dx \tag{42}$$

$n = 1, 2, \ldots,$ are called the *Fourier coefficients of f relative to the system* (41). The *Fourier series of f on* $[-c,c]$ is

$$\frac{a_0}{2} + \sum_{k=1}^{\infty}\left(a_k\cos\frac{k\pi x}{c} + b_k\sin\frac{k\pi x}{c}\right) \tag{43}$$

Since the intervals $[-c,c]$, $[-\pi,\pi]$ are related by a linear transformation, it is evident that the theorems on the convergence of Fourier series extend to the series (43). In particular if the series converges on $[-c,c]$, it converges for all x and outside of $[-c,c]$ represents the periodic extension, with period $2c$, of the function to which it converges on $[-c,c]$.

If f is integrable on $[-c,c]$ and is an *even function* on this interval, that is, $f(-x) = f(x)$, $-c \leq x \leq c$, then from (42) it follows that $b_n = 0$, $n = 1, 2, \ldots$, and

$$a_n = \frac{2}{c}\int_{0}^{c} f(x)\cos\frac{n\pi x}{c}\,dx \qquad n = 0, 1, 2, \ldots \tag{44}$$

In this case the Fourier series of f is

$$\frac{a_0}{2} + \sum_{k=1}^{\infty} a_k\cos\frac{k\pi x}{c} \tag{45}$$

Thus the Fourier series of an even function involves only cosines. If f is an *odd function* in the interval, that is, $f(-x) = -f(x)$, $-c \leq x \leq c$, then the coefficients $a_n = 0$, $n = 0, 1, 2, \ldots$, and

$$b_n = \frac{2}{c}\int_{0}^{c} f(x)\sin\frac{k\pi x}{c}\,dx \qquad k = 1, 2, \ldots \tag{46}$$

and the Fourier series of f is

$$\sum_{k=1}^{\infty} b_k \sin \frac{k\pi x}{c} \tag{47}$$

Let f be integrable on $[0,c]$. Then the series (45), with coefficients a_n defined by Eq. (44), is called the *Fourier cosine series* of f on $[0,c]$. Note that each term $\cos(n\pi x/c)$ is an even function. Accordingly if the cosine series converges on $[0,c]$ to f, then it converges on $[-c,0]$ and represents the *even extension* of f

$$F(x) = f(x) \qquad 0 \leq x \leq c \qquad\qquad F(-x) = F(x) \qquad -c \leq x \leq c$$

In fact the series converges for all x and represents the periodic extension, of period $2c$, of F. Similarly the series (47) is called the *Fourier sine series* of f on $[0,c]$. In this series each term $\sin(n\pi x/c)$ is an odd function. Thus if the series converges on $[0,c]$ to f, then it converges on $[-c,0]$, and represents the *odd extension* of f

$$F(x) = f(x) \qquad 0 \leq x \leq c \qquad\qquad F(-x) = -F(x) \qquad -c \leq x \leq c$$

The series in this case converges for all x to the periodic extension, with period $2c$, of F.

Often the orthogonal system of complex exponentials

$$1, e^{\pi ix/c}, e^{2\pi ix/c}, \ldots, e^{n\pi ix/c}, \ldots$$
$$e^{-\pi ix/c}, e^{-2\pi ix/c}, \ldots, e^{-n\pi ix/c}, \ldots \tag{48}$$

is useful. These are the complex-valued eigenfunctions of the Sturm-Liouville problem which yields the real system (41). The system (48) constitutes an orthogonal set on the interval $[-c,c]$. The *Fourier coefficients* of a function f relative to (48) are the complex numbers

$$\alpha_n = \frac{1}{2c} \int_{-c}^{c} f(x) e^{-n\pi ix/c} \, dx \qquad n = 0, \pm 1, \pm 2, \ldots$$

The complex form of the Fourier series of f on $[-c,c]$ is

$$\sum_{k=-\infty}^{\infty} \alpha_k e^{k\pi ix/c} \tag{49}$$

The coefficients α_n and the Fourier coefficients a_n, b_n in (42) are related by the equations

$$\alpha_n = \frac{a_n - ib_n}{2} \qquad \alpha_{-n} = \frac{a_n + ib_n}{2}$$

If f is real-valued on $[-c,c]$, then $\alpha_{-n} = \bar{\alpha}_n$ (complex conjugate), and the series (49) reduces to (40).

3   FOURIER-LEGENDRE SERIES

An important instance of a singular Sturm-Liouville problem is

$$[(1 - x^2)y']' + \lambda y = 0 \qquad -1 < x < 1 \tag{50}$$

$$B_1(y): y \text{ is bounded at } x = -1 \qquad B_2(y): y \text{ is bounded at } x = 1 \tag{51}$$

Equation (50) is Legendre's equation, and here the coefficient $p(x) = 1 - x^2$. Since $p(1) = p(-1) = 0$, the end points of the interval $[-1,1]$ are singular points of the differential equation. Suppose y is a solution of the problem. Then

$$\lim_{\substack{x \to 1 \\ x < 1}} [(1 - x^2)y'(x)] = 0 \qquad \lim_{\substack{x \to -1 \\ x > -1}} [(1 - x^2)y'(x)] = 0 \tag{52}$$

To establish the first relation note that since y is continuous on $(-1,1)$ and bounded on $[-1,1]$, it is integrable on $[0,1]$. Hence the function

$$g(x) = -\lambda \int_0^x y(\xi)\, d\xi + y'(0) = (1 - x^2)y'(x)$$

is continuous on $[0,1]$, and so

$$\lim_{\substack{x \to 1 \\ x < 1}} g(x) = g(1)$$

exists. Suppose $g(1) \neq 0$. Then

$$y(x) - y(0) = \int_0^x \frac{g(\xi)}{1 - \xi^2}\, d\xi$$

is unbounded near $x = 1$. This contradicts the assumption that y is bounded at $x = 1$. Thus $g(1) = 0$. The second relation in Eq. (52) is established in the same manner.

Let

$$L = \frac{d}{dx}\left[(1 - x^2)\frac{d}{dx}\right] \tag{53}$$

Then Green's formula in the form

$$\lim_{\substack{h \to 0 \\ h > 0}} \int_{-1+h}^{1-h} (vLu - uLv)\, dx = \lim_{\substack{h \to 0 \\ h > 0}} [(1 - x^2)(vu' - uv')]\Big|_{-1+h}^{1-h} \tag{54}$$

holds for the operator L. Equation (54) is valid for every pair of real-valued twice continuously differentiable functions u, v on the open interval (a,b). If in addition, u and v satisfy the boundary conditions (51) and (52), then

$$\lim_{\substack{h \to 0 \\ h > 0}} \int_{-1+h}^{1-h} vLu\, dx = \lim_{\substack{h \to 0 \\ h > 0}} \int_{-1+h}^{1-h} uLv\, dx \tag{55}$$

The proof that the eigenvalues are real and that real-valued eigenfunctions corresponding to distinct eigenvalues λ, μ are orthogonal on $[-1,1]$ in the sense

$$\int_{-1}^{1} \varphi_\lambda(x)\varphi_\mu(x)\, dx = \lim_{\substack{h \to 0 \\ h > 0}} \int_{-1+h}^{1-h} \varphi_\lambda(x)\varphi_\mu(x)\, dx = 0 \tag{56}$$

follows along the same lines as the proof given in Sec. 1.

Legendre's equation is usually written

$$(1 - x^2)y'' - 2xy' + \alpha(\alpha + 1)y = 0 \tag{57}$$

where $\alpha(\alpha + 1) = \lambda$. If a power-series solution of the form

$$y = x^r \sum_{k=0}^{\infty} c_k x^k$$

is assumed, then the method of Frobenius shows that the exponent r must satisfy $r(r-1) = 0$ (the indicial equation) and yields two linearly independent solutions

$$y_1 = 1 - \frac{\alpha(\alpha+1)}{2!} x^2 + \frac{\alpha(\alpha-2)(\alpha+1)(\alpha+3)}{4!} x^4 - \cdots \tag{58}$$

$$y_2 = x - \frac{(\alpha-1)(\alpha+2)}{3!} x^3 + \frac{(\alpha-1)(\alpha-3)(\alpha+2)(\alpha+4)}{5!} x^5 - \cdots$$

on $(-1,1)$.   Each series converges for $|x| < 1$ and defines an analytic function there.   If α is not an integer or zero, neither y_1 nor y_2 is defined and bounded on the closed interval $[-1,1]$.   Hence if a solution which is bounded on $[-1,1]$ is desired, then $\alpha = n$ is necessary, where n is an integer or zero.   If $\alpha = n$ is a positive integer or zero, one of the series terminates and the other does not.   In this event a polynomial solution of Legendre's equation is obtained.   The polynomial solution is analytic for all x and so satisfies the prescribed conditions on $[-1,1]$.   A constant multiple of the polynomial also satisfies the boundary conditions, and it is customary to choose the constant such that the resulting polynomial has the value 1 at $x = 1$.   Explicitly the polynomial solution of Legendre's equation when $\alpha = n$ is

$$P_n(x) = \frac{1}{2^n} \sum_{k=0}^{N} (-1)^k \frac{(2n-2k)!}{k!\,(n-k)!\,(n-2k)!} x^{n-2k} \tag{59}$$

where $N = n/2$ if n is even and $N = (n-1)/2$ if n is odd.   The function P_n is called *Legendre polynomial* (or Legendre coefficient) of order n.   Observe that P_n is an even or odd function of x according as n is even or odd.   The first few Legendre polynomials are

$$P_0(x) = 1 \qquad P_1(x) = x \qquad P_2(x) = \frac{3x^2 - 1}{2}$$

$$P_3(x) = \frac{5x^3 - 3x}{2} \qquad P_4(x) = \frac{35x^4 - 30x^2 + 3}{8}$$

If $\alpha = n$, n a positive integer or zero, the remaining series solution does not terminate. When this solution is multiplied by a suitable constant, the resulting function is called the *Legendre function of the second kind* of order n.   These are denoted by $Q_n(x)$.   By the method of reduction of order it can be shown that Q_n is expressible in terms of P_n and a function having a logarithmic singularity at $x = 1$.

$$Q_n(x) = \tfrac{1}{2}P_n(x) \log \frac{1+x}{1-x} - W_{n-1}(x) \tag{60}$$

where

$$W_{n-1}(x) = \frac{2n-1}{n} P_{n-1}(x) + \frac{2n-5}{3(n-1)} P_{n-3} + \frac{2n-9}{5(n-2)} P_{n-5}(x) + \cdots$$

and it is understood that the series terminates with either P_1 or P_0 according as n is even or odd.   The first few Q_n are

$$Q_0(x) = \tfrac{1}{2} \log \frac{1+x}{1-x} \qquad Q_1(x) = \frac{x}{2} \log\left(\frac{1+x}{1-x}\right) - 1$$

$$Q_2(x) = \tfrac{1}{2}P_2(x) \log \frac{1+x}{1-x} - \tfrac{3}{2}x$$

Thus, if $\alpha = n$, the general solution of Legendre's equation is

$$y = c_1 P_n(x) + c_2 Q_n(x) \tag{61}$$

where c_1, c_2 are arbitrary constants.

The eigenvalues of the original problem are

$$\lambda_n = n(n+1) \qquad n = 0, 1, \ldots$$

with corresponding eigenfunctions $P_n(x)$, $n = 0, 1, \ldots$.   Since $\lambda_n \neq \lambda_m$ whenever $n \neq m$, it follows from Eq. (56) that

$$\int_{-1}^{1} P_n(x)P_m(x)\, dx = 0 \qquad n \neq m \tag{62}$$

The Legendre polynomials have many interesting and useful properties.   For convenience several of these are listed below.

$$P_n(x) = \frac{1}{n!2^n} \frac{d^n}{dx^n} (x^2 - 1)^n \tag{63}$$

$$(1 - 2xh + h^2)^{-\frac{1}{2}} = \sum_{n=0}^{\infty} h^n P_n(x) \qquad |x| < 1; |h| < 1 \tag{64}$$

$$(n+1)P_{n+1}(x) - (2n+1)xP_n(x) + nP_{n-1}(x) = 0 \tag{65}$$

$$xP_n'(x) - P_{n-1}'(x) = nP_n(x) \tag{66}$$

$$P_{n+1}'(x) - P_{n-1}'(x) = (2n+1)P_n(x) \tag{67}$$

$$\int_{-1}^{1} f(x)P_n(x)\, dx = 0 \qquad \text{if } f \text{ is a polynomial of degree } m < n \tag{68}$$

$$\|P_n\|^2 = \int_{-1}^{1} |P_n(x)|^2\, dx = \frac{2}{2n+1} \tag{69}$$

Equation (63) is called *Rodrigues' formula*.   The function on the left in Eq. (64) is called a *generating function* for the Legendre polynomials.   These together with the recursion relations (65) to (67) are established in Ref. 1, for example.   Equation (69) can be proved with the aid of (63) and integration by parts.

$$\frac{1}{n!2^n} \int_{-1}^{1} P_n(x) \frac{d^n(x^2-1)^n}{dx^n}\, dx = \frac{1}{n!2^n} \left\{ \left[P_n(x) \frac{d^{n-1}}{dx^{n-1}} (x^2-1)^n \right]\Big|_{-1}^{1} - \int_{-1}^{1} P_n'(x) \right.$$

$$\left. \frac{d^{n-1}}{dx^{n-1}} (x^2-1)^n\, dx \right\} = -\frac{1}{n!2^n} \int_{-1}^{1} P_n'(x) \frac{d^{n-1}}{dx^{n-1}} (x^2-1)^n\, dx$$

Thus integration by parts n times together with (63) yields

$$\int_{-1}^{1} |P_n(x)|^2\, dx = \frac{(-1)^n}{(n!)^2 2^{2n}} \int_{-1}^{1} (x^2-1)^n \frac{d^{2n}}{dx^{2n}} [(x^2-1)^n]\, dx$$

But

$$\frac{d^{2n}}{dx^{2n}} [(x^2-1)^n] = (2n)! \qquad \int_{-1}^{1} (x^2-1)^n\, dx = \frac{(-1)^n 2^{2n+1}(n!)^2}{(2n+1)(2n)!}$$

and Eq. (69) follows.

The sequence $\{P_n\}$ of Legendre polynomials is complete in the space of all continuous functions on $[-1,1]$. Thus, if f is continuous on $[-1,1]$ and

$$\int_{-1}^{1} f(x)P_n(x)\, dx = 0 \qquad n = 0, 1, \ldots$$

then $f(x) = 0$, $-1 \le x \le 1$. This property is not proved in this text (see Ref. 10).

Let f be integrable on $[-1,1]$. The numbers

$$a_n = \frac{2n+1}{2} \int_{-1}^{1} f(x)P_n(x)\, dx \qquad n = 0, 1, 2, \ldots \tag{70}$$

are called the *Fourier-Legendre coefficients* of f. The formal series

$$\sum_{k=0}^{\infty} a_k P_k(x) \tag{71}$$

is called the *Fourier-Legendre* series of f. The expansion theorems for Fourier-Legendre series are quite analogous to those for Fourier series. It can be shown (see Ref. 10) that if f is piecewise smooth on $[-1,1]$, then at each point x in $(-1,1)$ where f is continuous the series (71) converges to $f(x)$ and at each point x where f is discontinuous the series converges to the average value $[f(x^+) + f(x^-)]/2$. At the end points $x = 1$, $x = -1$, the series converges to the values $f(1^-)$ and $f(-1^+)$, respectively.

The Legendre polynomial P_n is an even or odd function accordingly as the integer n is even or odd. Hence if it is an even function on $[-1,1]$, then

$$a_n = (2n+1)\int_{0}^{1} f(x)P_n(x)\, dx \tag{72}$$

provided n is even, while $a_n = 0$ if n is odd. In this case the Fourier-Legendre series of f is

$$\sum_{k=0}^{\infty} \left[(4k+1)\int_{0}^{1} f(\xi)P_{2k}(\xi)\, d\xi \right] P_{2k}(x) \tag{73}$$

Similarly if f is an odd function on $[-1,1]$, then a_n is given by Eq. (72) whenever n is odd, while $a_n = 0$ when n is even. The Fourier-Legendre series of an odd function is

$$\sum_{k=1}^{\infty} \left[(4k-1)\int_{0}^{1} f(\xi)P_{2k-1}(\xi)\, d\xi \right] P_{2k-1}(x) \tag{74}$$

4 FOURIER-BESSEL SERIES

Another singular problem which occurs in the applications is

$$(xy')' - \frac{\alpha^2 y}{x} + \lambda xy = 0 \qquad 0 < x < c \tag{75}$$

$$B_1(y) \colon y \text{ is bounded at } x = 0 \tag{76}$$

$$B_2(y) = y(c) = 0 \tag{77}$$

Here c is a given fixed positive number, and α is a real nonnegative parameter. The operator

$$L = \frac{d}{dx}\left(x\frac{d}{dx} \right) - \frac{\alpha^2}{x} \tag{78}$$

is singular at the left-hand end point of the interval [0,c].   An argument similar to that used in Sec. 3 shows that if y is a solution of the problem, then

$$\lim_{\substack{x \to 0 \\ x > 0}} xy' = 0 \tag{79}$$

Thus the boundary conditions are such that whenever u, v are twice continuously differentiable functions on [0,c] which satisfy the boundary conditions, then

$$\lim_{\substack{h \to 0 \\ h > 0}} \int_h^c (vLu - uLv)\, dx = \lim_{\substack{h \to 0 \\ h > 0}} [x(vu' - uv') \,|_h^c] = 0$$

Accordingly if φ_λ, φ_μ are eigenfunctions corresponding to distinct eigenvalues λ, μ, then

$$\int_0^c x\varphi_\lambda(x)\varphi_\mu(x)\, dx = \lim_{\substack{h \to 0 \\ h > 0}} \int_h^c x\varphi_\lambda(x)\varphi_\mu(x)\, dx = 0 \tag{80}$$

The eigenvalues are real.   Moreover they are positive.   To show this let u be an eigenfunction corresponding to the eigenvalue λ.   Choose h such that $0 < h < c$, multiply both sides of Eq. (75) by u, and integrate over the interval [h,c].   Then

$$\int_h^c uLu\, dx + \lambda \int_h^c xu^2\, dx = 0$$

Since

$$\int_h^c uLu\, dx = \int_h^c (xu')'u\, dx - \alpha^2 \int_h^c \frac{u^2}{x}\, dx$$

$$= xuu' \,|_h^c - \int_h^c x(u')^2\, dx - \alpha^2 \int_h^c \frac{u^2}{x}\, dx$$

it follows that

$$xuu' \,|_h^c - \int_h^c x(u')^2\, dx + \lambda \int_h^c xu^2 = \alpha^2 \int_h^c \frac{u^2}{x}\, dx \tag{81}$$

Let $h \to 0$.   Then the limit of the left side of this equation exists since each term has a limit.   Accordingly the limit of the right side as $h \to 0$ exists.   Now xuu' vanishes at $x = 0$ and at $x = c$.   Hence

$$\lambda \int_0^c xu^2\, dx = \int_0^c x(u')^2\, dx + \alpha^2 \int_0^c \frac{u^2}{x}\, dx \tag{82}$$

$$\lambda = \frac{\int_0^c x(u')^2\, dx + \alpha^2 \int_0^c (u^2/x)\, dx}{\int_0^c xu^2\, dx} \tag{83}$$

The denominator is positive, and the numerator is nonnegative.   Hence $\lambda \geq 0$.   Note that if $\alpha > 0$, then $\lambda > 0$.   If $\alpha = 0$, then $\lambda = 0$ if, and only if, $u(x) = $ const, $0 < x \leq a$. But $u(a) = 0$.   Hence u is the trivial function, and $\lambda = 0$ is not an eigenvalue in any event.

To show the existence of eigenvalues of the problem one can proceed as follows.   In the differential equation make the change of independent variable $t = \sqrt{\bar{\lambda}}\, x$.   The transformed differential equation is

$$t^2 \frac{d^2 y}{dt^2} + t \frac{dy}{dt} + (t^2 - \alpha^2)y = 0 \tag{84}$$

Bessel's equation of order α. This equation has a regular singular point at $t = 0$. If a power-series solution

$$y = t^r \sum_{k=0}^{\infty} c_k t^k$$

is assumed, the method of Frobenius shows that the exponent r must satisfy $r^2 - \alpha^2 = 0$ (the indicial equation) and yields a pair of linearly independent solutions. Two distinct cases arise according as α is or is not a nonnegative integer. If α is not a nonnegative integer, the general solution is

$$y = a_1 J_\alpha(t) + a_2 J_{-\alpha}(t) \tag{85}$$

where

$$J_\alpha(t) = \left(\frac{t}{2}\right)^\alpha \sum_{k=0}^{\infty} \frac{(-1)^k}{k! \, \Gamma(k + \alpha + 1)} \left(\frac{t}{2}\right)^{2k} \tag{86}$$

and $\Gamma(s)$ is the *gamma function*. The function J_α is called the *Bessel function of the first kind of order* α. If α is not real, then J_α is complex-valued. If α is real and $\alpha \geq 0$, then J_α is real-valued and continuous at $t = 0$.

If $\alpha = n$, n a positive integer or zero, then the *Bessel function of the first kind of integral order* n,

$$J_n(t) = \sum_{k=0}^{\infty} \frac{(-1)^k}{k! \, (k + n)!} \left(\frac{t}{2}\right)^{2k+n} \tag{87}$$

is an analytic solution of Bessel's equation (with $\alpha = n$) for all t. Since $\Gamma(k + n + 1) = (k + n)!$, one obtains J_n from J_α by setting $\alpha = n$. It can be shown that

$$J_{-n}(t) = (-1)^n J_n(t) \tag{88}$$

for $n = 0, 1, 2, \ldots$, and so J_{-n} is not a second linearly independent solution in this case. The general solution of Bessel's equation for $\alpha = n$ is

$$y = c_1 J_n(t) + c_2 Y_n(t) \tag{89}$$

where the second linearly independent solution is

$$Y_n(t) = \frac{2}{\pi} \left(\log \frac{t}{2} + \gamma\right) J_n(t)$$
$$- \frac{1}{\pi} \sum_{k=0}^{n-1} \frac{(n - k - 1)!}{k!} \left(\frac{2}{t}\right)^{n-2k} - \frac{1}{\pi} \sum_{k=0}^{\infty} \left\{\frac{(-1)^k (t/2)^{2k+n}}{k! \, (k + n)!} \left[\varphi(k + n) - \varphi(k)\right]\right\} \tag{90}$$

called the *Weber's Bessel function of the second kind of order* n, where γ is a constant (called Euler's constant; an approximate value is 0.5772157), and $\varphi(k) = 1 + \frac{1}{2} + \cdots + 1/k$, $k = 1, 2, \ldots$, $\varphi(0) = 0$. The Bessel function Y_n is not bounded in any neighborhood which includes the origin.

If α is not an integer, the Bessel function of the second kind is defined by

$$Y_\alpha(t) = \frac{\cos \alpha\pi J_\alpha(t) - J_{-\alpha}(t)}{\sin \alpha\pi}$$

This is a solution for $x \neq 0$, and the pair J_α, Y_α are linearly independent for all α. Accordingly, for arbitrary α a general solution of Bessel's equation is

$$y = c_1 J_\alpha(t) + c_2 Y_\alpha(t) \tag{91}$$

If α is real and $t > 0$, then $J_\alpha(t)$ and $Y_\alpha(t)$ are real-valued. Note that the Bessel function of the second kind is always unbounded in a neighborhood which includes the origin.

There are many relations and useful formulas involving the Bessel functions. A few are listed here for reference in the text.

$$J_0'(x) = -J_1(x)$$

$$\frac{d}{dx}[x^\alpha J_\alpha(x)] = x^\alpha J_{\alpha-1}(x)$$

$$xJ_\alpha'(x) + \alpha J_\alpha(x) = xJ_{\alpha-1}(x)$$
$$\alpha J_\alpha(x) - xJ_\alpha'(x) = xJ_{\alpha+1}(x)$$

$$J_{\alpha-1}(x) + J_{\alpha+1}(x) = \frac{2\alpha J_\alpha(x)}{x}$$

$$J_{1/2}(x) = \sqrt{\frac{2}{\pi x}} \sin x$$

$$J_{-1/2}(x) = \sqrt{\frac{2}{\pi x}} \cos x$$

$$J_{3/2}(x) = \sqrt{\frac{2}{\pi x}} \left(\frac{\sin x}{x} - \cos x \right)$$

Since in the original problem α is assumed real and nonnegative, and since the differential equation is equivalent to Bessel's equation, it follows from the preceding discussion that an eigenfunction must be of the form

$$y = a_1 J_\alpha(\sqrt{\lambda}\, x) \tag{92}$$

where a_1 is a nonzero real constant. From Eq. (86) it is clear that such a function satisfies the first boundary condition $B_1(y)$. In order to satisfy the second boundary condition the value must be such that

$$J_\alpha(\sqrt{\lambda}\, c) = 0 \tag{93}$$

Conversely, if λ is a positive number which satisfies Eq. (93), the function y in Eq. (92) is an eigenfunction. Accordingly λ is an eigenvalue of the problem if, and only if, $\lambda > 0$ and satisfies Eq. (93). It can be shown that $J_\alpha(t)$ has a denumerable set of real positive zeros ξ_k, such that

$$0 < \xi_1 < \cdots < \xi_m < \xi_{m+1} < \cdots \qquad \lim_{k \to \infty} \xi_k = +\infty$$

Hence the eigenvalues of the problem are

$$\lambda_k = \frac{\xi_k^2}{c^2} \qquad k = 1, 2, \ldots \tag{94}$$

where ξ_k is the kth positive zero of $J_\alpha(t)$. A corresponding set of real-valued eigenfunctions is defined by

$$J_\alpha\left(\frac{\xi_k x}{c} \right) \qquad k = 1, 2, \ldots \tag{95}$$

Equation (80) implies the orthogonality relations

$$\int_0^c x J_\alpha\left(\frac{\xi_j x}{c}\right) J_\alpha\left(\frac{\xi_k x}{c}\right) dx = 0 \qquad \neq k \tag{96}$$

If $j = k$, then

$$\int_0^c x \left[J_\alpha\left(\frac{\xi_k x}{c}\right)\right]^2 dx = \frac{c^2}{2} J_{\alpha+1}^2(\xi_k) \qquad k = 1, 2, \ldots \tag{97}$$

To derive Eq. (97) let k be a fixed positive integer, and set $\lambda = \lambda_k$, $y = J_\alpha(\xi_k x/c)$ in Eq. (75). Multiply both sides of this equation by $2xy'$ and integrate over $[0,c]$.

$$2\int_0^c (xy')'(xy')\,dx = 2\int_0^c (\alpha^2 - \lambda_k x^2) yy'\,dx$$

Integration by parts performed on the integral on the right yields

$$(xy')^2\big|_0^c = (\alpha^2 - \lambda_k x^2)y^2\big|_0^c + 2\lambda_k \int_0^c xy^2\,dx \tag{98}$$

Since y satisfies the boundary conditions, it follows that

$$2\lambda_k \int_0^c xy^2\,dx = (xy')^2\big|_{x=c} + \alpha^2 y^2\big|_{x=0}$$

If $\alpha = 0$, the second term on the right vanishes. If $\alpha > 0$, then $y^2(0) = J_\alpha^2(0)$, and again this term vanishes. Also

$$y' = \frac{dy}{dx} = \frac{\xi_k}{c} J_\alpha'\left(\frac{\xi_k x}{c}\right)$$

where the prime on J_n denotes differentiation with respect to the argument $t = \xi_k x/c$. Hence

$$2\frac{\xi_k^2}{c^2}\int_0^c x\left[J_\alpha\left(\frac{\xi_k x}{c}\right)\right]^2 dx = \xi_k^2[J'(\xi_k)]^2$$

and Eq. (97) follows from the relation

$$J'(\xi_k) = \frac{\alpha}{\xi_k} J_\alpha(\xi_k) - J_{\alpha+1}(\xi_k) = -J_{\alpha+1}(\xi_k)$$

Let f be an integrable function on $[0,c]$. The values

$$a_k = \frac{2}{c^2 J_{\alpha+1}^2(\xi_k)} \int_0^c x f(x) J_\alpha\left(\frac{\xi_k x}{c}\right) dx \qquad k = 1, 2, \ldots \tag{99}$$

are called the *Fourier-Bessel coefficients* of f relative to the orthogonal system (95). The series

$$\sum_{k=1}^\infty a_k J_\alpha\left(\frac{\xi_k x}{c}\right) \tag{100}$$

where the coefficients a_k are defined by Eq. (99), is called the *Fourier-Bessel* series of f. Conditions under which the Fourier-Bessel series converges to f are very similar to those stated for ordinary Fourier series. In particular, if f is piecewise smooth on $[0,c]$, the Fourier-Bessel series of f converges to $f(x)$ whenever x is a point of continuity of f in $(0,c)$, and converges to the average value $[f(x^+) + f(x^-)]/2$ at a point of discontinuity of f. At $x = c$ the series converges to zero. If $\alpha > 0$, then evidently the series converges to zero also at $x = 0$, since $J_\alpha(0) = 0$. It can be shown that the orthogonal sequence

$$\varphi_n(x) = \sqrt{x}\, J_\alpha\left(\frac{\xi_n x}{c}\right) \qquad n = 1, 2, \ldots$$

is complete in the class of all functions which are continuous on $[0,c]$. In turn this implies that the sequence (94) constitutes all the eigenfunctions of the problem; however, these properties are not proved in this text.

Let u, v be real-valued twice continuously differentiable functions on $(0,c)$ which satisfy the first boundary condition $B_1(y)$. Then

$$\lim_{h \to 0} x(vu' - uv')\big|_h^{c-h} = c[v(c)u'(c) - u(c)v'(c)]$$

Accordingly any homogeneous boundary condition $B_2(y)$ which implies

$$v(c)u'(c) - u(c)v'(c) = 0 \qquad (101)$$

whenever u, v satisfy $B_2(y)$, when adjoined to $B_1(y)$ will force Eq. (80) to hold and yield orthogonal eigenfunctions. For example, consider the singular problem with Eqs. (75) and (76) and the boundary condition

$$B_2(y) = \gamma y(c) + y'(c) = 0 \qquad (102)$$

where γ is a real constant. It is easy to verify that Eq. (101) holds whenever u, v satisfy $B_2(y) = 0$. Hence all the eigenvalues of this problem are real, and Eq. (80) holds. Moreover

1. If $\gamma > 0$, every eigenvalue is positive.
2. If $\gamma = 0$ and $\alpha > 0$, every eigenvalue is positive.
3. If $\gamma = \alpha = 0$, every eigenvalue $\lambda \geq 0$, and $\lambda = 0$ is an eigenvalue with corresponding eigenfunction $y = $ const.

These statements are established in the same manner as in the previous problem.

Assume either $\alpha > 0$ or $\gamma > 0$. Let λ be an eigenvalue, and make the change of independent variable $t = \sqrt{\lambda}\, x$. As before, this transforms the differential equation into Bessel's equation (84) of order α. Again it follows that the eigenfunction must have the form $y = J_\alpha(\sqrt{\lambda}\, x)$ in order to satisfy the first boundary condition. The boundary condition $B_2(y)$ implies that an eigenvalue λ must satisfy the equation

$$\gamma J_\alpha(\sqrt{\lambda}\, c) + \sqrt{\lambda}\, J_\alpha'(\sqrt{\lambda}\, c) = 0 \qquad (103)$$

where the prime denotes differentiation with respect to the argument $t = \sqrt{\lambda}\, x$. Under the assumptions made above on α and γ it can be shown that the positive zeros of the function $c\gamma J_\alpha(t) + t J_\alpha'(t)$ constitute a real sequence such that

$$0 < \xi_1 < \xi_2 < \cdots < \xi_m < \xi_{m+1} < \cdots \qquad \lim_{k \to \infty} \xi_k = +\infty$$

Accordingly the eigenvalues of the problem are

$$\lambda_k = \frac{\xi_k^2}{c^2} \qquad k = 1, 2, \ldots \qquad (104)$$

where ξ_k is the kth positive root of the function $c\gamma J_\alpha(t) + tJ'_\alpha(t)$, and the corresponding eigenfunctions are

$$J_\alpha\left(\frac{\xi_k x}{c}\right) \qquad k = 1, 2, \ldots \tag{105}$$

From the manner of their construction these functions satisfy the orthogonality property stated in Eq. (96). Also

$$\int_0^c x\left[J_\alpha\left(\frac{\xi_k x}{c}\right)\right]^2 dx = \frac{c^2}{2}\frac{\xi_k{}^2 + \gamma^2 c^2 - \alpha^2}{\xi_k{}^2}[J_\alpha(\xi_k)]^2 \tag{106}$$

The derivation of Eq. (106) is similar to the derivation of Eq. (97).

If f is an integrable function, the formal series

$$\sum_{k=1}^\infty a_k J_\alpha\left(\frac{\xi_k x}{c}\right) \tag{107}$$

is called the *Fourier-Bessel* series of f relative to the orthogonal system (105) whenever the coefficients are given by

$$a_k = \frac{2}{c^2}\frac{\xi_k{}^2}{\xi_k{}^2 + \gamma^2 c^2 - \alpha^2}\left[\frac{1}{J_\alpha(\xi_k)}\right]^2 \int_0^c xf(x)J_\alpha\left(\frac{\xi_k x}{c}\right) dx \tag{108}$$

where ξ_k is the kth positive zero of $c\gamma J_\alpha(t) + tJ'_\alpha(t), k = 1, 2, \ldots$. The expansion theorems for the Fourier-Bessel series (107) are similar to those for the series (100).

Finally, suppose $\alpha = \gamma = 0$. In this case the eigenvalues are $\lambda_0 = 0$, and

$$\lambda_k = \frac{\xi_k{}^2}{c^2} \qquad k = 1, 2, \ldots \tag{109}$$

where ξ_k is the kth positive root of $J'_0(t)$. The Fourier-Bessel series is

$$a_0 + \sum_{k=1}^\infty a_k J_0\left(\frac{\xi_k x}{c}\right) \tag{110}$$

where

$$a_0 = \frac{2}{c^2}\int_0^c xf(x)\, dx$$

$$a_k = \frac{2}{c^2[J_0(\xi_k)]^2}\int_0^c xf(x)J_0\left(\frac{\xi_k x}{c}\right) dx \qquad k = 1, 2, \ldots \tag{111}$$

Appendix

3

REFERENCES AND BIBLIOGRAPHY

REFERENCES

1 CHURCHILL, R. V.: "Fourier Series and Boundary Value Problems," 2d ed., McGraw-Hill Book Company, New York, 1963.

2 CODDINGTON, E. A., and N. LEVINSON: "Theory of Ordinary Differential Equations," McGraw-Hill Book Company, New York, 1955.

3 COURANT, R., and D. HILBERT: "Methods of Mathematical Physics," vol. 1, Interscience Publishers, Inc., New York, 1953.

4 COURANT, R., and D. HILBERT: "Methods of Mathematical Physics," vol. 2, "Partial Differential Equations," by R. Courant, Interscience Publishers, Inc., New York, 1962.

5 GELFAND, I., and G. SHILOV: "Generalized Functions," vol. 1, Academic Press Inc., New York, 1964.

6 HORNICH, H.: "Existenzprobleme bei linearen partiellen Differentialgleichungen," VEB Deutscher Verlag der Wissenschaften, Berlin, 1960.

7 KOLMOGOROV, A., and S. FOMIN: "Elements of the Theory of Functions and Functional Analysis," vol. 1, Graylock Press, Rochester, N.Y., 1957.

8 MOON, P., and D. SPENCER: "Field Theory for Engineers," D. Van Nostrand Company, Inc., Princeton, N.J., 1961.

9 PETROVSKY, I.: "Lectures on Partial Differential Equations," Interscience Publishers, Inc., New York, 1954.

10 SANSONE, G.: "Orthogonal Functions," Interscience Publishers, Inc., New York, 1959.

11 TAYLOR, A.: "Advanced Calculus," Ginn and Company, Boston, 1955.

BIBLIOGRAPHY

BATEMAN, H.: "Partial Differential Equations of Mathematical Physics," Dover Publications, Inc., New York, 1944.

BERGMAN, S., and M. SCHIFFER: "Kernel Functions and Elliptic Differential Equations," Academic Press Inc., New York, 1953.

BERNSTEIN, D.: "Existence Theorems in Partial Differential Equations," Princeton University Press, Princeton, N.J., 1950.

BERS, L., F. JOHN, and M. SCHECHTER: "Partial Differential Equations," Interscience Publishers, Inc., New York, 1964.

BIRKHOFF, G., and G. ROTA: "Ordinary Differential Equations," Ginn and Company, Boston, 1962.

BUDAK, B., A. SAMARSKII, and A. TIKHONOV: "A Collection of Problems on Mathematical Physics," The Macmillan Company, New York, 1964.

CARSLAW, H., and J. JAEGER: "Conduction of Heat in Solids," Oxford University Press, Fair Lawn, N.J., 1950.

CARSLAW, H.: "Introduction to the Theory of Fourier's Series and Integrals," Dover Publications, Inc., New York, 1930.

CHURCHILL, R.: "Fourier Series and Boundary Value Problems," 2d ed., McGraw-Hill Book Company, New York, 1963.

COLLATZ, L.: "Eigenwertprobleme und ihre numerische Behandlung," Chelsea Publishing Company, New York, 1948.

DETTMAN, J.: "Mathematical Methods in Physics and Engineering," McGraw-Hill Book Company, New York, 1962.

DUFF, G.: "Partial Differential Equations," University of Toronto Press, Toronto, 1956.

EPSTEIN, B.: "Partial Differential Equations," McGraw-Hill Book Company, New York, 1962.

FRIEDMAN, A.: "Generalized Functions and Partial Differential Equations," Prentice-Hall, Inc., Englewood Cliffs, N.J., 1963.

FRIEDMAN, A.: "Partial Differential Equations of Parabolic Type," Prentice-Hall, Inc., Englewood Cliffs, N.J., 1964.

FRIEDMAN, B.: "Principles and Techniques of Applied Mathematics," John Wiley & Sons, Inc., New York, 1956.

GARABEDIAN, P.: "Partial Differential Equations," John Wiley & Sons, Inc., New York, 1964.

GOULD, S.: "Variational Methods for Eigenvalue Problems," University of Toronto Press, Toronto, Canada, 1957.

GOURSAT, E.: "A Course in Mathematical Analysis," vol. 2, pt. 2, Differential Equations, Dover Publications, Inc., New York, 1955.

GOURSAT, E.: "A Course in Mathematical Analysis," vol. 3, pt. 1, Variation of Solutions, Partial Differential Equations of the Second Order, Dover Publications, Inc., New York, 1964.

GOURSAT, E.: "A Course in Mathematical Analysis," vol. 3, pt. 2, Integral Equations, Calculus of Variations, Dover Publications, Inc., New York, 1964.

GRAY, A., and G. MATHEWS: "A Treatise on Bessel Functions and Their Applications to Physics," 2d ed. prepared by A. Gray and T. M. MacRobert, The Macmillan Company, New York, 1952.

GREENSPAN, D.: "Introduction to Partial Differential Equations," McGraw-Hill Book Company, New York, 1961.

HADAMARD, J.: "Lectures on Cauchy's Problem in Linear Partial Differential Equations," Dover Publications, Inc., New York, 1952.

HALMOS, P.: "Finite Dimensional Vector Spaces," D. Van Nostrand Company, Inc., Princeton, N.J., 1958.

HELLWIG, G.: "Partial Differential Equations," Blaisdell Publishing Co., New York, 1964.

HILDEBRAND, F.: "Advanced Calculus for Applications," Prentice-Hall, Inc., Englewood Cliffs, N.J., 1962.

HORMANDER, L.: "Linear Partial Differential Operators," Academic Press Inc., New York, 1963.

HORN, J.: "Partial Differentialgleichungen," Walter de Gruyter & Co., Berlin, 1949.

INCE, E.: "Ordinary Differential Equations," Dover Publications, Inc., New York, 1950.

JEFFREYS, H., and B. JEFFREYS: "Methods of Mathematical Physics," Cambridge University Press, New York, 1950.

JOHN, F.: "Plane Waves and Spherical Means Applied to Partial Differential Equations," Interscience Publishers, Inc., New York, 1955.

KELLOGG, O.: "Foundations of Potential Theory," Frederick Ungar Publishing Co., New York, 1930.

KNESCHKE, A.: "Differentialgleichungen und Randwertprobleme," VEB Verlag Technik, Berlin, 1960.

KOSHLYAKOV, N., M. SMIRNOV, and E. GLINER: "Differential Equations of Mathematical Physics," Interscience Publishers, Inc., New York, 1964.

LANCZOS, C.: "Linear Differential Operators," D. Van Nostrand Company, Inc., Princeton N.J., 1961.

LENSE, J.: "Reihenentwicklungen in der mathematischen Physik," Walter de Gruyter & Co., Berlin, 1953.

LOVITT, W.: "Linear Integral Equations," Dover Publications, Inc., New York, 1950.

MACKE, W.: "Wellen," Akademische Verlagsgesellschaft Geest & Portig KG, Leipzig, 1958.

MACROBERT, T.: "Spherical Harmonics," Dover Publications, Inc., New York, 1947.

MORSE, P., and H. FESHBACK: "Methods of Theoretical Physics," pts. 1 and 2, McGraw-Hill Book Company, New York, 1953.

MURNAGHAN, F.: "Introduction to Applied Mathematics," John Wiley & Sons, Inc., New York, 1948.

PAGE, C.: "Physical Mathematics," D. Van Nostrand Company, Inc., Princeton, N.J., 1955.

RIESZ, F., and B. NAGY: "Functional Analysis," Frederick Ungar Publishing Co., New York, 1955.

SAGAN, H.: "Boundary and Eigenvalue Problems in Mathematical Physics," John Wiley & Sons, Inc., New York, 1961.

SAUER, R.: "Anfangswertprobleme bei partiellen Differentialgleichungen," Springer-Verlag OHG, Berlin, 1952.

SMIRNOV, V.: "A Course of Higher Mathematics," vol. 4, "Integral Equations and Partial Differential Equations," Addison-Wesley Publishing Company, Inc., Reading, Mass., 1964.

SNEDDON, I.: "Elements of Partial Differential Equations," McGraw-Hill Book Company, New York, 1957.

SOBOLEV, S.: "Applications of Functional Analysis in Mathematical Physics," Translations of Mathematical Monographs, vol. 7, American Mathematical Society, Providence, R.I., 1963.

SOBOLEV, S.: "Partial Differential Equations of Mathematical Physics," Addison-Wesley Publishing Company, Inc., Reading, Mass., 1964.

SOKOLNIKOFF, I., and R. REDHEFFER: "Mathematics of Physics and Modern Engineering," McGraw-Hill Book Company, New York, 1958.

SOMMERFELD, A.: "Partial Differential Equations in Physics," Academic Press Inc., New York, 1949.

STERNBERG, W., and T. SMITH: "The Theory of Potential and Spherical Harmonics," University of Toronto Press, Toronto, Canada, 1944.

TAYLOR, A.: "Introduction to Functional Analysis," John Wiley & Sons, Inc., New York, 1958.

TIKHONOV, A., and A. SAMARSKII: "Equations in Mathematical Physics," Pergamon Press, New York, 1965.

TITCHMARSH, E.: "Eigenfunction Expansions Associated with Second-order Differential Equations," pts. I and II, Oxford University Press, Fair Lawn, N.J., 1958.

TOLSTOV, G.: "Fourier Series," Prentice-Hall, Inc., Englewood Cliffs, N.J., 1962.

VULIKH, B.: "Introduction to Functional Analysis for Scientists and Technologists," Pergamon Press, New York, 1963.

WEBSTER, A.: "Partial Differential Equations of Mathematical Physics," Dover Publications, Inc., New York, 1957.

WEINBERGER, H.: "A First Course in Partial Differential Equations," Blaisdell Publishing Co., New York, 1965.

YOSIDA, K.: "Lectures on Differential and Integral Equations," Interscience Publishers, Inc., New York, 1960.

4 ANSWERS

CHAPTER 1

1 **(a)** Linear second order, two independent variables.
(c) Nonlinear, almost linear, second order, two independent variables.
(e) Nonlinear, quasilinear, third order, three independent variables.

3 **(a)** $xp - z = 0$ **(c)** $xq - yp = 0$
(e) $yp - xq = 2xyz$ **(g)** $u_{tt} - c^2 u_{xx} = 0$

6 **(a)** $xp + yq = z \log z$ **(c)** $xp + yq = z$
(e) $u_{xx} + u_{yy} = 0$ **(g)** $u_t = u_{xx}$
(i) $z = xp + yq + p^2 + q^2$

7 **(a)** $(x - 1)p + yq = z$ **(c)** $xp + yq - pq = z$

8 **(a)** $z = x^2 y + \dfrac{y^3}{3} + f(x)$ **(c)** $z = f(y)e^{-x^2/2} + x^2 - 2 + 3y$

10 **(a)** $z = f(4x + 3y) + \dfrac{x^3}{9}$

(c) $z = e^{-x/5} f(4x - 5y) + x^3 - 15x^2 + 150x - 749 + \dfrac{2e^{3y}}{13}$

(e) $z = f(ax + y) + e^{mx} \dfrac{m \cos by - ab \sin by}{m^2 + a^2 b^2}$

11 **(a)** $z = f(x^2 y^4)$ **(c)** $z = f(x^7 y) - \dfrac{x^2 y}{5}$

(e) $z = x^{-c/a} F(x^b y^{-a}) + \dfrac{x^2}{2a + c} + \dfrac{y^2}{2b + c}$

12 **(a)** $z = \dfrac{f(x^2 + y^2)}{x}$ **(c)** $z = (x + a)^{-c} f\left(\dfrac{x + a}{y + b}\right)$

13 **(a)** $u = e^{-x/2} f(x - 2y - z)$

14 **(a)** $F(2y - x^2, ze^{-x}) = 0$ **(c)** $F\left(z, \dfrac{x + z}{y + z}\right) = 0$

(e) $F\left(x^2 - z^2, \dfrac{x + z}{y}\right) = 0$ **(g)** $z = \dfrac{x^2(x^2 + y^2)}{2} + f(x^2 + y^2)$

(i) $z = \dfrac{axy \log (x/y)}{x - y} + f\left(\dfrac{x - y}{xy}\right)$ **(k)** $xyz = f(x + y + z)$

(m) $z = x^3 \sin(y + 2x) + f(y + 2x)$ **(o)** $z = \sin y + f(\sin x - \sin y)$

(q) $\sin^{-1}\dfrac{z}{a} = xy + f\left(\dfrac{y}{x}\right)$

15 (a) $z = yf\left(\dfrac{x^2 + y^2 + z^2}{y}\right)$

17 $z = a + xf\left(\dfrac{x}{y}\right)$

18 (a) $u = e^{-Rx/P_1}f(P_2x - P_1y, P_3x - P_1z)$ $P_1 \neq 0$
(c) $u = f(5x - 3y, 5z + y) + \frac{1}{5}\sin y - 2e^{-z}$
(e) $u = f[z + 2\log y, z + 2\log(x + z - 2)] - 2e^z$
(g) $u = f(yz, x + y + z) + xy$ **(i)** $u = x^n f\left(\dfrac{x}{y}, \dfrac{z}{x}\right)$
(k) $u = f(x_1{}^2 - 2x_2, x_2{}^2 - 2x_3, x_3{}^2 - 2x_4)$

19 (a) $f(x^2 - 2y, y^2 - 2z, ue^{-z^2/2}) = 0$ **(c)** $u = xf\left(\dfrac{y}{x}, \dfrac{z}{x}\right) + \dfrac{xy \log x}{z}$

20 (a) $z = e^y \cos(x - y)$ **(c)** $z^2 = xy$

(e) $z = \dfrac{xy}{2x - y + xy}$ **(g)** $(x^2 + y^2 + z^2)^2 = 2a^2(x^2 + y^2 + xy)$

(i) $(x^2 + y^2)(a^2z^2 - h^2y^2) = a^2hx^2z$ **(k)** $(x + y + z)^3 = 27xyz$

(m) $z = \left(\dfrac{x^2 + y^2}{2}\right)^{\frac{1}{2}} \exp\dfrac{x^2 - y^2}{2}$

(o) $a \sin(y - a \sin x) - \log[\sin(y - a \sin x)] = az - \log \sin y$

22 (a) $z = 1 - \dfrac{3(x - 1)}{2} + \dfrac{3(y - 1)}{2}$

$$+ \dfrac{15(x - 1)^2/4 - 9(x - 1)(y - 1)/2 + 3(y - 1)^2/4}{2} + \cdots$$

(c) No solution exists

23 (a) $u = (x + y + z)(xy/z)^{n-1}$

CHAPTER 2

4 (a) $z = f(x + y) + g(9x + y)$ **(c)** $z = f(y) + e^{-x}g(y) - \dfrac{x^2}{2} - xy$

(e) $z = f(x)\cos y + g(x)\sin y + \dfrac{e^{x+y}}{2}$

(g) $z = f(x) + e^{x/2}g(3x - 2y) + 2\sin(2x - 3y)$

(i) $z = f(2x + y) + g(2x - y) + x^4 + 2x^2 + \dfrac{\cos y}{4}$
(k) $z = f(x + y) + xg(x + y) + \frac{4}{9}e^{3y} - \cos x$

5 (a) $z = f(y - 2x) + g(y - 3x) + 7x^2 - 6xy + \dfrac{5y^2}{4} - \left(4x^2 - 3xy + \dfrac{y^2}{2}\right)\log(y - 2x)$

(c) $z = f(y - 2x) + g(y + 2x) + x\log y + y\log x$

6 (a) $z = f(x) + yg(x) + \dfrac{x^3 y^2}{2} + \dfrac{y^5}{20}$ (c) $z = e^{xy}g(x) - \dfrac{f(x)}{x} - xy$

(e) $z = f(y) + \dfrac{g(x)}{y} + \sin(x + y)$ (g) $z = \dfrac{f(x) + g(y)}{x - y}$

8 (a) $z = f(xy) + x^3 g\dfrac{y}{x}$ (c) $z = y^2 f(x) + \dfrac{g(xy)}{x}$

(e) $z = xy\log x + f(xy) + xg\left(\dfrac{x}{y}\right)$ (g) $z = xf(y) + yg(x)$

10 (a) $z = f(e^x - e^y) + g(e^x + e^y)$ (c) $z = f(x + y) + g\left(\dfrac{y}{x}\right)$

11 (a) $u = xf(x - y, x + z) + g(x - y, x + z) + 2\sin x - x\cos x$
$\qquad\qquad\qquad + 2\sin y - y\cos y + 2\sin z - z\cos z$

(c) $u = e^x f(x + z, y + z) + e^{-x}g(x - z, y - z) + \dfrac{e^{2x}}{3} - \cos y - z$

(e) $u = f(x + \sqrt{A}\,t, y + \sqrt{C}\,t) + g(x - \sqrt{A}\,t, y - \sqrt{C}\,t)$

12 (a) $z = \exp\left[hx + \dfrac{(1 - h^2)y}{1 - 2h}\right]$

(c) $z = \exp[hx + (h^2 - 2h)y] - \dfrac{x^2 + x}{4} - \dfrac{y^2}{2}$

(e) $u = \exp[h(x \pm iy)]$ $i = \sqrt{-1}$ (g) $u = \exp[hx \pm i(h^2 + k^2)^{1/2}y]$

(i) $u = \exp[\alpha x + \beta y \pm c(\alpha^2 + \beta^2)^{1/2}t]$

20 (a) $z = e^{-y}f(x) + e^{-x}g(y)$ (c) $z = \dfrac{xy}{x - y} + \dfrac{f(x) + g(y)}{x - y}$

22 (a) Parabolic; $z = f(x + y) + xg(x + y) + \dfrac{x^2}{8}$

(c) Parabolic; $z = \exp[\lambda_1(x + 2y)]f(x + y)$

$\qquad\qquad\qquad + \exp[\lambda_2(x + 2y)]g(x + y) + \dfrac{(x + 2y + a/c)^2}{c} + \dfrac{a^2 - 2c}{c^3}$

where λ_1, λ_2 are the roots of $\lambda^2 - a\lambda + c = 0$

(e) Parabolic; $z = f\left(\dfrac{y}{x}\right) + yg\left(\dfrac{y}{x}\right) + 2x^2$

(g) Hyperbolic wherever $xy \neq 0$; $z = f(xy) + xg\left(\dfrac{x}{y}\right) + xy\log x$

(i) Hyperbolic; $z = f(x + y - \cos x) + g(x - y + \cos x)$

(k) Hyperbolic wherever $x^2 + y^2 \neq 0$; $z = f(x^2 + y^2) + g\left(\dfrac{y}{x}\right) - xy$

(m) Hyperbolic wherever $x \neq h$; $z = \dfrac{f(x - at) + g(x + at)}{h - x}$

23 (a) Hyperbolic   (c) Elliptic   (e) Elliptic

24 (a) Unique solution is $u = b + \sin x \sin t$

(c) If $f(\xi)$ is an arbitrary twice differentiable function such that $f'(0) = 1$, then $u = f(x - t) - f(0)$ is a solution.

(e) Unique solution is $u = x^2 + t - \dfrac{t^2}{9}$ (g) No solution

(i) Unique solution is $z = \dfrac{x^2 + y^2}{2} + y - 1 + e^{x-y}(1 - x)$

(k) Unique solution is $z = 4xy + \dfrac{y^2}{2} - \dfrac{2y^3}{3} - e^{-y}$

(m) Unique solution is $z = \dfrac{x^4}{12} + \dfrac{x^2y^2}{2} + 1$

25 (a) $u = x + y - 1$ (c) $u = \dfrac{x^2}{2} + 2y - 2$

(e) $u = (\log x + y - x)^2 + 2(\log x + y) + x^2(e^y - 1) - xe^y$

CHAPTER 3

1 (c) $a_{11} + a_{22} + a_{33} = 0$ (e) $b = \pm a$

3 In three dimensions $f(r) = a + b/r$, a, b constants; two dimensions $f(r) = a + b \log r$.

8 (a) $\max u(x,y) = 1$, $\min u(x,y) = -1$; max occurs at the points $(1,0)$, $(-\tfrac{1}{2}, \sqrt{3}/2)$, $(-\tfrac{1}{2}, -\sqrt{3}/2)$; min occurs at the points $(\tfrac{1}{2}, \sqrt{3}/2)$, $(-1,0)$, $(\tfrac{1}{2}, -\sqrt{3}/2)$.

17 (a) $u = \dfrac{400}{\pi} \displaystyle\sum_{k=1}^{\infty} \dfrac{\sin [(2k - 1)\pi x/a] \sinh [(2k - 1)\pi(b - y)/a]}{(2k - 1) \sinh [(2k - 1)\pi b/a]}$

(c) $u = -\dfrac{8a^2 T}{\pi^3} \displaystyle\sum_{k=1}^{\infty} \dfrac{\sin [(2k - 1)\pi x/a] \sinh [(2k - 1)\pi(b - y)/a]}{(2k - 1)^3 \sinh [(2k - 1)\pi b/a]}$

21 (a) $u = \dfrac{2T \tan^{-1} [(\sin \pi x/a)(\sinh \pi y/a)]}{\pi}$

(b) $u = 2Th \displaystyle\sum_{k=1}^{\infty} \dfrac{1 - \cos (a\xi_k)}{\xi_k(ah + \cos^2 a\xi_k)} \sin (\xi_k x)e^{-\xi_k y}$

where $\{\xi_k\}$ denotes the sequence of positive roots of the transcendental equation $\tan a\xi = -\xi/h$.

23 $u = v + w_1 + w_2$ where $v = \dfrac{cx(x - a)}{2} + \dfrac{dy(y - b)}{2}$

$w_1 = \displaystyle\sum_{n=1}^{\infty} A_n \dfrac{\sinh [(2n - 1)\pi(a - x)/b] + \sinh [(2n - 1)\pi x/b]}{\sinh [(2n - 1)\pi a/b]} \sin \dfrac{(2n - 1)\pi y}{b}$

$w_2 = \displaystyle\sum_{n=1}^{\infty} B_n \dfrac{\sinh [(2n - 1)\pi(b - y)/a] + \sinh [(2n - 1)\pi y/b]}{\sinh [(2n - 1)\pi b/a]} \sin \dfrac{(2n - 1)\pi x}{b}$

where

$$A_n = \frac{4db^2}{\pi^3(2n-1)^3} \qquad B_n = \frac{4ca^2}{\pi^3(2n-1)^3}$$

27 $u = 64Ta^2b^2 \displaystyle\sum_{n=1}^{\infty} \sum_{m=1}^{\infty} \frac{\sin(\mu_n x/a)\sin(\mu_m x/b)e^{-\gamma_{nm}z}}{\mu_n^3 \mu_m^3}$

where

$$\mu_k = (2k-1)\pi \qquad k = 1, 2, \ldots \qquad \gamma_{nm} = \left[\left(\frac{\mu_n}{a}\right)^2 + \left(\frac{\mu_m}{b}\right)^2\right]^{1/2}$$

29 $u = \dfrac{c}{\pi}\left[\dfrac{\alpha}{2} + \displaystyle\sum_{n=1}^{\infty} \left(\dfrac{r}{a}\right)^n \dfrac{\sin n(\alpha - \theta) + \sin n\theta}{n}\right]$

31 $u = \dfrac{\pi - 20}{\pi} - \dfrac{2}{\pi}\displaystyle\sum_{n=1}^{\infty}\left(\dfrac{r}{a}\right)^{2m}\dfrac{\sin 2n\theta}{n}$

33 $u = \dfrac{8T}{\pi}\displaystyle\sum_{n=1}^{\infty}\dfrac{n}{4n^2-1}\dfrac{(r/a)^n - (a/r)^n}{(b/a)^n - (a/b)^n}\sin n\theta$

35 $u = 100\dfrac{\theta}{\alpha} + \dfrac{200}{\pi}\displaystyle\sum_{n=1}^{\infty}\dfrac{(-1)^n}{n}\left(\dfrac{a^{\mu_n}}{a^{\mu_n} + b^{\mu_n}}\right)\left[\left(\dfrac{r}{a}\right)^{\mu_n} + \left(\dfrac{b}{r}\right)^{\mu_n}\right]\sin \mu_n\theta$

where

$$\mu_n = \frac{n\pi}{\alpha} \qquad n = 1, 2, \ldots$$

37 (a) $u = 200 \displaystyle\sum_{k=1}^{\infty} \dfrac{J_0(\xi_k/a)\sinh[\xi_k(h-z)/a]}{\xi_k \sinh(\xi_k h/a)J_1(\xi_k)}$

where $\{\xi_k\}$ denotes the sequence of positive zeros of $J_0(\xi)$

39 $u = A_{00} + \displaystyle\sum_{n=0}^{\infty} \sum_{k=1}^{\infty} A_{nk}\cosh(\xi_{nk}h/a)J_{\nu_n}(\xi_{nk}r/a)\cos \nu_n\theta$ where, for each fixed n, $\{\xi_{nk}\}$
is the sequence of positive zeros of $J'_n(\xi)$, $\nu_n = n\pi/\beta$, and

$$A_{00} = \frac{2}{a^2\beta}\int_0^a \int_0^\beta f(r,\theta)r\,dr\,d\theta$$

$$A_{nk} = \frac{2\epsilon_n\xi_{nk}^2}{a^2\beta(\xi_{nk}^2 - \nu_n^2)\cosh(\xi_{nk}h/a)J_{\nu_n}^2(\xi_{nk})}\int_0^a \int_0^\beta f(r,\theta)J_{\nu_n}(\xi_{nk}r/a)\cos \nu_n\theta r\,dr\,d\theta$$

$n = 0, 1, \ldots \qquad k = 1, 2, \ldots \qquad \epsilon_0 = 2; \epsilon_n = 1 \text{ if } n \geq 1$

41 $\psi = 4\pi(b-a) \qquad 0 \leq r \leq a \qquad \cdot \qquad \psi = 4\pi b - \dfrac{2\pi(a^2 + r^2)}{r} \qquad a \leq r \leq b$

$\psi = \dfrac{M}{r} \qquad r \geq b$

where total mass $M = 2\pi(b^2 - a^2)$

43 $u = u_1 + \dfrac{ab^2h(u_2 - u_1)}{a + hb(b-a)}\left(\dfrac{1}{a} - \dfrac{1}{r}\right)$

47 $u = \dfrac{T}{3}\left[7 - 4\left(\dfrac{r}{a}\right)^2 P_2(\cos\theta)\right]$

49 $u = \dfrac{T}{2} + T \displaystyle\sum_{n=0}^{\infty} \dfrac{(4n+1)P_{2n}(0)}{(2n-1)(2n+2)}\left(\dfrac{r}{a}\right)^{2n} P_{2n}(\cos\theta)$

51 $u = \displaystyle\sum_{n=1}^{\infty}\sum_{m=1}^{\infty} A_{nm}\sin\dfrac{n\pi x}{a}\sin\dfrac{m\pi y}{b}$

$A_{nm} = -\dfrac{4}{nm\pi^2}\{ac(-1)^n[1-(-1)^m] + bd(-1)^m[1-(-1)^n]\}$

55 $\Psi = 4a^2 \displaystyle\sum_{k=1}^{\infty} \dfrac{J_0(\xi_k r(a))}{\xi_k^3 J_1(\xi_k)}$

where $\{\xi_k\}$ denotes the sequence of positive zeros of $J_0(\xi)$.

CHAPTER 4

1 **(d)** **(i)** $u = \sin kx \cos \omega t$ $\omega = kc$

(iii) $u = \exp[-(kx - \omega t)^2] - \exp[-(kx + \omega t)^2]$

(v) $u = (k\cos kx - \alpha \sin kx)\cos \omega t$ $\omega = kc$

3 **(a)** $u = e^{-x}\cosh ct$ **(c)** $u = A\sin \omega x \cos \omega ct + \dfrac{B}{c\mu}\cos \mu x \sin \mu ct$

(e) $u = \dfrac{U(1-x+ct) - U(-1-x+ct) + U(1-x-ct) - U(-1-x-ct)}{2}$

where $U(\xi)$ is the *unit step function* defined by

$U(\xi) = 1$ $\xi \geq 0$ $U(\xi) = 0$ $\xi < 0$

(g) $u = \dfrac{[U(\pi/2 - x + ct) - U(-\pi/2 - x + ct)]\cos(x - ct)}{2}$

$\qquad\qquad + \dfrac{[U(\pi/2 - x - ct) - U(-\pi/2 - x - ct)]\cos(x + ct)}{2}$

5 **(a)** $u = \sin \omega x \cos \omega ct + \dfrac{t^2}{2}$ **(c)** $u = \dfrac{\cosh bx \sinh bct}{bc} + 2xt^2 + \dfrac{t^3}{6}$

(e) $u = A\left[\dfrac{t\cos(kx - \omega t)}{2\omega} - \dfrac{\cos kx \sin \omega t}{2\omega^2}\right]$ $\omega = kc$

7 **(a)** $u = 1 - \dfrac{1}{c}\cos x \sin ct$ $0 \leq ct \leq x$

$\qquad u = \left(\dfrac{-1}{c}\right)\sin x \cos ct$ $0 \leq x \leq ct$

8 **(a)** $u = \begin{cases} e^{-x}\cosh ct + t & 0 \leq ct \leq x \\[2mm] -e^{-ct}\sinh x + \dfrac{x}{c} + 1 & 0 \leq x \leq ct \end{cases}$

19 (a) $u = \dfrac{32h}{\pi^3} \displaystyle\sum_{n=1}^{\infty} \dfrac{\sin [(2n-1)\pi x/b] \cos [(2n-1)\pi ct/b]}{(2n-1)^3}$

(c) $u = \dfrac{2A \sin \omega b}{b} \displaystyle\sum_{n=1}^{\infty} [(-1)^{n+1}\mu_n/(\mu_n{}^2 - \omega^2)] \sin \mu_n x \cos \mu_n ct$

$$+ \frac{4}{bc} \sum_{n=1}^{\infty} \frac{\sin \mu_{2n-1}x \cos \mu_{2n-1}ct}{\mu_{2u-1}^2} \qquad \mu_n = \frac{n\pi}{b}$$

21 (a) For the case of nonresonance

$$u = \frac{4cF_0}{b} \sum_{n=1}^{\infty} \frac{\omega \sin \omega_{2n-1}t - \omega_{2n-1} \sin \omega t}{\omega_{2n-1}^2(\omega^2 - \omega_{2n-1}^2)} \sin \frac{(2n-1)\pi x}{b}$$

If $\omega = \omega_m$ (resonance case), m even, solution is given by the above series. If $\omega = \omega_{2q-1}$, then in the series the term involving ω_{2q-1} is replaced by the term

$$\frac{\sin \omega_{2q-1}t - \omega_{2q-1}t \cos \omega_{2q-1}t}{\omega_{2q-1}^3} \sin \frac{(2n-1)\pi x}{b}$$

23 $u = -\dfrac{4F_0}{\pi c} \displaystyle\sum_{n=1}^{\infty} \dfrac{\cos (n\pi/2) \sin (n\pi/6) \sin (n\pi x/b) \sin (n\pi ct/t)}{n}$

25 (d) If $\omega \neq n\pi c/b$, $n = 1, 2, \ldots$, then $u = v + w$, where

$v = X(x) \cos \omega t$

$$X = \frac{h_0 \sin [\omega(b-x)/c] + h_1 \sin (\omega x/c) + X_p(x) \sin (\omega b/c) - \sin (\omega x/c)X_p(b)}{\sin (\omega b/c)}$$

$$X_p(x) = -F_0 \frac{[cv \sin (\omega x/c) - \omega \sin vx]/(c^2 v^2 - \omega^2)}{\omega}$$

$$w = \sum_{n=1}^{\infty} A_n \sin \frac{n\pi x}{b} \cos \frac{n\pi ct}{b}$$

$$A_n = \frac{2}{b} \int_0^b [x(b-x) - X(x)] \sin \frac{n\pi x}{b} dx$$

27 Let $\{\mu_n\}$ be the sequence of positive roots (in increasing order of magnitude) of the transcendental equation

$$\tan b\mu = \frac{\mu(h_1 + h_2)}{\mu^2 - h_1 h_2}$$

Let $\omega_n = c\mu_n$, $n = 1, 2, \ldots$. Then

$$u = \sum_{n=1}^{\infty} (A_n \cos \omega_n t + B_n \sin \omega_n t) \sin (\mu_n x + \theta_n)$$

$$A_n = \frac{1}{\alpha_n} \int_0^b f(x) \sin (\mu x_n + \theta_n) dx \qquad B_n = \frac{1}{\alpha_n \omega_n} \int_0^b g(x) \sin (\mu x_n + \theta_n) dx$$

$$\theta_n = \tan^{-1}\frac{\mu_n}{h_1} \qquad \alpha_n = \frac{1}{2}\left[b + \frac{(\mu_n{}^2 + h_1 h_2)(h_1 + h_2)}{(\mu_n{}^2 + h_1{}^2)(\mu_n{}^2 + h_2{}^2)} \right]$$

29 **(a)** $u = \left(1 - \dfrac{x}{b}\right) h_0 \sin \omega t + w_1 + w_2$

$$w_1 = \frac{2\omega h_0}{\pi} \sum_{n=1}^{\infty} \frac{(-1)^{n+1} \sin (n\pi x/b) \sin (n\pi ct/b)}{n\omega_n}$$

$$w_2 = -\frac{2}{\pi} \sum_{n=1}^{\infty} \left\{ \frac{g[1 - (-1)^n](1 - \cos \omega_n t)}{n\omega_n{}^2} \right.$$

$$\left. + h_0\omega^2 \frac{\omega_n \sin \omega t - \omega \sin \omega_n t}{n(\omega^2 - \omega_n{}^2)\omega_n} \right\} \sin \frac{n\pi x}{b} \qquad \omega_n = \frac{n\pi c}{b}$$

31 $u = v + w$, where

$$v = \frac{8b^2 e^{-\gamma t}}{\pi^3} \sum_{n=1}^{\infty} \sin \frac{(2n-1)\pi x}{b} \left[\frac{\cos \omega_{2n-1} t + (\gamma/\omega_{2n-1}) \sin \omega_{2n-1} t}{(2n-1)^3} \right]$$

$$w = \frac{2g}{\pi} \sum_{n=1}^{\infty} C_n(t) \sin \frac{n\pi x}{b}$$

$$C_n(t) = [1 - (-1)^n] \frac{e^{-\gamma t}(\gamma \sin \omega_n t + \omega_n \cos \omega_n t) - \omega_n}{n\omega_n(\gamma^2 + \omega_n{}^2)}$$

$$\omega_n = \left(\frac{n^2 \pi^2 c^2}{b^2} - \gamma^2 \right)^{\frac{1}{2}} \qquad n = 1, 2, \ldots$$

35 Let $\mu_n = (2n - 1)\pi/2b$, $\omega_n = (c^2\mu_n{}^2 - \gamma^2)^{\frac{1}{2}}$, $n = 1, 2, \ldots$.   Then

$$u = h_1 x \sin \omega t + v$$

$$v(x,t) = e^{-\gamma t} \left[\frac{2h_1\omega}{b} \sum_{n=1}^{\infty} \frac{(-1)^n}{\omega_n\mu_n{}^2} (\sin \omega_n t \sin \mu_n x) + w \right]$$

$$w(x,t) = \sum_{n=1}^{\infty} \left\{ \frac{1}{\omega_n} \int_0^t e^{\gamma \xi} \sin [\omega_n(t - \xi)] F_n(\xi)\, d\xi \right\} \sin \mu_n x$$

$$F_n(t) = \frac{2}{b\mu_n{}^2} [\mu_n F_0 \sin \nu t + \omega h_1(\omega \sin \omega t - 2\gamma \cos \omega t)(-1)^{n+1}]$$

37 $\xi = \displaystyle\sum_{n=1}^{\infty} A_n X_n(x) \cos \omega_n t$

where $X_n(x) = \sin \mu_n x$ and $\{\mu_n\}$ denotes the sequence of positive roots of

$$\tan b\mu = -\frac{EA}{Mc^2\mu}$$

$$A_n = \frac{b_1 - b}{b\alpha_n\mu_n{}^2} \sin \mu_n b \qquad \alpha_n = \frac{1}{2}\left(b - \frac{Mc^2 \sin^2 \mu_n b}{EA} \right) \qquad \omega_n = c\mu_n; \, n = 1, 2, \ldots$$

39 **(a)** Let $\omega_n = (n^2\pi^2 c^2/b^2 - \omega^2)^{\frac{1}{2}}$, $n = 1, 2, \ldots$.   Then $u = v + w$ where

$$v(x,t) = \sum_{n=1}^{\infty} (A_n \cos \omega_n t + B_n \sin \omega_n t) \sin \frac{n\pi x}{b}$$

$$w(x,t) = \frac{4F_0}{\pi} \sum_{n=1}^{\infty} \left[\frac{\nu \sin \omega_{2n-1} t - \omega_{2n-1} \sin \nu t}{(2n-1)\omega_{2n-1}(\nu^2 - \omega_{2n-1}^2)} \right] \sin \left[(2n-1) \frac{\pi x}{b} \right]$$

$$A_n = \frac{2}{b} \int_0^b f(x) \sin \frac{n\pi x}{b}\, dx \qquad B_n = \frac{2}{b\omega_n} \int_0^b g(x) \sin \frac{n\pi x}{b}\, dx$$

41 $u = \sum\limits_{n=1}^{\infty} J_0\left(\xi_n\sqrt{\dfrac{x}{b}}\right)(A_n \cos \omega_n t + B_n \sin \omega_n t)$

$A_n = \dfrac{1}{bJ_1{}^2(\xi_n)} \int_0^b J_0\left(\xi_n\sqrt{\dfrac{x}{b}}\right) f(x)\, dx$

$B_n = \dfrac{1}{b\omega_n J_1{}^2(\xi_n)} \int_0^b J_0\left(\xi_n\sqrt{\dfrac{x}{b}}\right) g(x)\, dx$

where $\{\xi_n\}$ denotes the positive zeros of $J_0(\xi)$ arranged in increasing order, and $\omega_n = c\xi_n/2\sqrt{b}$, $n = 1, 2, \ldots$.

53 (c) $u = v + w$, where

$\varphi_{nm} = \sin\dfrac{n\pi x}{a}\sin\dfrac{m\pi y}{b}$

$v = -\dfrac{16g}{\pi^2} \sum\limits_{n\text{ odd}} \sum\limits_{m\text{ odd}} \left(\dfrac{1 - \cos\omega_{nm}t}{nm\omega_{nm}{}^2}\right)\varphi_{nm}(x,y)$

$w = \dfrac{256}{\pi^2} \sum\limits_{n\text{ odd}} \sum\limits_{m\text{ odd}} \dfrac{\varphi_{nm}(x,y)\cos\omega_{nm}t}{nm(n^2 - 4)(m^2 - 4)}$

$\omega_{nm} = c\pi\left[\left(\dfrac{n}{a}\right)^2 + \left(\dfrac{m}{b}\right)^2\right]^{1/2}$

57 If $F(r,t) = F_0 \sin \omega t$, F_0 constant, and $\omega \neq \omega_n$, all n, then

$u = 2F_0 \sum\limits_{n=1}^{\infty}\left[(\omega \sin\omega_n t - \omega_n \sin\omega t)\dfrac{J_0(\xi_n r/a)}{\xi_n \omega_n(\omega^2 - \omega_n{}^2)J_1(\xi_n)}\right]$

where $\{\xi_n\}$ denotes the positive zeros of $J_0(\xi)$ and $\omega_n = c\xi_n/a$, $n = 1, 2, \ldots$.

59 If $\omega \neq \omega_{nm} = \xi_{nm}/a$, where $\{\xi_{nm}\}$ is the sequence of positive zeros of $J_n(\xi)$, $n = 0$, $1, \ldots$, then $u = \varphi(\theta)\sin\omega t + w(r,\theta,t)$, where

$w = \sum\limits_{n=0}^{\infty} \sum\limits_{m=1}^{\infty} \dfrac{\omega \sin\omega_{nm}t - \omega_{nm}\sin\omega t}{\omega_{nm}(\omega^2 - \omega_{nm}{}^2)}[\alpha_{nm}^{(e)}\varphi_{mn}^{(e)}(r,\theta) + \alpha_{nm}^{(0)}\varphi_{nm}^{(0)}(r,\theta)]$

$\alpha_{nm}^{(a)} = \dfrac{1}{\|\varphi_{nm}^{(e)}\|^2}\int_0^a\int_0^{2\pi}\left[-\omega^2\varphi(\theta) + \dfrac{c^2\varphi''(\theta)}{r^2}\right]\varphi_{nm}^{(e)}(r,\theta)r\, dr\, d\theta$

with $\alpha_{nm}^{(0)}$ defined similarly by replacing (e) with (0) throughout,

$\varphi_{nm}^{(e)} = J_n\left(\dfrac{\xi_{nm}r}{a}\right)\cos n\theta \qquad \varphi_{nm}^{(0)} = J_n\left(\dfrac{\xi_{nm}r}{a}\right)\sin n\theta$

$\|\varphi_{nm}^{(e)}\|^2 = \dfrac{\pi a^2 J_{n+1}^2(\xi_{nm})}{2} = \|\varphi_{nm}^{(0)}\|^2 \qquad n = 1, 2, \ldots ; m = 1, 2, \ldots$

$\|\varphi_{om}^{(e)}\|^2 = \pi a^2 J_1{}^2(\xi_{0m}) \qquad m = 1, 2, \ldots$

62 Eigenvalues $\lambda_{nm} = \xi_{nm}^2/a^2$, where $\{\xi_{nm}\}$ denotes the positive zeros of $J_{n\pi/\beta}(\xi)$; corresponding real-valued eigenfunctions are

$$\varphi_{nm} = J_{n\pi/\beta}\left(\frac{\xi_{nm}r}{a}\right)\sin\frac{n\pi\theta}{\beta} \qquad n = 0, 1, \ldots ; m = 1, 2, \ldots$$

$$\|\varphi_{nm}\|^2 = a^2\beta\frac{[J_{n\pi/\beta+1}(\xi_{nm})]^2}{4}$$

Solution is $u = v + w$, where

$$\omega_{nm} = \frac{c\xi_{nm}}{a}$$

$$v = \sum_{n=0}^{\infty}\sum_{m=1}^{\infty}\varphi_{nm}(r,\theta)(A_{nm}\cos\omega_{nm}t + B_{nm}\sin\omega_{nm}t)$$

$$A_{nm} = \frac{1}{\|\varphi_{nm}\|^2}\int_0^a\int_0^\beta f(r,\theta)\varphi_{nm}(r,\theta)r\,dr\,d\theta$$

$$B_{nm} = \frac{1}{\omega_{nm}\|\varphi_{nm}\|^2}\int_0^a\int_0^\beta g(r,\theta)\varphi_{nm}(r,\theta)r\,dr\,d\theta$$

$$w = \sum_{n=0}^{\infty}\sum_{m=1}^{\infty}\left\{\frac{1}{\omega_{nm}}\int_0^t F_{nm}(\xi)\sin[\omega_{nm}(t-\xi)]\,d\xi\right\}\varphi_{nm}(r,\theta)$$

$$F_{nm}(t) = \frac{1}{\|\varphi_{nm}\|^2}\int_0^a\int_0^\beta F(r,\theta,t)\varphi_{nm}(r,\theta)r\,dr\,d\theta$$

$$G(r,\theta,t;r_0,\theta_0,\tau) = \frac{4}{\beta a^2}\sum_{n=0}^{\infty}\sum_{m=1}^{\infty}\frac{\varphi_{nm}(r,\theta)\varphi_{nm}(r_0,\theta_0)\sin[\omega_{nm}(t-\tau)]}{\omega_{nm}\|\varphi_{nm}\|^2}$$

63 Eigenvalues $\lambda_{nm} = \xi_{nm}^2$, where, for each $n = 0, 1, \ldots$, $\{\xi_{nm}\}$ denotes the sequence of positive roots of the equation

$$J_n(a\xi)Y_n(b\xi) - J_n(b\xi)Y_n(a\xi) = 0$$

Corresponding real-valued eigenfunctions are

$$\varphi_{nm}^{(e)} = [J_n(\xi_{nm}r)Y_n(\xi_{nm}a) - J_n(\xi_{nm}a)Y_n(\xi_{nm}r)]\cos n\theta \qquad n = 0, 1, \ldots$$

$$\varphi_{nm}^{(0)} = [J_n(\xi_{nm}r)Y_n(\xi_{nm}a) - J_n(\xi_{nm}a)Y_n(\xi_{nm}r)]\sin n\theta \qquad n = 1, 2, \ldots$$

$$\|\varphi_{nm}\|^2 = 2\epsilon_n\frac{J_n^2(\xi_{nm}a) - J_n^2(\xi_{nm}b)}{\pi\xi_{nm}^2 J_n^2(\xi_{nm}b)}$$

where $\epsilon_0 = 2$, $\epsilon_n = 1$, $n \geq 1$. Characteristic frequencies $\omega_{nm} = c\xi_{nm}$, normal modes

$$u_{nm} = [A_{nm}\varphi_{nm}^{(e)}(r,\theta) + B_{nm}\varphi_{nm}^{(0)}(r,\theta)]\cos\omega_{nm}t + [C_{nm}\varphi_{nm}^{(e)}(r,\theta) + D_{nm}\varphi_{nm}^{(0)}(r,\theta)]\sin\omega_{nm}t$$

where it is understood that $\varphi_{0m}^{(e)} = 0$, $m = 1, 2, \ldots$. The formal series solution is identical in form to the solution obtained in Example 4-2 with the φ_{nm} defined above and the region of integration the annulus $a \leq r \leq b$, $0 \leq \theta \leq 2\pi$.

69 (a) $u = \dfrac{A}{r} + Bt$

75 $G = 0$ if $t \leq \tau$, and if $t > \tau$, then

$$G(r,\theta,t;r_0,\theta_0,\tau) = \frac{3(t-\tau)}{a^3} + \sum_{k=0}^{\infty}\sum_{m=1}^{\infty}\frac{\psi_{2k,m}(r,\theta)\psi_{2k,m}(r_0,\theta_0)\sin[\omega_{2k,m}(t-\tau)]}{\omega_{2k,m}\,\|\psi_{2k,m}\|^2}$$

where

$$\psi_{2k,m} = j_{2k,m}\left(\frac{\xi_{2k,m}r}{a}\right)P_{2k}(\cos\theta) \qquad k = 0,1,\ldots\,;\; m = 1,2,\ldots$$

$$\|\psi_{2k,m}\|^2 = \frac{a^3[1 - 2k(2k+1)/\xi_{2k,m}^2]\,j_{2k}^2(\xi_{2k,m})}{2(4k+1)}$$

and, for each $k = 0,1,\ldots$, $\{\xi_{2k,m}\}$ denotes the positive zeros of $j_{2k}'(\xi)$, $\omega_{2k,m} = c\xi_{2k,m}/a$ are the characteristic frequencies. If $F(r,\theta,t) = F_0 \sin \omega t$, $f(r,\theta) = h(r)\cos\omega t$, $g = 0$, then the solution is

$$u = \frac{F_0}{\omega^2}(\omega t - \sin\omega t) + v$$

$$v = A_0 + \sum_{k=0}^{\infty}\sum_{m=1}^{\infty} A_{2k,m}\psi_{2k,m}(r,\theta)\cos\omega_{2k,m}t$$

$$A_0 = \frac{3}{2a^3}\int_0^a r^2 h(r)\,dr$$

$$A_{2k,m} = -\frac{P_{2k}(0)}{(2k-1)(2k+2)\,\|\psi_{2k,m}\|^2}\int_0^a r^2 h(r)j_{2k}\left(\frac{\xi_{2k,m}r}{a}\right)dr$$

$$P_0(0) = 1 \qquad P_{2k}(0) = (-1)^k\frac{1\cdot3\cdot5\cdots(2k-1)}{2\cdot4\cdots(2k)} \qquad k \geq 1$$

79 $u = V(r,z)\cos p\theta\cos\omega t + W(r,z,t)\cos p\theta$

$$V(r,z) = \sum_{m=1}^{\infty} A_m J_p\left(\frac{\xi_m r}{a}\right)\cosh\mu_m z$$

$$W(r,z,t) = \sum_{m=1}^{\infty}\sum_{q=0}^{\infty} B_{mq} J_p\left(\frac{\xi_m r}{a}\right)\cos\frac{q\pi z}{h}\cos\omega_{mq}t$$

where $\{\xi_m\}$ denotes the positive zeros of $J_p'(\xi)$ arranged in order of increasing magnitude, $\mu_m = (\xi_m^2/a^2 - \omega^2/c^2)^{1/2}$, $\omega_{mq} = c(\xi_m^2/a^2 + q^2\pi^2/h^2)^{1/2}$,

$$A_m = \frac{2\xi_m^2}{a^2(\xi_m^2 - p^2)\mu_m\sinh(\mu_m h)J_p^2(\xi_m)}\int_0^a J_p\left(\frac{\xi_m r}{a}\right)\varphi(r)r\,dr$$

$$B_{mq} = \frac{2A_m(-1)^{q+1}\mu_m\sinh(\mu_m h)}{h(\mu_m^2 + q^2\pi^2/h^2)}$$

$$B_{m0} = -\frac{A_m\sinh\mu_m h}{h\mu_m} \qquad m = 1,2,\ldots\,;\; q = 1,2,\ldots$$

CHAPTER 5

3 $u = (u_1 - u_2) \operatorname{erf} \delta_1 + u_2 \operatorname{erf} \delta_2$

5 **(a)** $u = u_0 \operatorname{erf} \dfrac{x}{\sqrt{4\kappa t}}$ **(c)** $u = u_1 + (u_2 - u_1) \operatorname{erf} \dfrac{x}{\sqrt{4\kappa t}}$

7 $u = u_0 e^{-bt} \operatorname{erf} \dfrac{x}{\sqrt{4\kappa t}}$

11 **(a)** $u = \dfrac{4L}{\pi^2} \displaystyle\sum_{k=1}^{\infty} \dfrac{(-1)^{k+1}}{(2k-1)^2} \sin \dfrac{(2k-1)\pi x}{L} e^{-(2k-1)^2 \pi^2 \kappa t / L^2}$

(c) $u = u_1 + (u_2 - u_1) \dfrac{x}{L} + v(x)$

$$v(x) = \sum_{n=1}^{\infty} \left[B_n + 2 \frac{(-1)^n u_2 - u_1}{n\pi} \right] \sin \frac{n\pi x}{L} e^{-n^2 \pi^2 \kappa t / L^2}$$

$$B_n = \frac{2}{L} \int_0^L f(x) \sin \frac{n\pi x}{L} \, dx$$

(e) $u = \dfrac{q_0 x}{K} + \displaystyle\sum_{n=1}^{\infty} A_n \sin \dfrac{(2n-1)\pi x}{2L} e^{-(2n-1)^2 \pi^2 \kappa t / 4L^2}$

$$A_n = \frac{-4Lq_0}{\pi^2 K (2n-1)^2} + \frac{2}{L} \int_0^L f(x) \sin \frac{(2n-1)\pi x}{2L} \, dx$$

13 $u = A[1 + h(L - x)] \dfrac{\sin \omega_1 t}{1 + hL} + v(x,t)$

$$v(x,t) = \sum_{n=1}^{\infty} [C_n(t) + B_n] \sin \mu_n x$$

where $\{\mu_n\}$ denotes the positive roots of $\tan \mu L = -\mu/h$, and

$$C_n(t) = \left[\frac{F_0 \sin \mu_n x_0}{\kappa^2 \mu_n^4 + \omega^2} (\kappa \mu_n^2 \cos \omega t + \omega \sin \omega t - \kappa \mu_n^2 e^{-\kappa \mu_n^2 t}) \right.$$

$$\left. - \frac{A\omega_1}{\mu_n (\kappa^2 \mu_n^4 + \omega_1^2)} (\kappa \mu_n^2 \cos \omega_1 t + \omega_1 \sin \omega_1 t - \kappa \mu_n^2 e^{-\kappa \mu_n^2 t}) \right] \Big/ \|X_n\|^2$$

$$\|X_n\|^2 = \frac{L}{2} + \frac{\cos^2 \mu_n L}{2h} \qquad B_n = \frac{1}{\|X_n\|^2} \int_0^L f(x) \sin \mu_n x \, dx$$

15 $u = 8u_0 \displaystyle\sum_{n=1}^{\infty} \dfrac{J_0(\xi_n r/a)}{\xi_n^3 J_1(\xi_n)} e^{-\kappa \xi_n^2 t / a}$

where $\{\xi_n\}$ denotes the positive zeros of $J_0(\xi)$.

17 $u = u_0 + \displaystyle\sum_{q=1}^{\infty} A_q \sin \dfrac{(2q-1)\pi z}{2b} e^{\kappa \lambda_q t}$

$$+ \sum_{n=0}^{\infty} \sum_{m=1}^{\infty} \sum_{q=1}^{\infty} [A_{nmq} \psi_{nmq}^{(e)}(r,\theta,z) + B_{nmq} \psi_{nmq}^{(0)}(r,\theta,z)] e^{-\kappa \lambda_{nmq} t}$$

where

$$\lambda_q = \frac{(2q-1)^2\pi^2}{4b^2} \qquad \lambda_{nmq} = \frac{\xi_{nm}^2}{a^2} + \frac{(2q-1)^2\pi^2}{4b^2} \qquad n = 0, 1, \dots; m = 1, 2, \dots;$$
$$q = 1, 2, \dots$$

and, for each n, $\{\xi_{nm}\}$ denotes the sequence of positive zeros of $J_n'(\xi)$,

$$A_q = \frac{-4u_0}{(2q-1)\pi^2 a^2} + \frac{2}{\pi a^2 b} \int_0^a \int_0^{2\pi} \int_0^b f(r,\theta,z) \sin\frac{(2q-1)\pi z}{2b} r \, dr \, d\theta \, dz$$

$$A_{nmq} = \frac{1}{\|\psi_{nmq}\|^2} \int_0^a \int_0^{2\pi} \int_0^b f(r,\theta,z)\psi_{nmq}^{(e)}(r,\theta,z) r \, dr \, d\theta \, dz \qquad n = 0, 1, \dots; m = 1, 2, \dots;$$
$$q = 1, 2, \dots$$

$$B_{nmq} = \frac{1}{\|\psi_{nmq}\|^2} \int_0^a \int_0^{2\pi} \int_0^b f(r,\theta,z)\psi_{nmq}^{(0)}(r,\theta,z) r \, dr \, d\theta \, dz \qquad n = 1, 2, \dots; m = 1, 2, \dots;$$
$$q = 1, 2, \dots$$

$$\psi_{nmq}^{(e)} = J_n\left(\frac{\xi_{nm}r}{a}\right) \cos n\theta \sin\frac{(2q-1)\pi z}{2b}$$

$$\psi_{nmq}^{(0)} = J_n\left(\frac{\xi_{nm}r}{a}\right) \sin n\theta \sin\frac{(2q-1)\pi z}{2b}$$

$$\|\psi_{nmq}^{(e)}\|^2 = \|\psi_{nmq}^{(0)}\|^2 = \frac{\pi a^2 b}{4\xi_{nm}^2}(\xi_{nm}^2 - n^2)J_n^2(\xi_{nm}) \qquad n = 1, 2, \dots$$

$$\|\psi_{0mq}\|^2 = \frac{\pi a^2 b}{2} J_0^2(\xi_{0m}) \qquad m = 1, 2, \dots; q = 1, 2, \dots$$

21 $u = \dfrac{1}{r}\displaystyle\sum_{n=1}^{\infty} A_n \sin(\mu_n r)e^{-\kappa\mu_n^2 t}$ where $\{\mu_n\}$ denotes the sequence of positive roots of

$\tan a\mu = a\mu/(1-ah)$ and

$$A_n = 4\frac{a^2\mu_n + (ah-1)^2}{a^2\mu_n^2 + ah(ah-1)} \frac{3h - a\mu_n^2}{\mu_n^4} \sin a\mu_n$$

23 $u = \dfrac{q_0 a^2}{K}\left(\dfrac{bh-1}{b^2h} - \dfrac{1}{r}\right) + w(r,t)$

$$w(r,t) = \sum_{m=1}^{\infty} A_m \psi_m(r)e^{-\kappa\xi_m^2 t}$$

where $\{\xi_m\}$ denotes the sequence of positive roots of the equation

$$\xi[j_0'(a\xi)y_0'(b\xi) - y_0'(a\xi)j_0'(b\xi)] + h[j_0'(a\xi)y_0(b\xi) - y_0'(a\xi)j_0(b\xi)] = 0$$

$$\varphi_m(r) = f_0(\xi_m r)y_0'(\xi_m a) - j_0'(\xi_m a)y_0(\xi_m r)$$

$$A_m = \frac{1}{\alpha_m}\left[u_0\int_a^b \psi_m(r)r^2 \, dr - \frac{q_0 a^2}{K}\int_a^b\left(\frac{bh-1}{b^2h} - \frac{1}{r}\right)\psi_m(r)r^2 \, dr\right]$$

$$\alpha_m = \int_a^b \psi_m^2(r)r^2 \, dr \qquad j_0(x) = \sqrt{\frac{\pi}{2x}}J_{1/2}(x) \qquad y_0(x) = \sqrt{\frac{\pi}{2x}}Y_{1/2}(x)$$

INDEX

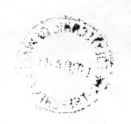